To John Baron with complements and best wishes

Berlin, June 12, 2013

Karsten [illegible]

Karsten Schrör

Acetylsalicylic Acid

Related Titles

Nicolaou, K. C., Montagnon, T.

Molecules that Changed the World

2008
ISBN: 978-3-527-30983-2

Corey, E. J., Czakó, B., Kürti, L.

Molecules and Medicine

2007
ISBN: 978-0-470-22749-7

Behme, S.

Manufacturing of Pharmaceutical Proteins

2008
ISBN: 978-3-527-32444-6

Curtis-Prior, P. (ed.)

The Eicosanoids

2004
ISBN: 978-0-471-48984-9

Karsten Schrör

Acetylsalicylic Acid

WILEY-VCH Verlag GmbH & Co. KGaA

The Author

Prof. Dr. Karsten Schrör
Institut für Pharmakologie und Klinische Pharmakologie
Heinrich-Heine-Universität Düsseldorf
Universitätsstr 1
40225 Düsseldorf
Germany

Cover
Reproduction of the first commercial Aspirin preparation with kind permission of Bayer Vital GmbH, Cologne.

Library of Congress Card No.: applied for

British Library Cataloguing-in-Publication Data
A catalogue record for this book is available from the British Library.

Bibliographic information published by the Deutsche Nationalbibliothek
The Deutsche Nationalbibliothek lists this publication in the Deutsche Nationalbibliografie; detailed bibliographic data are available on the Internet at http://dnb.d-nb.de.

Typesetting Thomson Digital, Noida, India
Printing betz-Druck GmbH, Darmstadt
Binding Litges & Dopf GmbH, Heppenheim

Printed in the Federal Republic of Germany
Printed on acid-free paper

ISBN: 978-3-527-32109-4

Contents

Preface *XI*

1 **General Aspects** *3*
1.1 History *5*
1.1.1 Willow Bark and Leaves as Antipyretic, Anti-Inflammatory Analgesics *5*
1.1.2 Salicylates as the Active Ingredient of Willow Bark and Other Natural Sources *6*
1.1.3 Synthesis of Acetylsalicylic Acid and First Clinical Studies *7*
1.1.4 Mode of Aspirin Action *12*
1.1.5 Anti-Inflammatory/Analgesic Actions of Aspirin *14*
1.1.6 Aspirin in the Cardiovascular System *15*
1.1.7 Current Research Topics *17*
1.2 Chemistry *25*
1.2.1 Structures and Chemical Properties of Salicylates *25*
1.2.1.1 Salicylates in Clinical Use *25*
1.2.1.2 Aspirin Formulations *27*
1.2.2 Determination of Salicylates *30*
1.2.2.1 Gas–Liquid Chromatography *30*
1.2.2.2 High-Performance Liquid Chromatography *30*
1.2.2.3 Spectrophotometry *30*

2 **Pharmacology** *33*
2.1 Pharmacokinetics *35*
2.1.1 Absorption and Distribution *36*
2.1.1.1 Absorption *36*
2.1.1.2 Particular Aspirin Formulations *40*
2.1.1.3 Distribution *40*
2.1.1.4 Modifying Factors *42*
2.1.2 Biotransformation and Excretion *46*
2.1.2.1 Biotransformation of Aspirin *46*
2.1.2.2 Biotransformations of Salicylic Acid *48*
2.1.2.3 Excretion of Salicylates *49*

Acetylsalicylic Acid. Karsten Schrör

ISBN: 978-3-527-32109-4

2.1.2.4 Modification of Biotransformations and Excretion of Salicylates *50*
2.2 Cellular Modes of Action *53*
2.2.1 Inhibition of Cyclooxygenases *54*
2.2.1.1 Acetylation of Cyclooxygenases and Traditional NSAIDs *55*
2.2.1.2 Modulation of COX-2 Gene Expression *61*
2.2.1.3 Further Actions of Salicylates on Arachidonate Metabolism *62*
2.2.2 COX-Independent Actions on Cell Function *66*
2.2.2.1 Kinases *67*
2.2.2.2 Transcription Factors *69*
2.2.2.3 Oxidative Stress and Nitric Oxide *73*
2.2.2.4 Immune Responses *75*
2.2.3 Energy Metabolism *79*
2.2.3.1 Fatty Acid β-Oxidation *80*
2.2.3.2 Uncoupling of Oxidative Phosphorylation *82*
2.2.3.3 Toxic Actions of Salicylates on the Liver *84*
2.3 Actions on Organs and Tissues *88*
2.3.1 Hemostasis and Thrombosis *89*
2.3.1.1 Platelets *90*
2.3.1.2 Endothelial Cells *95*
2.3.1.3 Plasmatic Coagulation *98*
2.3.1.4 Fibrinolysis *98*
2.3.2 Inflammation, Pain, and Fever *105*
2.3.2.1 Inflammation *106*
2.3.2.2 Pain *110*
2.3.2.3 Fever *113*
2.3.3 Aspirin and Malignancies *120*
2,3.3.1 COX-Related Antitumor Effects of Aspirin *120*
2.3.3.2 Nonprostaglandin-Related Antitumor Actions of Aspirin *122*
2.3.3.3 Nonspecific Actions of Salicylates *124*

3 Toxicity and Drug Safety *129*
3.1 Systemic Side Effects *131*
3.1.1 Acute and Chronic Toxicity *132*
3.1.1.1 Occurrence and Symptoms *132*
3.1.1.2 Treatment *136*
3.1.1.3 Habituation *139*
3.1.2 Bleeding Disorders *142*
3.1.2.1 Prolongation of Bleeding Time by Aspirin *142*
3.1.2.2 Aspirin-Related Bleeding Risk in Surgical Interventions *144*
3.1.2.3 Aspirin, Other Drugs, and Alcohol *146*
3.1.2.4 Prevention and Treatment of Bleedings *146*
3.1.3 Safety Pharmacology in Particular Life Situations *150*
3.1.3.1 Pregnancy and Fetal Development *150*
3.1.3.2 The Elderly Patient *153*
3.2 Organ Toxicity *157*

3.2.1 Gastrointestinal Tract *157*
3.2.1.1 Pathophysiology of GI Injury *158*
3.2.1.2 Prostaglandins and Gastric Mucosal Protection *160*
3.2.1.3 Mode of Aspirin Action *161*
3.2.1.4 Helicobacter pylori *164*
3.2.1.5 Clinical Studies *166*
3.2.1.6 Aspirin and Other Drugs *170*
3.2.2 Kidney *179*
3.2.2.1 Analgesic Nephropathy *179*
3.2.2.2 Mode of Aspirin Action *180*
3.2.2.3 Clinical Studies *181*
3.2.3 Liver *187*
3.2.3.1 Drug-Induced Liver Injury *187*
3.2.3.2 Mechanisms of Aspirin Action *187*
3.2.3.3 Aspirin and Other Drugs *188*
3.2.4 Audiovestibular System *191*
3.2.4.1 Pathophysiology of Hearing and Equilibrium Disturbances *191*
3.2.4.2 Mode of Aspirin Action *191*
3.2.4.3 Aspirin, Arachidonic Acid, and Prostaglandins *193*
3.2.4.4 Clinical Trials *193*
3.3 Non-Dose-Related (Pseudo)allergic Actions of Aspirin *197*
3.3.1 Aspirin Hypersensitivity (Widal's Syndrome) *197*
3.3.1.1 Pathophysiology and Clinics *197*
3.3.1.2 Mode of Aspirin Action *199*
3.3.1.3 Clinical Trials *201*
3.3.2 Urticaria/Angioedema, Stevens–Johnson and Lyell Syndromes *205*
3.3.2.1 Urticaria/Angioedema *205*
3.3.2.2 Stevens–Johnson Syndrome and Toxic Epidermal Necrolysis (Lyell Syndrome) *206*
3.3.3 Reye's Syndrome *208*
3.3.3.1 Clinics, Laboratory, and Morphological Findings *209*
3.3.3.2 Etiology *209*
3.3.3.3 Clinical Studies *212*
3.3.3.4 Actual Situation *217*

4 Clinical Applications of Aspirin *223*
4.1 Thromboembolic Diseases *227*
4.1.1 Coronary Vascular Disease *228*
4.1.1.1 Thrombotic Risk and Mode of Aspirin Action *229*
4.1.1.2 Clinical Trials: Primary Prevention in Individuals Without Risk Factors *231*
4.1.1.3 Clinical Trials: Primary Prevention in Individuals with Risk Factors *235*
4.1.1.4 Clinical Trials: Stable Angina Pectoris *237*
4.1.1.5 Clinical Trials: Secondary Prevention *238*
4.1.1.6 Clinical Trials: Percutaneous Coronary Interventions *243*
4.1.1.7 Clinical Trials: Coronary Artery Bypass Graft Surgery (CABG) *244*
4.1.1.8 Aspirin and Other Drugs *246*

4.1.1.9 Actual Situation *250*
4.1.2 Cerebrovascular Diseases *260*
4.1.2.1 Thrombotic Risk and Mode of Aspirin Action *261*
4.1.2.2 Clinical Trials: Primary Prevention *263*
4.1.2.3 Clinical Trials: Secondary Prevention *265*
4.1.2.4 Aspirin and Other Drugs *267*
4.1.2.5 Actual Situation *270*
4.1.3 Peripheral Arterial Disease *276*
4.1.3.1 Thrombotic Risk and Mode of Aspirin Action *276*
4.1.3.2 Clinical Trials: Primary Prevention *278*
4.1.3.3 Clinical Trials: Secondary Prevention *279*
4.1.3.4 Clinical Trials: Peripheral Transluminal Angioplasty (PTA) *281*
4.1.3.5 Aspirin and Other Drugs *282*
4.1.3.6 Actual Situation *283*
4.1.4 Venous Thrombosis *287*
4.1.4.1 Thrombotic Risk and Mode of Aspirin Action *287*
4.1.4.2 Clinical Trials *287*
4.1.4.3 Actual Situation *288*
4.1.5 Preeclampsia *290*
4.1.5.1 Thrombotic Risk and Mode of Aspirin Action *290*
4.1.5.2 Clinical Trials *293*
4.1.5.3 Clinical Trials in Pregnancy-Induced Hypertension: Reasons for Data Variability *294*
4.1.5.4 Actual Situation *298*
4.1.6 Aspirin "Resistance" *303*
4.1.6.1 Definition and Types of Pharmacological Aspirin "Resistance" *304*
4.1.6.2 Detection of Aspirin "Resistance" *304*
4.1.6.3 Pharmacological Mechanisms of Aspirin "Resistance" *307*
4.1.6.4 Clinical Trials *310*
4.2 Pain, Fever, and Inflammatory Diseases *322*
4.2.1 Aspirin as an Antipyretic Analgesic *322*
4.2.1.1 Fever, Pain, and Antipyretic/Analgesic Actions of Aspirin *323*
4.2.1.2 Clinical Trials *326*
4.2.1.3 Aspirin and Other Drugs *327*
4.2.1.4 Actual Situation *328*
4.2.2 Arthritis and Rheumatism *332*
4.2.2.1 Pathophysiology and Mode of Aspirin Action *332*
4.2.2.2 Clinical Trials *334*
4.2.2.3 Aspirin and Other Drugs *335*
4.2.2.4 Actual Situation *336*
4.2.3 Kawasaki Disease *339*
4.2.3.1 Pathophysiology and Mode of Aspirin Action *339*
4.2.3.2 Clinical Trials *340*
4.2.3.3 Aspirin and Other Drugs *341*
4.2.3.4 Actual Situation *341*
4.3 Further Clinical Indications *343*

4.3.1 Colorectal Cancer *343*
4.3.1.1 Etiology and Pathophysiology of Intestinal Adenomas, Colorectal Cancer, and Mode of Aspirin Action *344*
4.3.1.2 Clinical Trials: Epidemiological Studies *346*
4.3.1.3 Clinical Trials: Randomized Prospective Prevention Trials *349*
4.3.1.4 Aspirin and Other Drugs *352*
4.3.1.5 Actual Situation *352*
4.3.2 Alzheimer's Disease *359*
4.3.2.1 Pathophysiology and Mode of Aspirin Action *359*
4.3.2.2 Clinical Trials *361*
4.3.2.3 Aspirin and Other Drugs *363*
4.3.2.4 Actual Situation *363*

Appendix A *367*

Appendix B *369*

Index *371*

Preface

Acetylsalicylic acid, best known by its first trade name aspirin, belongs to the small and steadily decreasing number of therapeutic agents that are well known to both the health professional and the layman and enjoy a good reputation with both. Aspirin is not only one of the most intensively studied but also one of the most frequently used and cheapest drugs in the world, with an immense annual production rate. Ironically, the first clinical investigator to study the drug more than 100 years ago did so with "not little distress."

What are the reasons behind this exciting development? Certainly in the beginning, there was an urgent need for an effective and well-tolerated antipyretic analgesic that became invaluable during the large flue pandemia in 1918. In the following decades, aspirin soon became very popular as a household remedy for almost any condition associated with flu, feverish discomfort, or any kind of malaise – "take an aspirin." Another breakthrough was the discovery of its novel mode of action – inhibition of injury-induced prostaglandin biosynthesis. This offered a plausible mechanistic explanation for its multitude of pharmacological activities. Later, the antiplatelet/antithrombotic properties of aspirin came in focus and opened a new and broad therapeutic field for the prevention of atherothrombotic events. Aspirin is still a treatment of first choice in many of these indications, most notably the prevention of myocardial infarction and stroke. Experimental and clinical research on aspirin is still ongoing with current issues including investigation of transcriptional effects of aspirin on gene regulation, specifically in inflammation and malignancies, as well as possible new clinical indications, such as colorectal carcinoma.

This book provides an overview of all aspects of therapeutically relevant aspirin actions and the underlying mechanisms with specific focus on the unique structural property of the compound consisting of two bioactive molecules, that is, a reactive acetyl group and the salicylate moiety, each with different pharmacokinetic and pharmacodynamic properties that synergize in many cases. This and the resulting possible benefits and risks in clinical use are discussed in four main chapters, each divided into several subsections. A list of selected references is provided in each section and includes reports that have been published up to 2007.

The preparation of this manuscript would not have been possible without the continuous help and assistance of a number of individuals. The first person to mention is my wife Elke. I thank her for her patience and continuous support during the last 4 years while this manuscript was prepared. I also like to thank all of my friends and colleagues in Australia, Germany, and the rest of Europe, Taiwan, and the United States for critical reading of parts of the manuscript and providing many valuable suggestions.

Acetylsalicylic Acid. Karsten Schrör

ISBN: 978-3-527-32109-4

Many thanks also to Erika Lohmann and Petra Kuger in Düsseldorf for typing and carefully reading the manuscript as well as preparing figures and tables.

As a young research scientist in the mid-1970s, I had the privilege to work for some time in the laboratory of Sir John Vane in Beckenham (England). This was an exciting period, when several important discoveries were made in this laboratory: detection of prostacyclin and the demonstration that aspirin inhibits injury-induced prostaglandin biosynthesis. The significance of these findings was acknowledged when Sir John Vane was awarded the Nobel Prize in Medicine in 1982. When this book project was still in progress, I asked Sir John whether he would be willing to write a preface and he kindly agreed to do so. Unfortunately, he passed away before the project was finished. He would have celebrated his 80th birthday in March 2007. This book is dedicated to him.

Düsseldorf, October 2008 *Karsten Schrör*

1 General Aspects

- 1.1 History
- 1.2 Chemistry

2 Pharmacology

- 2.1 Pharmacokinetics
- 2.2 Cellular Modes of Action
- 2.3 Actions on Organs and Tissues

3 Toxicity and Drug Safety

- 3.1 Systemic Side Effects
- 3.2 Organ Toxicity
- 3.3 Non-Dose-Related (Pseudo)allergic Actions of Aspirin

4 Clinical Applications of Aspirin

- 4.1 Thromboembolic Diseases
- 4.2 Pain, Fever, and Inflammatory Diseases
- 4.3 Further Clinical Indications

1 General Aspects

1.1 History
1.1.1 Willow Bark and Leaves as Antipyretic, Anti-Inflammatory Analgesics
1.1.2 Salicylates as the Active Ingredient of Willow Bark and Other Natural Sources
1.1.3 Synthesis of Acetylsalicylic Acid and First Clinical Studies
1.1.4 Mode of Aspirin Action
1.1.5 Anti-Inflammatory/Analgesic Actions of Aspirin
1.1.6 Aspirin in the Cardiovascular System
1.1.7 Current Research Topics
1.2 Chemistry
1.2.1 Structures and Chemical Properties of Salicylates
1.2.2 Determination of Salicylates

Acetylsalicylic Acid. Karsten Schrör

ISBN: 978-3-527-32109-4

1
General Aspects

1.1
History

1.1.1
Willow Bark and Leaves as Antipyretic, Anti-Inflammatory Analgesics

Medical Effects of Willow Bark Treatment of diseases by plants or extracts thereof is as old as the history of mankind. This is also true for fever and pain, two particularly frequent and inconvenient symptoms of acute illnesses, as well as arthritis and rheumatism, two examples for chronic painful diseases. Rheumatism already existed in old Egypt, as seen from cartilage alterations in Egyptian mummies. The Egyptians were aware of the pain-relieving effects of potions made from myrtle and willow leaves. Clay tablets from the Sumerian period also described the use of willow leaves for medical purposes. *Hippocrates* recommended leaves of the willow tree for medical purposes about 400 BC. *Pliny* (compilations) and *Dioscurides* (*Materia Medica*) also recommended decocts of willow leaves or ash from the willow bark to treat sciatica (lumbago) and gout in about 100 AD. Outside Europe, it was the Nama (Hottentots) in Southern Africa who had "for a long time" used the bark of willow trees to treat rheumatic diseases (cited after Ref. [1]). This comment was made by Dr Ensor at Cape of Good Hope in his reply to a publication of Dr MacLagan in 1876, describing the use of salicylates for treating rheumatism [2].

The first known communication on the medical use of willow bark extracts in modern times came from Reverend *Edward Stone* from Chipping Norton (Oxfordshire, England). In 1763, he treated some 50 cases of "aigues, fever, and intermitting disorders" with a powdered dry bark preparation of willow tree [3]. The doses were about "20 gr(ains) [1.3 g] to a dram of water every 4 h." On June 2, 1763, he wrote a letter entitled "An account of the success of the bark of the willow in the cure of aigues" to the Earl of Macclesfield, the then president of the Royal Society of London. In this letter, he summarized his opinion about this treatment as follows:

> ". . . As this tree delights in moist or wet soil where agues chiefly abound, the general maxim, that many natural maladies carry their cure along with them or that their remedies lie not far from their causes, was so very apposite to this particular case, that I could not help applying it; and this might be the intention of providence here, I must own had some little weight with me . . .."

After claiming to have obtained good results he concluded:

> ". . . I have no other motives for publishing this valuable specific than that it may have a fair and full trial in all its variety of circumstances and situations, and that the world may reap the benefits accruing from it."

Acetylsalicylic Acid. Karsten Schrör

ISBN: 978-3-527-32109-4

1.1.2
Salicylates as the Active Ingredient of Willow Bark and Other Natural Sources

Plants as Natural Sources of Salicylates In 1828, German pharmacist *Buchner* was the first to prepare a yellowish mash of bitter taste from boiled willow bark, which he named salicin, after the Latin word for willow (*salix*). He considered salicin as the antipyretic component of willow bark and recommended its use for treatment of fever. A similar conclusion had earlier been reached by the Italians *Brugnatelli* and *Fontana* in 1826 using a less purified preparation of willow bark. They also considered salicin as the active principal component of willow bark (cited after Ref. [4]). In 1830, Frenchman *Leroux* was the first to obtain salicin in crystalline form. Only 3 years later, in 1833, the pharmacist *Merck* in Darmstadt (Germany) announced highly purified salicin from willow bark as an antipyretic for half of the price of quinine (cited after Ref. [5]) – at the time a really attractive offer.

Salicin is not only the antipyretic ingredient of willow bark but also the reason for its bitter taste and irritation of stomach mucosa. Salicin hydrolyzes in aqueous media to glucose and salicylic alcohol (saligenin). Saligenin has no bitter taste and can be easily oxidized to salicylic acid. *Raffaele Piria,* an Italian, was the first to successfully synthesize salicylic acid (acide salicylique ou salicylique) in 1839 from salicin and correctly determined the empirical formula $C_7H_6O_3$. This increased the possibility of replacing the poorly palatable salicin with salicylic acid, for example, as a good water-soluble sodium salt. This became practically significant after the new rich natural sources for salicylates were detected. This included wintergreen oil obtained from the American evergreen (*Gaultheria procumbens*) and spireic acid (acidum salicylicum) from the American teaberry (*Spirea ulmaria*). Gaultheria oil (wintergreen oil) contains large amounts of methylsalicylate from which free salicylic acid can easily be obtained.

Chemical Synthesis of Salicylic Acid The modern pharmaceutical history of salicylate and its derivatives begins with the synthetic production of the compound. In 1859, *Hermann Kolbe,* a German from Marburg, produced the first fully synthetic salicylic acid from the already known decomposition products phenol and carbonic acid. Kolbe later improved the technology by using sodium phenolate and carbon dioxide under high-pressure conditions at 140 °C; he was the first to receive a patent for this procedure. Kolbe stimulated his student *Friedrich von Heyden* to further improve the technique to make the compound on an industrial scale. Von Heyden did this in the kitchen of his villa in Dresden (Saxony). The development of a technology to synthesize large amounts of salicylate, independent of the limited availability of natural sources with varying contents and seasonal variations of the active ingredient, opened the door for its broader clinical use and caused a massive drop in price: the price of 100 g of salicylic acid prepared from salicin (gaultheria oil) dropped from 10 to 1 Taler/100 g (Dollar = American for Taler) for the chemical product made through Kolbe's synthesis (cited after Ref. [6]). In 1874, von Heyden founded a factory Salizylsäurefabrik Dr. von Heyden in Radebeul, today part of Dresden (Germany). This factory was extremely successful: after making 4000 kg of salicylic acid in the first year, the annual production was increased to 25 tons only 4 years later. Thus, salicylate became available and known to all civilized countries (cited after Ref. [7]). Interestingly, after solving some legal issues, von Heyden also produced the salicylic acid that was later used by Bayer to make aspirin [8].

Practical Use of Salicylates After salicylate as a cheap chemical became available in essentially unlimited amounts, it was tested for several practical applications. For example, salicylic acid was soon found to have antiseptic properties that could be useful to preserve milk and meat. The compound was also recommended as an alternative to carbolic acid, which was the antiseptic of choice in surgery those days. The antipyretic action of salicylate in infectious diseases was for a time attributed to its antiseptic activity, until it was shown that the sodium salt with little antiseptic properties was an equally effective antipyretic (cited after Ref. [1]). Importantly, salicylic acid was also studied as a potential drug in a large variety of diseases. In 1875, *Ebstein* and *Müller* detected the blood sugar-lowering action of the compound [9]. Shortly thereafter, the uricosuric action of salicylate was described. Thus, salicylates appeared to be useful for treatment of diabetes and gout.

Salicylic Acid as an Antirheumatic Agent Of the several discoveries regarding practical applications of salicylates, the most significant was the finding that synthetic salicylates were potent anti-inflammatory analgesics and useful for treating rheumatic diseases. *Franz Stricker* was the first to publish that sodium salicylate was not only an antipyretic remedy but also an effective drug for treatment of rheumatic fever. He introduced salicylate in 1876 as an analgesic antirheumatic drug at the Charité in Berlin [10]. Two months later, Scottish physician *T.J. MacLagan* published the first of a series of articles showing that administration of salicylate to patients with rheumatic fever resulted in the disappearance of pain and fever. Similar results were reported by Frenchman *Germain Sée* 1 year later [11]. These three studies marked the beginning of the widespread therapeutic use of salicylates, here sodium salicylate, as analgesic anti-inflammatory drugs in clinical practice.

1.1.3 Synthesis of Acetylsalicylic Acid and First Clinical Studies

The Invention of Acetylsalicylic Acid Despite the undoubted benefits of sodium salicylate in the treatment of pain, fever, and inflammatory disorders, there were several problems with the practical handling of the compound. These included an unpleasant sweetish taste and, in particular, irritation of the stomach, often associated with nausea and vomiting. Another side effect was hearing disorders (tinnitus). These side effects were frequent at the high doses of 4–6 g per day, which had to be taken regularly by patients suffering from chronic (rheumatic) pain. Thus, after having an effective technology to generate large amounts of synthetic salicylate, several efforts were made to improve the efficacy of the compound by appropriate chemical modifications, eventually resulting in reduced dosage and improved tolerance. Several groups of researchers addressed this issue with different results [12, 13] (discussed subsequently) until scientists at the Bayer Company in Elberfeld, today in Wuppertal (Germany), succeeded in synthesizing acetylsalicylic acid (ASA) in a chemically pure and stable form. The compound was soon found to be at least as effective as sodium salicylate as an antipyretic analgesic, but was considerably better tolerated.

The History of Bayer Aspirin Several individuals at Bayer Company had markedly influenced the development of aspirin. The first was *Arthur Eichengrün* (Figure 1.1). He joined the Bayer Company in 1895 and became head of the newly founded Pharmaceutical Research Department [14]. According to a report published 50 years later [15], it was he who had the idea to acetylate salicylate to improve its efficacy and tolerance. This concept of acetylation of drugs to improve their efficacy was not new at the time and had already been successfully used to synthesize phenacetin from *p*-aminophenol, a considerably more powerful analgesic.

Arthur Eichengrün (1867 –1949)

Felix Hoffmann (1868 –1946)

Figure 1.1 Arthur Eichengrün and Felix Hoffmann.

Felix Hoffmann (Figure 1.1) was the chemist working on this issue in Eichengrün's group, and it was he who originally worked out a new technology of the acetylation reaction of salicylate (and other natural products such as guaiacole, cinchonine, and morphine). He was the first to produce chemically pure and stable ASA on August 10, 1897, according to his handwritten notes in the laboratory diary (Figure 1.2).

Two other individuals have to be mentioned in this context: *Heinrich Dreser*, head of the Department of Pharmacology at Bayer, and *Carl Duisberg*, the then head of the research and later president of the Bayer Company. Dreser was not interested in this kind of research. Initially, he was also not informed by the pharmacists about the successful clinical testing of the new compound, although according to his contract with Bayer, the pharmacists should have had reported this finding to him. He was probably not amused to hear about these results. According to Eichengrün [15] and other sources, he did everything to block the further development of the compound. Duisberg, the key manager in charge, emphatically supported the activities of Eichengrün and Hoffmann. The further development and clinical introduction of aspirin as an antipyretic analgesic, eventually resulting in the worldwide spread of the compound, is his merit. The new drug received the trade name "aspirin," which is composed from "acetic" and "spireic acid," a former name of *o*-hydroxybenzoic acid (salicylic acid), originally prepared from *S. ulmaria*, the richest natural source of salicylates.

The first description of the pharmacology of aspirin was published in 1899 by Dreser [16]. The names of the inventors Hoffmann and Eichengrün were not mentioned in this paper. Dreser considered aspirin as a better tolerable prodrug of the active principal salicylic acid. This was basically correct even from today's viewpoint for the symptoms the substance was supposed to be used at the time. According to Eichengrün [15], Dreser had nothing to do with the invention. Interestingly, it was neither Eichengrün nor Hoffmann but Dreser who had the financial benefits from the discovery. According to a contract with Bayer, the products invented under the direction of Eichengrün had to be patented in Germany to get a royalty for the inventor from the company [15]. Acetylsalicylic acid was registered on February 1, 1899 under the trade name "Aspirin" by the Imperial Patent Bureau (Kaiserliches Patentamt) in Berlin and a few weeks later introduced as a tablet (1 tablet = 5 gr(ain) 325 mg) in Germany. However, the compound did not receive recognition as a substance to be patented in Germany. Aspirin was patented in 1900 exclusively in the United States (Figure 1.3). This patent expired in 1917. In 1918, after the World War I ended, Bayer's assets were considered enemy property, confiscated by the US government, and auctioned in the United States the same year [17]. It took Bayer about 80 years to buy back the rights of the trade name "Aspirin."

A recent article raised doubts about Hoffmann as the inventor of aspirin and ascribed this merit to Eichengrün [18]. Eichengrün was Jewish and lost the fruits of his scientific research, including the invention of several other products in addition to aspirin, such as acetate silk, because of political reasons during the Nazi regime after 1933. Eichengrün was interned in 1944 in a concentration camp and remained there until the end of World War II. In 1949, the year of his death, Eichengrün stated in an article, published in the German scientific journal *Die Pharmazie*, that it was he and Felix Hoffmann who should be considered as the inventors of aspirin. Though Hoffmann was probably the person who designed and did the experiments eventually resulting in pure and chemically stable ASA – additionally evidenced by the (obviously undisputed) sole mention of his name on the patent application in 1900 (Figure 1.3) – one should also ascribe a significant contribution to Eichengrün.

Further Attempts to Make Acetylsalicylic Acid At this point, it should be noted that Hoffmann was not the first person who tried to chemically synthesize ASA. In 1853, Frenchman *Charles Frédéric Gerhardt* from Straßburg (Alsace) described the synthesis of a new compound from acetyl chloride and sodium salicylate, which he named Salicylate acétique (see Ref. [19]).

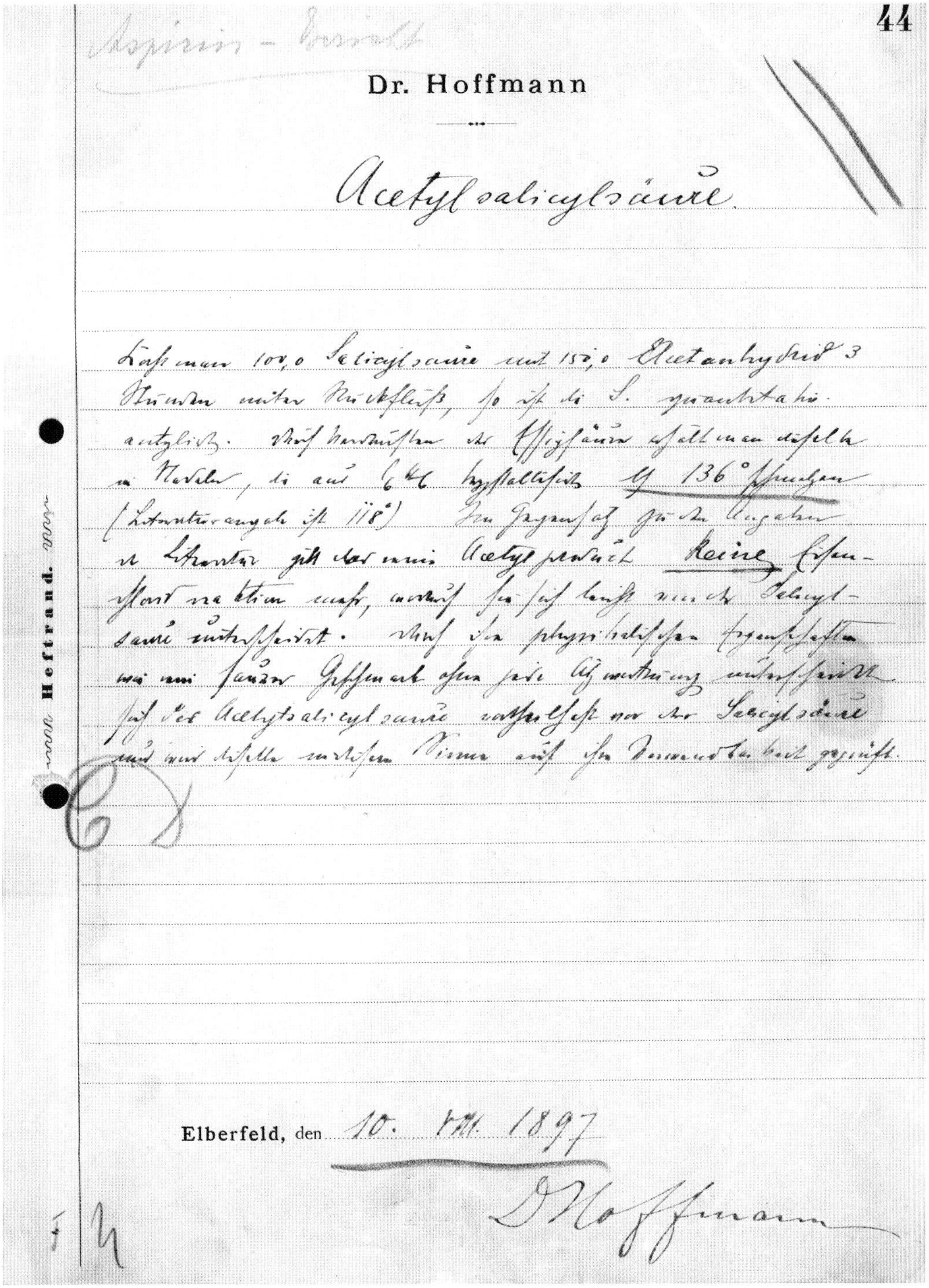

44

Dr. Hoffmann

Acetylsalicylsäure.

Läßt man 100,0 Salicylsäure mit 150,0 Acetanhydrid 3 Stunden unter Rückfluß, so ist die S. grösstentheils acetylirt. Nach Verdünsten der Essigsäure erhält man dieselbe in Nadeln, die aus C_6H_6 krystallisirt bei 136° schmelzen (Literaturangabe ist 118°). Im Gegensatz zu den Angaben der Literatur gibt das reine Acetylprodukt keine Eisenchloridreaktion mehr, wodurch sie sich leicht von der Salicylsäure unterscheidet. Nach den physikalischen Eigenschaften wie einem sauren Geschmack ohne jede Aetzwirkung unterscheidet sich die Acetylsalicylsäure vortheilhaft von der Salicylsäure und wird dieselbe in diesem Sinne auf ihre Verwendbarkeit geprüft.

Heftrand.

Elberfeld, den 10. VIII. 1897

Dr Hoffmann

Figure 1.2 Laboratory record of Dr Felix Hoffmann for August 10, 1897.

This publication of Gerhardt was used by several authors to ascribe the invention of ASA to him (e.g., [1, 20]). This appears not to be correct for several reasons. His preparation of ASA was impure, due to the insufficient technical procedure used by him [12], and rather a labile, intermediate raw product of the reaction between acetyl chloride (prepared by him by a suboptimal procedure) and sodium salicylate. The chemical structure was not determined. Because of inappropriate processing of the raw product, Gerhardt only obtained salicylic acid as a stable end product. He concluded that acetylated salicylic acid is unstable and in water immediately breaks down to salicylic acid and acetate. Both statements are wrong and do not qualify Gerhardt for the claim to have invented the synthesis of ASA [7].

In 1859, *H. von Gilm*, a pharmacist from Innsbruck (Austria), reported the synthesis of ASA as did *Karl Kraut* 10 years later in 1869. Again, these preparations were impure – see also comments in the patent application of Hoffmann (Figure 1.3). During the following 20 years, there were apparently no further attempts to improve the synthetic procedure to obtain ASA as a pure, chemically stable compound.

Thus, the origin of ASA, in contrast to the natural product salicylic acid, was exclusively in organic chemistry. From the point of view of an organic chemist, the substance had no obvious practical benefit, and there were definitely no ideas or even concepts about its possible use as a therapeutic agent. Thus, ASA probably would have suffered the fate of several hundreds of chemicals before and many thousands thereafter – a product of chemical synthesis, principally easy to make but more difficult to generate in pure and chemically stable form and without any practical significance. On the contrary, Hoffmann and Eichengrün, combined their knowledge about the chemistry of a natural product with synthetic chemistry with the intention to make a new drug out of it with improved pharmacological properties. These studies would probably not have been done without the support of the Bayer Company. Therefore, the company had good reason to celebrate the 100th anniversary of the compound, which in the meantime became the most popular drug in the world [18].

In this context, an interesting comparison with the discovery of prostacyclin can be made. Its chemical structure and a suggested (later confirmed) enzymatic synthetic pathway were originally described in 1971 by Pace-Asciak and Wolfe. These authors considered this (labile) product as just another prostaglandin, in addition to dozens of already known compounds, which was possibly overlooked by earlier investigators because of its low biological activity. This was tested at that time in bioassay experiments using the rat stomach strip. It also remained uncertain whether the compound was synthesized at all in the intact stomach wall and, if so, was released in biologically active amounts [21].

A completely different approach was followed by the group around Sir John Vane. The group's work on prostacyclin started with the discovery of a biological effect – inhibition of platelet aggregation – of an enzymatic product made from prostaglandin endoperoxides on artery walls. This prostaglandin, originally named PGX, differed in its biological behavior from all other known prostaglandins [22]. PGX was later identified as the already known enzymatic product of prostaglandin endoperoxides, described by Pace-Asciak and Wolfe, and was renamed as prostacyclin (PGI_2). Despite the originality and merits of Pace-Asciak and Wolfe regarding the detection of biosynthetic pathways of natural prostacyclin and its chemical structure, the medical history of prostacyclin starts with the work of Vane's group, which was the first to discover the biological significance of prostacyclin for control of hemostasis.

The Introduction of Acetylsalicylic Acid into the Clinics *Kurt Witthauer*, a specialist in internal medicine in a city hospital (Diakonie Krankenhaus – still existent!) in Halle/Saale (Germany), and Julius Wolgemuth [23] from Berlin published the first clinical investigations on aspirin in 1899. Witthauer began his report as follows:

> "... Nowadays, certain courage is necessary to recommend a new drug. Almost every day those are thrown on the market and one has to have an excellent memory to keep all the new names and brands in mind. Many drugs appear, are praised and recommended by companies and certain authors but after a short time have disappeared without any further comments [24]."

The author also did not forget to instruct his readers that he did this study with "not little distrust."

UNITED STATES PATENT OFFICE.

FELIX HOFFMANN OF ELBERFELD GERMANY ASSIGNOR TO THE FARBEN-FABRIKEN OF ELBERFELD COMPANY OF NEW YORK.

ACETYL SALICYLIC ACID.

SPECIFICATION forming part of Letters Patent No. 644,077, dated February 27, 1900.

Application filed August 1, 1898 Serial No. 087,385 (Specimens.)

To all whom it may concern:

Be it known that I, FELIX HOFFMANN, doctor of philosophy, chemist, (assignor to the FARBENFABRIKEN OF ELBERFELD COMPANY, of New York) residing at Elberfeld, Germany, have invented a new and useful Improvement in the Manufacture or Production of Acetyl Salicylic Acid; and I hereby declare the following to be a clear and exact description of my invention.

In the *Annalen der Chemie und Pharmacie,* Vol. 150, pages 11 and 12, Kraut has described that he obtained by the action of acetyl chlorid on salicylic acid a body which he thought to be acetyl salicylic acid. I have now found that on heating salicylic acid with acetic anhydride a body is obtained the properties of which are perfectly different from those of the body described by Kraut. According to my researches the body obtained by means of my new process is undoubtedly the real acetyl salicylic acid

$$C_6H_4\begin{cases} OCO.CH_3 \\ COOH. \end{cases}$$

Therefore the compound described by Kraut cannot be the real acetyl salicylic acid but is another compound. In the following I point out specifically the principal differences between my new compound and the body described by Kraut.

If the Kraut product is boiled even for a long while with water, (according to Kraut's statement,) acetic acid is not produced, while my new body when boiled with water is readily split up, acetic and salicylic acid being produced. The watery solution of the Kraut body shows the same behavior on the addition of a small quantity of ferric chlorid as a watery solution of salicylic acid when mixed with a small quantity of ferric chlorid—that is to say, it assumes a violet color. On the contrary, a watery solution of my new body when mixed with ferric chlorid does not assume a violet color. If a melted test portion of the Kraut body is allowed to cool it begins to solidify (according to Kraut's statement) at from 118° to 118.5° centigrade while a melted test portion of my product solidifies at about 70° centigrade. The melting-points of the two compounds cannot be compared because Kraut does not give the melting-point of his compound. It follows from those details that the two compounds are absolutely different.

In producing my new compound I can proceed as follows, (without limiting myself to the particulars given:) A mixture prepared from fifty parts of salicylic acid and seventy-five parts of acetic anhydride is heated for about two hours at about 150° centigrade in a vessel provided with a reflex condenser. Thus a clear liquid is obtained, from which on cooling a crystalline mass is separated, which is the acetyl salicylic acid. It is freed from the acetic anhydride by pressing and then recrystallized from dry chloroform. The acid is thus obtained in the shape of glittering white needles melting at about 135°centigrade, which are easily soluble in benzene, alcohol, glacial acetic acid, and chloroform, but difficultly soluble in cold water. It has the formula

$$C_6H_4\begin{cases} OCOCH_3 \\ COOH \end{cases}$$

and exhibits therapeutical properties.

Having now described my invention and in what manner the same is to be performed, what I claim as new, and desire to secure by Letters Patent, is—

As a new article of manufacture the acetyl salicylic acid having the formula:

$$C_6H_4\begin{cases} O.COCH_3 \\ COOH \end{cases}$$

being when crystallized from dry chloroform in the shape of white glittering needles, easily soluble in benzene, alcohol and glacial acetic acid, difficultly soluble in cold water being split by hot water into acetic acid and salicylic acid, melting at about 135° centigrade substantially as herein before described.

In testimony whereof I have signed my name in the presence of two subscribing witnesses.

Witnesses:
R. E. JAHN,
OTTO KÖNIG.

FELIX HOFFMANN.

Figure 1.3 The acetyl salicylic acid patent.

However, his impressions about the results were obviously quite positive and he came to the conclusion:

> "... According to my positive results, the company is now prepared – after waiting for a long while – to introduce the new compound on the market. I sincerely hope that the difficult technology to make it will not cause a too high price, to allow the broad general use of this valuable new drug."

Aspirin as a Household Remedy Against Fever, Inflammation, and Pain Soon after the introduction of ASA into medical use under the brand name "aspirin," the new drug became a very popular remedy against fever, inflammation, and pain. A local German newspaper, *Kölner Stadtanzeiger*, published the following recommendation for treatment of flu on March 6, 1924:

> "... As soon as you feel yourself ill, you should go to bed and have a hot-water bottle at your feet. You should drink hot chamomilae tea or grog in order to sweat and should take 3 tablets of aspirin a day. If you follow these instructions you will recover within a few days, in most cases ...."

This extract is remarkable for several reasons: during the past 25 years of practical use, aspirin had become a drug whose name was not only well known to health professionals but also to the general public. Certainly, the flu pandemia with millions of victims alone in Europe at the beginning of the last century as well as the limited availability of antipyretic analgesics other than aspirin contributed to this. However, the compound was generally recommended – and accepted – by the lay man and doctors – as a "household remedy" for treating pain, fever, inflammation, and many other kinds of "feeling bad," although essentially nothing was known about the mechanisms of action behind these multiple activities of the drug. It was only in the 1950s, when the first report was published, that salicylates including aspirin at anti-inflammatory doses uncouple oxidative phosphorylation in a number of organs and tissues [25]. However, this at the time was considered to be mainly of toxicological interest. Whether this contributes to the clinical efficacy of aspirin as anti-inflammatory agent is still unknown.

1.1.4 Mode of Aspirin Action

Aspirin and Prostaglandins In 1971, the journal *Nature* published three articles of the group of *John Vane* at the Royal College of Surgeons of England. These articles demonstrated for the first time a mechanism of action of aspirin that explained the multiple biological activities of the compound by one single pharmacological effect: inhibition of prostaglandin biosynthesis [26]. In his pioneering paper, the later Sir John Vane showed by elegant bioassay experiments that aspirin – and salicylate – inhibited prostaglandin formation in cell-free systems after tissue injury (Figure 1.4). This finding and his later discovery of prostacyclin were honored with the Nobel Prize for Medicine in 1982.

Inhibition of Prostaglandin Synthesis as a Mechanism of Action for Aspirin-like Drugs

J. R. VANE

Department of Pharmacology, Institute of Basic Medical Sciences, Royal College of Surgeons of England, Lincoln's Inn Fields, London WC2A 3PN

NATURE NEW BIOLOGY VOL. 231 JUNE 23 1971

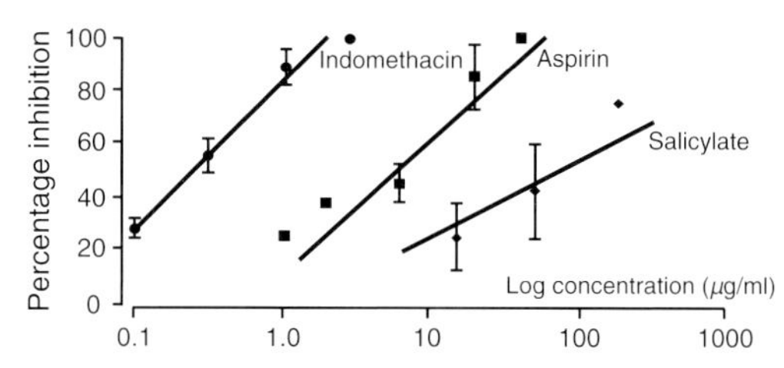

Figure 1.4 First description of inhibition of prostaglandin biosynthesis by aspirin and salicylate and the reference compound indomethacin by John Vane. Note the dose dependency of this reaction by all compounds including aspirin (modified after [26]).

Prostaglandins, thromboxane A_2, and leukotrienes are members of a group of natural lipid mediators that are generated by oxidation from arachidonic acid. Because of this origin, they all have a 20-carbon backbone and are summarized as "eicosanoids" (Greek: *eikos* = twenty). Today, more than 150 eicosanoids are known and have been structurally identified. Arachidonic acid, the precursor fatty acid, is a constituent of the cell membrane phospholipids and is released from them by phospholipases. Eicosanoid synthesis starts with the availability of free arachidonic acid.

The first oxidation step of arachidonic acid to generate prostaglandins is catalyzed by cyclooxygenases (COXs). These enzymes are widely distributed throughout the body. The primary products – the prostaglandin endoperoxides – are then converted to the terminal products of this pathway, that is, prostaglandins and thromboxane A_2, in a more cell-specific manner. The active products are not stored but released, act on their cellular target, and are afterward degraded enzymatically.

Prostaglandins exert their multiple actions via specific G-protein-coupled receptors at the cell surface. The direction and intensity of these actions is determined by the kind and the number of available prostaglandin receptors from which today about 10 are known. Prostaglandins act as local mediators that dispatch signals between cells. Thus, prostaglandin generating cells, and these are probably all cells of the body, do not require prostaglandin biosynthesis for survival. Consequently, prostaglandins are not essential for vital functions, such as energy metabolism or maintenance of the cell cytoskeleton.

The cellular prostaglandin synthesis can be markedly increased in response to disturbed homoeostasis (injury) to adapt cellular functions to changes in the environmental conditions. An increased prostaglandin synthesis "on demand," therefore, reflects a tissue-specific response to increased needs. Examples for physiological stimuli are hemostasis and pregnancy, whereas the increased prostaglandin production in inflammation, atherosclerosis, and tumorigenesis rather reflects the response to pathological stimuli.

Thus, any change in generation of prostaglandins or the related thromboxanes *per se* is neither good nor bad but rather reflects a functioning cell-based adaptation or defense mechanism. Functional disorders may arise, when prostaglandins become limiting factors for control of cell and organ function, respectively. Thus, any pharmacological interference with these processes may be either positive or negative but in most cases is not associated with any measurable functional change at the organ level as long as other mediator systems can compensate for it.

Aspirin and Cyclooxygenases Aspirin blocks the biosynthesis of prostaglandins and thromboxane A_2 at the level of prostaglandin endoperoxides or cyclooxygenase(s) by irreversible acetylation of a critical serine in the substrate channel of the COX enzyme (Section 2.2.1). This limits the access of substrate (arachidonic acid) to the catalytic active site of the enzyme [27] and explains the antiplatelet action of the substance, first described by the group of *Philip Majerus* [28]. The group of *William Smith* and *David DeWitt* detected this unique mechanism of action and has made major other contributions to this issue. The contributions of William Smith in elucidating the molecular reaction kinetics of aspirin were acknowledged with the Aspirin Senior Award in 1997.

Two genes have been identified that encode for cyclooxygenases: COX-1 and COX-2. In addition, there is a steadily increasing number of splice variants of these two genes. They are also transcriptionally regulated and might cause synthesis of gene products. Both COX isoforms are molecular targets for aspirin. However, the inhibition of COX-1 appears to dominate at lower concentrations of the compound, whereas aspirin and its primary metabolite salicylate are about equipotent inhibitors of COX-2 (Section 2.2.1). Thus, aspirin contains two pharmacologically relevant groups: the reactive acetyl moiety and salicylate. Both components are biologically active and act independently of each other at different sites. The molecular interaction of aspirin with COX-1 was further elucidated after the crystal structure of the enzyme became clarified by Michael Garavito and his group. The contribution of *Patrick Loll* to this work was acknowledged with the Aspirin Junior Award [29].

The detection of inhibition of prostaglandin synthesis was the first plausible explanation for the multiple pharmacological actions of aspirin via an ubiquitary class of endogenous mediators, prostaglandins and thromboxanes. With increasing knowledge of the complex nature of these reactions, specifically the multiple interactions of prostaglandins with other mediator systems, some details of these findings are now interpreted in a different way. This is particularly valid for the anti-inflammatory activities of aspirin, which are mainly due to the formation of the more stable metabolite salicylate [30] and, possibly, the generation of aspirin-triggered lipoxin (ATL), resulting from the interaction of aspirin-treated (acetylated) COX-2 and the 5-lipoxygenase from white cells (Sections 2.2.1 and 2.3.2). This eventually resulted in the detection of resolvins, a new class of anti-inflammatory mediators, also involved in aspirin action by *Charles Serhan* and his group. The contributions of *Jose Claria* to this work [31] were acknowledged with the Aspirin Junior Award in 1996.

Aspirin and Gene Transcription After the discovery of the inducible isoform of COX-2, it became rapidly clear that aspirin was rather ineffective on this enzyme at low antiplatelet concentrations, and higher doses were required to suppress COX-2-dependent prostaglandin formation. This finding also confirmed the clinical experience that the analgesic and anti-inflammatory effects of the compound require substantially higher doses than those necessary for inhibition of platelet function. *Kenneth Wu* and colleagues were the first to show that aspirin and salicylate interact with the binding of transcription factors to the promoter region of the COX-2 gene [32]. These factors regulate the gene expression level after stimulation by inflammatory mediators. Later work of *Xiao-Ming Xu* and others of this group eventually identified the binding of CCAAT/enhancer-binding protein-β (C/EBP-β or NF/IL-6) as one critical control mechanism [33, 34]. It becomes now increasingly evident that the anti-inflammatory and antineoplastic actions of salicylates and aspirin, respectively, involve inhibition of COX-2 gene transcription because COX-2 overexpression is an important permissive factor in these disorders and also might generate compounds other than prostaglandins (Section 4.3.1). The molecular mechanisms of control of inducible COX activity are still under intense research. However, it is interesting to note that the upregulation of salicylate biosynthesis by plants – the natural sources of salicylates – represents a most effective, transcriptionally regulated defense system that becomes activated in response to about all kinds of noxious stimuli and exhibits a number of similarities to the prostaglandin pathway in animals and men (Section 2.2.2).

1.1.5
Anti-Inflammatory/Analgesic Actions of Aspirin

The disclosure of a causal relationship between inhibition of prostaglandin synthesis and anti-inflammatory/analgesic actions of aspirin was not only a satisfactory explanation for its mode of action but also a stimulus for the mechanism-based drug research. These new compounds should be able to block prostaglandin biosynthesis via COX inhibition and should also be more potent than aspirin to allow lower dosing at increased efficacy and reduced side effects. Indomethacin was the first of these so-called aspirin-like drugs [35] and was already used as a reference compound in the pioneering experiments of John Vane (Figure 1.4). Many others followed too. In 2005, there were more than 20 chemically defined substances on the German market, designed and developed as inhibitors of prostaglandin biosynthesis and approved for clinical use as antipyretic/anti-inflammatory analgesics. There were more than 40 (!) brands containing ibuprofen, 35 containing diclofenac, and 12 containing indomethacin, most of them available in several different galenic preparations. However, there were less than 20 preparations containing aspirin as the only active ingredient. Thus, the invention of aspirin did significantly stimulate basic research for new anti-inflammatory analgesics and still does so as is evident from the detection of ATLs and resolvins. Nevertheless, the use of aspirin has also remarkably increased after the discovery of its prostaglandin-related mode of action (Figure 1.5). Today, ASA in its different commercial preparations is still among the most frequently used antipyretic analgesics for self-medication of headache, flu, and other acute inflammatory/painful states. According to a recent survey in Germany (MONICA registry), salicylates

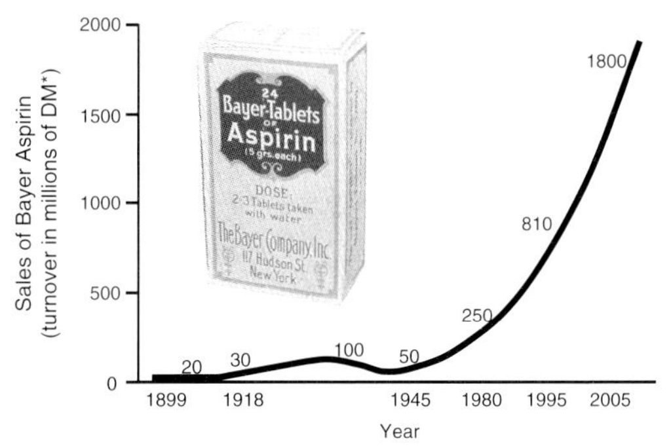

Figure 1.5 From 1899 to 2005 (projected) turnover of Bayer aspirin in millions of DM (Deutsche Mark). (*): 1 DM (Deutsche Mark) is equivalent to 0.51 € or 0.67 USD (January 2007) [37].

still keep the key position (70–80%) in OTC drugs for these symptoms [36].

1.1.6 Aspirin in the Cardiovascular System

Bleeding Time and Platelet Function Aspirin was in clinical use for about half a century, when the first reports about disturbed hemostasis were published. In 1945, *Singer*, an ETN specialist, reported late bleedings after tonsillectomies [38]. He attributed this to his prescription of aspirin for analgesic purposes. Withdrawal of aspirin or its replacement by metamizol (dipyrone) resulted in disappearance of bleeding. A relationship between aspirin intake and bleeding was also considered for tooth extractions and epistaxis [39, 40]. *Beaumont* and *Willie* [41] reported that aspirin prolonged bleeding time in patients with cardiac diseases. *Quick* and *Clesceri* [42] suggested that this effect of high-dose (6 g) aspirin might be caused by the reduction of a stable procoagulatory factor in plasma, probably prothrombin. In 1967, Quick demonstrated that a prolonged bleeding time was specific for aspirin and was not seen after salicylate was administered [43]. Similar results were obtained by the group of *Mustard* [44]. These authors confirmed the results of Quick and extended them also to several animal species. In addition, they showed that the antiplatelet effect of aspirin depends on the kind of platelet stimulus. Specifically, aspirin did not inhibit ADP-induced primary platelet aggregation, a finding that was largely ignored later in many platelet function assays *in vitro* (Section 2.3.1). During 1967–1968, *Weiss* and colleagues, *Zucker* and *Peterson*, and *O'Brien* published the first more systematic studies on the action of aspirin on platelet function in healthy men [45–48]. O'Brien found a significant inhibition of platelet function at the "subclinical" dose of 150 mg and strongly recommended a clinical trial on the compound in patients at elevated thrombotic risk.

Mechanisms of Antiplatelet Action of Aspirin The elucidation of the mechanism of action of aspirin as an antiplatelet drug starts with a study by *Bryan Smith* and *Al Willis* [49], both working at the time in the laboratory of John Vane. These authors were the first to show that inhibition of platelet function by aspirin was associated with the inhibition of prostaglandin biosynthesis and concluded that this explains the antiplatelet effects of the compound. At this time, thromboxane was still unknown. A few years later, Roth and Majerus [28] showed that aspirin causes irreversible acetylation and inhibition of platelet cyclooxygenase. The group of *Garret FitzGerald* [50] confirmed this finding as well as serine $_{529\ (530)}$ in the COX substrate channel as acetylation site for the cloned enzyme from human platelets. These studies provided the rationale for the use of aspirin as an antiplatelet drug in the secondary and primary prevention of atherothrombotic vessel occlusion, specifically myocardial infarction and ischemic stroke (Sections 4.1.1 and 4.1.2). Preeclampsia is another indication for prophylactic aspirin administration after the description of a positive effect of aspirin in high-risk patients by *Crandon* and *Isherwood* [51] and its confirmation in a randomized trial by *Beaufils* and colleagues 6 years later [52] (Section 4.1.5).

Aspirin and Prevention of Myocardial Infarction and Stroke In 1949, *Paul C. Gibson* reported for the first time the successful use of aspirin for prevention of anginal pain and coronary thrombosis [53]. This report was based on a questionnaire sent by him to 20 doctors: fifteen of them had already successfully used the compound for this condition and all of them considered aspirin as "valuable" or "very valuable," specifically with respect to its analgesic properties. Gibson explained these beneficial effects by a combination of anticoagulatory and analgesic properties of the compound. The recommended doses were 1300 mg (20 grain) or 650 mg (10 grain) aspirin.

The first larger and more systematic investigation of the significance of antithrombotic effects of aspirin for the prevention of myocardial infarctions was published in 1950 by *Lawrence Craven* [54] (Figure 1.6), a suburban general practitioner from Glendale (California) [55].

Figure 1.6 Photograph of Dr Lawrence L. Craven in 1914, at the age of 31, when he graduated from the University of Minnesota College of Medicine and Surgery (with permission of the University of Minnesota).

His finding reads in the original contribution as follows:

> ". . . during the past two years, I have advised all of my male patients between the ages of 40 and 65 to take from 10–30 grains [650–1950 mg] of acetylsalicylic acid daily as a possible preventive of coronary thrombosis. More than 400 have done so, and of these none has suffered a coronary thrombosis. From past experience, I should have expected at least a few thrombotic episodes among this group. There would appear to be enough evidence of the antithrombotic action of acetylsalicylic acid to warrant further study under more carefully controlled conditions . . .."

In the following years, Craven increased the number of his patients to about 8000 – still without having seen any myocardial infarction – and recommended the agent also for prevention of stroke [56, 57]. Unfortunately, he died in 1957, 1 year after publication of his last study, at the age of 74 from a heart attack – despite regularly using aspirin [58].

Craven's study was a stroke of luck in several aspects: first, he treated exclusively males at an age of increased risk for myocardial infarction who, according to the current knowledge, benefit most from aspirin prophylaxis. He used a dose of aspirin that was high, but in comparison to anti-inflammatory doses at the time for treatment of chronic inflammatory diseases, was rather low. Thus, not too many side effects were to be expected, which was good for the compliance of his patients. Finally, he had no problems with statistics because there were no infarctions in the patient group.

Unfortunately, these data did not find the necessary attention during the following 20 years – possibly due to the low impact factor of the journals where they were published and the fact that Craven himself died of heart attack despite regularly taking aspirin. Until the 1970s of the last century, the significance of thrombosis against spasm for the genesis of myocardial infarction was also in question. Until 1988, more than 15 000 patients were studied in seven placebo-controlled trials for the secondary prevention of myocardial infarction at the cost of many millions of dollars. None of these studies was significant on its own, possibly because, from today's viewpoint, of poor study design, the highly variable aspirin doses (300–1500 mg/day), the apparently absent systematic control of patient compliance, and a highly variable time point when aspirin treatment was started, in one study (AMIS) up to 5 years (!) after the acute event [59].

These data finished the discussion on the possible use of aspirin for the prevention of myocardial infarctions. In addition, infrequent though severe side effects such as GI or cerebral bleeding and a suggested though never established relationship with Reye's syndrome (Section 3.3.3) have tainted its reputation and resulted in its removal from the list of essential drugs by the WHO in 1988. Ironically, at about the same time, that is, 1988/1989, the first prospective randomized placebo-controlled trial – the US American Physicians' Health Study (Section 4.1.1) – and the subsequent meta-analyses by the Antiplatelet Trialists' Collaboration were published and showed a significant reduction of the incidence of myocardial infarction or other atherothrombotic events in both healthy volunteers and patients at elevated cardiovascular risk [60]. The ISIS-2 study of 1989 convincingly demonstrated for the first time a significant reduction in infarct mortality by aspirin and resulted in the first official guideline recommendation of aspirin use in these patients [61]. The prevention of atherothrombotic vessel occlusions by aspirin in patients at increased vascular risk is now a therapeutic standard. The medical decision to use aspirin in the individual patient is determined by the individual benefit/risk ratio. This issue is particularly relevant in the primary prevention in patients with a low risk profile (Section 4.1.1).

1.1.7 Current Research Topics

Clinical Applications In 1988, *Gabriel Kune* from Melbourne (Australia) published the first report on reduced incidence of colorectal carcinoma by about 40% in regular (daily) aspirin users as compared to those who did not regularly take the drug [62]. These data were generated in a retrospective, exploratory case–control study, which also noted a significant risk reduction in patients using nonsteroidal anti-inflammatory drugs (NSAIDs) other than aspirin. These data were later confirmed in a large epidemiological trial [63]. However, there are still not sufficient prospective randomized trials to calculate the individual benefit/risk ratio, considering the long treatment period of at least 10 years before significant improvements can be expected. Furthermore, all of the three available randomized trials have determined the (re)occurrence of colorectal adenomas as surrogate parameters for colorectal malignancies (Section 4.3.1). Another mutually interesting therapeutic option is Alzheimer's disease [64] (Section 4.3.2).

Basic Research One of the issues of interest in basic research with a considerable clinical concern is the so-called aspirin resistance, that is, a reduced antiplatelet activity of aspirin at antithrombotic doses. Any reduced pharmacological activity will also limit its clinical efficacy in thrombosis prophylaxis. Recent evidence, however, suggests that the clinical significance of this phenomenon might be much less than originally anticipated and, in contrast to "resistance" to clopidogrel, will not require systematic screening (Section 4.1.6).

A particular pharmacological property of aspirin that is not shared with either natural salicylate or coxibs is the acetylation of COX-2 and the subsequent generation of "aspirin-triggered lipoxin" ATL, an anti-inflammatory mediator. This may help better understand and explain clinically well-known phenomena, such as adaptation of stomach mucosa to long-term (high-dose) aspirin use (Section 3.2.1) [65] and the inhibition of neutrophil recruitment to an inflamed site [31] (Section 2.3.2). These activities might be relevant to all clinical situations with an upregulated COX-2, including acute and chronic inflammation, tumorigenesis, and advanced atherosclerosis.

Finally, the newly discovered actions of aspirin on gene regulation are of considerable interest. Control of gene expression by preventing binding of transcription factors to selected areas in the promoter region of genes is not limited to COX-2, but may also occur with other genes that are regulated by the same transcription factors, such as inducible NO synthase or other "immediate early genes" that are involved in rapid adaptation of cell function to changes in the environment. Targeted control of gene expression and function appears to be pharmacologically much more attractive – and efficient – than just the inhibition of activity of selected enzymes. Salicylates in plants are transcriptionally regulated "resistance genes" and form an essential part of cellular defense mechanisms. Research on these pleiotropic actions of aspirin in man is also a pharmacological challenge [66] and, eventually, might result in the design and development of new class(es) of "aspirin-like" drugs.

Summary

Extracts or other preparations from willow bark or leaves have been used since ancient times for the treatment of fever, inflammation, and pain. These ancient uses have been rediscovered in modern times. The identification of salicylates as the active fraction eventually resulted in its chemical synthesis, allowing broad-spectrum practical use.

The availability of synthetic salicylate was also the precondition for chemical modification of its structure to increase the activity and to reduce side effects. In this respect, the first successful synthesis of chemically pure and stable ASA by Felix Hoffmann, working in the group of Arthur Eichengrün at Bayer in 1897, was the key event. The new compound was introduced into the market in 1899 under the trade name "Aspirin" and soon became a well-known and widely accepted household remedy for the treatment of pain, fever, inflammation, and almost every kind of "feeling bad."

The first pharmacological explanation for these multiple actions was provided by the discovery of Sir John Vane in 1971 that aspirin blocked prostaglandin biosynthesis. This explained the analgesic/anti-inflammatory properties of aspirin and other salicylates, which according to the current knowledge, are probably due to the inhibition of COX-2. Interestingly, acetylation of salicylates added a new property to the compound, which is not shared by any natural salicylate – transacetylation of target proteins, most notably COX-1 in platelets – with subsequent inhibition of platelet function. Inhibition of platelet function is the rationale for the widespread use of aspirin in the prevention of thromboembolic events. Acetylation of COX-2 might result in the generation of ATL, an inhibitor of leukocyte recruitment that facilitates resolution of inflammation.

Current areas of interest in basic and clinical research on aspirin include "aspirin resistance," its definition, measurement, and significance and the possible clinical benefit of aspirin in the prevention of malignant disorders such as colorectal carcinomas and Alzheimer's disease. The mode of action of aspirin appears here to be more complex and involves, in addition to inhibition of enzyme activity, actions on gene regulation. Targeted modulation of cytokine- and tumor promoter-induced upregulation of "early response" genes, such as COX-2 or iNOS and probably others, by aspirin and salicylates appears to be much more attractive and promising than just inhibition of enzyme activity. In this way, salicylate acts in plants as a transcriptionally regulated "resistance gene." Transfer of this principle to the animal kingdom and men, eventually, might result in the design and development of new and even more effective class(es) of "aspirin-like" drugs.

Table 1.1 The history of salicylates and acetylsalicylic acid.

Date	Event
400 BC–100 AD	*Hippocrates* recommends bark and leaves of the willow tree (*Salix alba*) for medical use. This recommendation is later encyclopedized by *Pliny* and *Dioscurides* as popular medical knowledge of the time.
1763	Rev. *Edward Stone* recommends the use of willow bark extracts for treatment of "Aigues and intermitting disorders."
1826–1830	*Brugnatelli* and *Fontana* as well as *Buchner* identifiy salicin as the active antipyretic ingredient of the willow bark. *Leroux* in 1830 is the first to isolate salicin in crystalline form. The compound, prepared from willow barks, is later sold by Ernst *Merck* as an antipyretic drug for half the price of quinine.
1839	*Piria* prepares salicylic acid from salicin and correctly determines the brutto formula $C_7H_6O_3$.
1835–1843	New rich sources of natural salicylates are detected, most notably wintergreen oil from the American evergreen (*G. procumbens*), containing about 99% methylsalicylate. This finding markedly improves the availability of salicylates for practical use.
1859–1874	*Kolbe* synthesizes for the first time pure salicylate from the already known decomposition products phenol and carbonic acid. His student *von Heyden* improves the technology of synthesis and founds the first salicylic acid producing factory in Radebeul (Dresden) in 1874. The plant rapidly produces tons of salicylic acid. This provides unlimited amounts of the compound independent of natural sources for medical use.
1875	*Ebstein* and *Müller* detect the blood sugar-lowering action of salicylates.
1876	*Stricker* introduces salicylate as an analgesic/antirheumatic drug at the Charité in Berlin. Shortly thereafter, Scottish physician *MacLagan* and Frenchman *Germain Sée* from Strassburg (Alsace) also describe an antipyretic/analgesic activity of the compound.
1897	*Felix Hoffmann*, working in the pharmaceutical research group at Bayer laboratories in Elberfeld under direction of *Arthur Eichengrün*, synthesizes for the first time acetylsalicylic acid as a chemically pure and stable compound.
1899	*Heinrich Dreser*, head of the pharmacological research laboratories at Bayer, publishes the first report on the pharmacology of acetylsalicylic acid. He considers the compound as a prodrug of the active metabolite salicylic acid. The first clinical studies by *Witthauer* and *Wolgemuth* are published the same year.
1899	Introduction of acetylsalicylic acid for the treatment of fever and pain under the trade name "Aspirin."
Since 1899	Worldwide use of aspirin as a household remedy for treatment of fever, pain, and inflammation.
1945–1952	*Singer* describes a bleeding tendency after surgical interventions if aspirin was used for analgesic purposes. This observation is confirmed in other case reports. Singer explains this by a reduction of prothrombin levels.
1949–1950	*Gibson* reports of the positive results with aspirin for the treatment of anginal pain, according to a survey by 20 physicians. *Craven*, a general practitioner from Glendale (California), publishes shortly thereafter his first study on antithrombotic effects of aspirin. According to his data, daily administration of 650–1950 mg aspirin completely prevented myocardial infarctions in 400 male, medium-aged patients during an observation period of 2 years.

(*Continued*)

Table 1.1 (*Continued*)

Date	Event
1967–1968	*O'Brien, Zucker, Weiss*, and coworkers publish the first mechanistic studies on the antiplatelet activity of aspirin. Daily doses of 75 mg cause an inhibition of platelet function over several days. According to his data, O'Brien recommends a clinical trial with aspirin for thrombosis prevention in patients at elevated vascular risk.
1971	*Sir John Vane* detects the inhibition of prostaglandin synthesis by aspirin (and salicylate) and considers this as the mechanism of the anti-inflammatory action. This work and the later detection of prostacyclin are acknowledged with the Nobel Prize for medicine in 1982.
1971	*Brian Smith* and *Al Willis*, both working in John Vane's laboratory, detect the inhibition of prostaglandin synthesis by aspirin in platelets and explain its antiplatelet action by this property.
Since 1971	Systematic search for and development of cyclooxygenase inhibitors with prospective use as symptomatic anti-inflammatory analgesics.
1975	The group of Philip *Majerus* detects the acetylation of platelet cyclooxygenase by aspirin and explains by this mechanism the inhibition of thromboxane formation and platelet function.
1979	*Crandon* and *Isherwood* report that regular intake of aspirin in pregnancy reduces the risk of preeclampsia. *Beaufils* confirms this finding in high-risk patients in1985 in a randomized study.
1983	Publication of the first placebo-controlled randomized double-blind trial (Veterans Administration Study) on prophylactic use of aspirin (324 mg/day) in men with acute coronary syndromes by *Lewis* and colleagues. The study shows a 50% reduction of the incidence of myocardial infarctions and death within an observation period of 3 months.
1988	Publication of the first clinical findings of a relationship between aspirin intake and prevention of colon cancer by *Gabriel Kune* and colleagues from Melbourne. In this retrospective, exploratory case–control study, regular (daily) use of aspirin reduced the risk of (incident) colon cancer by 40%. These findings were later confirmed and extended by *Michael Thun* and colleagues (1991) in a large prospective epidemiological trial (CPS-II Study) in the United States.
1988–1989	Publication of the first two prospective, placebo-controlled long-term trials on primary prevention of myocardial infarctions in apparently healthy men in the United States (USPHS) and the United Kingdom (BMDS), respectively. The results are controversial. The American study suggests a beneficial effect of aspirin on the prevention of a first myocardial infarction. *Charles Hennekens* (USPHS) receives the Aspirin Senior Award in 1999 for his significant contributions to the use of aspirin for prevention of atherothrombotic events.
1989	The ISIS-2 trial, a prospective, placebo-controlled randomized trial in patients with acute myocardial infarction, demonstrates a remarkable protective action of aspirin (162 mg/day), alone and a doubling of the effect in combination with streptokinase, on prevention of recurrent myocardial infarctions and death for an observation period of 5 weeks. This study leads to guideline recommendation of aspirin in secondary prevention.
1988–1990	*William Smith, David De Witt*, and colleagues demonstrate that the molecular mechanism of aspirin action is due to steric hindering of access of the substrate (arachidonic acid) to the enzyme (cyclooxygenase) and does not involve direct binding of the agent to the active center. *William Smith* receives the Aspirin Senior Award in 1997 for this and other major contributions to the better understanding of the molecular mechanism of aspirin action.

Table 1.1 (*Continued*)

Date	Event
1991	Publication of the prospective, randomized, placebo-controlled study (SALT trial) on low-dose (75 mg/day) aspirin in patients with transient ischemic attacks (TIA). The number of strokes, TIA, and myocardial infarctions are markedly reduced by aspirin, whereas the number of hemorrhagic infarctions is increased. The benefit–risk ratio is clearly in favor of prevention.
1991	*Kenneth Wu* and colleagues show inhibition of cytokine-induced expression of cyclooxygenase (-2) in human endothelial cells by aspirin and salicylate but not by indomethacin. This suggests salicylate-mediated inhibition of gene transcription. Later work of this group [33] identifies inhibition of binding of the C/EBP-β transcription factor as (one) molecular mechanism of action.
1995	Patrick Loll, Daniel Picot, and R. Michael Garavito describe the crystal structure of COX-1 inactivated by an aspirin analogue. *Patrick Loll* receives the Aspirin Junior Award in 2000 for his significant contributions to this research.
1995	Jose Claria and *Charles Serhan* detect the generation of ATL by the interaction of acetylated COX-2 with the 5-lipoxygenase of white cells. The contributions of *Claria*, working in Serhan's group, to this research are acknowledged with the Young Researchers' Aspirin Award in 1996.
Since 1998	Search for new fields of clinical use of aspirin and aspirin-"like" drugs, including the prevention of progression of Alzheimer's disease.
2000	The Oxford group around *Sir Richard Peto, Rory Collins,* and *Peter Sleight* receives the Aspirin Senior Award for their outstanding contributions to developing and conducting meta-analyses in studies with antiplatelet drugs. The latest edition dates to 2002.
2005	Publication of the prospective, randomized, placebo-controlled Women's Health Study (WHS). The study demonstrates the usefulness of aspirin (100 mg each second day) for the prevention of atherothrombotic events in apparently healthy women during a 10-year observation period. There is significant protection from ischemic cerebral infarctions but not from myocardial infarctions. The protective action of aspirin increases with increasing vascular risk.
2006	The CHARISMA study compares low-dose aspirin (75–162 mg) alone and in combination with clopidogrel (75 mg) in primary prevention of vascular events in high-risk populations. Although the combination is useful in secondary prevention, similar to CAPRIE, the comedication of clopidogrel with aspirin did not reduce the vascular risk but increased bleeding in patients with risk factors but without preexisting event.
2010	?

References

1 Gross, M. and Greenberg, L.A. (1948) *The Salicylates. A Critical Bibliographic Review*, Hillhouse Press, New Haven, CT.

2 MacLagan, T. (1876) The treatment of rheumatism by salicin and salicylic acid. *British Medical Journal*, **1** (803), 627.

3 Stone, E. (1763) An account of the success of the bark of the willow in the cure of agues. *Transactions of the Royal Entomological Society of London*, **53**, 195–200.

4 Sharp, G. (1915) The history of the salicylic compounds and of salicin. *The Pharmaceutical Journal*, **94**, 857.

5 Horsch, W. (1979) Die Salicylate. *Die Pharmazie*, **34**, 585–604.
6 Bekemeier, H. (1977) *On the History of the Salicylic Acid, Vol. 42 (R34)*, Wissenschaftliche Beitrage der Martin-Luther-Universität Halle-Wittenberg, pp. 6–13.
7 Schlenk, O. (1947) *Die Salizylsäure. Arzneimittel forschungen 3*, Verlag Dr. Werner Saenger, Berlin.
8 Schreiner, C. (1997) *100 Years Aspirin. The Future has Just Begun*, Bayer AG.
9 Ebstein, W. and Müller, J. (1875) Weitere Mittheilungen über die Behandlung des Diabetes mellitus mit Carbolsäure nebst Bemerkungen über die Anwendung von Salicylsäure bei dieser Krankheit. *Berliner Klinische Wochenschrift*, **12**, 53–56.
10 Stricker, S. (1876) Über die Resultate der Behandlung der Polyarthritis rheumatica mit Salicylsäure. *Berliner Klinische Wochenschrift*, **13**, 1–2, 15–16, 99–103.
11 Sée, G. (1877) Études sur l'acide salicyliquè et les salicylates; traitement du rhumatisme aigu et chronique, de la goutte, et de diverses affections du système nerveux sensitif par les salicylates. *Bulletin de l'Academie Nationale de Medecine (Paris)*, **6**, 689–706. 717–754.
12 Gordonoff, T. (1965) Zur Geschichte der Antipyrese. *Wiener Medizinische Wochenschrift*, **115**, 45–46.
13 Hangarter, W. (1974) Herkommen, Geschichte, Anwendung und weitere Entwicklung der Salizylsäure, in *Die Salizylsäure und ihre Abkömmlinge* (ed. W. Hangarter), F.K. Schattauer, Stuttgart, pp. 3–11.
14 Stadlinger, H. (1947) Arthur Eichengrün 80 Jahre. *Die Pharmazie*, **2**, 382–384.
15 Eichengrün, A. (1949) 50 Jahre Aspirin. *Die Pharmazie*, **4**, 582–584.
16 Dreser, H. (1899) Pharmakologisches über Aspirin (Acetylsalizylsäure). *Pflügers Archiv. European Journal of Physiology*, **76**, 306–318.
17 Mueller, R.I. and Scheidt, S. (1994) History of drugs for thrombotic diseases. Discovery, development and directions for the future. *Circulation*, **89**, 432–449.
18 Sneader, W. (2000) The discovery of aspirin: a reappraisal. *British Medical Journal*, **321**, 1591–1594.
19 Gerhardt, C.F. (1855) *Lehrbuch der Organischen Chemie. Deutsche Originalausgabe unter Mitwirkung von Prof. Dr. Rudolf Wagner. 3. Band*, Verlag Otto Wigand, Leipzig, p. 350.
20 Collier, H.O.J. (1963) Aspirin. *Scientific American*, **169**, 1–10.
21 Pace-Asciak, C. and Wolfe, L.S. (1971) A novel prostaglandin derivative formed from arachidonic acid by rat stomach homogenates. *Biochemistry*, **10**, 3657–3664.
22 Moncada, S., Gryglewski, R., Bunting, S. and Vane, J.R. (1976) An enzyme isolated from arteries transforms prostaglandin endoperoxides to an unstable substance that inhibits platelet aggregation. *Nature*, **263**, 663–665.
23 Wolgemuth, J. (1899) Über Aspirin (Acetylsalicylsäure). *Therapeutische Monatshefte*, **13**, 276–278.
24 Witthauer, K. (1899) Aspirin, ein neues Salicylpräparat. *Heilkunde*, **3**, 396–398.
25 Brody, T.M. (1956) Action of sodium salicylate and related compounds on tissue metabolism *in vitro*. *The Journal of Pharmacology and Experimental Therapeutics*, **117**, 39–51.
26 Vane, J.R. (1971) Inhibition of prostaglandin biosynthesis as a mechanism of action of aspirin-like drugs. *Nature – New Biology*, **231**, 232–235.
27 DeWitt, D.L., El-Harith, E.A., Kraemer, S.A. *et al.* (1990) The aspirin- and heme-binding sites of ovine and murine prostaglandin endoperoxide synthases. *The Journal of Biological Chemistry*, **265**, 5192–5198.
28 Roth, R.G. and Majerus, P.W. (1975) The mechanism of the effect of aspirin on platelets. I. Acetylation of a particulate fraction protein. *The Journal of Clinical Investigation*, **56**, 624–632.
29 Loll, P.J., Picot, D. and Garavito, R.M. (1995) The structural basis of aspirin activity inferred from the crystal structure of inactivated prostaglandin H_2 synthase. *Nature Structural Biology*, **2**, 637–643.
30 Higgs, G.A., Salmon, J.A., Henderson, B. *et al.* (1987) Pharmacokinetics of aspirin and salicylate in relation to inhibition of arachidonate cyclooxygenase and antiinflammatory activity. *Proceedings of the National Academy of Sciences of the United States of America*, **84**, 1417–1420.
31 Claria, J. and Serhan, C. (1995) Aspirin triggers previously undescribed bioactive eicosanoids by human endothelial cell-leukocyte interactions. *Proceedings of the National Academy of Sciences of the United States of America*, **92**, 9475–9479.
32 Wu, K.K., Sanduja, R., Tsai, A.L. *et al.* (1991) Aspirin inhibits interleukin 1-induced prostaglandin H synthase expression in cultured endothelial cells. *Proceedings of the National Academy of Sciences of the United States of America*, **88**, 2384–2387.
33 Xu, X.-M., Sansores-Garcia, L., Chen, X.-M. *et al.* (1999) Suppression of inducible cyclooxygenase-2 gene

transcription by aspirin and sodium salicylate. *Proceedings of the National Academy of Sciences of the United States of America*, **96**, 5292–5297.

34 Saunders, M.A., Sansores-Garcia, L., Gilroy, D. *et al.* (2001) Selective expression of C/EBPβ binding and COX-2 promoter activity by sodium salicylate in quiescent human fibroblasts. *The Journal of Biological Chemistry*, **276**, 18897–18904.

35 Shen, T.Y. (1981) Burger Award address. Toward more selective antiarthritic therapy. *Journal of Medicinal Chemistry*, **24**, 1–5.

36 Schenkirsch, G., Heier, M., Stieber, J. *et al.* (2001) Trends der Schmerzmitteleinnahme über einen Zeitraum von 10 Jahren. *Deutsche Medizinische Wochenschrift*, **126**, 643–648.

37 Jack, D.B. (1997) One hundred years of aspirin. *The Lancet*, **350**, 437–439.

38 Singer, R. (1945) Acetylsalicylic acid, a probable cause for secondary post-tonsillectomy hemorrhage. A preliminary report. *Archives of Otolaryngology, Chicago*, **42**, 19–20.

39 Smith, J.M. and MacKinnon, J. (1951) Aetiology of aspirin bleeding. *The Lancet*, **1**, 569.

40 Wising, P. (1952) Haematuria, hypoprothrombinemia and salicylate medication. *Acta Medica Scandinavica*, **141**, 256.

41 Beaumont, J.L. and Willie, A. (1955) Influence sur l'hémostase, de l'hypertension artérielle, des antivitamines K, de l'héparine et de l'acide acétyl salicylique. *Sang*, **26**, 880.

42 Quick, A.J. and Clesceri, L. (1960) Influence of acetylsalicylic acid and salicylamide on the coagulation of blood. *The Journal of Pharmacology and Experimental Therapeutics*, **128**, 95–98.

43 Quick, A.J. (1967) Salicylates and bleeding. The aspirin tolerance test. *The American Journal of the Medical Sciences*, **252**, 265–269.

44 Evans, G., Packham, M.A., Nishizawa, E.E. *et al.* (1968) The effect of acetylsalicylic acid on platelet function. *The Journal of Experimental Medicine*, **128**, 877–894.

45 Weiss, H.J. and Aledort, L.M. (1967) Impaired platelet–connective tissue reaction in man after aspirin ingestion. *The Lancet*, **2**, 495–497.

46 O'Brien, J.R. (1968) Effects of salicylates on human platelets. *The Lancet*, **1**, 779–783.

47 Weiss, H.J., Aledort, L.M. and Kochwa, S. (1968) The effects of salicylates on the hemostatic properties of platelets in man. *The Journal of Clinical Investigation*, **47**, 2169–2180.

48 Zucker, M.B. and Peterson, J. (1968) Inhibition of adenosine diphosphate-induced secondary aggregation and other platelet functions by acetylsalicylic acid ingestion. *Proceedings of the Society for Experimental Biology and Medicine*, **127**, 547–551.

49 Smith, J.B. and Willis, A.L. (1971) Aspirin selectively inhibits prostaglandin production in human platelets. *Nature – New Biology*, **231**, 235–236.

50 Funk, C.D., Funk, L.B., Kennedy, M.E. *et al.* (1991) Human platelet/erythroleukemia cell prostaglandin G/H synthase: cDNA cloning, expression, and gene chromosomal assignment. *The FASEB Journal*, **5**, 2304–2312.

51 Crandon, A.J. and Isherwood, D.M. (1979) Effect of aspirin on incidence of preeclampsia. *The Lancet*, **1**, 1356.

52 Beaufils, M., Uzan, S., Donsimoni, R. and Colau, J.C. (1985) Prevention of preeclampsia by early antiplatelet therapy. *The Lancet*, **1**, 840–842.

53 Gibson, P.C. (1949) Aspirin in the treatment of vascular diseases. *The Lancet*, **2**, 1172–1174.

54 Craven, L.L. (1950) Acetylsalicylic acid, possible prevention of coronary thrombosis. *Annals of Western Medicine and Surgery*, **4**, 95–99.

55 Miner, J. and Hoffhines, A. (2007) The discovery of aspirin's antithrombotic effects. *Texas Heart Institute Journal*, **34**, 179–186.

56 Craven, L.L. (1953) Experiences with aspirin (acetylsalicylic acid) in the non-specific prophylaxis of coronary thrombosis. *Mississippi Valley Medicinal Journal*, **75**, 38–40.

57 Craven, L.L. (1956) Prevention of coronary and cerebral thrombosis. *Mississippi Valley Medicinal Journal*, **78**, 213–215.

58 Mann, C.C. and Plummer, M.L. (1991) *The Aspirin Wars*, Alfred A. Knopf, New York.

59 Reilly, I.A. and FitzGerald, G.A. (1988) Aspirin in cardiovascular disease. *Drugs*, **35**, 154–176.

60 Antiplatelet Trialists' Collaboration (1988) Secondary prevention of vascular disease by prolonged antiplatelet treatment. *British Medical Journal*, **296**, 320–331.

61 ISIS-2 (1988) Randomized trial of intravenous streptokinase, oral aspirin, both or neither among 17,187 cases of suspected acute myocardial infarction. *The Lancet*, **2**, 349–360.

62 Kune, G.A., Kune, S. and Watson, L.F. (1988) Colorectal cancer risk, chronic illnesses, operations, and medications: case control results from the Melbourne

Colorectal Cancer Study. *Cancer Research*, **48**, 4399–4404.

63 Thun, M.J., Namboodiri, M.N. and Heath, C.W., Jr (1991) Aspirin use and reduced risk of fatal colon cancer. *The New England Journal of Medicine*, **325**, 1593–1596.

64 Breitner, J.C. and Zandi, P.P. (2001) Do non-steroidal antiinflammatory drugs reduce the risk of Alzheimer's disease? *The New England Journal of Medicine*, **345**, 1567–1568.

65 Fiorucci, S., de Lima, O.M., Jr, Mencarelli, A. *et al.* (2002) Cyclooxygenase-2-derived lipoxin A4 increases gastric resistance to aspirin-induced damage. *Gastroenterology*, **123**, 1598–1606.

66 Amin, A.R., Attur, M.G., Pillinger, M. and Abramson, S.B. (1999) The pleiotropic functions of aspirin: mechanisms of action. *Cellular and Molecular Life Sciences*, **56**, 305–312.

1.2 Chemistry

This chapter on chemistry and measurement of salicylates is focused on pharmaceutical aspects of aspirin and other salicylates. It includes a description of the physicochemical behavior of this class of compounds and methods for their measurement in biological media. This chapter is not written with the intention to provide a complete overview of all chemical and analytical aspects of salicylates but rather to inform about those issues that are relevant to the understanding of their pharmacology and toxicology in biological systems.

The first part of this chapter describes chemical structures and physicochemical properties of aspirin and other salicylates (Section 1.2.1). The physicochemical properties of salicylates are unique to this class of compounds, specifically the mesomeric structures of salicylate, eventually resulting in chelating properties and allowing incorporation of salicylate into the cell membrane phospholipids. This physicochemical property is of outstanding importance to understand the actions of salicylates on cellular energy metabolism, specifically the uncoupling of oxidative phosphorylation. Another aspect is the (poor) water solubility of aspirin and salicylic acid (in contrast to sodium salicylate), eventually resulting in local irritations of stomach mucosa after oral administration (Section 3.2.1). Finally, the particular crystal structure of aspirin and the recent discovery of two polymorphic forms, coexisting in one and the same aspirin crystal, is an issue of considerable pharmacological interest.

The second part of this chapter describes analytical methods of salicylate determination in biological media (Section 1.2.2). Several techniques are available for simultaneous measurement of aspirin and its major metabolites.

1.2.1 Structures and Chemical Properties of Salicylates

The glucoside salicin was the first active ingredient of the willow bark, which was isolated as a crystalline agent by Leroux in 1830 (Section 1.1.2). Leroux obtained 1 ounce (about 28.3 g) salicin from 3 pounds of willow bark (*Salix helix*) (cited after Ref. [67]). Salicin was later used by Piria as starting material for the preparation of salicylic acid (Section 1.1.2). Salicylic acid (*o*-hydroxybenzoic acid) is a relatively strong acid, having a pK_a of 2.9 and is poorly water soluble (0.2%). Its solubility can be considerably improved by converting the compound into the sodium salt, which is approximately 50% water soluble. Salicylates for systemic use are either esters with substitutions in the carboxyl group, such as methylsalicylate, or esters of organic acids with substitutions in the phenolic *o*-hydroxyl group, such as aspirin. Aspirin is the acetate ester of salicylic acid (Figure 1.7). The crystalline and molecular structure of aspirin has been elucidated [68, 69].

Computer calculations, however, have provided evidence for another even more stable crystalline form of aspirin with a close relationship with the already known form I [70]. Most recent experimental studies were able to confirm the real existence of form II and, in addition, showed that a polymorphism exists between these two forms. Importantly, the two different polymorphs can coexist within one and the same crystal [71, 72]. This new and unexpected finding with aspirin as the first compound to show this unique property raises a number of principal questions regarding the definition of crystal polymorphism, which is outside the further discussion about pharmacological and toxicological properties of the compound. There are also the legal issues that are important in this context – each polymorph of each compound can be patented separately.

1.2.1.1 Salicylates in Clinical Use

Salicylic Acid Salicylic acid (molecular weight 138.1) in the form of sodium salt (molecular weight 160.1) was the first purely synthetic salicylate in clinical use (Section 1.1.3). It is no more used for systemic internal administration because of its unpleasant, sweetish taste and irritation of the stomach mucosa. Not only as an antipyretic analgesic but also as an anti-inflammatory drug, it has been replaced by better tolerable agents, such as

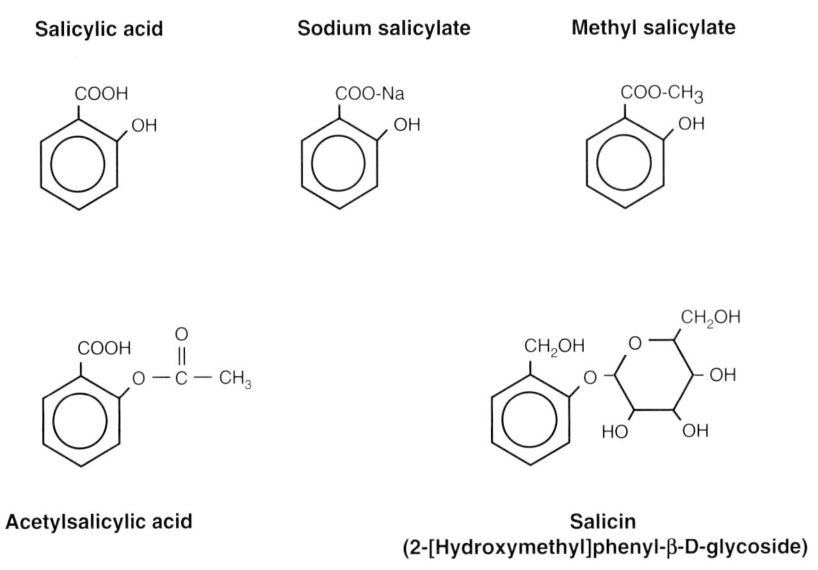

Figure 1.7 Chemical structures of selected salicylates.

aspirin or the structurally different, more potent anti-inflammatory compounds, NSAID (Section 4.2.2) and acetaminophen (paracetamol). Nevertheless, salicylate is still being used as an external medication, for example, in ointments because of its antiseptic and keratolytic properties.

Despite its disappearance from internal medicine, the pharmaceutical and biological properties of salicylate are of considerable pharmacological interest because this compound is the primary metabolite of aspirin and responsible for many of its biological actions including salicylate poisoning (Section 3.1.1). Salicylate shows a peculiar physicochemical behavior because of the formation of a ring structure by hydrogen bridging. This requires a hydroxy group in a close neighborhood of the carboxyl group and is only seen with the *o*-hydroxybenzoic acid salicylic acid (Figure 1.8) but not with its *m*- and *p*-analogues. The *o*-position of the hydroxyl group facilitates the release of a proton with decreasing pH by increasing the mesomery of the resulting anion. These properties are biologically relevant for the protonophoric actions of salicylates in the uncoupling of oxidative phosphorylation by eliminating the impermeability of cell membranes to protons (Section 2.2.3). In addition, they help understand the local irritation of the stomach mucosa subsequent to direct contact with the compound and its incorporation into mucosal cells (Section 3.2.1). The *m*- and *p*-hydroxy analogues of benzoic acid do not share these properties with salicylate and are biologically largely inactive.

Structure–activity studies of 80 salicylate-type compounds for uncoupling oxidative phosphorylation in isolated mitochondria showed that the essential pharmacophore for this activity is a compound with a negatively charged (carboxyl) group at the *o*-position, that is, acetylsalicylate. The *m*- and *p*-hydroxybenzoate analogues of salicylic acid failed to do so. This suggested

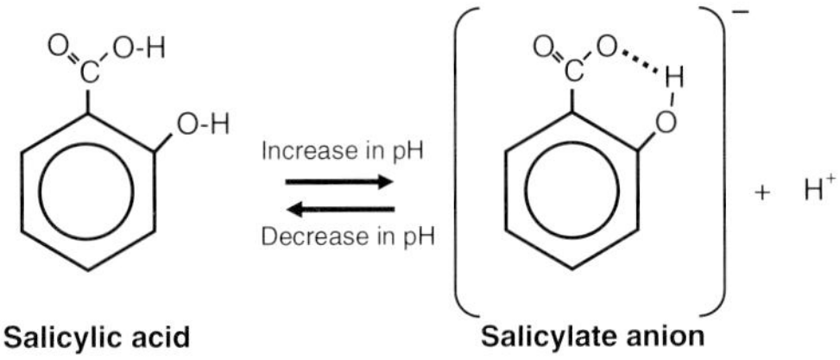

Increase in pH
Decrease in pH
Salicylic acid
Salicylate anion

Figure 1.8 pH-dependent equilibrium of ionized and nonionized forms of salicylate.

the *o*-position of the hydroxyl group as an essential steric requirement for this activity. Mechanistically, this was explained by the unique proton bridging between the oxygen of the carboxyl group and the proton in the hydroxy group, allowing for a mesomeric state that promotes a nondissociated configuration and facilitates tissue penetration [73].

Acetylsalicylic Acid Acetylsalicylic acid (molecular weight 180.2) or aspirin is the acetate ester of salicylic acid. The pharmacological properties are similar to those of salicylate. However, aspirin also has activities of its own, which are added by the reactive acetate group – the (nonselective) acetylation of cellular targets, including proteins and DNA. This results in biological effects that are not shared by salicylate. Examples are the inhibition of platelet function by irreversible acetylation of COX-1 and the acetylation of COX-2 with subsequent formation of 15-(R)-HETE and generation of ATL by white-cell 5-lipoxygenases (Section 2.2.1).

Aspirin is a white powder with a pleasant acidic taste. The compound is poorly soluble in water (0.3%) and somewhat better soluble in ethanol (20%). The solubility in aqueous media depends on pH. It amounts to only 60 μg/ml at pH 2, but increases dramatically with increasing pH (Figure 1.9). The solubility in aqueous media is also markedly improved after its conversion into the sodium salt, specifically at acidic pH (Table 1.2).

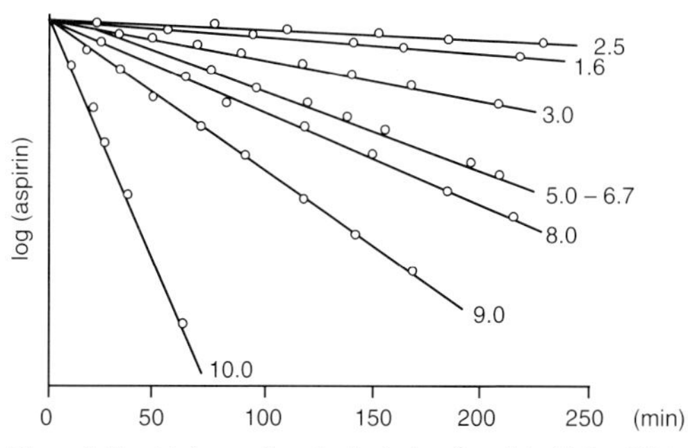

Figure 1.9 pH-dependent hydrolysis of aspirin (1.5 mM) in aqueous solution at 42 °C. Note the high stability (poor solubility) of aspirin at acidic pH and the significant increase in solubility (rapid degradation) at alkaline pH [74].

Table 1.2 Dissolution rates of various salicylates in 0.1 N hydrochloric acid (modified after Levy and Leonards, 1966).

Compound	Dissolution rate (units at current stirring and 37 °C)	
	Absorption	Release
Salicylic acid	38	1.0
Acetylsalicylic acid (aspirin)	49	1.3
Sodium acetylsalicylate	4900	130.0
Sodium salicylate	5200	140.0

This pH-dependency of solubility of aspirin is one of the reasons for its irritation of stomach mucosa in the acidic gastric juice (Section 3.2.1) as well as the increasing absorption in the upper intestine because of markedly improved solubility that also dominates the increased dissociation rate.

Methylsalicylate Methylsalicylate is the active ingredient of wintergreen oil from the American teaberry (*G. procumbens*). This oil was used as a natural source of salicylates since the early nineteenth century because it contained 98% methylsalicylate, the active constituent of numerous drug combinations for external use in rheumatic diseases. It is also of toxicological interest because of its much higher toxicity in comparison to other salicylates. Particularly dangerous is the erroneous ingestion (by children) of methylsalicylate-containing ointments and other products for external use (Section 3.1.1).

1.2.1.2 Aspirin Formulations

Several galenic preparations of aspirin have been developed for practical use with the intention to improve the solubility and stability of the compound and to reduce or even avoid gastric irritations. The first property is particularly relevant if short-term or immediate action of aspirin is desired, for example, in the treatment of migraine or tension-related headache. Alternatively, aspirin can be administered intravenously or orally as the well water-soluble lysine salt. This is of particular interest if high levels of unmetabolized aspirin have to be obtained in the systemic circulation, for example, to immediately block platelet function in the acute coronary syndromes (Section 4.1.1). Another approach is an enteric-coated formulation, predominantly for long-term use. A detailed discussion of these and other formulations in clinical use is found elsewhere (Section 2.1.1).

Generics A frequently discussed issue in clinical pharmacology is whether all formulations of all manufacturers containing the same active ingredient at the same dose are bioequivalent – with the consequence of preferential use of the cheapest formulation. In the case of salicylates, this has been refused already many years ago [74]. A more recent study additionally suggested that even aspirin analogues that passed pharmaceutical *in vitro* dissolution specifications may not be bioequivalent *in vivo* [75]. Moreover, a comparative study of selected aspirin formulations in Germany has found large differences regarding the pharmaceutical quality of aspirin-containing monopreparations with suggested use as antipyretic analgesics.

In 1986, a comparison of all pharmaceutical OTC formulations of aspirin was performed. Included were all products containing aspirin as active ingredient, which were on the German market for suggested use as antipyretic analgesics. Only tablets and no other galenic preparations or tablets for other symptoms were included.

All 11 tablets fulfilled general requirements regarding, for example, the content of the active ingredient. However, according to the authors, marked and unacceptable differences existed with respect to the individual *in vitro* release kinetics. These criteria were determined according to an US Standard and were not met by 5 out of the 11 preparations tested (Table 1.3) [76].

Table 1.3 *In vitro* release kinetics of ASA under standard conditions from commercially available ASA preparations in Germany [76].

ASA preparation	Declared ASA content (mg)	Percentage of declaration found	ASA release kinetics (% in 30 min ± SD)
Acetylin	500	98.9	89.6 ± 2.8
Aspirin	500	99.4	100.7 ± 1.0
Aspirin junior	100	102.9	103.5 ± 2.7
Aspro	320	98.1	96.4 ± 4.7
Ass 500 Dolormin	500	98.0	77.2 ± 9.2
ASS-Dura			
Ch.-B. 18613	500	98.2	65.4 ± 16.5
Ch.-B. 074035	500	102.9	72.4 ± 9.5
ASS-Fridetten			
Ch.-B. 019026	500	98.3	68.8 ± 10.0
Ch.-B. 020047	500	100.1	75.9 ± 5.7
ASS-ratiopharm	500	96.6	89.3 ± 9.3
ASS-Woelm	500	100.9	77.9 ± 8.8
Temagin ASS 600			
Ch.-B. 212142	600	100.9	50.2 ± 11.4
Ch.-B. 212143	600	100.4	44.0 ± 9.9
Trineral	600	99.2	90.8 ± 3.9

According to predefined quality standards, the content of the active ingredient should be 95–105% of declaration and at least 80% of the compound should be released within 30 min under the conditions chosen.

The authors repeated this study 2 years later on 62 different aspirin preparations. They found that of the tested formulations, 20 (!) still did not meet the quality standards mentioned above [77]. Thus, despite containing ASA as the active ingredient at identical amounts, not all aspirin preparations might be the same in terms of bioavailability of the compound.

Summary

Aspirin and its main metabolite salicylic acid are poorly water soluble at acidic and neutral pH. The solubility increases markedly in alkaline pH and is more than 100-fold higher for the sodium salts as compared to the free acids. This is also valid for strong acidic pH.

Salicylates, in contrast to aspirin, have unique physicochemical properties. These are caused by the close steric neighborhood of the acetate hydroxyl group to the carboxyl group. This allows the formation of a chelate ring structure and facilitates the release of protons. The major functional consequence is the action of salicylate as protonophore, for example, in mitochondrial membranes, to uncouple oxidative phosphorylation because of the abolition of the membrane impermeability to protons (Section 2.2.3). Neither aspirin nor other salicylates exhibit comparable physicochemical properties.

Aspirin is commercially available in many different galenic formulations. *In vitro* studies on bioequivalence provided different results for different formulations. However, the functional consequences for clinical use have not been studied sufficiently.

References

67 Büge, A. (1977) Zur Chemie der Salicylsäure und ihrer wichtigsten Derivate, in *100 Years of the Salicylic Acid as an Antirheumatic Drug, Vol. 42 (R34)* (ed. H. Bekemeier), Martin-Luther-Universität Halle-Wittenberg, Wissenschaftliche Beiträge, pp. 14–38.

68 Wheatley, P.J. (1964) The crystal and molecular structure of aspirin. *Journal of Chemical Society*, 6036–6048.

69 Kim, Y. and Machida, K. (1986) Vibrational spectra, normal coordinates and infrared intensities of aspirin crystal. *Chemical & Pharmaceutical Bulletin*, **34**, 3087–3096.

70 Ouvrard, C. and Price, S.L. (2004) Toward crystal structure prediction for conformationally flexible molecules: the headaches illustrated by aspirin. *Crystal Growth Design*, **4**, 1119–1127.

71 Bond, A.D., Boese, R. and Desiraju, G.R. (2007a) On the polymorphism of aspirin. *Angewandte Chemie, International Edition*, **46**, 615–617.

72 Bond, A.D., Boese, R. and Desiraju, G.R. (2007b) On the polymorphism of aspirin: crystalline aspirin as intergrowth of two "polymorphic" domains. *Angewandte Chemie, International Edition*, **46**, 618–622.

73 Whitehouse, M.W. (1964) Biochemical properties of antiinflammatory drugs. III. Uncoupling of oxidative phosphorylation in a connective tissue (cartilage) and liver mitochondria by salicylate analogues: relationship of structure to activity. *Biochemical Pharmacology*, **13**, 319–336.

74 Horsch, W. (1979) Die Salicylate. *Die Pharmazie*, **34**, 585–604.

75 Gordon, M.S., Ellis, D.J., Molony, B. *et al.* (1994) *In vitro* dissolution versus *in vivo* evaluation of four different aspirin products. *Drug Development and Industrial Pharmacy*, **20**, 1711–1723.

76 Blume, H. and Siewert, M. (1986) Zur Qualitätsbeurteilung von acetylsalizylsäurehaltigen Fertigarzneimitteln. 1. Mitteilung: Vergleichende Reihenuntersuchung zur pharmazeutischen Qualität handelsüblicher ASS-Monopräparate. *Pharmazeutische Zeitung*, **47**, 2953–2958.

77 Siewert, M. and Blume, H. (1988) Zur Qualitätsbeurteilung von acetylsalizylsäurehaltigen Fertigarzneimitteln. 2. Mitteilung: Untersuchung zur Chargenkonformität biopharmazeutischer Eigenschaften handelsüblicher ASS-Präparate. *Pharmazeutische Zeitung Wissenschaft*, **131**, 21.

1.2.2
Determination of Salicylates

Measurement of salicylates in biological fluids, that is, mainly plasma and urine, is of interest for several purposes. The purpose also determines the selection of the method. Most frequent is the control of plasma levels to verify that the plasma concentrations are within the therapeutic range. Determination of plasma levels is also necessary in the case of intoxication and for controlling the efficacy of detoxification procedures. Plasma or urinary levels of salicylates allow checking patients' compliance, an important issue for long-term aspirin use in cardiovascular prophylaxis and a frequent explanation of the so-called aspirin resistance (Section 4.1.6). Finally, measurements are of interest to study the pharmacokinetics of salicylates in research, in particular, drug metabolism and interactions.

The therapeutic plasma levels of salicylate differ, depending on the indication. They are in the range of 100–200 μg/ml at anti-inflammatory doses and 50–100 μg/ml for analgetic purposes (see Figure 2.23). The plasma levels of unmetabolized ASA are approximately one order of magnitude lower. Thus, assay methods should be sensitive enough to detect amounts above 1 μg/ml [78]. In most cases, no separate determination of aspirin and salicylate is necessary because there is a rapid and complete conversion of aspirin into salicylate *in vivo*.

1.2.2.1 Gas–Liquid Chromatography

Gas–liquid chromatography (GLC) is the reference standard. The technique allows separate determination of ASA, salicylic acid, and their metabolites. The detection limit is 1 μg/ml.

1.2.2.2 High-Performance Liquid Chromatography

High-performance liquid chromatography (HPLC) is an alternative to GLC but more complex and time consuming. Reverse-phase HPLC techniques with photometric detection are the methods of choice [79]. This has the advantage that the complete spectrum of aspirin and its metabolites can be measured simultaneously. However, one major problem is associated with this type of assay. This is the spontaneous hydrolysis of aspirin to salicylate in protic solvents, including water, methanol, and plasma (Section 2.1.1). Thus, some degradation of aspirin may occur *ex vivo*. A modification of this technique for human plasma, including extraction of salicylates in organic solvents, allows simultaneous determination of aspirin and its metabolites down to the levels of 100 ng/ml with an interassay variation of less than 10% (Figure 1.10). This technique combines simplicity in sample treatment with stability of aspirin over several days (!) without significant decomposition [80].

1.2.2.3 Spectrophotometry

Spectrophotometry is the earliest and most widely used method for measuring serum salicylate levels. The classical assays are colorimetric assays, taking advantage from the intense red color of salicylate/Fe^{3+} complexes. The technology is simple and particularly suitable for compliance measurements.

Trinder Method The Trinder method [81] is a colorimetric test where salicylic acid is determined by measuring the absorbance of the ferric ion–salicylate complex after total serum protein is precipitated by mercuric chloride and allowed to react with ferric iron supplied by ferric nitrate. This method involves the generation of a complex between salicylate and ferric ion. A_{max} of the ferric complex is 540 nm. Quantification is done by measuring light absorbance at this wavelength by a spectrophotometer.

The Trinder method is simple, inexpensive, rapid, and very reliable. Spectrophotometry lowers the detection limit to about 100 μg/ml. This is sufficient for therapeutic and toxic purposes. However, the Trinder method measures salicylate rather than ASA. (False) Positive results may be obtained with salicylamide or methylsalicylate. Conversely, the method can also be used to measure these compounds, for example, in the case of poisoning. The Trinder test is also rather nonspecific and sensitive to a large number of other acids and amines [82]. This also includes compounds and their metabolites, which are increased in patients with Reye-like symptoms because of hepatic (metabolic) failure [83]

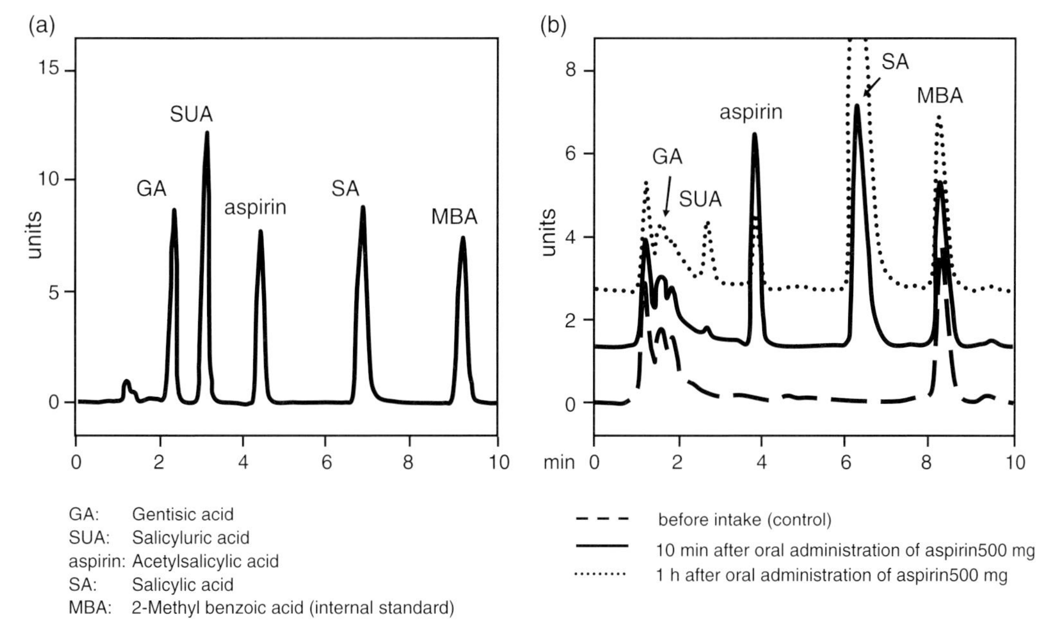

Figure 1.10 Chromatograms of a standard mixture of acetylsalicylic acid and major metabolites (50 ng each) (a) and plasma levels of a volunteer before, 10 min after, and 1 h after oral administration of 500 mg aspirin (b) (modified after Ref. [80]).

(Section 3.3.3). More recently, Morris *et al.* [84] have described a two-step colorimetric method that, however, so far has only been used in the research.

Second-Derivative Synchronous Fluorescence Spectrometry Another method that allows simultaneous determination of ASA and its major metabolites in one assay is second-derivative synchronous fluorescence spectrometry (SDSFS) [85]. This method appears to be the first nonchromatographic technology for the simultaneous determination of aspirin and its major metabolites in one single-serum sample. The technique is not sensitive to several other drugs, found frequently in the sera of healthy subjects (antipyrine, ibuprofen, indomethacin, theophylline, and others).

Summary

Several methods are available to determine aspirin and its major metabolites in biological fluids, including plasma (serum), liquor, synovial fluid, and urine. Most of them have the necessary sensitivity (detection limit 1 µg/ml or less).

HPLC separation and subsequent identification of the spots by appropriate standards is the most frequently used technology. Advantages are the simplicity and reproducibility of the method, a high sensitivity (detection limit about 100 ng/ml), and the possibility of simultaneous determination of several aspirin metabolites together with aspirin itself in one sample. Disadvantages of this and some other technologies include the spontaneous (pH-dependent) and enzymatic hydrolysis of aspirin. However, this problem can be solved by appropriate sample processing.

The Trinder method, a colorimetric assay, determines salicylate and is also useful and simple, though less sensitive. It exhibits a number of cross-reactions with other compounds, which might become relevant, for example, in hepatic failure. GC/MS is clearly the most reliable technology. However, it needs expensive equipment and experienced investigators.

References

78 Svirbely, J. (1987) *Salicylates. Methods in Clinical Chemistry* (eds A.J. Pesce and L.A. Kaplan), C.V. Mosby Company, St. Louis, MO, pp. 417–424.

79 Klimes, J., Sochor, J., Zahradnicek, M. *et al.* (1992) Simultaneous high-performance liquid chromatographic determination of salicylates in whole blood, plasma and isolated erythrocytes. *Journal of Chromatography*, **584**, 221–228.

80 Kees, F., Jehnich, D. and Grobecker, H. (1996) Simultaneous determination of acetylsalicylic acid and salicylic acid in human plasma by high-performance liquid chromatography. *Journal of Chromatography B*, **677**, 172–177.

81 Trinder, P. (1954) Rapid determination of salicylate in biological fluids. *The Biochemical Journal*, **57**, 301–303.

82 Kang, E.S., Todd, T.A., Capaci, M.T. *et al.* (1983) Measurement of true salicylate concentrations in serum from patients with Reye's syndrome. *Clinical Chemistry*, **29**, 1012–1014.

83 Chu, A.B., Nerurkar, L.S., Witzel, N. *et al.* (1986) Reye's syndrome. Salicylate metabolism, viral antibody levels, and other factors in surviving patients and unaffected family members. *American Journal of Diseases of Children*, **140**, 1009–1012.

84 Morris, M.C., Overton, P.D., Ramsay, J.R. *et al.* (1990) Development and validation of an automated, enzyme-mediated, colorimetric assay of salicylate in serum. *Clinical Chemistry*, **36**, 131–135.

85 Konstantianos, D.G. and Ioannou, P.C. (1992) Simultaneous determination of acetylsalicylic acid and its major metabolites in human serum by second-derivative synchronous fluorescence spectrometry. *Analyst*, **117**, 877–882.

2 Pharmacology

2.1 Pharmacokinetics
2.1.1 Absorption and Distribution
2.1.2 Biotransformation and Excretion

2.2 Cellular Modes of Action
2.2.1 Inhibition of Cyclooxygenases
2.2.2 COX-Independent Actions on Cell Function
2.2.3 Energy Metabolism

2.3 Actions on Organs and Tissues
2.3.1 Hemostasis and Thrombosis
2.3.2 Inflammation, Pain, and Fever
2.3.3 Aspirin and Malignancies

Acetylsalicylic Acid. Karsten Schrör

ISBN: 978-3-527-32109-4

2 Pharmacology

The pharmacology of aspirin is unique and characterized by the action of two compounds as parts of the same molecule: reactive acetate and salicylic acid (salicylate). In addition, unmetabolized aspirin itself has biological activities similar to salicylate on many cellular targets. Reactive acetate and salicylate as primary metabolites of aspirin are released at an equimolar ratio by enzymatic or spontaneous hydrolytic cleavage. Hydrolysis starts already in the stomach lumen, within minutes after intake of the drug, but reaches its maximum only after the passage of the intestinal epithelium within the blood of the presystemic and systemic circulations, respectively. This reaction is catalyzed by "aspirin" esterases that are ubiquitously present throughout the body. The resulting reactive acetate moiety can be transferred to any acceptor structure in the neighborhood, which has the most important amino acids in several proteins, such as $serine_{530}$ in the cyclooxygenase (COX) of blood platelets. The other primary metabolite, salicylate, has also a broad spectrum of cellular targets. Because of the much longer half-life and an anti-inflammatory and analgesic potential comparable to aspirin, it is probably salicylate that causes most if not all of the analgesic, antipyretic, anti-inflammatory, and antiproliferative actions of aspirin. An exception is the antiplatelet effect that is specific for aspirin.

This chapter describes first the pharmacokinetics of aspirin and salicylate, focusing on the bioavailability of the active drug(s), plasma and tissue distribution, and clearance from blood and other body fluids (Section 2.1). This is followed by a discussion of the multiple modes of action of aspirin and salicylate at the cellular and subcellular levels (Section 2.2). This involves the description of pharmacological properties of the compounds on mediator systems, cellular signaling, and energy metabolism. This discussion of the pharmacological properties of aspirin is entirely mechanism based without paying too much attention to the salicylate concentrations and the question, whether they can be achieved *in vivo* and, therefore, is of potential value for the use of aspirin as a drug.

Another issue is the functional consequences of these pharmacological mechanisms at the tissue and organ levels (Section 2.3). Here, only those concentrations of the compounds are interesting that can also be obtained with therapeutic doses *in vivo* or, at least, do not cause severe tissue injuries. These pharmacological actions in "real life" should help to understand the clinical actions of aspirin as discussed in detail in Chapter 4. Safety aspects with respect to toxicology are discussed separately in Chapter 3.

2.1 Pharmacokinetics

The pharmacokinetics of aspirin, like the pharmacokinetics of all drugs, involves all processes between uptake and excretion including tissue penetration, storage, and metabolic transforma-

Acetylsalicylic Acid. Karsten Schrör

ISBN: 978-3-527-32109-4

tions. The following steps can be separated: *absorption*, in most cases from the gastrointestinal (GI) tract, passage into the blood compartment and the liver, and subsequent *distribution* throughout the body. The distribution then allows the active compound to reach the cellular targets inside the organs and tissues and to interact with specific binding sites on macromolecules. This interaction is the prerequisite for any biological response and, according to the nonspecific nature of transacetylation, will cause numerous biological actions in different cells and tissues, respectively, which are discussed in Section 2.2.

The drug interaction with biological targets is terminated by the dissolution of the compound from its binding site and the release into the extracellular space. This allows several types of *biotransformations*, eventually resulting in the generation of phase I and phase II metabolites, mainly by the liver. These metabolites as well as unmetabolized salicylates are then cleared by *excretion*, in the case of salicylates almost entirely by the kidney.

For formal purposes, the pharmacokinetic reasons for the generation of a biological signal, that is, all processes that culminate in the binding of salicylates to the cellular target after absorption and distribution, can be distinguished from the pharmacokinetic reasons for the disappearance of the signal, that is, all processes involved in inactivation and excretion of the compound, mainly in the form of inactive metabolites. Therefore, absorption and distribution (Section 2.1.1) are discussed separately from biotransformation and excretion (Section 2.1.2).

2.1.1 Absorption and Distribution

2.1.1.1 Absorption

In daily practice, it is the oral ingestion that is the predominant form of application and is, therefore, discussed in more detail here. Several factors determine in sequence the speed and extent of absorption, and systemic bioavailability of aspirin after oral administration. The first is the solubility of the compound in aqueous media that is mainly determined by physicochemical properties of aspirin (Section 1.2.2), the kind of formulation, and the pH of the solvent. The second is the speed of passage through the stomach. This determines the contact time with the stomach mucosa and can vary considerably depending on the stomach filling state. The retention time in the stomach lumen is also important for side effects, that is, GI intolerance, and determines the passage time to reach the most significant absorption area – the mucosa of the small intestine.

Solubility of Aspirin in Aqueous Media The classical and first variable for action of any drug is its solubility in the (aqueous) medium of interest: *corpora non agunt nisi soluta*. Dissolution of a conventional aspirin tablet by 50% in 0.1 N HCl in a stirred sample *in vitro* under standard conditions is quite long and requires 30–60 min. This suggests a poor solubility of the drug at the acidic conditions of stomach juice. The stability of the compound is inversely related to pH and has a maximum at pH 2.5 ([1]; Section 1.2.1). Thus, the acidic pH in the stomach favors the stability of aspirin and prevents hydrolysis. However, it also largely prevents dissolution. Both factors contribute to the poor gastric tolerance of plain aspirin (Section 3.2.1). In the case of ingestion of high (toxic) doses of aspirin, absorption can be additionally retarded and reduced by the formation of concretions (insoluble aggregates), facilitated by the poor solubility of the drug under these conditions. Standard doses of 300 mg aspirin will result in millimolar concentrations of the compound in 50–100 ml gastric juice (Section 3.2.1).

Absorption in the Stomach The data about the extent of absorption of aspirin within the stomach are variable. According to Cooke and Hunt [2], about 10% of a predissolved 250 mg dose of aspirin is absorbed from an acid solution in the stomach though the main absorption site is the upper intestine. This is partially because of not only the already mentioned poor solubility of the compound at strong acidic pH but also the comparably small

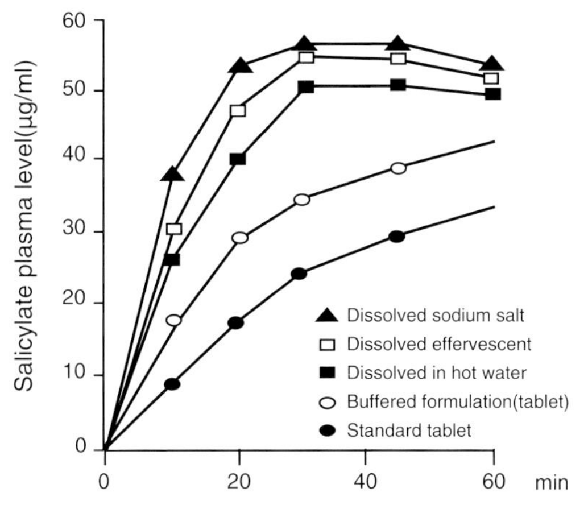

Figure 2.1 Plasma total salicylate concentrations in healthy volunteers after oral intake of aspirin in various galenic formulations on an empty stomach. All preparations contained 640 mg of acetylsalicylic acid (modified after [4]).

absorption surface of the stomach mucosa: 0.2–0.3 m^2 as opposed to the 100–200 m^2 surface of the small intestine. The use of buffered aspirin preparations further reduces the small absorption in the stomach because of an increased proportion of the ionized form [2]. Thus, buffer capacity of the aspirin formulation and its maintenance over time are important variables for gastric tolerance. On the contrary, use of predissolved preparations or water-soluble sodium salts will improve absorption and increase systemic bioavailability [3], though the use of buffered preparations has little effect on these parameters [4] (Figure 2.1).

Though the absorption of aspirin in the stomach is small, it can, nevertheless, have important functional consequences, most important a potentiation of bleeding in connection with alcohol (Section 3.2.1).

Human gastric mucosal epithelial cells have a significant alcohol dehydrogenase activity that oxidizes alcohol and is inhibited by aspirin in a noncompetitive way [5]. Intake of 1 g oral aspirin results in a significant increase, by about 15%, of systemic bioavailability of alcohol in blood. No such effect is obtained after i.v. administration of alcohol, suggesting that it is not due to an inhibition – by aspirin – of alcohol metabolism by the liver. Interestingly, this effect is seen only in men but not in women, possibly due to the low or even absence of first pass metabolism of alcohol in the female stomach. Similar findings were obtained in rats. Though not all investigators could confirm this finding [6], social drinkers should be made aware of the possibility that aspirin may potentiate the effects of alcohol consumed postprandial [7].

The extent and velocity of absorption of aspirin from the stomach are also influenced by the speed of stomach emptying. Addition of antacids or buffering of stomach juice stimulates gastric emptying. This increases initially the plasma levels of aspirin and salicylate. Delayed gastric emptying, for example, by proton-pump inhibitors or atropine, has the opposite effect. In this respect, it is interesting to note that plasma salicylate levels in patients, who underwent total gastrectomy (Billroth II), were not significantly different from those in healthy controls after oral aspirin intake. All important pharmacokinetic variables (absorption kinetics, plasma half-life, elimination kinetics) were also unchanged [8].

Only the nondissociated, lipophilic aspirin can penetrate into the epithelial cells of the stomach mucosa [9]. According to a pK_a of 3.5 for aspirin, 95% of the substance are not dissociated at a pH of 2 in the stomach lumen and, therefore, might exert direct erosive actions on the mucosa epithelial cells [2, 9, 10] (Section 3.2.1).

The pK_a for aspirin is 3.5. This means that 50% of the compound is ionized at this pH and almost all of it at pH 6, that is, it is negatively charged within the stomach lumen. In this ionized form, the molecules are lipid insoluble and can (theoretically) penetrate cell membranes only via special channels. At pH levels below 3.5, the majority of aspirin molecules are nondissociated, that is, lipid soluble and can penetrate cell membranes independent of specialized channels.

Orally taken aspirin exists in the acid stomach lumen primarily in the nonionized form. However, the totally dissolved amount is very small because of the poor solubility of the compound. After diffusion into the superficial stomach mucosal cells, there is a dissociation of aspirin within these cells: pH 7 versus pH 2, equivalent to an ionic gradient of 10^5 (!). This prevents rediffusion of salicylates into the stomach lumen ("ionic trap") and results in an intracellular (intramucosal) accumulation of aspirin and subsequent cytotoxic effects on mucosal cells. Similar considerations apply to salicylate with a pK_a value of 3.0 (Figure 2.2).

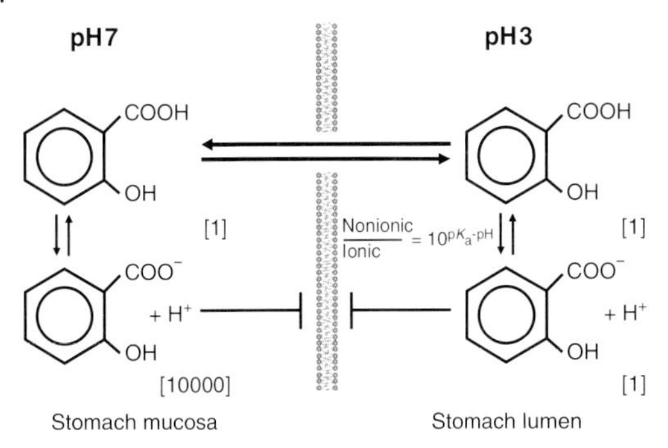

Figure 2.2 Local accumulation of salicylate ($pK_a = 3.0$) in the stomach mucosa: ion trapping hypothesis (for further explanation, see the text).

A pH-dependent kinetics of absorption and distribution is relevant not only for the stomach but also for local accumulation of salicylates at sites of inflammation with acidic pH. Similar considerations apply to the acceleration of urinary salicylate excretion after aspirin overdosing by alkalinization of urine (Section 3.1.1).

Taken together – not only because of pharmacodynamic actions of aspirin (Section 3.2.1) but also for pharmacokinetic reasons – the stomach is not required for aspirin absorption and is entirely a site of salicylate-related side effects.

Absorption in the Intestine Like most other drugs, aspirin is mainly absorbed in the upper intestine by passive diffusion of the nonionized form. The pH in the upper duodenum is about 2–4 and then increases gradually to about 7–8 in the distal small intestine and colon. The much larger surface of the (small) intestine, amounting to 100–200 m^2, as well as the improved solubility of aspirin with increasing pH result in an increase of the totally absorbed amount as a net response, despite the higher degree of dissociation.

Absorption from Other Application Sites Cutaneous administration of salicylates in the form of ointments as an external medication is well known. This stimulated the idea of percutaneous administration of aspirin for systemic application after a significant absorption of salicylates had been shown [11]. Percutaneous application will improve gastric tolerance by avoiding the gastrointestinal passage. This might be useful in patients at an elevated risk for gastrointestinal bleeding or toxicity. In addition, the antiplatelet effects of aspirin might be enhanced by using skin patches as a drug reservoir from which the compound is slowly released. Avoiding high peak levels might additionally result in less inhibition of vascular prostacyclin production.

Cutaneous aspirin (750 mg/day) was reported to inhibit serum thromboxane formation by 95 ± 3% in a small group (6) of healthy volunteers without inhibition of basal or bradykinin-stimulated vascular prostacyclin production [12]. However, a more systematic follow-up study in a larger group of volunteers [10] indicated that this approach may not always work as suggested. In this study, aspirin at the same cutaneous dose (750 mg/day to 29 volunteers for 10 days) had a systemic bioavailability of only 4–8% and did reduce serum thromboxane by only <90%, which is probably of borderline clinical significance.

It was also shown that aspirin applied by skin patches undergoes rapid hydrolysis to salicylate [13], thus eliminating the antiplatelet activity. However, cutaneous aspirin also reduced the prostaglandin content of the gastroduodenal mucosal and caused mucosal injury. Thus, cutaneous aspirin has no advantages in comparison with other forms of application for systemic use.

Bioavailability Aspirin, applied as a (predissolved) standard oral single dose (600 mg), is essentially completely absorbed in the GI tract. There is an appreciable "first-pass" metabolism of aspirin to salicylate during intestinal uptake and liver passage. Thus, the duration of passage through the intestine, that is, the duration of exposition of aspirin to esterases of the intestinal wall and the presystemic circulation (Section 2.1.2), are critical for systemic bioavailability of the uncleaved compound. These factors do not play any role for the bioavailability of the primary metabolite salicylic acid. The deacetylation follows a dose-independent, zero-order kinetics and reduces the systemic bioavailability of standard plain aspirin to about 50% at clinically relevant doses between 40 and 1300 mg [14–17]. The esterases that catalyze this reaction are nonspecific and are located in the intestinal mucosa, blood of the portal vein (red cells, platelets, plasma), and liver parenchyma.

The first-pass effect is less relevant for the anti-inflammatory actions of salicylates because in this indication aspirin and salicylate are about equipotent. For acetylation reactions, such as inhibition of platelet function and, perhaps, generation of 15-hydroxy-eicosatetraenoic acid (15-(*R*)-HETE) by acetylation of COX-2, it is highly relevant. During presystemic circulation, platelets will be exposed to the total absorbed amount of aspirin in portal vein blood whereas the organs in the systemic circulation, including vessel walls and gut will be exposed to only about 50% of total administered aspirin (Figure 2.3). Theoretically, even complete inhibition of the platelet cyclooxygenase will not require any circulating aspirin in the systemic circulation [18]. This will reduce systemic side effects and improve compliance, specifically during long-term prophylactic use.

These data are valid for standard preparations of plain aspirin but not for formulations with delayed release. The high potency and abundance of aspirin esterases in the intestinal mucosa and presystemic circulation allow for sustained hydrolysis of slow-release aspirin formulations during the prolonged GI passage time. Consequently, the systemic bioavailability of unchanged aspirin is markedly reduced [18, 19] whereas the bioavailability of salicylic acid remains unchanged [20, 21].

A 75 mg "slow-release" aspirin formulation was designed to take advantage from this particular pharmacokinetics for cardiocoronary prophylaxis. The intention was to obtain selective inhibition of platelet COX-1 in the presystemic circulation but not the inhibition of prostacyclin in the systemic circulation because the levels of unmetabolized aspirin were too low in the systemic circulation to block vascular prostaglandin production. With this particular preparation, the C_{max} of aspirin was 15-fold lower than that for a standard formulation with the same dose [22]. This hypothesis worked in a proof of concept study [23] but was never introduced clinically.

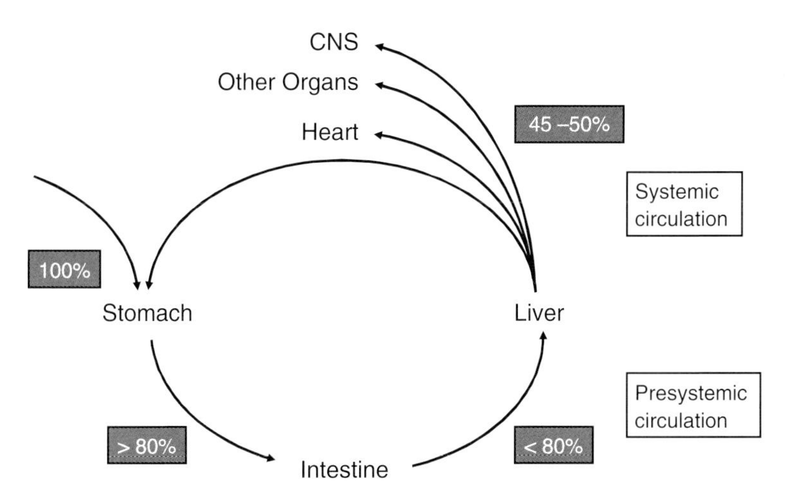

Figure 2.3 Systemic bioavailability of aspirin after oral administration. After oral administration (100%), the majority (>80%) of aspirin enters unchanged the upper intestine. Aspirin passes the intestinal wall and enters the portal vein blood. In this presystemic circulation, about 50% of standard aspirin are deacetylated to salicylic acid. The remaining 45–50% of aspirin enter the systemic circulation via the liver and act on organs and the vessel wall. These conversion rates of aspirin refer to standard plain, uncoated preparations, and standard doses. Slow-release formulations will undergo a significantly greater presystemic deacetylation because of the prolonged passage time through the intestine. In all cases the deacetylated aspirin is converted 1 : 1 to salicylic acid, that is, the bioavailability of salicylate is not affected by the deacetylation procedure.

2.1.1.2 Particular Aspirin Formulations

Several galenic preparations have been developed to improve the pharmacokinetics of the compound and to reduce gastric side effects. These include buffered (i.e., easily soluble or predissolved) formulations and enteric-coated forms. Another option is i.v. administration of aspirin as water-soluble lysine salt or administration as sodium salt. Intravenous administration is preferentially used to obtain rapidly effective plasma levels of undegraded aspirin, for example, as a first measure in the treatment of acute coronary syndromes. Alternatively, effervescent formulations have been developed for rapid absorption and early start of action. These formulations are valuable galenic improvements above the poorly soluble plain standard preparation (Figure 2.1).

Buccal The administration of aspirin as a chewing tablet allows for buccal absorption, that is, a direct, rapid access of the active compound to the systemic circulation, avoiding first-step metabolism by the liver. This approach is thought to be particularly useful in the treatment of headache.

Intravenous Animal experiments suggested that i.v. aspirin did not cause gastric injury despite of nearly complete inhibition of serum thromboxane formation [24]. In clinical use, no overt signs of mucosal injury (microbleedings, gastric potential differences) did occur after i.v. application of aspirin, possibly because of avoidance of direct contacts of the (undissolved) substance with the stomach mucosa (Section 3.2.1). Intravenous aspirin (250–500 mg) is a widely used first-line measure to obtain fast inhibition of platelet function in acute coronary syndromes.

Buffered In one of the first systematic studies on the influence of galenics on aspirin plasma levels and GI blood loss, Stubbé *et al.* [25] found an apparently complete disappearance of occult blood from stool with an appropriately buffered aspirin preparation (Alka-Seltzer). In addition, peak plasma levels of salicylate were obtained earlier and were higher than after plain preparations [25]. This was assumed to be due to faster emptying of the drug into the intestine, and improved solubility.

Erosions of the stomach mucosa, occasionally seen with buffered plain aspirin, are probably due to a disturbed passage ("concretions") of large tablets and not due to direct mucosal injury [26]. On the contrary, buffering may shorten the half-life of action because of a more rapid renal excretion due to the alkaline pH [3, 26].

Enteric Coated Historically the first and most promising approach to improve gastric tolerance of aspirin was the introduction of enteric-coated formulations. The theoretical considerations to introduce this formulation were the insignificant (10%) absorption of aspirin in the stomach and the large body of evidence that gastric injury by aspirin requires direct contact of the active compound with the stomach mucosa. Enteric-coated formulations are resistant against stomach juice. This will avoid any physical interaction of the drug with the stomach mucosa, allowing for the release of the active drug only into the upper intestine. This is the major site of salicylate absorption at a markedly improved solubility because of the more alkaline pH.

2.1.1.3 Distribution

Similar to absorption, distribution of salicylates within body fluids and tissues is determined mainly by pH-dependent passive diffusion of the unbound fraction. As already seen with the stomach, there is a balance between the free, nondissociated acid at both sites of the cell membrane. Any decrease in pH, for example, during acute salicylate intoxication, enhances the accumulation of the substance in the tissue and increases the symptoms of poisoning (Section 3.1.1).

Dose Dependency of the Distribution Volume The apparent distribution volume of salicylates is dose dependent. At low therapeutic doses it amounts to about 0.2 l/kg. This is equivalent to a predominant distribution in the extracellular space because of high (80–95%) binding to plasma albumin [27].

Protein binding of aspirin and salicylate occurs via the phenolic hydroxyl group of the substances [28, 29]. At high (anti-inflammatory) doses or salicylate poisoning, the apparent distribution volume increases to about 0.5 l/kg. This is because of a saturation of salicylate binding to plasma proteins, subsequent diffusion into the intracellular space, and binding to tissue proteins. In addition, some metabolizing enzymes also become saturated (Section 2.1.2). An increase of the volume of distribution and reduced binding to plasma proteins also probably explain the prolonged bleeding time after aspirin in uremic patients [30] and the overproportional high levels of free salicylate after high-dosage aspirin treatment, for example, in children with Kawasaki's disease (Section 4.3.2). It is possible that the intracellular accumulation of (free) salicylic acid that has to be expected at high anti-inflammatory doses is a precondition for its pharmacological efficacy. High tissue levels of salicylates might be a requirement for not only the inhibition of cytokine or tumor promoter-induced gene expression, including transcriptional COX-2 regulation (Section 2.2.2) but also non-COX-related effects, for example, on cell energy metabolism (Section 2.2.3).

Salicylate Levels in Plasma The plasma salicylate levels differ markedly, depending on the dose and the distribution equilibrium, including the protein (albumin) content. At low antiplatelet doses essentially all salicylate is albumin bound. However, this is not relevant for the antiplatelet effects of the compound since these effects are entirely due to the acetylation potential and independent of salicylate. At analgesic single doses of 0.6–1 g, mean salicylate plasma levels of about 30 μg/ml and aspirin levels of 3 μg/ml are obtained. The protein binding (ultrafiltration) of a 600 mg single oral dose amounts to 82 and 58% for salicylate and aspirin, respectively [31]. The level of analgesia appears to correlate well with the salicylate plasma level [32]. Thus, at these concentrations, 10% of total salicylate is free, that is, able to move out from plasma into cells and tissues. This percentage of free salicylate at inflammatory daily doses of about 5 g, equivalent to total plasma levels of 300 μg/ml and more (1–3 mM) amounts to about 10% and to 30% at toxic doses (Table 2.1). Thus, higher doses of aspirin result not only in higher levels of total plasma salicylate but additionally also in a marked increase of the unbound fraction [33, 34].

Salicylate Levels in Selected Tissues and Body Fluids The maximum tissue levels of salicylates in the synovial fluid amount to about 50% of plasma level [35]. Salicylate concentrations in the cerebrospinal fluid are about 10–25% of the plasma level, and there is no tight correlation to the plasma level [36]. Similar low percentages of plasma level, about 30%, are found in the perilymph. In contrast, salicylate levels in the fetal circulation are only slightly lower than in the maternal circulation [37]. This is particularly relevant for newborn prior to

Table 2.1 Plasma salicylate levels and percentage of free salicylate in dependency on dosing (modified after [33]).

			Max. plasma salicylate (mM)		
Clinical use	ASA dose	mg/kg (70 kg BW)	Total	Free	Free salicylate
Antiplatelet	100 mg/day (maintenance dose)	<2	<0.1	[a]	
Analgesic	1 g single dose	15	0.5	0.005	1%
Anti-inflammatory	6–8 g/day	100	1.5–2.5	0.15–0.60	10%
Toxic	30 g single dose and more	≥400	3.0–10.0	1.0–5.0	30%

[a] Irrelevant since clinical efficacy depends on acetylation and is independent of salicylate.

delivery whose renal and hepatic clearance systems are not yet fully developed (see Section 3.1.2).

2.1.1.4 Modifying Factors

Food Intake The compliance for regular intake of drugs is improved if this is coupled with food intake. This is particularly valid for patients at older age and the intake of drugs that might irritate the stomach, such as aspirin. However, simultaneous eating might prolong the passage time through the stomach, allow for adsorption of the drug to food particles and allow a reduced velocity of absorption in the small intestine [38].

Aspirin is not recommended to be taken on an empty stomach. The reason is that the substance is a stronger irritant for the mucosa in the empty state and will also stay there for a longer period of time. In addition, absorption occurs predominantly in the upper small intestine and is independent of the stomach filling state. However, the absorption of aspirin is quantitatively optimal if the drug is ingested in a relatively large volume of water on an empty stomach [39].

Ferner and colleagues compared the plasma levels of aspirin and salicylate after oral intake of 1200 mg standard aspirin in healthy volunteers either starved for 37 h or after having a standard breakfast. The maximum plasma levels of aspirin were $17 \pm 3\,\mu g/ml$ at 22 min in starved and $24 \pm 4\,\mu g/ml$ in nonstarved persons. The maximum salicylate plasma levels were $57 \pm 7\,\mu g/ml$ in starved and $65 \pm 8\,\mu g/ml$ in nonstarved subjects. None of these differences were significant. There were also no differences in the pharmacodynamics of aspirin as measured in terms of metabolic parameters. The conclusion was that eating, that is, the filling state of the stomach, does affect neither the bioavailability nor the efficacy of standard aspirin and salicylic acid [40].

It has, however, to be considered that delayed gastric emptying or prolonged passage through the intestine will also prolong the exposure time of aspirin against hydrolyzing enzymes. This, eventually results in a reduced systemic bioavailability of aspirin (Section 2.1.2) whereas the bioavailability of salicylic acid remains unchanged. Similar data were published with controlled-release formulations [22]. However, marked delays in the absorption of aspirin from enteric-coated tablets may be observed when they are consumed with food. This effect appears to be particularly pronounced in women [41]. Finally, the delayed and prolonged absorption of aspirin from enteric-coated formulations has to be considered in the treatment of salicylate poisoning (Section 3.1.1).

Vegetables as a Natural Source of Salicylates Many fruits and vegetables contain different forms of salicylates, in particular, the salicylic acid ester salicin. These salicylates or metabolites thereof might circulate in plasma. Their level might be increased by an appropriate (vegetarian) diet and, eventually, add to the therapeutic benefit of exogenous aspirin. There is mixed information, whether clinically relevant amounts of salicylates can be obtained in plasma after intake of salicylate-rich diets [42, 43]. One recent study has, however, shown that the low circulating salicylate levels in plasma [43] can be significantly, though not impressively, increased by a vegetarian diet [44] (Figure 2.4). This has been taken as evidence to explain beneficial effects of certain vegetables in chemoprevention of colorectal carcinoma and is discussed in detail elsewhere (Section 4.3.1). Al-

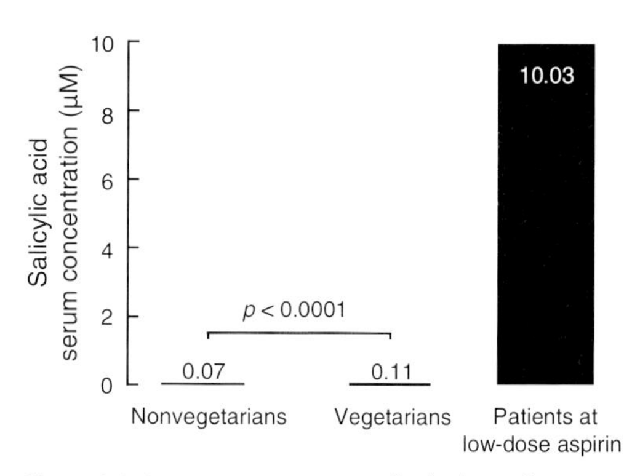

Figure 2.4 Serum concentrations of salicylic acid (µM) in nonvegetarians, vegetarians (Buddhist monks of European origin), and patients taking low-dose aspirin (75 mg daily) (modified after Blacklock *et al.*, [44]).

though this possibility certainly exists, the alternative explanation, namely the abdication of (red) meat, the most important natural source not only of cholesterol but also of arachidonic acid (for thromboxane formation), by vegetarians, appears at least as likely.

Gender The possible influence of gender on the pharmacokinetics of aspirin was studied after single (1000 mg) oral, intramuscular or intravenous administration of the (water soluble) lysine salt of aspirin to healthy volunteers. No differences were detected with respect to half-life and distribution volume. The bioavailability after oral and intramuscular administration of standard aspirin was not different between genders, amounting to 54 and 89% in both groups, respectively [45]. The results for low-dose (100 mg) [19] and high-(antirheumatic) dose [46] aspirin were similar though two other small studies found a reduced metabolic capacity for aspirin in women [47, 48]. Taken together, there might be some reduced plasma aspirin hydrolysis rate in women. However, whether these differences in aspirin pharmacokinetics are clinically relevant is questionable. A separate issue is gender-dependent variations in blood alcohol levels after aspirin intake (see above).

Age Bioavailability and metabolism of aspirin are similar in healthy men at the age of 21–40 as compared to those at the age of 55–75. However, in the elderly, there were reduced serum peak levels and a prolonged half-life as opposed to young individuals. This is probably due to the larger volume of distribution for aspirin in elderly persons [49]. However, these age-dependent differences in aspirin bioavailability and metabolism were not considered to play a major role for its therapeutic use [50].

Summary

Standard plain aspirin is poorly soluble in aqueous media at acidic pH. The galenic formulation, the speed of tablet dispersion, the local pH, and velocity of gastric emptying determine the passage time through the stomach and, therefore, stomach (in)tolerance (Section 3.2.1). Absorption of aspirin occurs predominantly in the upper small intestine and is nearly complete there. The systemic bioavailability of standard plain aspirin is about 50% and is reduced to 25–15% or even less after administration as "controlled-release" formulations.

During and after absorption, aspirin undergoes hydrolytic cleavage by esterases in the intestine, portal vein blood, and liver. This results in an equimolar generation of salicylic acid, the primary metabolite. Salicylic acid has a larger volume of distribution, in particular, at acidic pH, and accounts for many of the pharmacological actions of aspirin – except inhibition of platelet function. The percentage of free salicylate is dose dependent and increases from about 1% at analgesic doses (1 g) to 30% and more at toxic doses.

Several galenic formulations have been developed to improve gastric tolerance and to modulate the systemic bioavailability of aspirin. For oral use, these are buffered (i.e., easily soluble or predissolved) formulations or (i.v.) administration of aspirin as water-soluble lysine salt. For long-term use, enteric-coated formulations appear to have some advantages.

The bioavailability of aspirin is independent of gender and age. Food intake can reduce the bioavailability of aspirin if this is associated with adsorption to food constituents or a prolonged exposure against esterases in the intestinal mucosa. Otherwise, there are no relevant pharmacokinetic interactions between aspirin and other compounds with respect to drug absorption and distribution.

References

1 Horsch, W. (1979) Die Salizylate. *Die Pharmazie*, **34**, 585–604.

2 Cooke, A.R. and Hunt, J.N. (1970) Absorption of acetylsalicylic acid from unbuffered and buffered gastric contents. *The American Journal of Digestive Diseases*, **2**, 95–102.

3 Mason, W.D. and Winer, N. (1983) Influence of food on aspirin absorption from tablets and buffered solutions. *Journal of Pharmaceutical Sciences*, **72**, 819–821.

4 Leonards, J.R. (1963) The influence of solubility on the rate of gastrointestinal absorption of aspirin. *Clinical Pharmacology and Therapeutics*, **4**, 476–479.

5 Roine, R., Gentry, R.T. and Hernández-Muňoz, R. *et al.* (1990) Aspirin increases blood alcohol concentrations in humans after ingestion of alcohol. *The Journal of the American Medical Association*, **264**, 2406–2408.

6 Melander, O., Lidén, A. and Melander, A. (1995) Pharmacokinetic interactions of alcohol and acetylsalicylic acid. *European Journal of Clinical Pharmacology*, **48**, 151–153.

7 Gentry, R.T., Baraona, E., Amir, I. *et al.* (1999) Mechanism of the aspirin-induced rise in blood alcohol levels. *Life Sciences*, **65**, 2595–2512.

8 Hurtado, C., Acevedo, C., Domecq, C. *et al.* (1988) Absorption kinetics of acetylsalicylic acid in gastrectomized patients. *Medical Science Research*, **16**, 1241–1243.

9 Graham, D.Y. and Smith, J.L. (1986) Aspirin and the stomach. *Annals of Internal Medicine*, **104**, 390–398.

10 Cryer, B. and Feldman, M. (1999) Effects of very low dose daily, long-term aspirin therapy on gastric duodenal and rectal prostaglandin levels and on mucosal injury in healthy humans. *Gastroenterology*, **117**, 17–25.

11 Feldmann, R.J. and Maibach, H.I. (1970) Absorption of some organic compounds through the skin of man. *Journal of Investigative Dermatology*, **54**, 399–404.

12 Keimowitz, R.M., Pulvermacher, G., Mayo, G. *et al.* (1993) Transdermal modification of platelet function. A dermal aspirin preparation selectively inhibits platelet cyclooxygenase and preserves prostacyclin biosynthesis. *Circulation*, **88**, 556–561.

13 McAdam, B., Leimowitz, R.M., Maher, M. *et al.* (1996) Transdermal modification of platelet function: an aspirin patch system results in marked suppression of platelet cyclooxygenase. *The Journal of Pharmacology and Experimental Therapeutics*, **277**, 559–564.

14 Rowland, M., Riegelman, S., Harris, P.A. *et al.* (1967) Kinetics of acetylsalicylic acid disposition in man. *Nature*, **215**, 413–414.

15 Rowland, M., Riegelman, S., Harris, P.A. *et al.* (1972) Absorption kinetics of aspirin in man following oral administration of an aqueous solution. *Journal of Pharmaceutical Sciences*, **61**, 379–385.

16 Harris, P.A. and Riegelman, S. (1969) Influence of the route of administration on the area under the plasma concentration–time curve. *Journal of Pharmaceutical Sciences*, **58**, 71–75.

17 Pedersen, A.K. and FitzGerald, G.A. (1984) Dose-related kinetics of aspirin. Presystemic acetylation of platelet cyclooxygenase. *The New England Journal of Medicine*, **311**, 1206–1211.

18 Siebert, D.J., Bochner, F., Imhoff, D.M. *et al.* (1983) Aspirin kinetics and platelet aggregation in man. *Clinical Pharmacology and Therapeutics*, **33**, 367–374.

19 Bochner, F., Somogyi, A.A. and Wilson, K.M. (1991) Bioinequivalence of four 100 mg oral aspirin formulations in healthy volunteers. *Clinical Pharmacokinetics*, **21**, 394–399.

20 Cummings, A.J. and Martin, B.K. (1962) Relationship of plasma salicylate concentration to urinary salicylate excretion rate. *Nature*, **195**, 1104–1105.

21 Cummings, A.J. and King, M.L. (1966) Urinary excretion of acetylsalicylic acid in man. *Nature*, **209**, 620–621.

22 Charman, W.N., Charman, S.A., Monkhouse, D.C. *et al.* (1993) Biopharmaceutical characterisation of a low-dose (75 mg) controlled-release aspirin formulation. *British Journal of Clinical Pharmacology*, **36**, 470–473.

23 Clarke, R., Mayo, G., Price, P. *et al.* (1991) Suppression of thromboxane A_2 but not of systemic prostacyclin by controlled-release aspirin. *The New England Journal of Medicine*, **325**, 1137–1141.

24 Ligumsky, M., Golanska, E.M., Hansen, D.G. *et al.* (1983) Aspirin can inhibit gastric mucosal cyclo-oxygenase without causing lesions in the rat. *Gastroenterology*, **84**, 756–761.

25 Stubbé, L.T., Pietersen, J.H. and Van Heulen, C. (1962) Aspirin preparations and their noxious effect on the gastro-intestinal tract. *British Medical Journal*, **5279**, 675–680.

26 Mason, W.D. (1984) Comparative aspirin absorption kinetics after administration of sodium- and

potassium-containing buffered solutions. *Journal of Pharmaceutical Sciences*, **73**, 998–999.

27 Dromgoole, S.H. and Furst, D.E. (1992) Salicylates, in *Applied Pharmacokinetics: Principles of Therapeutic Drug Monitoring* (eds W.E. Evans, J.J. Schentag and W.J. Jusko), Lippincott Williams & Wilkins.

28 Cohen, L.S. (1976) Clinical pharmacology of acetylsalicylic acid. *Seminars in Thrombosis and Hemostasis*, **2**, 146–175.

29 Green, F.A. and Young, C.Y. (1981) Acetylation of erythrocytotic membrane peptides by aspirin. *Transfusion*, **21**, 55–58.

30 Gaspari, F., Vigano, G., Orisio, S. *et al.* (1987) Aspirin prolongs bleeding time in uremia by a mechanism distinct from platelet cyclooxygenase inhibition. *The Journal of Clinical Investigation*, **79**, 1788–1797.

31 Ghahramani, P., Rowland-Yeo, K., Yeo, W.W. *et al.* (1998) Protein binding of aspirin and salicylate measured by *in vivo* ultrafiltration. *Clinical Pharmacology and Therapeutics*, **68**, 285–295.

32 Bromm, B., Rundshagen, I. and Scharein, E. (1991) Central analgesic effects of acetylsalicylic acid in healthy men. *Arzneimittel-Forschung/Drug Research*, **41**, 1123–1129.

33 Smith, M.J.H. and Dawkins, P.D. (1971) Salicylate and enzymes. *The Journal of Pharmacy and Pharmacology*, **23**, 729–744.

34 Insel, P.A. (1990) Analgesic, antipyretic and antiinflammatory agents: drugs employed in the treatment of rheumatoid arthritis and gout, in *Goodman and Gilman's The Pharmacological Basis of Therapeutics*, 8th edn, Pergamon Press, New York, pp. 638–681.

35 Sholkoff, S.D., Eyring, J.E., Rowland, M. *et al.* (1967) Plasma and synovial fluid concentrations of acetylsalicylic acid in patients with rheumatoid arthritis. *Arthritis and Rheumatism*, **10**, 348–351.

36 Bannwarth, B., Netter, P., Pourel, J. *et al.* (1989) Clinical pharmacokinetics of nonsteroidal anti-inflammatory drugs in the cerebrospinal fluid. *Biomedicine & Pharmacotherapy*, **43**, 121–126.

37 Palmisano, P.A. and Cassady, G. (1969) Salicylate exposure in the perinate. *The Journal of the American Medical Association*, **209**, 556–558.

38 Winstanley, P.A. and Orme, M.L. (1989) The effects of food on drug bioavailability. *British Journal of Clinical Pharmacology*, **28**, 621–628.

39 Koch, P.A., Schultz, C.A., Wills, R.J. *et al.* (1978) Influence of food and fluid ingestion on aspirin bioavailability. *Journal of Pharmaceutical Sciences*, **67**, 1533–1535.

40 Ferner, R.E., Williams, F.M., Graham, M. *et al.* (1989) The metabolic effects of aspirin in fasting and fed subjects. *British Journal of Clinical Pharmacology*, **27**, 104.

41 Mojaverian, P., Rocci, M.L., Jr, Conner, D.P. *et al.* (1987) Effect of food on the absorption of enteric-coated aspirin: correlation with gastric residence time. *Clinical Pharmacology and Therapeutics*, **41**, 11–17.

42 Janssen, P.L.T.M., Katan, M.B., van Staveren, W.A. *et al.* (1997) Acetylsalicylate and salicylates in foods. *Cancer Letters*, **114**, 163–164.

43 Baxter, G.J., Lawrence, J.R., Graham, A.B. *et al.* (2002) Identification and determination of salicylic acid and salicyluric acid in urine of people not taking salicylate drugs. *Annals of Clinical Biochemistry*, **39**, 50–55.

44 Blacklock, C.J., Lawrence, J.R., Wiles, D. *et al.* (2001) Salicylic acid in the serum of subjects not taking aspirin. Comparison of salicylic acid concentrations in the serum of vegetarians, non-vegetarians, and patients taking low-dose aspirin. *Journal of Clinical Pathology*, **54**, 553–555.

45 Aarons, L.K., Hopkins, M., Rowland, S. *et al.* (1989) Route of administration and sex differences in the pharmacokinetics of aspirin administered as its lysine salt. *Pharmacological Research*, **6**, 660–666.

46 Rainsford, K.D., Ford, N.L.V., Brooks, P.M. *et al.* (1980) Plasma aspirin esterases in normal individuals, patients with alcoholic liver disease and rheumatoid arthritis: characterization and the importance of the enzymic components. *European Journal of Clinical Investigation*, **10**, 413–420.

47 Ho, P.C., Triggs, E.J., Bourne, D.W. *et al.* (1985) The effects of age and sex on the disposition of acetylsalicylic acid and its metabolites. *British Journal of Clinical Pharmacology*, **19**, 675–684.

48 Miners, J.O., Grgurinovich, N., Whitehead, A.G. *et al.* (1986) Influence of gender and oral contraceptive steroids on the metabolism of salicylic acid and acetylsalicylic acid. *British Journal of Clinical Pharmacology*, **22**, 135–142.

49 Mason, W.D., Falbe, J.W., Fu, C.H.J. *et al.* (1989) Comparative aspirin bioavailability in young and old men. *Pharmaceutical Research*, **6** (Suppl. 9), S233.

50 Montgomery, P.R., Berger, L.G., Mitenko, P.A. *et al.* (1986) Salicylate metabolism: effect of age and sex in adults. *Clinical Pharmacology and Therapeutics*, **39**, 571–576.

2.1.2
Biotransformation and Excretion

The biotransformation of aspirin involves two principally distinct processes that occur in sequence but independent of each other and at different reaction kinetics. They are independently controlled by different enzymes and have a different biological significance: (i) generation of salicylic acid after hydrolytic removal of the acetyl moiety by "Aspirin" esterases and transfer of the acetyl moiety to macromolecules; (ii) biotransformation(s) of salicylic acid and its metabolites.

Deacetylation of aspirin with subsequent generation of salicylic acid occurs within minutes and, because of the ubiquitous presence of carboxyesterases, is independent of the aspirin dosage (Section 2.1.1). In contrast, the pharmacokinetics of salicylate, including half-life, volume of distribution, the composition of salicylate metabolites and their fractional excretion are largely determined by the aspirin dosage. The pharmacokinetic parameters of salicylic acid are of particular clinical interest because many of the therapeutic and toxic effects of aspirin, including analgesic, antipyretic effects, and anti-inflammatory actions (Section 2.3.2) as well as actions on cell metabolism (Section 2.2.3) and organ toxicity, for example, stomach (Section 3.2.1) and ototoxicity (Section 3.2.4) are caused by salicylic acid. However, the transacetylation of macromolecules by aspirin becomes now increasingly important after the detection of new acetylation targets. The most prominent example is COX-2 that, unlike COX-1, generates a new product 15-(*R*)-HETE, a precursor of several lipoxins, after acetylation by aspirin. This may be relevant for tissue-protective, anti-inflammatory actions of aspirin as well as nitric oxide (NO) generation via the endothelial NO synthase (eNOS) in the cardiovascular system (Section 2.3.2). The elimination of salicylates occurs via the kidney, at therapeutic doses predominantly in the form of the glycine conjugation product salicyluric acid [51–54]. Major metabolic pathways of aspirin and salicylic acid, respectively, are summarized in Figure 2.5.

2.1.2.1 Biotransformation of Aspirin

"Aspirin" Esterases Enzymatic hydrolysis of aspirin to the primary metabolite salicylic acid starts already in the gastrointestinal mucosa [55]. It continues in the portal vein blood and liver [56] and eventually reduces the bioavailability of aspirin to about 50% after oral administration of standard preparations. Deacetylation occurs at zero-order kinetics, that is, is independent of the dosage (Section 2.1.1). The plasma half-life of aspirin in men after i.v. administration is about 15–20 min [57].

Aspirin hydrolysis in the systemic circulation is catalyzed by different "aspirin esterases," most notably those in red cells (cytosol) [58, 59] and plasma [46]. This enzyme is located intracellularly in the erythrocyte cytosol, is independent of Ca^{2+} and other bivalent cations, and unrelated to acetylcholine esterase [58]. In men, red-cell aspirin esterase accounts for about half of the aspirin hydrolysis activity *in vitro* [59]. This activity also explains the more than twofold faster aspirin hydrolysis in whole blood *in vitro* as opposed to plasma [59], –0.5 h versus –1.9 h (Table 2.2).

The other major enzyme activity is plasmatic and associated with the plasma (pseudo)-choline esterase. The aspirin esterase activity and pseudo cholinesterase activities in plasma (serum) are positively correlated in men but exhibits a skewed distribution [60]. It hydrolyzes aspirin in a manner, different from choline esters [61], requires Ca^{2+} for optimal activity, and accounts for about 80% of aspirin esterase activity in plasma. The remaining esterase activity in plasma can be attributed to (aryl)-albumin esterase that is acetylated by aspirin [46]. Aspirin esterase activity has also been detected in liver microsomes and other organs, and there it could be separated from cholinesterases and other nonspecific carboxyesterases [62]. Thus, a specific "aspirin" esterase is likely not to be the only enzyme that cleaves aspirin but additional esterases for circulating carboxyesters with broad spectrum substrate specificity might be involved. Studies with radioactive labeled aspirin have shown that a variety

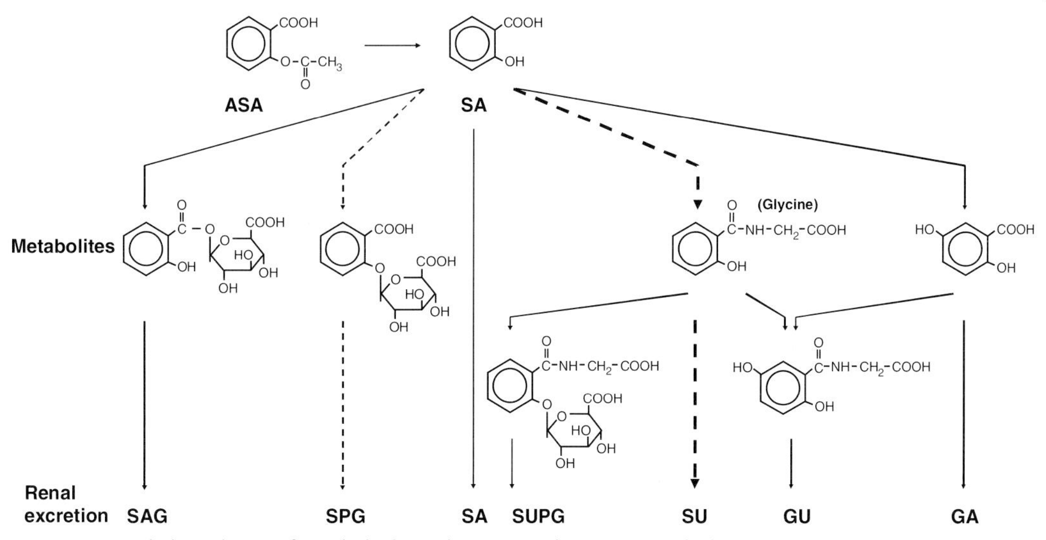

Figure 2.5 Metabolic pathways of acetylsalicylic acid (ASA) at a therapeutic single dose (500 mg) in men. ASA is initially hydrolyzed to salicylic acid (SA) as the primary metabolite and the reactive acetyl group (not shown). SA is excreted in urine either unchanged (10%) or as SA phenol glucuronide (SPG) and SA acyl glucuronide (SAG) (5–10%). The dominant metabolic pathway is in conjugation with glycine to salicyluric acid (SU) (70–75%). SU is mainly excreted as such or as SU phenol glucuronide (SUPG) (≤1%). SA can also be hydroxylated to gentisic acid (GA) (<5%) and gentisuric acid (GU) (<1%) that can also be formed from SU by glycine conjugation. Dashed lines mark metabolic pathways with limited (saturated) capacity at this dose (≥500 mg). The elimination of SA at higher therapeutic and toxic doses occurs increasingly as unchanged SA with increasing half-life of the substance (for further explanation, see the text).

of proteins, glycoproteins, and lipids can be acetylated by the compound *in vivo* [63].

As a result, biotransformation of aspirin generates two active metabolites: a reactive acetyl moiety that transacetylates macromolecules and modifies their activity, most notably COX-1 in platelets and COX-2 elsewhere, and salicylate. Transacetylation is rather nonspecific and is also seen with various other macromolecules, for example, albumin [64], hemoglobin, and red-cell membrane peptides [65] as well as hormones and DNA.

Table 2.2 Hydrolysis half-life of aspirin at 37 °C in different body fluids compared to a physiological buffer solution (pH 7.4) (modified after [15]).

Medium	Aspirin concentration (μg/ml)	Aspirin half-life (h)
Krebs buffer	10	15.5
Gastric juice	10	16.0
Duodenal juice	10	17.0
Blood	13	0.5
Plasma	13	1.9

Influence of Gender, Age, and Diseases The plasma aspirin esterase activity is independent of age and gender [50] and is not induced by regular aspirin intake [56, 66–68]. It is also unchanged in patients with rheumatoid arthritis but depressed in patients with alcoholic liver cirrhosis, probably because of the reduced plasma albumin levels [46]. In addition, gene polymorphisms in plasma cholinesterase with subsequently reduced enzymatic activity [61] might result in increased aspirin relative to salicylate levels. However, there are only anecdotal reports regarding possible changes in plasmatic aspirin esterase activity in patients with a heterozygous defect in plasma cholinesterase [69] and no data on patients with homozygous defects in this enzyme. Interestingly, aspirin-sensitive asthma (Section 3.3.1) and aspirin-sensitive urticaria (Section 3.3.2) appear to be associated with a re-

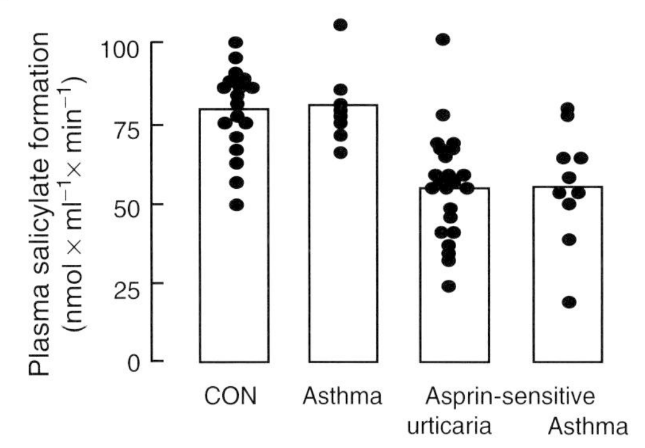

Figure 2.6 Plasma "aspirin esterase" activities in patients with aspirin-sensitive urticaria and aspirin-sensitive asthma (Widal trial) patients as compared to non-aspirin-sensitive asthmatics and nonasthmatic controls [69].

duced plasmatic aspirin esterase and cholinesterase activity, respectively, in a small clinical trial [69] (Figure 2.6).

Plasma aspirin esterase activity is transiently reduced, by about 20–25%, in patients shortly, 2 days, after surgery and returns to normal within 2 weeks. These changes are paralleled by changes in plasma cholinesterase activity [70]. Transient decreases followed by an increase in aspirin esterase activity were also seen in patients subjected to coronary artery bypass surgery but appear not to be related to aspirin "resistance" in these patients [71] (Section 4.1.6).

Changes in aspirin esterase activity appear not to have major consequences for the analgesic/anti-inflammatory actions of aspirin. In this indication, aspirin and salicylate are nearly equipotent. This is not surprising because both compounds are almost equipotent inhibitors of COX-2 (Section 2.2.1). Whether changes in aspirin esterase activity modify the antiplatelet effect of aspirin has not been studied systematically, some anecdotic reports do not exclude this possibility, especially at low doses [72].

Pharmacokinetic Modification by Drugs Drugs that induce the cytochrome P450 (CYP) system in the liver will stimulate aspirin esterase activity. This has been shown for phenobarbitone, phenytoin, carbamazepine, and valproic acid.

Aspirin esterase activity in serum of healthy controls amounted to 181 ± 34 μg/ml/h and was approximately doubled to 332 ± 93 μg/ml/h in epileptics, treated with phenobarbitone, phenytoin, carbamazepine, or valproic acid. This suggests a drug-induced induction of esterase activity, possibly via the P450 system. Further support for this rather nonspecific effect came from the increase in serum-cholinesterase activity that was stimulated in epileptic patients for about the same amount [73].

Changes in aspirin esterase activity might become clinically relevant in situations when drugs, affecting the activity of the enzyme, are employed concomitantly with aspirin. For example, cholinesterase inhibitors and succinylcholine might be expected to increase the proportion of aspirin relative to salicylate in the circulation [46].

2.1.2.2 Biotransformations of Salicylic Acid

The biotransformations of salicylic acid are dose dependent as is the excretion rate. A schematic overview is shown in Table 2.2.

Plasma Half-Life The plasma half-life of salicylic acid at analgesic doses (0.6–1.2 g) amounts to about 3 h [57]. At these doses, excretion of salicylic acid occurs mainly – 70–75% – via conjugation with glycine as salicyluric acid. A smaller part is excreted as unchanged salicylic acid or conjugated with glucuronic acid to form acyl and phenolic glucuronides. Another small fraction of salicylic acid is hydroxylated to gentisic acid and further converted to gentisuric acid. In addition, there are some minor metabolites, amounting to <10% of total salicylic acid (Figure 2.5).

The plasma half-life of salicylic acid becomes markedly prolonged at doses above 500 mg, in particular, if the drug is given repeatedly (Figure 2.7) and may increase up to 20 h and more in the case of salicylate poisoning (Section 3.1.1). The reasons are saturation of metabolizing enzymatic pathways, such as glycine conjugation to salicyluric acid, and an increased free percentage of free salicylic acid (Table 2.3) that can easily penetrate into tissues, in particular, at the acidic pH during intoxication.

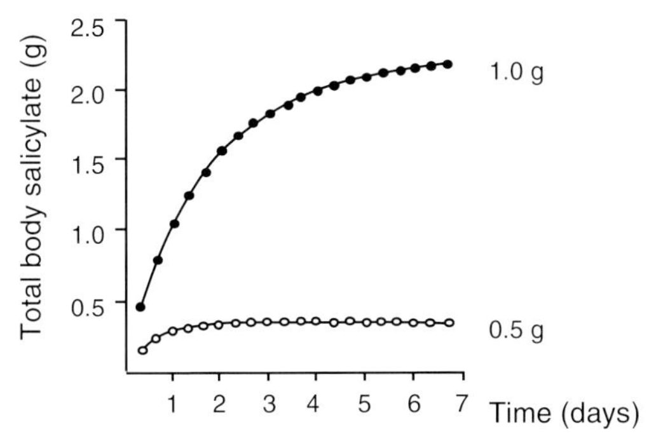

Figure 2.7 Accumulation of salicylate in healthy adults within 1 week after oral intake of 0.5 or 1 g aspirin in 8 h intervals (modified after [1]).

2.1.2.3 Excretion of Salicylates

The elimination of aspirin occurs completely (>98%) as salicylic acid and salicylic acid metabolites in urine. As mentioned above, the composition of the spectrum of metabolites is dependent on the aspirin dose. The absorption/secretion balance within the kidney tubules depends on the pH of tubulus fluid: alkalinization of the urine stimulates the dissociation of the acid(s) and increases the excretion of salicylate (metabolites) 5–10-fold. This effect can be used therapeutically in the treatment of salicylate poisoning (Section 3.1.1).

At an analgesic dose of 1.5 g aspirin, the approximate recovery rates of salicylates in urine are as follows: 70–75% salicyluric acid, including glucuron-conjugated products, 10% salicylic acid, 1–2% gentisic acid, <1% gentisuric acid [75–77]. Several of the metabolic pathways of salicylic acid are pH-dependent and have a limited capacity [78]. This, in quantitative terms, is particularly relevant to the conversion of salicylic acid to salicyluric acid that becomes markedly reduced at high and toxic doses of aspirin whereas the excretion of nonmetabolized salicylate is increased (Table 2.4). The enzymatic generation of salicyluric acid is saturated after a single aspirin dose of >300 mg [79], possibly because of exhaustion of the glycine pool [80].

Consequently, there is also a large interindividual variability in urinary excretion of salicylate metabolites. In two studies, the urinary excretion of salicyluric acid after oral administration of 900 mg aspirin varied between 6 and 72 and 1–31% of dose within 12 h [82, 83]. It is interesting that despite these large interindividual variations, the intraindividual reproducibility was quite well, that is, +12%, suggesting that the metabolic pattern of salicylate is a genetically fixed individual feature [82]. This hypothesis is supported by studies on salicylate metabolism in twins [84]. A genetically fixed expression pattern of CYP2E1 [85] or other cytochromes that are involved in salicylate metabolism is an option [86]. Possible consequences of a variable pharmacokinetic of salicylates for the

Table 2.3 Pharmacokinetic parameters of aspirin and salicylate (adapted from [74]).

Parameter	Aspirin	Salicylate
Bioavailability (%)	68	100
Urinary excretion (%)	~1	2–30[a]
Protein binding (%)	50–60	80–95[b]
Clearance (ml/min/kg)	9.3	~0.2[c]
Volume of distribution (ml/kg)	~150	150
Half-life (h)	0.25	Dose dependent[d]
Effective concentrations (μg/ml or mM)	See salicylate	50 to ≥200 (0.5 to ≥1.0 mM)
Toxic concentrations (μg/ml or mM)	See salicylate	>200 (>1 mM)

[a] Dependent on dose and pH of urine.
[b] 95% at 14 g/ml, further decrease at higher doses.
[c] ~0.2 h at 300 g/ml.
[d] ~2 h at 300 mg to 20 h and more in intoxication.

Table 2.4 Metabolites of aspirin (salicylic acid) recovered in the urine after intake of therapeutic or toxic doses.

Metabolite	Therapeutic (aspirin 600 mg) ($n = 45$)	Overdose (plasma SA 240–600 μg/ml) ($n = 24$)	Overdose (plasma SA 715–870 μg/ml) ($n = 13$)
Salicylic acid	9 ± 1	32 ± 4	65 ± 4
Salicyluric acid	75 ± 1	47 ± 3	22 ± 4
Salicylic acid phenol glucuronide	11 ± 1	23 ± 2	15 ± 4
Gentisic acid	5 ± 1	10 ± 2	7 ± 2
Total salicylate recovered (mg, as SA equivalents)	246 ± 8	2999 ± 374	8092 ± 1470

Metabolites are given as percent of the amount of total salicylate (SA) (adapted from [81]).

analgesic and anti-inflammatory effect of the compounds have not been studied systematically.

2.1.2.4 Modification of Biotransformations and Excretion of Salicylates

Dose Metabolism, biotransformations, and excretion of salicylates are largely dose dependent. After saturation of the major metabolic routes, generation of salicyluric acid by conjugation with glycine, levels of unchanged salicylate increase in circulating blood and the compound also becomes the main metabolite in urine (Table 2.4). This results in a prolongation of the plasma half-life of salicylate from 2–3 to 20–30 h.

Changes in urinary pH have also significant effects on salicylate excretion. The renal clearance is about four times as great at pH 8.0 as at pH 6.0, mainly because of the inhibition of tubular reabsorption. Again, this can be used for the treatment of salicylate poisoning by alkalinization of urine (Section 3.1.1).

Drug Interactions Salicylates compete with a variety of drugs for plasma protein binding. This includes bilirubin, uric acid, phenytoin, sulfinpyrazone, and thyroxin among others. Possible consequences are changes in the pharmacokinetics of the free, active fraction of the compound. Salicylates may also interfere with high protein bound anticoagulants.

High doses of aspirin to rat induce the CYP2E1 isoform of cytochrome enzymes, eventually resulting in increased salicylate degradation. This could also be relevant for interactions with acetaminophen, especially in mixed type intoxications, because acetaminophen is also metabolized by this enzyme (Section 3.2.3). However, in men, aspirin at low antiplatelet (50 mg/day for 2 weeks) or analgesic/anti-inflammatory doses (1 g tid for 6 days) [87] did not induce this cytochrome though there was induction of other cytochrome P450 enzymes, for example, CYP2C19 [88]. Thus, CYP2E1 induction might be mainly of toxicological interest though the consequences of pharmacokinetic interactions with drugs – that are substrate for CYP2C19, such as warfarin – are unknown.

Age Typical alterations in pharmacokinetics with increasing age are a reduced protein binding with an increased apparent volume of distribution as well as a reduced clearance. This is also valid for aspirin, salicylate, and their metabolites. The plasma levels of salicyluric acid are slightly higher in the elderly (age: >75 years) than in younger persons [47]. In general, the variations are small and probably clinically not relevant [89].

Summary

The biotransformation of aspirin to salicylic acid is complete within minutes and is independent of the aspirin dose over a wide range of therapeutic doses. In contrast, biotransformation and excretion of salicylic acid are largely dose dependent and involve capacity-limited pathways that become saturated already at therapeutic doses, such as a single analgesic dose of 0.5–1 g aspirin.

Excretion of aspirin at low-dose (<500 mg) occurs mainly (70–75%) as glycine conjugate (salicyluric acid). Further metabolites are salicylic acid glucuronides (5–10%), salicylic acid (10%) and gentisic acid (10%). The transformation of salicylic acid to salicyluric acid is capacity limited and cannot be further increased at doses ≥500 mg, probably because the available glycine pool in the liver becomes exhausted. This enzymatic conversion of salicylic acid is genetically controlled and exhibits a large interindividual variability despite of good intraindividual reproducibility.

At toxic doses, the limited salicylate metabolism results in accumulation of salicylic acid and an increase in its plasma half-life from 2–3 to up to 20–30 h. This results in an enhanced volume of distribution and increased tissue levels, associated with metabolic acidosis and uncoupling of oxidative phosphorylation (Section 2.2.3). The pH dependency of renal salicylate excretion can be used for enhanced elimination by alkalinization of the urine in the case of salicylate overdosing (Section 3.1.1).

References

51 Levy, G. (1965) Pharmacokinetics of salicylate elimination in man. *Journal of Pharmaceutical Sciences*, 54, 959–967.

52 Levy, G. (1978) Clinical pharmacokinetics of aspirin. *Pediatrics*, **62** (Suppl.), 867–872.

53 Levy, G. (1979) Pharmacokinetics of salicylate in man. *Drug Metabolism Reviews*, **9**, 3–19.

54 Shen, J., Wanwimolruk, S., Purves, R.D. *et al.* (1991) Model representation of salicylate pharmacokinetics using unbound plasma salicylate concentrations and metabolite urinary excretion rates following a single oral dose. *Journal of Pharmacokinetics and Biopharmaceutics*, **19**, 575–595.

55 Spenney, J.G. (1978) Acetylsalicylic acid hydrolase of gastric mucosa. *The American Journal of Physiology*, **234**, E606–610.

56 Williams, F.M., Mutch, E., Nicholson, E.M. *et al.* (1989a) Human liver and plasma aspirin esterase. *The Journal of Pharmacy and Pharmacology*, **41**, 407–409.

57 Rowland, M. and Riegelman, S. (1968) Pharmacokinetics of acetylsalicylic acid and salicylic acid after intravenous administration in man. *Journal of Pharmaceutical Sciences*, **57**, 1313–1319.

58 Costello, P.B. and Green, F.A. (1983) Identification and partial purification of the major aspirin hydrolyzing enzyme in human blood. *Arthritis and Rheumatism*, **26**, 541–547.

59 Costello, P.B., Caruana, J.A. and Green, F.A. (1984) The relative roles of hydrolases of the erythrocyte and other tissues in controlling aspirin survival *in vivo*. *Arthritis and Rheumatism*, **27**, 422–426.

60 Adebayo, G.I., Williams, J. and Healy, S. (2007) Aspirin esterase activity – evidence for skewed distribution in healthy volunteers. *European Journal of Internal Medicine*, **18**, 299–303.

61 La Du, B. (1971) Plasma esterase activity and the metabolism of drugs with ester groups. *Annals of the New York Academy of Sciences*, **179**, 684–694.

62 Ali, B. and Kaur, S. (1983) Mammalian tissue acetylsalicylic acid esterases. Identification, distribution and discrimination from other esterases. *The Journal of Pharmacology and Experimental Therapeutics*, **226**, 589–594.

63 Rainsford, K.D., Schweitzer, A. and Brune, K. (1983) Distribution of the acetyl compared with the salicyl moiety of acetylsalicylic acid. *Biochemical Pharmacology*, **32**, 1301–1308.

64 Hawkins, D., Pinckard, R.N., Crawford, I.P. *et al.* (1969) Structural changes in human serum albumin induced by ingestion of acetylsalicylic acid. *The Journal of Clinical Investigation*, **48**, 536–542.

65 Green, F.A. and Jung, C.Y. (1981) Acetylation of erythrocyte membrane peptides by aspirin. *Transfusion*, **21**, 55–58.

66 Stiel, D., Griffin, J. and Andrew, D.S. (1985) Plasma aspirin esterase activity: relationship to aspirin ingestion and peptic ulceration. *Australian and New Zealand Journal of Medicine*, **15**, 562.

67 Williams, F.M., Wynne, H., Woodhouse, K.W. *et al.* (1989b) Plasma aspirin esterase – the influence of old age and frailty. *Age and Ageing*, **18**, 39–42.

68 Yelland, C., Summerbell, J., Nicholson, E. *et al.* (1991) The association of age with aspirin esterase activity in human liver. *Age and Ageing*, **20**, 16–18.

69 Williams, F.M., Asad, S.I., Lessof, M.H. *et al.* (1987) Plasma esterase activity in patients with aspirin-sensitive asthma or urticaria. *European Journal of Clinical Pharmacology*, **33**, 387–390.

70 Puche, E., Gomez-Valverde, E., Garcia Morillas, M. *et al.* (1993) Postoperative decline in plasma aspirin-esterase and cholinesterase activity in surgical patients. *Acta Anaesthesiologica Scandinavica*, **37**, 20–22.

71 Hohlfeld, Th. Zimmermann, N., Gams, E. *et al.* (2000) Aspirin resistance coincides with platelet COX-2 expression but not with enhanced aspirin esterase activity. *Circulation*, **102** (Suppl. II), II-237.

72 Benedek, I.C.H., Joshi, A.S., Pieniaszek, H.J. *et al.* (1995) Variability in the pharmacokinetics and pharmacodynamics of low dose aspirin in healthy male volunteers. *Journal of Clinical Pharmacology*, **35**, 1181–1186.

73 Puche, E., Garcia de la Serrana, H., Mota, C. *et al.* (1989) Serum aspirin-esterase activity in epileptic patients receiving treatment with phenobarbital, phenytoin, carbamazepine and valproic acid. *International Journal of Clinical Pharmacology Research*, **9**, 55–58.

74 Insel, P.A. (1996) Analgesic-antipyretic and antiinflammatory agents and drugs employed in the treatment of gout, in *Goodman and Gilman's The Pharmacological Basis of Therapeutics*, 9th edn, McGraw-Hill, New York, pp. 617–657.

75 Forman, M.B., Davidson, E.D. and Webster, L.T. (1971) Enzymatic conversion of salicylate to salicylurate. *Molecular Pharmacology*, **7**, 247–259.

76 Wilson, J.T., Howell, R.L., Holladay, M.W. *et al.* (1978) Gentisuric acid: metabolic formation in animals and identification as a metabolite of aspirin in man. *Clinical Pharmacology and Therapeutics*, **23**, 635–643.

77 Cham, B.E., Bochner, F., Imhoff, D.M. *et al.* (1980) Simultaneous liquid-chromatographic quantitation of salicylic acid, salicyluric acid, and gentisic acid in urine. *Clinical Chemistry*, **26**, 111–114.

78 Bochner, F., Graham, G.G., Polverino, A. *et al.* (1987) Salicylic phenolic glucuronide pharmacokinetics in patients with rheumatoid arthritis. *European Journal of Clinical Pharmacology*, **32**, 153–158.

79 Bedford, C., Cummings, A.J. and Martin, B.K. (1965) A kinetic study of the elimination of salicylate in man. *British Journal of Pharmacology*, **24**, 418–431.

80 Notarianni, L.J., Ogunbona, F.A., Oldham, H.G. *et al.* (1983) Glycine conjugation of salicylic acid after aspirin overdose. *British Journal of Clinical Pharmacology*, **15**, 587.

81 Patel, D.K., Ogunbona, A., Notarianni, L.J. *et al.* (1990) Depletion of plasma glycine and effect of glycine by mouth on salicylate metabolism during aspirin overdose. *Human & Experimental Toxicology*, **9**, 389–395.

82 Caldwell, J., O'Gorman, J. and Smith, R.L. (1980) Inter-individual differences in the glycine conjugation of salicylic acid. *British Journal of Clinical Pharmacology*, **9**, 114.

83 Hutt, A.J., Caldwell, J. and Smith, R.L. (1986) The metabolism of aspirin in man: a population study. *Xenobiotica*, **16**, 239–249.

84 Furst, D.E., Gupta, N. and Paulus, H.E. (1977) Salicylate metabolism in twins. Evidence suggesting a genetic influence and induction of salicylurate formation. *The Journal of Clinical Investigation*, **60**, 32–42.

85 Pankow, D., Damme, B. and Schrör, K. (1994) Acetylsalicylic acid – inducer of cytochrome P-450 2E1? *Archives of Toxicology*, **68**, 261–265.

86 Damme, B., Darmer, D. and Pankow, D. (1996) Induction of hepatic cytochrome P4502E1 in rats by acetylsalicylic acid or sodium salicylate. *Toxicology*, **106**, 99–103.

87 Park, J.-Y., Kim, K.-A., Park, P.-W. *et al.* (2006) Effect of high-dose aspirin on CYP2E1 activity in healthy subjects measured using chlorzoxazone as a probe. *Journal of Clinical Pharmacology*, **46**, 109–114.

88 Chen, X.-P., Tan, Z.-R. and Huang, S.-L. *et al.* (2003) Isozyme-specific induction of low-dose aspirin on cytochrome P450 in healthy subjects. *Clinical Pharmacology and Therapeutics*, **73**, 264–271.

89 Woodhouse, K.W. and Wynne, H. (1987) The pharmacokinetics of non-steroidal anti-inflammatory drugs in the elderly. *Clinical Pharmacokinetics*, **12**, 111–122.

2.2 Cellular Modes of Action

The pharmacological actions of aspirin are determined by *two* compounds: the reactive acetate moiety and the primary metabolite salicylic acid (salicylate). Both are complex and involve different mechanisms of action that, for the most part, are independent of each other. This includes inhibition of cyclooxygenases by transcriptional regulation of protein expression and inhibition of enzyme activity (Section 2.2.1). Other actions of salicylates are prostaglandin independent. This involves interactions with cellular signaling cascades that control inflammation, pain, and tumorigenesis, and also apoptosis and its mediators as central events for cell injury (Section 2.2.2). In addition, there are direct but nonspecific actions of salicylates on cell membranes, most notably the action of salicylate as a protonophore in mitochondrial membranes. This eventually results in uncoupling of oxidative phosphorylation (Section 2.2.3) with multiple follow-up effects, for example, in salicylate poisoning (Section 3.3.1).

The description of pharmacological properties of aspirin covers all biochemical and molecular changes that are induced by aspirin and its metabolites at the cellular and subcellular level at doses that are compatible with cell survival. The pharmacological properties are described independent of their possible usefulness for medical treatment.

In this context, the fundamental differences regarding a drug as a chemical/*pharmaceutical product* as opposed to its use as a *medicine* for the treatment of diseases have to be considered. A pharmaceutical is a chemical entity, exerting pharmacological actions in biological systems, independent of their kind and consequences for this particular system, for example, the human organism or a diseased organ. In contrast to this, a medicinal drug is a compound (pharmaceutical) that is used with the intention to prevent or to treat a disease and to obtain a therapeutic benefit for the patient. Thus, a chemical is qualified as a medicinal drug not by certain pharmacological properties but by its clinical efficacy and the benefit/risk ratio in controlled, randomized trials. In other words, not all pharmaceuticals are medicinal drugs. With regard to the global broad-spectrum biological activities of aspirin, it is evident that aspirin has many more pharmacological actions than those that are currently used for prevention or treatment of diseases.

Biologically relevant consequences of these multiple molecular modes of action of aspirin at the cellular and subcellular level are transcriptional and posttranscriptional modifications of protein expression, inhibition of enzyme activity, particularly the well-known inhibition of cyclooxygenases, and translational and posttranslational modifications, for example, the generation of "aspirin-triggered lipoxin" (ATL) from the interaction of (acetylated) COX-2 with the 5-lipoxygenase of neutrophils. A selection of these multiple actions of aspirin on cell function is shown in Table 2.5 and graphically demonstrated in Figure 2.8. The consequences of these (sub)cellular actions for tissue and organ function *in vivo* are discussed in Section 2.3.

Table 2.5 Cellular targets of aspirin action.

Two bioactive components in one molecule
→Reactive acetate
Nonselective, nonenzymatic transacetylation of target proteins with subsequent inhibition of enzyme activity (COX-1) or (additional) generation of new products (15-(*R*)-HETE) as a substrate for new biologically active mediators – ATL (COX-2)
Acetylation of eNOS
Acetylation of other cellular targets
→Salicylate
Modulation of gene transcription (COX-2, iNOS) via inhibition of binding of transcription factors to the gene promoter
Nonselective inhibition of (phospho)kinases by different mechanisms: substrate binding, ATP binding, ATPase activity
Protonophoric actions on cell membranes with subsequent uncoupling of oxidative phosphorylation

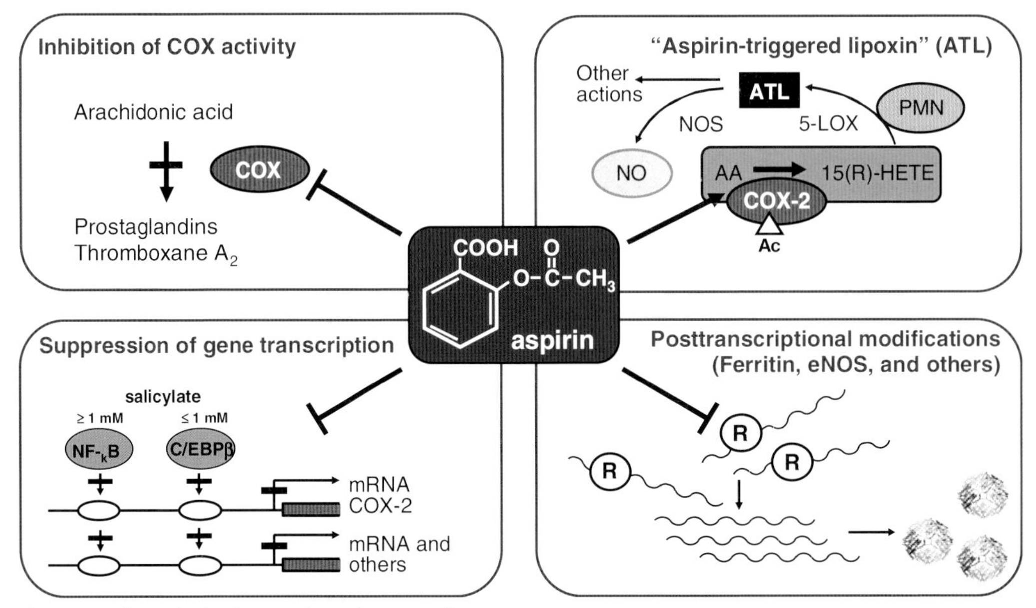

Figure 2.8 The multiple pharmacological actions of aspirin.

2.2.1
Inhibition of Cyclooxygenases

COX-1 and COX-2 There are two genes encoding for proteins with cyclooxygenase activity: COX-1, which is constitutively expressed and present ubiquitously in the organism and COX-2, which is also constitutively expressed in some tissues, including vascular endothelium, kidney, and CNS, but becomes markedly upregulated by inflammatory and mitogenic stimuli. Immunocompetent white cells, such as monocytes/macrophages, are rich sources for COX-2 [90] upon stimulation by inflammatory cytokines or tumor promoters. COX-1 is the dominating isoform in blood platelets, and its constant constitutive expression in platelets is an important prerequisite for the irreversibility of the antiplatelet action of aspirin. The gene for COX-2 belongs to the group of "early response genes" and encodes for a protein with high turnover rate and short half-life that synthesizes prostaglandins "on demand." COX-2-derived prostaglandins are, therefore, involved not only in pathologic situations, such as inflammation, pain, and immune reactions, but also in physiological functions, such as renal perfusion and sodium excretion, pregnancy (from fertility to labor), and neuronal signal transmission in the CNS. A significant proportion of endothelial prostacyclin and PGE_2 is also generated via COX-2. Prostacyclin contributes to the antithrombotic effects of endothelium, PGE_2 to the regulation of vessel tone. Both functions become relevant in advanced atherosclerosis, an inflammatory disease, where COX-2 is (constitutively) upregulated (Section 2.3.1).

The three-dimensional structures of COX enzymes are well known. They are integral membrane enzymes that catalyze the first step in the conversion of arachidonic acid to prostaglandins and thromboxanes. The entrance to the catalytic site lies at the apex of a long, narrow hydrophobic channel that runs from the membrane surface into the protein interior, providing an access for arachidonic acid to diffuse directly from the membrane inferior to the enzyme active site without traversing the aqueous department [91, 92]; Loll, personal communication, 2006). The generation of eicosanoids from arachidonic acid through the prosta-

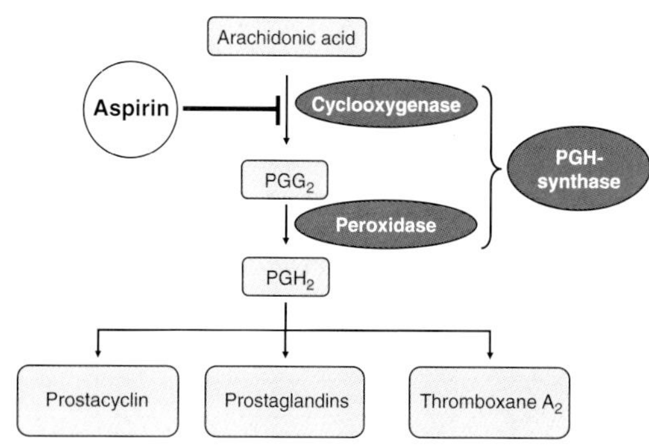

Figure 2.9 Generation of prostaglandins, prostacyclin and thromboxane A_2 via the cyclooxygenase pathway of arachidonic acid. Inhibition of cyclooxygenase activity by aspirin.

glandin H_2 synthase (PGHS) proceeds via several steps. The first is the introduction of two oxygen functions into arachidonic acid with subsequent cyclization and the formation of the prostaglandin hydroendoperoxide PGG_2 (cyclooxygenase reaction). This is followed by the reduction of the hydroperoxide at C15 to the corresponding hydroxy fatty acid, PGH_2 (peroxidase reaction). PGH_2 is then further converted to prostaglandins, prostacyclin and thromboxane A_2 by specific synthases and isomerases, respectively (Figure 2.9).

Aspirin and COX Isoforms Aspirin interacts with prostaglandin and thromboxane formation at several levels: most intensively studied is the interference with product formation via acetylation of the COX enzyme(s) [93]. However, sodium salicylate at therapeutic doses (3 g orally) also reduces PGE_2 excretion in urine to a comparable extent [94], which cannot be explained by target enzyme acetylation. An interference of both aspirin and salicylate with COX-2 gene regulation was detected, suggesting transcriptional effects of the compounds prior to expression of COX enzyme activity. Importantly, inhibition of binding of transcription factors to regulatory units (promoter region) of genes by salicylates might also change the regulation of other genes and gene products that are controlled by the same transcription factors. This is discussed in detail separately (Section 2.2.2) because this effect is not restricted to or even specific for cyclooxygenases.

2.2.1.1 Acetylation of Cyclooxygenases and Traditional NSAIDs

Aspirin modifies the activity of both COX-1 and COX-2 by acetylation. In addition, COX splice variants (COX-1b (COX-3), COX-2a and others) have been detected with yet poorly defined biological relevance and susceptibility to acetylation by aspirin. Aspirin inhibits the recombinant human COX-1b, an effect not seen with acetaminophen [95]. COX-2 is the molecular target of aspirin for inhibition of inflammatory and immune responses. COX-2 is also involved in aspirin-related actions on cell proliferation and apoptosis (Section 2.3.3). Prostaglandins, generated via both COX isoforms, contribute to inflammation and pain and both are inhibited by aspirin. However, inhibition of COX-2-dependent prostaglandin formation *in vivo* requires higher concentrations of aspirin and, in contrast to inhibition (acetylation) of platelet COX-1, is also sensitive to salicylates at about equal concentrations. An interesting suggestion by John Oates' group was that the inhibitory action of salicylate on COX-1 and COX-2 *in vitro* is controlled by the extent to which hydroperoxides are metabolized (reduced) by the peroxidase function of the enzyme. Thus, lipid hydroperoxides might also control the catalytic activity of the enzyme *in vivo* [96]. Most traditional nonsteroidal anti-inflammatory drugs (NSAIDs) inhibit both COX-1 and COX-2 at therapeutic concentrations [97–100], whereas acetaminophen fails to do so (Table 2.6).

The so-called "selectivity" of traditional NSAIDs for COX-1 and COX-2 isoenzymes and the different results found in dependency on the experimental setup to determine this "selectivity" have raised some uncertainties regarding their functional significance with respect to clinical use. These different results are in many cases due to a different methodology for measurement rather than a drug-related specificity.

The expression level of COX-1 protein is usually constant and determines the constitutive activity of the enzyme. In contrast, the expression of COX-2 protein has to be first induced by appropriate stimuli before inhibition of this activity can be studied at all. These stimuli (mostly tumor promoters, inflammatory interleukins, or endotoxin) have many biological effects in addition to upregulating COX-2. Thus, upregulation depends upon the duration and intensity of the stimulus. It is, therefore, not surprising that the sensitivity of COX-2 to inhibitors is much different and usually greater in purified enzymes or COX-2-transfected cells than to inhibitors in intact cells after upregulation by any of the stimuli mentioned above. In this context, the frequently used approach to upregulate COX-2 by 24 h *in vitro* incubation with higher concentrations of endotoxin appears not to be very physiological. However, it is closer to real life than the constitutive activity in any cell-free system. In other words, the potency and selectivity of COX-2 versus COX-1 inhibition *in vivo* may not always be identical with those determined *in vitro* in sophisticated assay systems.

Table 2.6 Inhibition of COX-1 (bovine aortic endothelial cells) and COX-2 (endotoxin-stimulated mouse macrophages)-dependent prostaglandin formation by aspirin, sodium salicylate, other nonsteroidal anti-inflammatory drugs, and acetaminophen (paracetamol).

Substance	COX-1	COX-2	COX-2/COX-1
Aspirin	0.3	50	166
Salicylate	35.0	100	2.8
Indomethacin	0.01	0.6	60
Ibuprofen	1.0	15	15
Diclofenac	0.5	0.4	0.7
Naproxen	2.2	1.3	0.6
Acetaminophen	2.7	>20	>7

The data are IC_{50} values (g/ml) of three to five independent experiments (modified after [97]).

Acetylation of COX-1 by Aspirin The first systematic studies on the molecular mechanism of inhibition of prostaglandin biosynthesis by aspirin were published in the 1970s by the group led by Philip Majerus, using the platelet enzyme. At the time, it was already known that aspirin transacetylates numerous proteins and DNA [64, 101, 102]. Roth and Majerus [103] detected the acetylation of platelet COX by aspirin. These authors and others additionally showed that the acetylation site was a serine close to the N-terminus of the enzyme [104–106]. The half-life of inactivation of the enzyme *in vitro* amounted to 10–20 min at 100 μM aspirin and was in the same time frame as the inhibition of platelet function. The acetylation reaction after oral aspirin was dose dependent after single doses of 20–650 mg (Table 2.7), lasted for several days [107], and was pH independent [106]. This led to the conclusion that the serine$_{530}$ side chain, lining the substrate binding site of the COX protein, is the target of acetylation by aspirin (Figure 2.10). It was also suggested that the acetylation of this serine side chain by aspirin, which is readily reversible under different conditions, becomes stabilized in the highly hydrophobic environment of the COX-active site. This probably explains why one single aspirin dose can irreversibly inhibit prostanoid synthesis in the platelet (Loll, personal communication, 2006). Aspirin does not modify the peroxidase activity of the synthase, pointing to a specific action of aspirin on the cyclooxygenase step.

Table 2.7 Dose-dependent acetylation of platelet cyclooxygenase by aspirin in healthy volunteers after single oral administration (modified after [107]).

Aspirin dose (mg)	Number of subjects	COX acetylation after aspirin (free cpm/mg protein)	Inactivation (%)
20	7	1.061	34
80	4	318	73
160	4	204	82
325	16	184	89
650	3	78	>95

This hypothesis of serine$_{530}$ as the molecular target of aspirin was further substantiated after elucidation of the amino acid sequence of COX protein. The primary structure of the gene that encoded for the enzyme was also clarified. It was shown that the enzyme from sheep seminal vesicle [108] and human platelets [109] was acetylated at the same serine$_{529}$ (sheep) or serine$_{530}$ (human), respectively. Acetylation resulted in a complete loss of enzyme activity, confirming the original observations of Roth *et al.* [104]. Replacement of serine by alanine, another bulky amino acid without an acet-

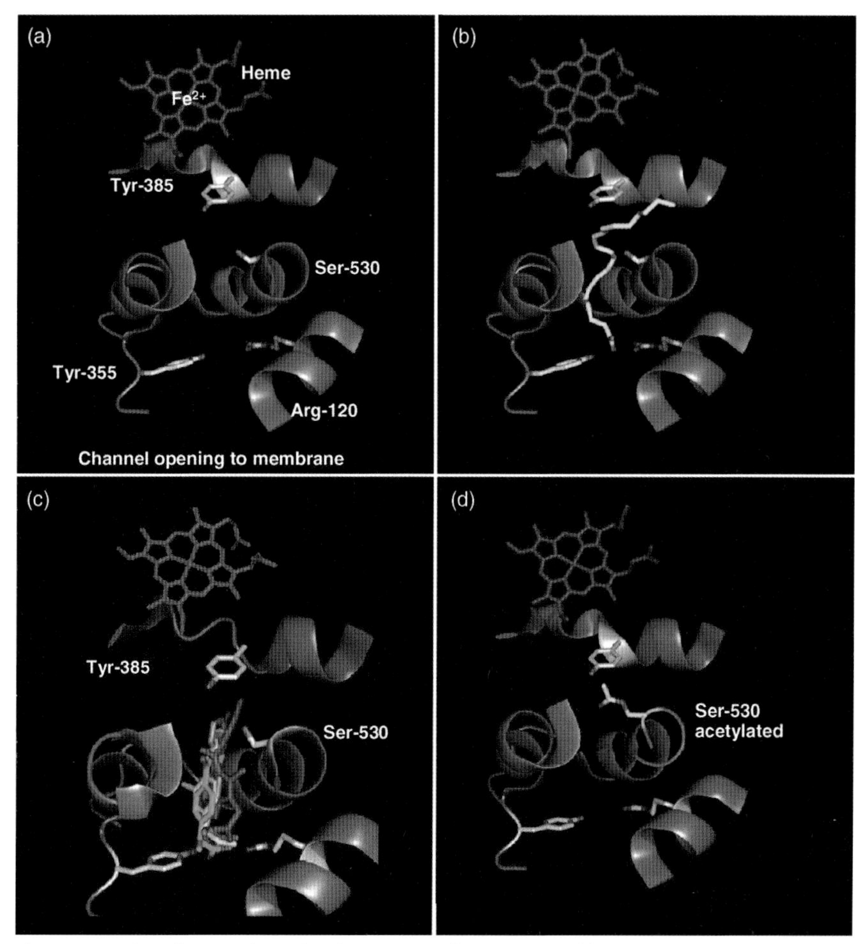

Figure 2.10 The substrate channel and active site of COX-1 (a) and binding of arachidonic acid (b). Salicylic acid and traditional NSAIDs prevent binding of arachidonic acid by mechanical channel blockade (c) whereas aspirin (d) blocks the access of arachidonic acid through acetylation of serine 530. This modification of serine side chain by aspirin is stabilized against hydrolysis by the hydrophobic environment of the channel (Loll, personal communication).

ylation site, by site-directed mutagenesis, prevented the inhibitory action of aspirin on enzyme activity [110]. However, this procedure did not reduce the catalytic activity of the enzyme, suggesting that serine$_{530}$ was not essential for catalysis or binding to the active site. Replacement of serine$_{530}$ with asparagine resulted in a complete loss of cyclooxygenase activity whereas the peroxidase activity was increased, indicating that these two properties of PGH_2 synthase could be changed independently (Table 2.8) [108]. Removal of Tyr$_{385}$ also was associated with a complete loss of enzymatic activity. These and other data clearly indicated that this amino acid is essential for the cyclooxygenase activity of the enzyme [111–113]. After the crystal structure of COX-1 was elucidated [114, 91], it also became morphologically apparent that aspirin blocked the access of the arachidonic acid substrate to the tyrosine$_{385}$ inside the protein, located at the end of a hydrophobic channel that arachidonic acid has to pass through to reach the catalytic site. Aspirin introduces a bulky compound into the

Table 2.8 Cyclooxygenase and hydroperoxidase activities of virally transformed mutants of PGG/PGH synthase from membrane preparations of sheep.

Mutated amino acids	COX Activity (nmol/min × g)	k_m-AA (μM)	Enzyme-half-life in the presence of aspirin (min)	Peroxidase activity (nmol/min × g)
None (wt)	450	7	30	70
$Serine_{530}$ acetylated	0	—	—	70
$Serine_{530} \leftrightarrow alanine_{530}$	388	8	Stable	79
$Serine_{530} \leftrightarrow asparagine_{530}$	0	—	—	222

Effects of aspirin: Acetylation of Ser_{530} in the COX protein (wt) results in complete loss of COX activity without inhibition of peroxidase activity of the complex. Replacement of Ser_{530} by alanine does not change the COX or peroxidase activity of the complex but makes the enzyme resistant against aspirin, suggesting that Ser_{530} is the acetylation target of aspirin. Replacement of Ser_{530} by asparagine inhibits the cyclooxygenase activity completely but increases the peroxidase activity, suggesting that both activities can be regulated independently (after DeWitt *et al.* [108] and Smith WL [personal communication]).

channel that causes steric hindrance of the access of arachidonic acid to the catalytic site. Salicylic acid and traditional NSAIDs prevent binding of arachidonic acid by a mechanical channel blockade as a result of their lipophilic properties. Arg_{120} fixes the salicylate group by hydrogen bounds in the correct steric position [115, 116]. Figure 2.10 demonstrates the current view on the COX-1 substrate channel and active site of COX-1, as well as the modification by aspirin.

Another recently discovered effect of acetylation of COX-1 in platelets is the inhibition of the release of sphingosine-1-phosphate from human platelets. The inhibition parallels the inhibition of thromboxane formation and persists for 3 days in healthy men after ingestion of one single dose (1 g) of aspirin. Sphingosine phosphate does apparently not modify the platelet function but might change the function of sphingosine-derived lipid signaling [117] (Figure 2.11).

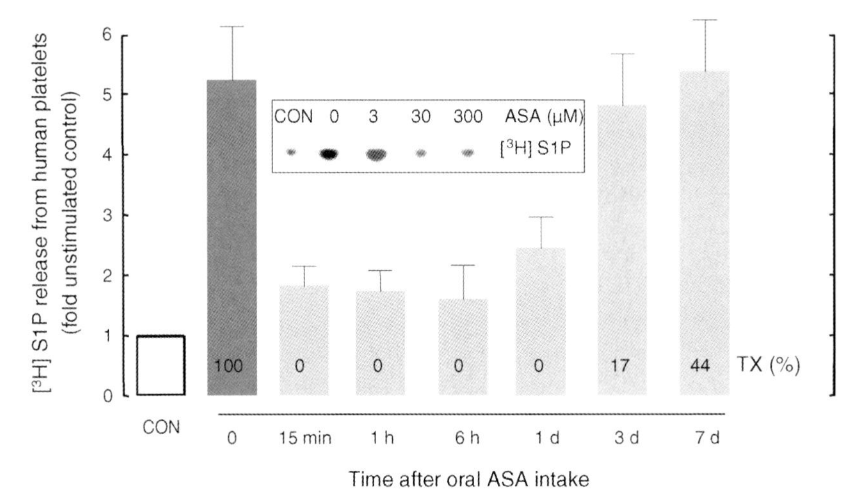

Figure 2.11 Inhibition of thrombin-induced release of sphingosine-1-phosphate (S1P) from human platelets by oral aspirin (ASA) (500 mg) and aspirin *in vitro*. The numbers in the columns indicate the remaining thromboxane (TX) formation. No inhibition is seen with salicylate (not shown) (modified after [117]).

Acetylation of COX-2 by ASA Transfer of the acetyl moiety toward one critical location in the COX enzyme(s) – the serine$_{530}$ in COX-1 and COX-2 (*Note*: the COX-1 numbering system is used by convention that is also for COX-2: Ser$_{516}$ in COX-2 is structurally identical with serine$_{530}$ in human COX-1) has chemically the same reaction but has different consequences for COX-1 and COX-2. Acetylation of COX-2 modifies the steric structure of COX-2 protein in a way to generate 15-(*R*)-HETE. As already seen with COX-1, replacement of Ser$_{530}$ with alanine not only prevented the aspirin acetylation reaction but also reduced the enzymatic activity to only 50% whereas replacement of Tyr$_{385}$ by phenylalanine completely prevented any COX activity. All of these procedures left the peroxidase activity unchanged [113]. This is presumably due to the cyclooxygenase side pocket of COX-2 that allows fatty acids and inhibitors to bind more tightly at the COX-2 active site [116]. In contrast to COX-1, Arg$_{120}$ plays only an accessory role for the binding of arachidonic acid and inhibitors in COX-2 – a deletion mutant was still active though at markedly reduced affinity.

Thus, in addition to the inhibition of prostaglandin formation, acetylation of COX-2 by aspirin results in the generation of a new product, 15-(*R*)-HETE [118–120]. 15-(*R*)-HETE is made by COX-2 in much higher proportions than the cyclooxygenase product PGE_2. Interestingly, the production of 15-(*R*)-HETE by aspirin-acetylated COX-2 was inhibited by most traditional NSAIDs [112] and selective COX-2 inhibitors [121]. The significance of this finding for the anti-inflammatory action of these compounds remains to be determined.

This particular property of aspirin led to the hypothesis that the generation of 15-(*R*)-HETE by acetylated COX-2 may not just be a removal of metabolic "waste" but might serve specific purposes. *Charles Serhan* was the first to show that 15-(*R*)-HETE was the precursor of a class of new products the "aspirin-triggered lipoxins" by synergistic interaction of acetylated COX-2 with 5-lipoxygenase from white cells [122] (Figure 2.12). ATL not only contributes to the anti-inflammatory actions of aspirin (Section 2.3.2) but might also be involved in endothelial protection (Section 2.3.1) and, possibly represents a new class of anti-inflammatory and proresolving lipid mediators [123]. The most recent finding that ATL contributes to the control of innate immunity by inducing proteasomal degradation of a tumor necrosis factor receptor-associated factor-6 (TRAF-6) requires particular attention [124]. In addition, ATL appears also to be gastroprotective during repeated or long-term aspirin use, possibly associated with enhanced COX-2 expression in the stomach mucosa (Section 3.2.1). However, most of the available evidence so far comes from *in vitro* or animal studies, and clinical data are still missing.

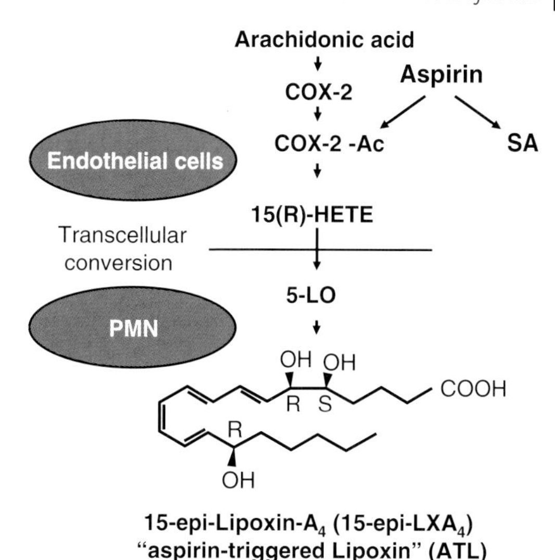

Figure 2.12 Generation of 15-(*R*)-HETE in endothelial cells by acetylated COX-2 (COX-2-Ac) and its transcellular conversion to 15-epi-lipoxin A_4 (15-epi-LXA_4) or aspirin-triggered lipoxin by 5-lipoxygenase (5-LO) of polymorphonuclear cells (modified after [122]).

The generation of (*R*)-precursors of biologically active eicosanoids by aspirin is not limited to arachidonic acid but was also seen with docosahexaenoic acid and eicosapentaenoic acid, two precursors of series "3" prostaglandins. Interaction with aspirin-treated COX-2 in human endothelial cells resulted in the generation of 17-(*R*)-hydroxydocosahexaenoic acid, 17-(*R*)-HDHA. Human neutrophils transform COX-2-aspirin-derived 17-(*R*)-HDHA into two sets of novel di- and trihydroxy products – the resolvins. Analogous pathways exist for eicosapentaenoic acid. The name resolvins was given because these products of transcellular biosynthesis, similar to ATL, dampen inflammatory events. They are generated within the inflammatory resolution phase and downregulate leukocytic exudate cell numbers to prepare for orderly and timely resolution [125, 123].

Generation of New "Aspirin-Like Drugs" After elucidation of the molecular mechanism of aspirin action, many drugs were developed with the intention to inhibit COX-dependent prostaglandin formation – the so-called "aspirin-like drugs" or nonsteroidal anti-inflammatory drugs, including indomethacin, diclofenac, ibuprofen, and many others. In contrast to aspirin, indomethacin-type compounds are reversible inhibitors of the enzyme and compete with the substrate arachidonic acid for binding at the catalytic site [108, 126].

COX-2 differs from COX-1 by the presence of a side pocket in the substrate channel. Compounds that fit into this side pocket, such as coxibs, are COX-2 selective. Attempts have been made to create "aspirin-like," that is, irreversibly acting compounds, with a higher COX-2 selectivity by modifying the side chain in a way that increases affinity for the COX-2 protein. These compounds bind covalent to the enzyme and act irreversibly, like aspirin, but have a certain COX-2 selectivity *in vitro* and in animal models [127]. However, their efficacy is lower than that of the specific coxibs, and no data about clinical testing have been published so far.

Inhibition of ASA-Induced COX-1 Inhibition by Reversibly Acting NSAIDs The binding of NSAIDs and aspirin to similar sites in the lining of the COX-active site may result in drug interactions with particular relevance to the antiplatelet effects of aspirin. This interaction, first shown for indomethacin [104, 128], was later confirmed for other NSAIDs (Section 2.3.1) and, most recently, pyrazoles, such as dipyrone (metamizol) [129]. If these compounds are given shortly before aspirin, they may occupy salicylate binding sites inside the COX protein and prevent the access of aspirin to its acetylation site (Figure 2.13). Since NSAIDs act reversibly, the duration of this interaction is determined by the half-life of the particular compounds, in most cases only a few hours. However, this time is sufficient for deacetylation of aspirin by esterases that occurs within 30 min. Thus, aspirin might have lost its antiplatelet activity because the active form is no longer present when the acetylation site at the enzyme becomes available again. Functionally, this results in a prevention of antiplatelet effects of aspirin (Section 2.3.1) or aspirin "resistance" (Section 4.1.6). Interestingly, the inhibitory effect of dipyrone on aspirin-induced inhibition of platelet aggregation is seen also during continuous aspirin intake, probably as result of a higher affinity of dipyrone than salicylic acid to the (nonspecific) hydrophobic binding sites in the COX-1 channel

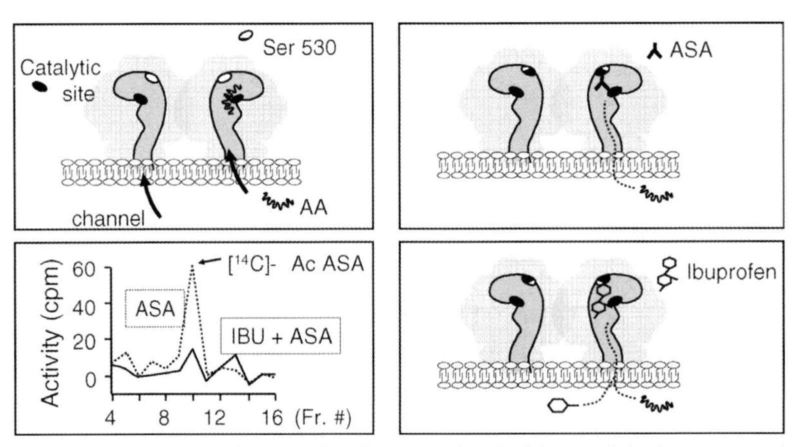

Figure 2.13 The inhibition by ibuprofen of aspirin-induced inhibition of platelet COX-1. Evidence for inhibition of aspirin binding by ibuprofen (IBU) (for further explanation, see the text) (modified after [130, 131]).

(cf. Figure 4.24). No such interactions exist for selective COX-2 inhibitors [130].

2.2.1.2 Modulation of COX-2 Gene Expression

Protein acetylation explains the inhibition of COX activity by aspirin. In the platelet without significant protein synthesis, this inhibition of COX-1 is only seen with aspirin whereas salicylic acid has no effect. On the contrary, both aspirin and salicylate are about equipotent inhibitors of COX-2 (Table 2.6). This corresponds to their efficacy in the treatment of fever, pain, and inflammation (Section 2.3.2) as well as the clinical experience that higher doses are needed for these effects as compared with those that inhibit platelet function. Overall, this suggests additional mechanism(s) of action beyond the acetylation of cyclooxygenases.

In 1991, *Kenneth Wu* and colleagues were the first to show that both aspirin and salicylate inhibit expression of a COX protein, later identified as COX-2 [132] at the level of gene transcription, an effect not shared with other "aspirin-like" drug.

Prostaglandin formation was stimulated in human vascular endothelial cells *in vitro* with interleukin-1 or phorbol ester. This resulted in enhanced COX mRNA expression and protein biosynthesis. Pretreatment of the cells with aspirin at low concentrations (0.1–10 μg/ml) reduced the increase in mass of enzyme protein by more than 60%. Similar effects were seen with salicylate but not with indomethacin. This inhibition of protein synthesis was preceded by a reduced cytokine-induced increase in COX mRNA, suggesting an action of salicylates but not of nonsteroidal anti-inflammatory drugs at the level of COX gene transcription [133, 134] (Figure 2.14).

Actions of Salicylates on Transcription Factors At the time, it was not known that human vascular endothelial cells express the inducible COX-2 isoform after stimulation by cytokines. Later work suggested that salicylate, after entering the cell, interacts with the CCAT/enhancer-binding protein-β (C/EBP-β) in the nucleus, prevents the binding of this transcription factor, and downregulates COX-2 gene expression. This effect is seen at therapeutic analgesic/anti-inflammatory concentrations (0.01–≤1 mM) of the compounds. At higher concentrations (>1 mM) salicylates additionally inhibit the binding of additional transcription factors, such as nuclear factor κB (NFκB) [135]. These effects of salicylates are not limited to COX-2 but are also seen with inducible NO synthase (iNOS) and probably many other genes that are regulated by the same transcription factors in their promoter region. This issue is discussed in more detail in Section 2.2.2. Similar to salicylates, naproxen was also found to transcriptionally downregulate COX-2 (and COX-1) protein expression [136]. Whether this also applies to other traditional NSAIDs is unknown.

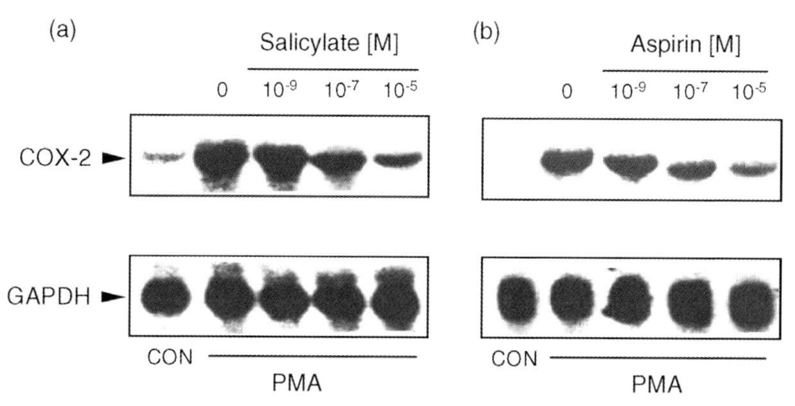

Figure 2.14 Equipotent inhibition of phorbol ester (PMA)-induced COX-2 gene expression in human umbilical vein endothelial cells (a) and fibroblasts (b) by aspirin and sodium salicylate in anti-inflammatory concentrations (GAPDH: internal standard) (modified after [134]).

2.2.1.3 Further Actions of Salicylates on Arachidonate Metabolism

In addition to modifying COX-1 and COX-2 activity, salicylates can also influence other metabolic pathways of arachidonic acid. In general, these actions are seen only at high concentrations of the compounds. Their significance for the clinical efficacy of aspirin is uncertain.

Arachidonic Acid Release Daily intake of aspirin at oral doses of ≥3 g reduces incorporation of arachidonic acid into isolated mononuclear cells by about 50%. Similar effects are seen after incubation of white cells with 0.3 mg/ml aspirin *in vitro* [137]. This suggests that aspirin at anti-inflammatory concentrations might modify the generation of arachidonic acid metabolites via alteration of the fatty acid composition of membrane phospholipids. An inhibition of arachidonic acid release by higher concentrations of aspirin (0.56 mM) was also shown for human platelets [138] and cytokine-induced expression of soluble phospholipase A_2 [139].

Generation of 12-H(P)ETE 12-Hydroxyeicosatetraenoic acid (12-HETE) is another major arachidonic acid metabolite in platelets, formed from the 12-hydroperoxy precursor (12-HPETE) via the 12-lipoxygenase pathway. 12-HPETE inhibits platelet function and thromboxane biosynthesis [140, 141]. Therefore, enhanced availability of 12-HPETE after inhibition of COX-1 in platelets could contribute to the antiplatelet actions of aspirin. However, because of the high concentrations and short half-life of 12-HPETE, this mode of action is not likely to be important *in vivo*. Another action of HPETE could be the control of COX activity by modulation of the hydroperoxide tone (see above).

Summary

Aspirin inhibits prostaglandin and thromboxane formation at different levels. Most intensively studied is the acetylation of COX-1 and COX-2 proteins. In addition both aspirin and salicylate might also interfere with COX-2 gene transcription by inhibition of binding of transcription factors to the promoter region of the gene.

The mechanism of COX inhibition by aspirin has been elucidated. Aspirin binds covalently to a serine ($serine_{530}$) in the substrate channel of COX-1 and COX-2. This inhibits the access of substrate (arachidonic acid) to the catalytic site of the enzyme at $tyrosine_{385}$. The peroxidase reaction of the enzyme is not affected but might be involved in the control of enzyme activity under certain conditions.

Acetylation of COX-1 is the mechanism of antiplatelet actions of aspirin. Acetylation of COX-2 not only inhibits prostaglandin formation but in addition also generates a new compound, 15-(*R*)-HETE. This hydroxy fatty acid is substrate for 5-lipoxygenases in white cells that generates 15-epi-lipoxin A_4 or ATL, an anti-inflammatory compound.

Competitive-type NSAIDs as well as pyrazoles (dipyrone) interact with aspirin by preventing its binding in the COX substrate channel. Because of the short half-life of aspirin (15–20 min) this may result in a complete deacetylation of the compound by plasma esterases and disappearance of the antiplatelet effects. Traditional NSAIDs and COX-2-selective inhibitors also antagonize the activity of acetylated COX-2 to synthesize 15-(*R*)-HETE and subsequent generation of ATL. This could explain the increase in GI toxicity of coxibs in the presence of aspirin (Section 3.2.1). Whether this mechanism is also relevant to the anti-inflammatory effects of aspirin remains to be determined.

References

90 DeWitt, D.L. and Smith, W.L. (1988) Primary structure of prostaglandin G/H synthase from sheep vesicular gland determined from the complementary DNA sequence. *Proceedings of the National Academy of Sciences of the United States of America*, **85**, 1412–1416.

91 Loll, P., Picot, D. and Garavito, R.M. (1995) The structural basis of aspirin activity inferred from the crystal structure of inactivated prostaglandin H_2 synthase. *Nature Structural Biology*, **2**, 637–643.

92 Gupta, K., Selinsky, B.S., Kaub, C.J. *et al.* (2004) The 2.0 Å resolution crystal structure of prostaglandin H_2 synthase-1: structural insights into an unusual peroxidase. *Journal of Molecular Biology*, **335**, 503–518.

93 Vane, J.R. (1971) Inhibition of prostaglandin synthesis as a mechanism of action of aspirin-like drugs. *Nature: New Biology*, **231**, 232–235.

94 Hamberg, M. (1972) Inhibition of prostaglandin synthesis in man. *Biochemical and Biophysical Research Communications*, **49**, 720–726.

95 Censarek, P., Freidel, K. and Hohlfeld, T. *et al.* (2006) Human cyclooxygenase-1b is not the elusive target of acetaminophen. *European Journal of Pharmacology*, **551**, 50–53.

96 Aronoff, D.M., Boutaud, O., Marnett, L.J. *et al.* (2003) Inhibition of prostaglandin H_2 synthases by salicylate is dependent on the oxidative state of the enzymes. *The Journal of Pharmacology and Experimental Therapeutics*, **304**, 589–595.

97 Mitchell, J.A., Akarasereenont, P., Thiemermann, Ch. *et al.* (1994) Selectivity of nonsteroidal antiinflammatory drugs as inhibitors of constitutive and inducible cyclooxygenase. *Proceedings of the National Academy of Sciences of the United States of America*, **90**, 11693–11697.

98 Mitchell, J.A., Saunders, M., Barnes, P.J. *et al.* (1997) Sodium salicylate inhibits cyclo-oxygenase-2 activity independently of transcription factor (nuclear factor κB) activation: role of arachidonic acid. *Molecular Pharmacology*, **51**, 907–912.

99 Warner, T.D., Giuilano, F., Vojnovic, I. *et al.* (1999) Nonsteroid drug selectivities for cyclooxygenase rather than cyclooxygenase-2 are associated with human gastrointestinal toxicity: a full *in vitro* analysis. *Proceedings of the National Academy of Sciences of the United States of America*, **96**, 7563–7568.

100 Warner, T.D. and Mitchell, J.A. (2004) Cyclooxygenases: new forms, new inhibitors, and lessons from the clinic. *The FASEB Journal*, **18**, 790–804.

101 Pinckard, R.N., Hawkins, D. and Farr, R.S. (1968) *In vitro* acetylation of plasma proteins, enzymes and DNA by aspirin. *Nature*, **219**, 68–69.

102 Bridges, K.R., Schmidt, G.J., Jensen, M. *et al.* (1975) The acetylation of haemoglobin by aspirin. *The Journal of Clinical Investigation*, **56**, 201–207.

103 Roth, G.J. and Majerus, P.W. (1975) The mechanism of the effect of aspirin on human platelets. I. Acetylation of a particulate fraction protein. *Journal of Clinical Investigation*, **56**, 624–632.

104 Roth, G.J., Stanford, N. and Majerus, P.W. (1975) Acetylation of prostaglandin synthetase by aspirin. *Proceedings of the National Academy of Sciences of the United States of America*, **72**, 3073–3076.

105 Roth, G.J. and Siok, C.J. (1978) Acetylation of the NH_2-terminal serine of prostaglandin synthetase by aspirin. *The Journal of Biological Chemistry*, **253**, 3782–3784.

106 Van der Oudera, F.J., Buytenhek, M., Nugteren, D.H. *et al.* (1980) Acetylation of prostaglandin endoperoxide synthetase with acetylsalicylic acid. *European Journal of Biochemistry*, **109**, 1–8.

107 Burch, J.W., Stanford, N. and Majerus, P.W. (1978) Inhibition of platelet prostaglandin synthetase by oral aspirin. *The Journal of Clinical Investigation*, **61**, 314–319.

108 DeWitt, D.L., El-Harith, E.A., Kraemer, S.A. *et al.* (1990) The aspirin and heme-binding sites of ovine and murine prostaglandin endoperoxide synthases. *The Journal of Biological Chemistry*, **265**, 5192–5198.

109 Funk, C.D., Funk, L.B., Kennedy, M.E. *et al.* (1991) Human platelet/erythroleukemia cell prostaglandin G/H synthase: cDNA cloning, expression and gene chromosomal assignment. *The FASEB Journal*, **5**, 2304–2312.

110 Shimokawa, T. and Smith, W.L. (1992) Prostaglandin endoperoxide synthase: the aspirin acetylation region. *The Journal of Biological Chemistry*, **267**, 12387–12392.

111 Shimokawa, T., Kulmacz, R.J., DeWitt, D.L. *et al.* (1990) Tyrosine 385 of prostaglandin endoperoxide synthase is required for cyclooxygenase catalysis. *The Journal of Biological Chemistry*, **265**, 20073–20076.

112 Meade, E.A., Smith, W.L. and DeWitt, D.L. (1993) Differential inhibition of prostaglandin endoperoxide synthase (cyclooxygenase) isozymes by aspirin and other nonsteroidal antiinflammatory drugs. *The Journal of Biological Chemistry*, **268**, 6610–6614.

113 Hochgesang, G.P., Jr, Rowlinson, S.W. and Marnett, L.J. (2000) Tyrosine-385 is critical for acetylation of cyclooxygenase-2 by aspirin. *Journal of the American Chemical Society*, **122**, 6514–6515.

114 Picot, D., Loll, P.J. and Garavito, R.M. (1994) The X-ray crystal structure of the membrane protein prostaglandin H_2 synthase-1. *Nature*, **367**, 243–249.

115 Mancini, J.A., Riendeau, D., Falgueyret, J.-P. *et al.* (1995) Arginine 120 of the prostaglandin G/H-synthase-1 is required for the inhibition by nonsteroidal anti-inflammatory drugs containing a carboxylic acid moiety. *The Journal of Biological Chemistry*, **270**, 29372–29377.

116 DeWitt, D.L. (1999) COX-2 selective inhibitors: the new super aspirins. *Molecular Pharmacology*, **55**, 625–631.

117 Rauch, B.H., Ulrych, T., Rosenkranz, A.C. *et al.* (2008) Aspirin inhibits release of sphingosine-1-phosphate from human platelets. *Arteriosclerosis, Thrombosis, and Vascular Biology*, **28**, E104.

118 Lecomte, M., Laneuville, O., Ji, C. *et al.* (1994) Acetylation of human prostaglandin endoperoxide synthase-2 (cyclooxygenase-2) by aspirin. *The Journal of Biological Chemistry*, **269**, 13207–13215.

119 Mancini, J.A., O'Neill, G.P., Bayly, C. *et al.* (1994) Mutation of serine 516 in human prostaglandin G/H-synthase-2 to methionine or aspirin acetylation of this residue stimulates 15-R-HETE synthesis. *FEBS Letters*, **342**, 33–37.

120 O'Neill, G.P. *et al.* (1994) Overexpression of human prostaglandin G/H synthase-1 and -2 by recombinant vaccinia virus: inhibition by nonsteroidal antiinflammatory drugs and biosynthesis of 15-hydroperoxyeicosatetraenoic acid. *Molecular Pharmacology*, **45**, 245–254.

121 Mancini, J.A., Vickers, P.J., O'Neill, G.P. *et al.* (1997) Altered sensitivity of aspirin-acetylated prostaglandin G/H-synthase-2 to inhibition by non-steroidal antiinflammatory drugs. *Molecular Pharmacology*, **51**, 52–60.

122 Clària, J. and Serhan, C.N. (1995) Aspirin triggers previously undescribed bioactive eicosanoids by human endothelial cell–leukocyte interactions. *Proceedings of the National Academy of Sciences of the United States of America*, **92**, 9475–9479.

123 Serhan, C.N. (2007) Resolution phase of inflammation: novel endogenous anti-inflammatory and proresolving lipid mediators and pathways. *Annual Review of Immunology*, **25**, 101–137.

124 Machado, F.S., Esper, L., Dias, A. *et al.* (2008) Native and aspirin-triggered lipoxins control innate immunity by inducing proteasomal degradation of TRAF6. *The Journal of Experimental Medicine*, **205**, 1077–1086.

125 Serhan, C., Hong, S., Gronert, K. *et al.* (2002) Resolvins: a family of bioactive products of omega-3 fatty acid transformation circuits initiated by aspirin treatment that counter proinflammatory signals. *The Journal of Experimental Medicine*, **196**, 1025–1037.

126 Kulmacz, R.J. (1989) Topography of prostaglandin H synthase. Antiinflammatory agents and the protease-sensitive arginine 253 region. *The Journal of Biological Chemistry*, **264**, 14136–14144.

127 Kalgutkar, A.S., Crews, B.C., Rowlinson, S.W. *et al.* (1998) Aspirin-like molecules that covalently inactivate cyclooxygenase-2. *Science*, **280**, 1268–1270.

128 Rome, L.H., Lands, W.E.M., Roth, G.J. and Majerus, P.W. (1976) Aspirin as a quantitative acetylating reagent for the fatty acid oxygenase that forms prostaglandins. *Prostaglandins*, **11**, 23–30.

129 Hohlfeld, T., Zimmermann, N., Weber, A.-A. *et al.* (2008) Pyrazolinone analgesics prevent the antiplatelet effect of aspirin and preserve human platelet thromboxane biosynthesis. *Journal of Thrombosis and Haemostasis*, **6**, 166–173.

130 Ouellet, M., Riendeau, D. and Percival, M.D. (2001) A high level of cyclooxygenase-2 inhibitor activity is associated with a reduced interference of platelet cyclooxygenase-1 inactivation by aspirin. *Proceedings of the National Academy of Sciences of the United States of America*, **98**, 14583–14588.

131 Catella-Lawson, F., Reilly, M.P., Kapoor, S.C. *et al.* (2001) Cyclooxygenase inhibitors and the antiplatelet effects of aspirin. *The New England Journal of Medicine*, **345**, 1809–1817.

132 Habib, A., Créminon, C., Frobert, Y. *et al.* (1993) Demonstration of an inducible cyclooxygenase in human endothelial cells using antibodies raised against the carboxyl-terminal region of the cyclooxygenase-2. *The Journal of Biological Chemistry*, **268**, 23448–23454.

133 Wu, K.K., Sanduja, R., Tsai, A.L. *et al.* (1991) Aspirin inhibits interleukin-1 induced prostaglandin H synthase expression in cultured endothelial cells. *Proceedings of the National Academy of Sciences of the United States of America*, **88**, 2384–2387.

134 Xu, X.-M., Sansores-Garcia, L. and Chen, X.-M. *et al.* (1999) Suppression of inducible cyclooxygenase-2 gene transcription by aspirin and sodium salicylate. *Proceedings of the National Academy of Sciences of the United States of America*, **96**, 5292–5297.

135 Saunders, M.A., Sansores-Garcia, L., Gilroy, D. *et al.* (2001) Selective suppression of C/EBPβ binding and COX-2 promoter activity by sodium salicylate in quiescent human fibroblasts. *The Journal of Biological Chemistry*, **276**, 18897–18904.

136 Zyglewska, T., Sanduja, R., Ohashi, K. *et al.* (1992) Inhibition of endothelial cell prostaglandin H synthase gene expression by naproxen. *Biochimica et Biophysica Acta*, **1131**, 78–82.

137 Bomalaski, J.S., Alvarez, J., Touchstone, J. *et al.* (1987) Alteration of uptake and distribution of eicosanoid precursor fatty acids by aspirin. *Biochemical Pharmacology*, **36**, 3249–3253.

138 Vedelago, H.R. and Mahadevappa, V.G. (1988) Mobilization of arachidonic acid in collagen-stimulated human platelets. *The Biochemical Journal*, **256**, 981–987.

139 Vervoordeldonk, M.J., Pineda Torra, I.M., Aarsman, A.J. *et al.* (1996) Aspirin inhibits expression of the interleukin-1beta-inducible group II phospholipase A_2. *FEBS Letters*, **397**, 108–112.

140 Siegel, M.I., McConnell, R.T., Porter, N.A. *et al.* (1980) Arachidonate metabolism via lipoxygenase and 12-L-hydro-peroxy-5,8,10,14-eicosatetraenoic acid peroxidase sensitive to anti-inflammatory drugs. *Proceedings of the National Academy of Sciences of the United States of America*, **77**, 308–312.

141 Aharony, D., Smith, J.B. and Silver, M.J. (1982) Regulation of arachidonate-induced platelet aggregation by the lipoxygenase product, 12-hydroperoxyeicosatetraenoic acid. *Biochimica et Biophysica Acta*, **718**, 193–200.

2.2.2
COX-Independent Actions on Cell Function

The most convincing evidence for prostaglandin-independent though biologically significant actions of salicylates is their biosynthesis by plants. Salicylates are natural plant constituents and part of a defense system that protects them from injury by exogenous noxes such as bacteria or viruses. Salicylate generation increases plant resistance and can be substantially upregulated at the transcriptional level in response to injury. The failure to do so, for example, after genetic manipulation, results in severe damage or cell death. Plants only can synthesize (polyunsaturated) fatty acids up to an 18 hydrocarbon (C18) backbone. Thus, neither arachidonic acid (C20) nor prostaglandins or other mediators derived from arachidonic acid peroxidation can be generated. However, plants can synthesize jasmonic acid via a lipid peroxidation pathway that has many structural and functional similarities to prostaglandins. Thus, generation of salicylates in plants represents a prostaglandin-independent protective mechanism that increases plant resistance and has provided several innovative approaches for the development of insecticides and pest management.

Aspirin and salicylate exhibit a broad spectrum of pharmacological actions on cell function [142] that is probably not completely elucidated yet. The transacetylation and nontransacetylation-related actions of aspirin are nonselective and nonspecific. Transacetylations may occur at any appropriate molecular site in any macromolecule, most notably plasma albumin and hemoglobin [102] as well as DNA [143], whereas salicylic acid will accumulate inside cell membranes, including those of mitochondria, with subsequent alterations in cell signaling and energy metabolism.

One of the earliest studies demonstrating COX-independent actions of aspirin on inflammatory cells came from Gerald Weissmann's group. They showed that aspirin and salicylate were about equipotent inhibitors of neutrophil aggregation and Ca^{2+} entry (signaling) whereas only aspirin but not salicylate inhibited platelet aggregation and thromboxane formation (signaling) (Figure 2.15). Since neutrophils do not synthesize prostaglandins, this finding suggested that aspirin and salicylate inhibit neutrophil reactions through mechanisms independent of the prostaglandin system. Later works showed a disturbed assembly of heterotrimeric G proteins within the lipid bilayer of cell membranes by salicylates. This suggests that membrane effects of salicylate, possibly related to its particular physicochemical properties (Section 2.2.3), might interfere with transmembrane signal transduction. These actions of salicylates require higher concentrations than the transacetylation reaction. However, the concentrations of 1–3 mM – used in this particular study – were still within the therapeutic range obtained with anti-inflammatory doses of aspirin [144, 145].

The modulation of enzymes of the cell energy metabolism and, most interestingly, heat-shock proteins and chaperones [146–148], probably, is

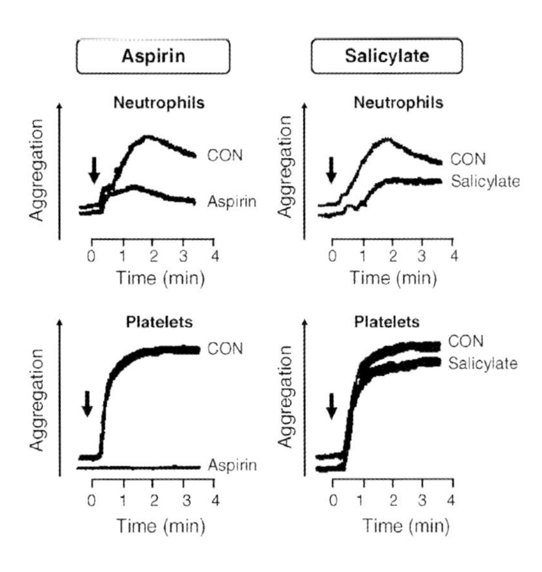

Figure 2.15 Different actions of aspirin and sodium salicylate on human platelets and granulocytes *in vitro*. Stimulation of platelet aggregation by arachidonic acid was blocked after pretreatment with aspirin but remained unchanged after pretreatment with an equimolar concentration of Na salicylate. In contrast, stimulation of neutrophil aggregation by the chemoattractant of MLP was inhibited to a similar extent by both aspirin and Na salicylate. Arrows mark the addition of the agonists (modified after [144]).

involved in regulatory actions of aspirin on enzymatic processes in inflammation, immune responses, and tumor defense. All of these actions are salicylate mediated.

Salicylate concentrations, necessary to exert analgesic/anti-inflammatory effects, are substantially higher (about 150–450 µg/ml or 1–3 mM) than those necessary to inhibit prostaglandin biosynthesis. This either suggests additional sites of action or indicates that enhanced prostaglandin production is rather an epiphenomenon than a causal factor of these disorders. After cell stimulation by inflammatory cytokines or tumor promoters, salicylates interact with cellular signal generation and signal transduction at both the transcriptional and posttranscriptional levels. Transcription factors and kinases appear to be the central cellular target and cytokines a major class of mediators involved [149, 150]. The functional consequences of these multiple activities of salicylates for inflammation, pain, and fever are discussed in more detail in Section 2.3.2, the consequences for cell proliferation with particular relevance to malignancy in Section 2.3.3. This section deals with the mechanistic aspect of these activities.

2.2.2.1 Kinases

Phosphorylation of proteins by kinases is a central biochemical mechanism to regulate enzyme activity. To become active, kinases have to be first phosphorylated. This process starts by binding of ATP to an ATP (substrate) binding site of the enzyme. Subsequently, the active site of the enzyme becomes phosphorylated by transfer of energy-rich phosphate. Finally, the activated phosphate group is transferred by the kinase reaction to a target substrate, such as another enzyme or transcription factor, respectively, eventually resulting in a biological response (Figure 2.16).

Principally, salicylates can interact with these events at several levels: one is the inhibition of substrate (ATP) binding because of chemical analogies between the ring structures of adenine and salicylic acid. This reaction is competitive, that is, reversible, stochiometric at a 1 : 1 relationship and requires higher concentrations of salicylates (>1 mM). The other is an interference with kinase activity by steric interaction with the transfer of the energy-rich phosphate from the ATP-binding site to the active site of the kinase, for example, after binding to another binding site, such as an arginine (by analogy with the arginine$_{120}$ in COX-1). This reaction is noncompetitive and nonstochiometric and probably requires lower concentrations of salicylates (<1 mM).

Inhibition of kinases by salicylates is both a simple and comprehensive explanation for the diversity of salicylate actions on cell function at higher concentrations. The consequences of kinase inhibition for cell function are then determined by the function of the particular phosphorylated target protein or transcription factor, respectively. Kinase activity might be differentially modified by salicylates in intact cells as opposed to cell homogenates [151], perhaps because of the high millimolar concentrations, which can be used in cell homogenates but not in intact cells or even tissues *in vivo*. Myriads of kinases can be inhibited by salicylates and the biological significance of this action is difficult to predict [152].

Nonselective inhibition of kinases as a mode of action of salicylates was first hypothesized by Frantz and O'Neill [153]. These authors showed that sodium salicylate caused a concentration-dependent, nonselective inhibition of a variety of different transcription factors at millimolar concentrations. This effect was probably due to nonselective inhibition of cellular kinases since an apparently identical response at the same concentrations was also seen in a cellular kinase preparation (Figure 2.17). The hypothesis was that activation of these transcription factors required phosphorylation by kinases that were prevented by salicylates.

Inhibition of ATP Binding to Kinases Yin *et al.* [154] suggested another more specific mechanism, namely a competitive and specific inhibition by salicylates of ATP-binding to the inhibitory kinase-β (IKK-β). This effect was reversible and could be antagonized by increasing the ATP concentra-

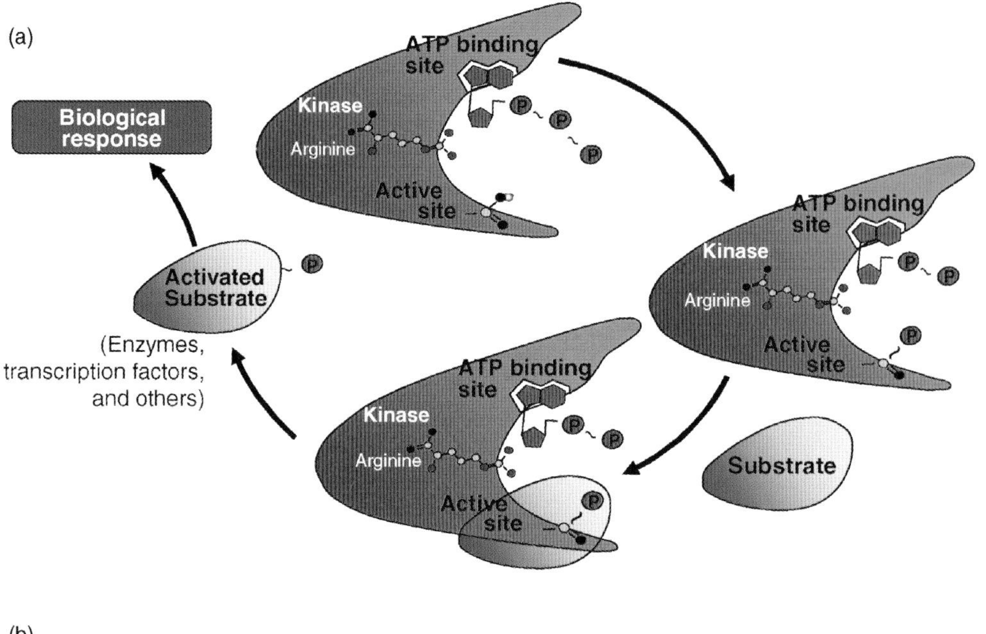

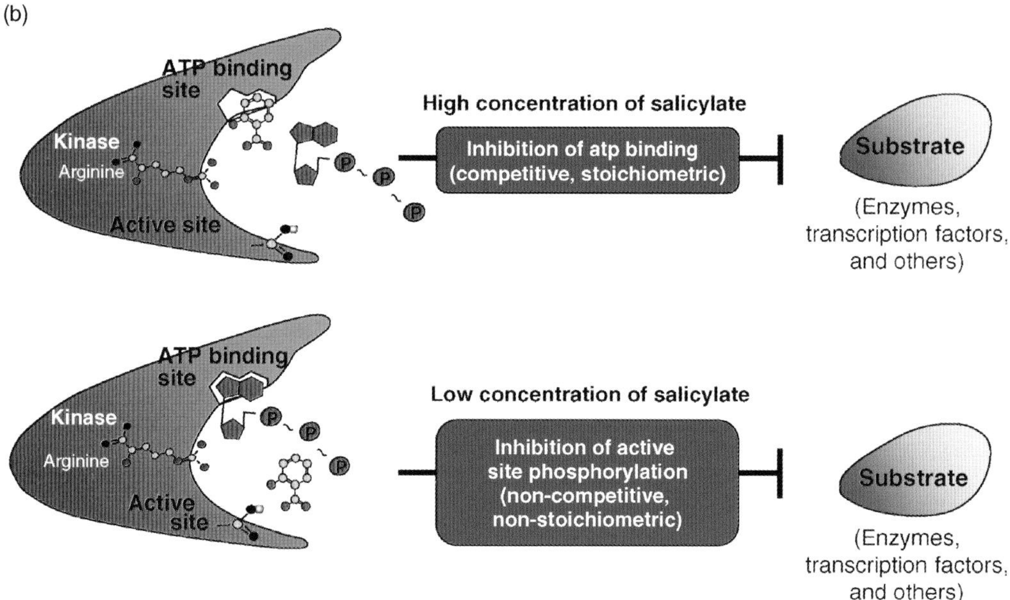

Figure 2.16 Hypothetic cellular mechanism of kinase activation and action (a) and possible sites of interaction of salicylates (b) via inhibition of ATP binding (top) or interaction with the active site (bottom) (for further explanation, see the text).

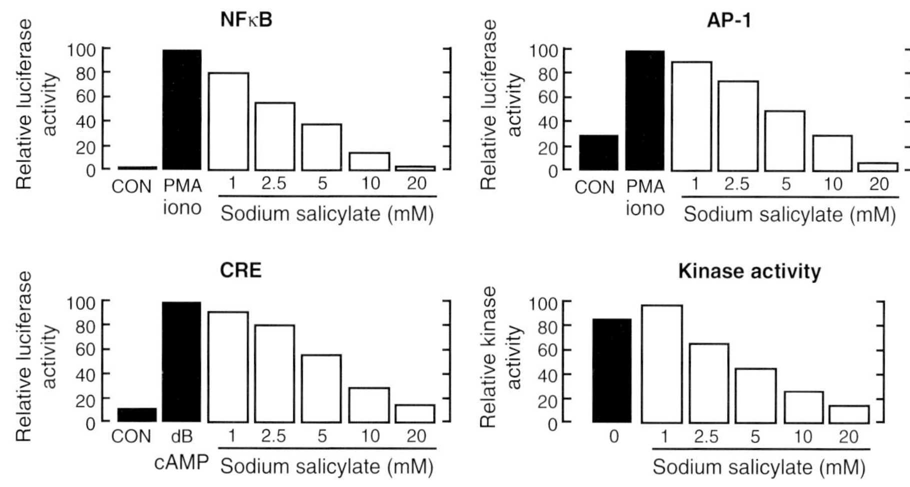

Figure 2.17 Effects of salicylates on transcriptional activation of three differently regulated transcription factors (NFκB, AP-1, and CRE) in transfected JURKAT cells. Note the apparently identical concentration response on three functionally different transcription factors after stimulation by phorbolester/ionomycin (PMA/iono) or cAMP (CRE) and the same inhibition in an acellular, nonselective kinase activity assay (modified after [153]).

tion. As mentioned above, phosphorylation is essential for enzymatic activity, that is, phosphate transfer. However, IKK-β is definitely not the only kinase that is inhibited by aspirin and salicylate, respectively [155].

Inhibition of Kinase Activity In search for a more detailed understanding of the cellular actions of aspirin with a possible relationship with kinase activity, the group of *Kenneth Wu* [156] made the interesting observation that salicylates not only interact with substrate (ATP) binding to kinases but also with kinase activity.

Specific cellular binding sites of salicylates were identified by incubation of homogenates of human foreskin fibroblasts with ^{14}C-sodium salicylate. The binding protein fraction was isolated and sequenced. This fraction contained a 15-amino acid sequence that was identical with a sequence in the heavy chain of human immunoglobulin binding protein (BiP). The k_D values of salicylate binding to the crude extract and to recombinant BiP were apparently identical: 45 and 55 μM, respectively, suggesting that salicylates may specifically interact with this sequence. Binding occurred via the *o*-hydroxy group of salicylate, leaving the carboxy function free for chemical reactions.

BiP (also known as GRP78) belongs to the heat-shock protein 70 (HSP70) family with important chaperone functions. These include binding of newly synthesized polypeptides, allowing for appropriate protein folding and transport across the membrane after binding to a polypeptide binding site. A synthetic heptapeptide containing this particular sequence displaced salicylate from its binding in a concentration-dependent manner, binding of the peptide-induced ATPase activity that was blocked by both aspirin and salicylate at micromolar concentrations. Neither aspirin nor salicylate did block ATP binding or modify BiP protein expression.

It was concluded that salicylates bind specifically to the polypeptide binding site of BiP in human cells, resulting in perturbation of a protein-involved inactivation of a specific kinase, such as ribosomal S6 kinase. In this way, salicylates may interfere with the chaperone function of BiP, that is, the processing of proteins important in inflammation [156] (Figure 2.18).

2.2.2.2 Transcription Factors

An enhanced expression of inducible genes is a regular component not only of inflammatory and ischemic diseases but also of malignant tumors. Gene expression is initiated by binding of transcription factors to specific positions in the promoter region of these genes. This results in gene activa-

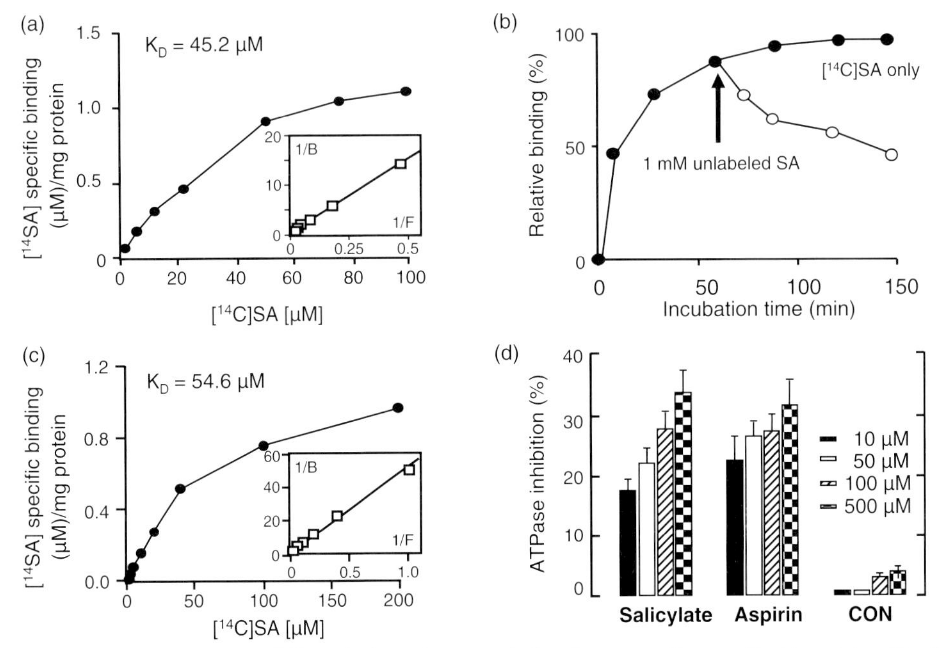

Figure 2.18 Specific binding of [^{14}C] salicylic acid (SA) to whole-cell extracts of human fibroblasts (a) and its displacement by addition of unlabeled salicylic acid (b). Similar binding kinetics of SA is seen in purified recombinant human immunoglobulin heavy chain binding protein (BiP) (c). (d) indicates the inhibition of basal ATPase activity of purified recombinant BiP by SA, aspirin, and the reference compound 3,4-dimethyoxy-γ-benzoic acid [156].

tion and transcription of the genetic code into messenger RNA. This message is then translated into a protein in the ribosomes within the cytosol. The genes for not only COX-2 but also iNOS belong to this group and are upregulated in response to tissue injury, eventually resulting in enhanced product formation, including not only prostaglandins and NO but also peroxynitrite and other products of (lipid)peroxidation.

Salicylates can modify the binding of a variety of transcription factors to the promoter region of genes. In many cases, modifications of gene regulation occur via inhibition of kinases that are necessary for activation of transcriptions factors, allowing for their subsequent binding to the promoter region. Alternatively, there might be a direct interaction with the binding of transcription factors, such as NFκB or nuclear factor of activated T cells (NFAT) to the promoter region, which is kinase independent [149, 157, 158]. In both cases, the result is essentially the same – prevention of gene activation and subsequent product formation.

The understanding of the biological significance of modulation of particular transcription factors by salicylates is often hampered by the fact that these changes are mainly found *in vitro* and, in many cases, require extremely high concentrations of salicylates, up to 100 mM (!), to become (statistically) significant. Thus, although from a pharmacological point of view it is interesting to know which genes can be modulated by salicylates, these studies do not necessarily suggest that these changes are biologically significant *in vivo*. It is also less likely that cell lines *in vitro*, expressing constitutively active (otherwise inducible) genes after gene transfer or stimulation by tumor promoters that can be directly compared with "normal" somatic nontransfected cells, being the subject of rather transient stimulation by cytokines or related mediators of inflammation, ischemia, or immune responses. Thus, transcription factors that are sensitive to salicylates at concentrations above 2 mM are of pharmacological interest but may have limited therapeutic value *in vivo*. This problem will be discussed in terms of two well known salicylate-sensitive transcription factors: NFκB and C/EBP-β.

Nuclear Factor (NF) NFκB The nuclear factor κB/RelA family of transcription factors (NFκB/RelA) regulates the expression of numerous genes involved in the control of immune and inflammatory responses, most notably tumor necrosis factor α (TNFα) and interleukin (IL)-1β. NFκB also controls cell survival either as a regulator of the apoptotic program for induction of apoptosis or, more commonly, as its inhibitor. Therefore, NFκB not only controls immediate inflammatory and immune responses [159] but also acts as a central regulator of longer lasting changes, including stress responses [160].

Intracellular NFκB resides inactive and bound to the inhibitory protein IkB in the cytosol of immunocompetent white cells, endothelial cells, and vascular smooth muscle cells as a heterotrimeric complex with IkB. Stimulation of IkB by IKK kinases results in phosphorylation, cleavage of the inhibitor, and translocation of the active NFκB heterodimer into the nucleus. IKK kinase activity is stimulated by cytokines, reactive oxygen species, and numerous other stimuli. The activation is mediated by increased activity of an IkB kinase (IKK) complex. The liberated heterodimer p50/p65 activates the genes of IL-1, IL-6, TNFα, ICAM-1, VCAM-1, and others, participating in the regulation of inflammation, immune responses, and cell survival. The net reaction is determined by signaling pathways, distal to NFκB.

Aspirin and salicylate inhibit NFκB activation via inhibition of IKK-β kinase activity in numerous cells and tissues *in vitro*, predominantly at millimolar concentrations [161, 162]. Therefore, these actions may not always be detected in intact cells *in vivo* [151]. The effects of aspirin on NFκB [161] are specific for salicylates and are not seen with indomethacin or other NSAIDs. Direct inhibition of IKK-β by salicylates [154, 163] probably explains the hypoglycemic actions of salicylates [164] and the inhibition of transcriptional activation of "tissue factor"(TF) [165, 166]. Another most interesting recent finding is the inhibition of influenza virus replication *in vitro* and *in vivo* by aspirin (not indomethacin) via inhibition of NFκB [167].

NFκB and Apoptosis Sodium salicylate (20 mM) produces a strong activation of p38 MAP kinases and cell death by apoptosis, suggesting that this kinase serves as a mediator of induced apoptosis in human fibroblasts and several cell lines including human colon adenocarcinoma cells. This activation of p38-MAPK and the subsequent induction of apoptosis might be important for the antineoplastic effect of the compound [168, 169] and is discussed in more detail in Section 2.2.3.

Other Transcription Factors As mentioned above, NFκB is not the only transcription factor that is modified by salicylates. C/EBP-β is another one that can be phosphorylated by several kinases, in particular, ribosomal p90 S6 kinase [170]. C/EBP-β controls transcriptional activation of COX-2, iNOS, and probably other genes that are involved in inflammatory and immune reactions [135, 171] (Section 2.2.1). The inhibition of the transcription factor C/EBP-β by salicylates – in contrast to inhibition of NFκB – is already seen at submillimolar concentrations and, therefore, probably significant also *in vivo* (Figure 2.19).

Further transcription factors that are potential targets of salicylates are activator protein-1 (AP-1) [153], STAT-6 [172] and NFAT. NFAT shares some homologies with NFκB and becomes activated after dephosphorylation by the phosphatase calcineurin. Salicylates inhibit DNA-binding and activation of this transcription factor without affecting the phosphorylation status or intracellular localization of NFAT [158]. Interestingly, a new pentafluoropropoxy derivative of salicylic acid (UR-1505) has recently been shown to block T-cell activation via inhibition of NFAT, eventually resulting in decreased T-cell proliferation and cytokine production [173]. An overview of selected transcription

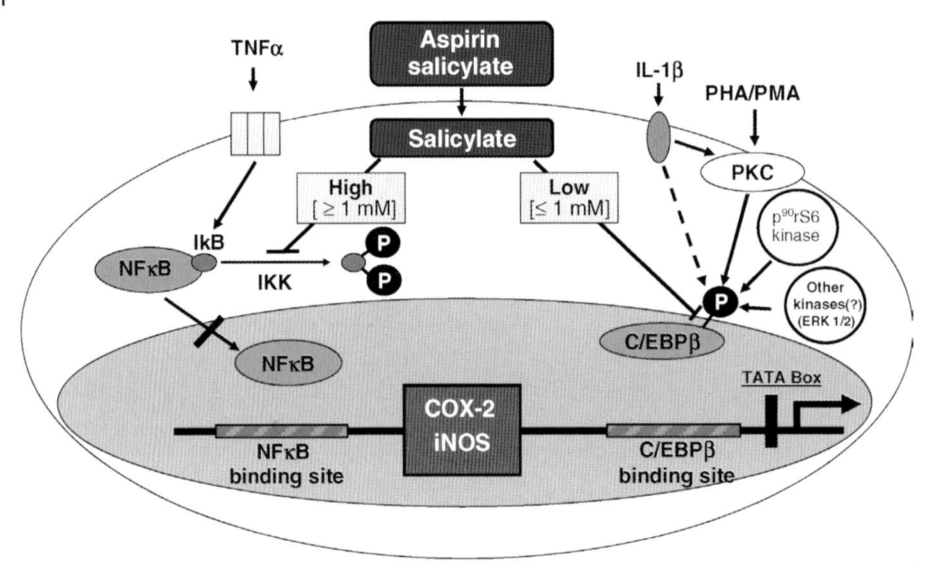

Figure 2.19 Different control of COX-2- and iNOS-promotor activity by salicylates via C/EBPβ and NFκB (for further explanation, see the text) (modified after [135, 170, 171]).

factors and affected genes that are modified by salicylates is summarized in Table 2.9.

NFκB and Neuroprotection Grilli *et al.* [177] showed that aspirin at low millimolar (IC_{50}: 1.7 mM) concentrations protected rat primary neuronal cultures and hippocampal slices from neurotoxicity elicited by the excitatory amino acid glutamate. Similar albeit less pronounced effects were seen with salicylate but not with indomethacin. The site of action was downstream to the glutamate receptor and involved salicylate-specific inhibition of NFκB activation via blocking phosphorylation of IkB [154, 178]. Another transcription factor, AP-1 remained unaffected [177]. This study was the first to show NFκB-mediated neuroprotective effects of salicylates. This action was not correlated with the anti-inflammatory activity or COX inhibition. More recent studies additionally indicated that glutamate-induced neuronal death involved inhibition of the

Table 2.9 Interactions of salicylates (aspirin and/or salicylate) with activation (act) or inhibition (inh) of transcription factors via kinase modulation.

Kinase	inh/act	TF	Salicylate (mM)	Affected genes	References
PKC, others	inh	C/EBP-β	0.01–1.0	iNOS, COX-2	Saunders *et al.* [135]
p79/p85S6	inh	C/EBP-β	0.01–0.5	COX-2, iNOS	Saunders *et al.* [135]; Cieslik *et al.* [171]
JNK	act	AP-1	0.1–1.0	Oncogenes	Ma *et al.* [174]
?	inh	Els family member	0.3–1.0	IL-4	Cianferoni *et al.* [175]
1src	inh	STAT 6	5–20	Many	Perez-G *et al.* [172]
IKK-β, others	inh	NFκB	1–10	Many	Kopp and Ghosh [161]
ERK 1/2	inh	AP-1	1–3	Many	Vartiainen *et al.* [176]
p38 MAPK	act	NFκB	2–20	Many	Schwenger *et al.* [168]

protein kinase PKCζ, an enzyme that controls apoptosis in neuronal cells. Aspirin directly inhibited the kinase activity of the purified enzyme, an action, which explained the prevention of neuronal death in these conditions [179]. Thus, inhibition of selected kinases by salicylates, possibly related to activation of transcription factors, may afford neuroprotection against excitatory, potentially toxic stimuli. A relationship to the analgesic action of salicylates (Section 2.3.2) is obvious.

In clinical reality, there are multiple reasons for neuronal injury and death, including vascular (cerebral infarction) and nonvascular (Alzheimer's disease) factors. These different etiologies of neuronal injury can only incompletely be mimicked in experimental setups. Consequently, *in vitro* data, obtained in isolated neurons, are not transferable to all types of neuronal injury *in vivo*. In a hypoxia-reoxygenation model of focal brain ischemia in the rat, inhibition of the sustained activation of cellular signaling pathways (p42/p44 MAP kinases) by salicylates (1–3 mM) was described independent of blockade of glutamate receptors or inhibition of NFκB activation [176, 180]. *In vivo*, iNOS and iNOS-derived products, such as peroxynitrite, appear to be important salicylate-sensitive proinflammatory and tissue-destructive mechanisms for brain injury. Salicylates inhibit iNOS-gene expression in submillimolar concentrations [171] and might afford neuroprotection via this mechanism. *In vivo*, iNOS expression is restricted to infiltrating leukocytes in focal brain ischemia. These cells are not present in neuronal cultures *in vitro*.

NFκB and End Organ Damage *In Vivo* Considering these limitations regarding the interpretation of *in vitro* data on the modulation of transcription factors by salicylates for *in vivo* conditions of tissue and organ injury, the question arises whether they have any consequences at all for tissue and organ survival *in vivo*. This issue was addressed in rats with severe genetically fixed arterial hypertension. The question was, whether treatment with aspirin could protect these animals from end organ damage and death and whether signaling pathways, operating via NFκB and AP-1, were involved.

Genetically modified rats, harboring both the human renin and angiotensinogen gene, generate large amounts of angiotensin II, develop hypertension, and die from heart and/or renal failure at about 7 weeks of age. These animals were treated with high (600 mg/kg/day) or low (25 mg/kg/day) dose aspirin intraperitoneally from week 4 to week 7. These (on a weight basis) high doses of aspirin – in comparison to those given to man – were necessary because of the high aspirin clearance rate in this species. The ED_{50} for analgesic effects ranges between 150 and 350 mg/kg, and the LD_{50} (strain-dependent) between 600 and 800 mg/kg.

High-dose aspirin significantly reduced mortality, cardiac hypertrophy, fibrosis, and albuminuria in these animals. These beneficial effects were independent of blood pressure changes and were not seen with low-dose aspirin, although both doses reduced COX activity to a comparable extent. High-dose aspirin inhibited NFκB and AP-1 activation and inflammatory reactions in heart and kidney.

The conclusion was that aspirin treatment leads to organ protection *in vivo* and that this effect is mediated via inhibition of NFκB- and AP-1-controlled signaling cascades in end organs, here predominantly the heart and kidney [181].

2.2.2.3 Oxidative Stress and Nitric Oxide

There are three isoforms of NO synthases (NOS) from which at least two are regulated by aspirin: the iNOS in macrophages, other inflammatory cells, fibroblasts and smooth muscle cells, and the eNOS, which, by definition, is mainly located in endothelial cells. iNOS generates large amounts of NO and oxygen-centered free radicals that amplify the inflammatory process whereas eNOS synthesizes small amounts of NO that regulate vessel tone via endothelium-dependent relaxation of vascular smooth muscle cells. Both enzymes are regulated by aspirin, though in opposite direction. Functionally, this corresponds to a synergistic action on endothelial cells by endothelial protection from oxidative stress, an effect also verified in other tissues by proteome analysis [148]. Neither iNOS- nor eNOS-dependent NO formation is modified by indomethacin and other NSAIDs. This suggests salicylate-specific actions on NO formation that are unrelated to prostaglandins.

Aspirin, iNOS, and Inflammation Aspirin and salicylate inhibit the cytokine-induced expression of iNOS gene via inhibition of transcription factors, such as NFκB and STAT-1 (see above). This is associated with an anti-inflammatory effect as seen from reduced release of TNFα [182–184]. Cieslik *et al.* [171] were the first to show that inhibition of iNOS (and COX-2) gene expression in macrophages is obtained at submillimolar, that is, therapeutic anti-inflammatory concentrations of salicylates and is mediated by inhibition of activation and binding of the transcription factor C/EBP-β (Figure 2.19). Other studies have shown that iNOS protein expression and activity in LPS-stimulated macrophages and cytokine-stimulated vascular smooth muscle cells can also be obtained at comparable aspirin concentrations (1 mM) at unchanged iNOS mRNA expression. This suggests translational/posttranslational effects of aspirin on iNOS-dependent NO formation that are independent of gene transcription [185] or interactions with COX-2 [186]. Overall, these findings agree well with the general concept of synergistic functions of iNOS and COX-2 in inflammation (Section 2.3.2) and the potent anti-inflammatory effect of salicylates.

Aspirin, eNOS, and Endothelial Protection In contrast to these anti-inflammatory effects of aspirin that are mediated via iNOS inhibition and modulation of COX-2 and, consequently, require anti-inflammatory doses of the compound, the modulation of eNOS-derived NO by aspirin in vascular cells is seen at lower concentrations (3–30 μM). The similarities to the different dosing of aspirin for inhibition of COX-1 and COX-2 are obvious. The enhanced NO generation by aspirin is considered endothelial protective [187, 188]. The expression of eNOS-protein remains unchanged [189]. Acetylation seems to be a crucial (posttranslational) mechanism because acetylating aspirin analogues have the same effect as aspirin itself whereas NSAIDs and nonacetylated salicylate are ineffective [188].

More recent works suggested that aspirin-induced upregulation of endothelial NO formation will not only improve local perfusion but in addition also increase the expression of two endothelial-protective proteins: heme oxygenase-1 (HO-1) and ferritin.

Heme oxygenases are rate-limiting enzymes of heme degradation. HO-1 is the inducible isoform. The enzyme catalyzes the formation of bilirubin, free iron ions, and carbon monoxide (CO). Bilirubin exerts strong antioxidant effects at physiological plasma concentrations. High-normal plasma levels of bilirubin were reported to be inversely related to atherogenic risk and to provide protection from endothelial damage.

The group of *Henning Schröder* was the first to show HO-1 induction by aspirin. HO-1 induction was followed by increased formation of bilirubin, CO, and ferritin, an another antioxidant protein. This was considered a novel, prostaglandin-independent vasoprotective action of aspirin [190, 189] (Figure 2.20). Generation of HO-1 by aspirin can also occur via ATL (Section 2.2.1) [191], another evidence for an endothelium-mediated anti-inflammatory effect of aspirin.

The cellular iron-binding protein ferritin protects from oxidative stress. It is also upregulated by aspirin in endothelial cells, probably at the translational and/or posttranslational level [192]. Again, this effect requires submillimolar con-

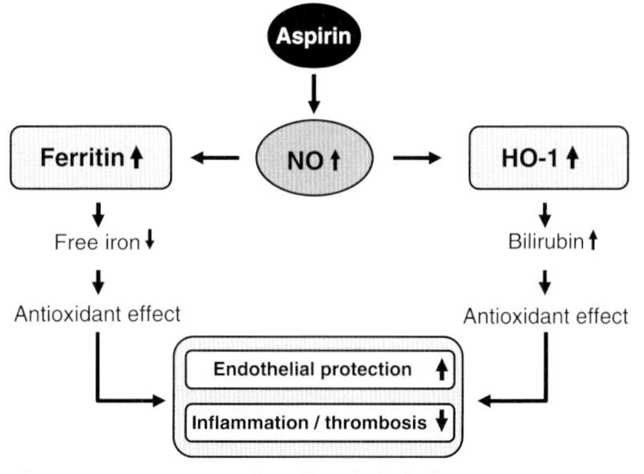

Figure 2.20 Aspirin-induced endothelial-protective pathways (modified after [190]).

centrations (0.03–0.3 mM) of aspirin and will functionally synergize with HO-1. It will be interesting to see whether similar actions of aspirin occur *in vivo*, in particular, in patients suffering from cardiovascular diseases. There is some evidence for an improvement of the (reduced) endothelium-dependent relaxation by aspirin in patients at advanced stages of atherosclerosis (Section 2.3.1).

2.2.2.4 Immune Responses

Anti-inflammatory and antimitogenic actions of aspirin and salicylate also involve modulations of immune reactions. Mechanistically these are often caused by inhibition of expression of inflammatory/immunogenic cytokines with subsequent modification of cytokine-induced signal generation, transduction, and perception. Aspirin was found to significantly reduce IL-4 mRNA expression and secretion in mitogen-primed human $CD4^+$ lymphocytes. This effect was not seen with traditional NSAIDs. It has been hypothesized that it was because of inhibition of binding of a yet unidentified inducible transcription factor in the IL-4 promoter region [172]. By this mechanism salicylates could selectively influence the nature of adaptive immune responses [175]. This kind of action might be involved in anti-inflammatory actions of aspirin in certain forms of asthma (Section 3.3.2).

Summary

The concept of a single mode of action of aspirin via inhibition of cyclooxygenase(s) is being increasingly challenged with improved knowledge of transcriptional and translational regulation of cell function. Aspirin and salicylates interact with these processes at several levels. These effects often require higher concentrations of the compounds than those that are necessary for inhibition of prostaglandin biosynthesis. These actions of aspirin are shared in many cases by salicylates but not by NSAID-type compounds that inhibit only prostaglandin synthesis via inhibition of COX enzyme activity.

At anti-inflammatory concentrations of about 200–300 μg/ml, aspirin has numerous effects on cellular signal generation and transmission, especially in consequence to cell stimulation by inflammatory cytokines, growth factors, or immunostimulants. Transcriptional, translational, and posttranslational levels of regulation might be affected that makes the net response difficult to predict. For example, there are different consequences of NFκB inhibition for cell functionality and survival in neuronal tissue as opposed to tumor cells. Inhibition of kinases is another general mode of action of high-dose salicylates. Although this effect is rather nonspecific, the sensitivity of different kinases to salicylates may not be the same, for example, ribosomal S6 kinase that phosphorylates (activates) the transcription factor C/EBP-β and subsequent gene expression of iNOS and COX-2. More work is necessary to establish the biological significance of these important new findings *in vivo*.

More recently, translational and posttranslational actions of salicylates have been found that protect the vascular endothelium from oxidative stress. Protection from generation of oxygen-centered radicals by enhanced HO-1 and eNOS activities has been described at medium concentrations of aspirin (30–300 μM) but not by salicylate. This suggests a different mode of action that requires target structure acetylation and, perhaps, generation of ATL (Section 2.3.1). It is unknown whether this mechanism operates *in vivo* and whether it is involved in antiatherosclerotic actions of aspirin, that is, improvement of endothelial dysfunction.

References

142 Wu, K.K. (2007) Salicylates and their spectrum of activity. *Anti-Inflammatory & Anti-Allergy Agents in Medicinal Chemistry*, **6**, 278–292.

143 Minchin, R.F., Illet, I.K., Teitel, C.H. *et al.* (1992) Direct *O*-acetylation of *N*-hydroxy arylamines by acetylsalicylic acid to form carcinogen–DNA adducts. *Carcinogenesis*, **13**, 663–667.

144 Abramson, S., Korchak, H., Ludewig, R. *et al.* (1985) Modes of action of aspirin-like drugs. *Proceedings of the National Academy of Sciences of the United States of America*, **82**, 7227–7231.

145 Abramson, S., Leszczynska-Piziak, J., Clancy, R.M. *et al.* (1994) Inhibition of neutrophil function by aspirin-like drugs (NSAIDS): requirement for assembly of heterotrimeric G proteins in bilayer phospholipid. *Biochemical Pharmacology*, **47**, 563–572.

146 Jurivich, D.A., Sistonen, L., Kroes, R.A. *et al.* (1992) Effect of sodium salicylate on the human heat shock response. *Science*, **255**, 1243–1245.

147 Fawcett, T.W., Xu, Q. and Holbrook, N.J. (1997) Potentiation of heat stress-induced hsp70 expression *in vivo* by aspirin. *Cell Stress & Chaperones*, **2**, 104–109.

148 Drew, J.E., Padidar, S., Horgan, G. *et al.* (2006) Salicylate modulates oxidative stress in the rat colon: a proteomic approach. *Biochemical Pharmacology*, **72**, 204–216.

149 Shackelford, R.E., Alford, P.B., Xue, Y. *et al.* (1997) Aspirin inhibits tumor necrosis factor-α gene expression in murine tissue macrophages. *Molecular Pharmacology*, **52**, 421–429.

150 Yoo, C.-G., Lee, S., Lee, C.-T. *et al.* (2001) Effect of acetylsalicylic acid on endogenous IκB kinase activity in lung epithelial cells. *American Journal of Physiology. Lung Cellular and Molecular Physiology*, **280**, L3–L9.

151 Alpert, D. and Vilček, J. (2000) Inhibition of IκB kinase activity by sodium salicylate *in vitro* does not reflect its inhibitory mechanism in intact cells. *The Journal of Biological Chemistry*, **275**, 10925–10929.

152 Wu, K.K. (2003) Control of COX-2 and iNOS gene expressions by aspirin and salicylate. *Thrombosis Research*, **110**, 273–276.

153 Frantz, B. and O'Neill, E.A. (1995) The effect of sodium salicylate and aspirin on NF-κB. *Science*, **270**, 2017–2019.

154 Yin, M.-J., Yamamoto, Y. and Gaynor, R.B. (1998) The anti-inflammatory agents aspirin and salicylate inhibit the activity of IkappaB kinase-beta. *Nature*, **396**, 77–80.

155 O'Neill, E.A. (1998) A new target for aspirin. *Nature*, **396**, 15–17.

156 Deng, W.-G., Ruan, K.-H., Du, M. *et al.* (2001) Aspirin and salicylate bind to immunoglobulin heavy chain binding protein (BiP) and inhibit its ATPase activity in human fibroblasts. *The FASEB Journal*, **15**, 2463–247.

157 Mazzeo, D., Panina-Bordignon, P., Recalde, H. *et al.* (1998) Decreased IL-12 production and Th1 cell development by acetyl salicylic acid-mediated inhibition of NF-κB. *European Journal of Immunology*, **28**, 3205–3213.

158 Aceves, M., Dueñas, A., Gómez, C. *et al.* (2004) A new pharmacological effect of salicylates: inhibition of NFAT-dependent transcription. *Journal of Immunology*, **173**, 5721–5729.

159 Holmes-McNary, M. (2002) Nuclear factor kappa B signaling in catabolic disorders. *Current Opinion in Clinical Nutrition and Metabolic Care*, **5**, 255–263.

160 Tegeder, I., Pfeilschifter, J. and Geisslinger, G. (2001) Cyclooxygenase-independent actions of cyclooxygenase inhibitors. *The FASEB Journal*, **15**, 2057–2072.

161 Kopp, E. and Ghosh, S. (1994) Inhibition of NF-κB by sodium salicylate and aspirin. *Science*, **265**, 956–959.

162 Pierce, J.W., Read, M.A., Ding, H. *et al.* (1996) Salicylates inhibit IκB-α phosphorylation, endothelial-leukocyte adhesion molecule expression, and neutrophil transmigration. *Journal of Immunology*, **156**, 3961–3969.

163 Kwak, Y.T., Guo, J., Shen, J. *et al.* (2000) Analysis of domains in the IKKalpha and IKKbeta proteins that regulate their kinase activity. *The Journal of Biological Chemistry*, **275**, 14752–14759.

164 Yuan, M., Konstantopoulos, N., Lee, J. *et al.* (2001) Reversal of obesity- and diet-induced insulin resistance with salicylates or targeted disruption of IKKβ· *Science*, **293**, 1673–1677.

165 Oeth, P. and Mackman, N. (1995) Salicylates inhibit lipopolysaccharide-induced transcriptional activation of the tissue factor gene in human monocytic cells. *Blood*, **86**, 4144–4152.

166 Osnes, L.T.N., Foss, K.B., Joø, G.B. *et al.* (1996) Acetylsalicylic acid and sodium salicylate inhibit LPS-induced NF-κB/c-Rel nuclear translocation, and synthesis of tissue factor (TF) and tumor necrosis factor alpha (TNF-α) in human monocytes. *Thrombosis and Haemostasis*, **76**, 970–976.

167 Mazur, I., Wurzer, W.J., Ehrhardt, C. *et al.* (2007) Acetylsalicylic acid (ASA) blocks influenza virus propagation via its NFκB-inhibiting activity. *Cellular Microbiology*, **9**, 1683–1694.

168 Schwenger, P., Bellosta, P., Vietor, I. *et al.* (1997) Sodium salicylate induces apoptosis via p38 mitogen-activated protein kinase but inhibits tumor necrosis factor-induced c-Jun N-terminal kinase/stress-activated protein kinase activation. *Proceedings of the National Academy of Sciences of the United States of America*, **94**, 2869–2873.

169 Schwenger, P., Alpert, D., Skolnik, E.Y. *et al.* (1998) Activation of p38 mitogen-activated protein kinase by sodium salicylate leads to inhibition of tumor necrosis factor-induced IκBα phosphorylation and degradation. *Molecular and Cellular Biology*, **18**, 78–84.

170 Cieslik, K.A., Zhu, Y., Shtivelband, M. *et al.* (2005) Inhibition of p90 ribosomal kinase-mediated CCAAT/enhancer-binding protein beta activation and cyclooxygenase-2 expression by salicylate. *The Journal of Biological Chemistry*, **280**, 18411–18417.

171 Cieslik, K., Zhu, Y. and Wu, K.K. (2002) Salicylate suppresses macrophage nitric oxide synthase-2 and cyclooxygenase-2 expression by inhibiting CCAAT/enhancer-binding protein-beta binding via a common signaling pathway. *The Journal of Biological Chemistry*, **277**, 49304–49310.

172 Perez-G, M., Melo, M., Keegan, A.D. *et al.* (2002) Aspirin and salicylates inhibit the IL-4- and IL-13-induced activation of STAT6. *Journal of Immunology*, **168**, 1428–1434.

173 Román, J., Fernández de Arrriba, A., Barrón, S. *et al.* (2007) UR-1505, a new salicylate, blocks T cell activation through nuclear factor of activated T-cells. *Molecular Pharmacology*, **72**, 269–279.

174 Ma, W.Y., Huang, C. and Dong, Z. (1998) Inhibition of ultraviolet C irradiation-induced AP-1 activity by aspirin is through inhibition of JNKs but not Erks or P38 MAP kinase. *International Journal of Oncology*, **12**, 565–566.

175 Cianferoni, A., Schroeder, J.T., Kim, J. *et al.* (2001) Selective inhibition of interleukin-4 gene expression in human T cells by aspirin. *Blood*, **97**, 1742–1749.

176 Vartiainen, N., Goldstein, G., Keksa-Goldstein, V. *et al.* (2003) Aspirin inhibits p44/42 mitogen-activated protein kinase and is protective against hypoxia/reoxygenation neuronal damage. *Stroke*, **34**, 752–757.

177 Grilli, M., Pizzi, M., Memo, M. *et al.* (1996) Neuroprotection by aspirin and sodium salicylate through blockade of NF-κB activation. *Science*, **274**, 1383–1385.

178 Moro, M.A., De Alba, H.J., Cardenas, A. *et al.* (2000) Mechanisms of the neuroprotective effect of aspirin after oxygen and glucose deprivation in rat forebrain slices. *Neuropharmacology*, **39**, 1309–1318.

179 Crisanti, P., Leon, A., Lim, D.M. *et al.* (2005) Aspirin prevention of NMDA-induced neuronal death by direct protein kinase Cς inhibition. *Journal of Neurochemistry*, **93**, 1587–1593.

180 De Cristobal, J., Cardenas, A., Lizasoain, I. *et al.* (2002) Inhibition of glutamate release via recovery of ATP levels accounts for a neuroprotective effect of aspirin in rat cortical neurons exposed to oxygen-glucose deprivation. *Stroke*, **33**, 261–267.

181 Muller, D.N., Heissmeyer, V., Dechend, R. *et al.* (2001) Aspirin inhibits NFκB and protects from angiotensin II-induced organ damage. *The FASEB Journal*, **15**, 1822–1824.

182 Farivar, R.S. and Brecher, P. (1996) Salicylate is a transcriptional inhibitor of the inducible nitric oxide synthase in cultured cardiac fibroblasts. *The Journal of Biological Chemistry*, **271**, 31585–31592.

183 Sanchez de Miguel, L., de Frutos, T., González-Fernández, F. *et al.* (1999) Aspirin inhibits inducible nitric oxide synthase expression and tumour necrosis factor-α release by cultured smooth muscle cells. *European Journal of Clinical Investigation*, **29**, 93–99.

184 Wang, Z. and Brecher, P. (1999) Salicylate inhibition of extracellular signal-regulated kinases and inducible nitric oxide synthase. *Hypertension*, **34**, 1259–1264.

185 Katsuyama, K., Shichiri, M., Kato, H. *et al.* (1999) Differential inhibitory actions by glucocorticoids and aspirin on cytokine-induced nitric oxide production in vascular smooth muscle cells. *Endocrinology*, **140**, 2183–2190.

186 Amin, A.R., Vyas, P., Attur, M. *et al.* (1995) The mode of action of aspirin-like drugs: effect on inducible nitric oxide synthase. *Proceedings of the National*

Academy of Sciences of the United States of America, **92**, 7926–7930.

187 Wolin, M.S. (1998) Novel antioxidant action of aspirin may contribute to its beneficial cardiovascular actions. *Circulation Research*, **82**, 1021–1022.

188 Taubert, D., Berkels, R., Grosser, N. *et al.* (2004) Aspirin induces nitric oxide release – a novel mechanism of action. *British Journal of Pharmacology*, **143**, 159–165.

189 Grosser, N. and Schröder, H. (2003) Aspirin protects endothelial cells from oxidant damage via the nitric oxide-cGMP pathway. *Arteriosclerosis, Thrombosis, and Vascular Biology*, **23**, 1345–1351.

190 Grosser, N., Abate, A., Oberle, S. *et al.* (2003) Heme oxygenase-1 induction may explain the antioxidant profile of aspirin. *Biochemical and Biophysical Research Communications*, **308**, 956–960.

191 Nascimento-Silva, V., Arruda, M.A., Barja-Fidalgo, C. *et al.* (2005) Novel lipid mediator aspirin-triggered lipoxin A_4 induces heme oxygenase-1 in endothelial cells. *American Journal of Physiology*, **289**, C557–C563.

192 Oberle, S., Polte, T., Abate, A. *et al.* (1998) Aspirin increases ferritin synthesis in endothelial cells – a novel antioxidant pathway. *Circulation Research*, **82**, 1016–1020.

2.2.3 Energy Metabolism

Changes in protein expression and enzyme activity as well as interactions with cellular signal transduction pathways are examples for aspirin-induced biological responses at the cellular level. In most cases, these actions are energy dependent and will proceed only if sufficient free energy, usually provided by ATP, is available. Although interactions of drugs with cellular energy metabolism are less specific than interactions with cellular signaling pathways, they are very effective since sufficient energy supply is essential for generation, receipt, processing, and dispatch of biological signals. Examples are the generation, release, and actions of cytokines or the expression of adhesion molecules in inflammatory or ischemic conditions. Thus, compounds that interact with cellular energy supply or utilization might also significantly interfere with cell signaling.

Mitochondria are the cellular power plants. Salicylates interact with the mitochondrial energy metabolism at two different levels: inhibition of β-oxidation of long-chain (LC) fatty acids and uncoupling of oxidative phosphorylation, that is, the generation of ATP from the energy-providing electron transport system of the respiratory chain. Both actions are dose dependent and typical for higher concentrations of the compounds. Clinically, they present as hyperventilation, that is, increased oxygen uptake, and increased heat production (hyperpyrexia) (Section 3.1.1) in acute and chronic salicylate overdosing [193]. Functionally, this indicates a "waste" of energy generated via the respiratory chain as heat instead of the generation of ATP.

The metabolic effects of aspirin and salicylate and their consequences for cell function become most apparent in the liver, the main organ of energy metabolism. The actual concentration of free salicylate as well as its maintenance over time also determine the metabolic actions and may differ markedly between *in vitro* and *in vivo* conditions. Another variable is the protein content, that is, the binding to albumin. At total salicylate plasma levels of 0.1–1.0 mM, which are obtained at therapeutic single doses of about 1 g of aspirin to adults, the percent free plasma salicylate varies between 1 and 10% and increases further to 20–30% at toxic doses of the compound (Table 2.1).

Salicylates are phenols and, like other phenolic compounds, such as the classical metabolic inhibitor 2,4-dinitrophenol, they interact with mitochondrial proteins that are involved in oxidative phosphorylation. This interaction is due to an allosteric effect that results in changes in mitochondrial protein configuration after salicylate binding and, eventually, results in uncoupling of oxidative phosphorylation. Albumin binds phenols via the phenolic hydroxy group and restores the capacity for oxidative phosphorylation in isolated mitochondria by removing it from mitochondrial proteins [194]. Consequently, hepatic metabolic failure by salicylate is particularly prominent *in vitro* in protein-free media and can be antagonized by supplementation with serum albumin [195].

According to these findings, *in vitro* data obtained at constant levels of salicylates over hours at low or absent protein in the incubation medium cannot be directly transferred to the *in vivo* situation with high albumin levels and a continuous metabolic degradation and transformation of salicylates into inactive metabolites (Section 2.1.2). It should also be noted that salicylate actions on energy metabolism are usually strictly competitive and reversible and not associated with any permanent mitochondrial injury [33].

Another kind of actions of salicylate on the liver, independent of energy metabolism, is the modulation of hepatic cytochrome P450 enzymes, specifically, an increased expression of CYP2E1 [85, 86]. Thus, long-term actions of aspirin on the liver *in vivo* must not only be caused by the substance itself but might be additionally modified by changes in selected hepatic P450 enzyme activities. This might also change the catabolism of the numerous other chemicals and xenobiotics that are substrate to CYP2E1-catabolism, most notably acetaminophen (Section 2.1.2). A combination with high-dose aspirin, for example, in intoxication, may markedly influence the susceptibility of certain individuals to acetaminophen promoted liver damage [196] (Section 3.2.3). Interestingly, aspirin might also induce a microsomal constitutive isoform of hepatic cytochrome P450 that catalyzes the omega-hydroxylation of fatty acids [197].

2.2.3.1 Fatty Acid β-Oxidation

Basic Mechanisms Mitochondrial β-oxidation of fatty acids is a principal source of generation of ATP, the conserved form of energy, in liver, heart, and kidney. Because of their amphiphilic nature, fatty acids become easily associated with mitochondrial membranes. In order to enter the mitochondrial β-oxidation process, they have to pass the outer and inner mitochondrial membrane before further processing in the mitochondrial matrix. Short and medium-chain fatty acids can cross the mitochondrial membranes without prior activation. Long-chain fatty acids (C14–C18) first require conversion into acyl-carnitine for translocation across the inner mitochondrial membrane ("carnitine shuttle"), before further processing after carnitine removal occurs. The resulting acyl-CoA then undergoes β-oxidation, resulting in the generation of reducing NADH equivalents that are subsequently oxidized by the mitochondrial respiratory chain. The energy thus produced is stored in the form of ATP by the oxidative phosphorylation system, coupled with the transfer of electrons along the respiratory chain [198]. A severe and long-lasting impairment of hepatic β-oxidation is a fundamental mechanism of metabolic organ failure. Morphologically, this results in microvesicular steatosis as the result of accumulation of nonmetabolized fatty acids and their re-esterification into triglycerides.

Actions of Salicylates Inhibition of hepatic mitochondrial β-oxidation of fatty acids, predominantly long-chain, is obtained at millimolar concentrations of aspirin or salicylate *in vitro* [199] and a typical feature of high-dose aspirin treatment or overdosing, These impairments may result from deficiency in cofactors such as coenzyme A (CoA) or carnitine that are essential for fatty acid transport and metabolism and become exhausted as a result of formation of acyl derivatives, such as salicylyl-CoA [200] (Figure 2.21). This will impair the activation of long-chain fatty acids by preventing their passage through the mitochondrial membranes, eventually resulting in intracellular but extramitochondrial accumulation and a number of secondary effects, including re-esterification into triglycerides and formation of dicarboxylic acids. Marked changes in liver fatty acid metabolism were found in liver biopsy specimens of patients with rheumatoid arthritis after long-term treatment with high-dose aspirin [201].

The hepatic lipid distribution pattern was studied in liver specimens obtained at autopsy from seven patients with rheumatoid arthritis. All patients had taken 3–6 g aspirin daily for many years. They were compared with seven age-matched controls who had not taken aspirin. All patients of both groups died from myocardial infarction, and there was no known functional liver abnormality at the time of death.

The total lipid content was significantly, >20%, higher in liver biopsy specimens of aspirin-treated patients as opposed to age-matched controls without aspirin intake. Most striking differences were seen in free fatty acids, which were more than doubled in aspirin-treated patients, whereas total hepatic phospholipids were reduced by >30%. The phospholipid depletion was due to a considerable, about 40–50%, decrease in phosphatidylethanolamine, phosphatidylcholine, and cardiolipin though other phospholipid classes remained unchanged.

It was concluded that major metabolic impairments of fatty acid oxidation occur in patients at long-term (years) high-dose aspirin treatment (Table 2.10). The increase in neutral lipids and free fatty acids in these patients suggest reduced oxidative capacity, indicating a relationship between abnormalities in fatty acid oxidation and aspirin intake [201].

Unfortunately, this study did not analyze the composition of the free fatty acid fraction, specifically the percentage of long-chain fatty acids or the occurrence of dicarboxylic acids. Nor was there any morphological data of the liver specimens. Thus, there was no information about microvesicular steatosis. It is also interesting that despite the markedly elevated free fatty acids, there was no

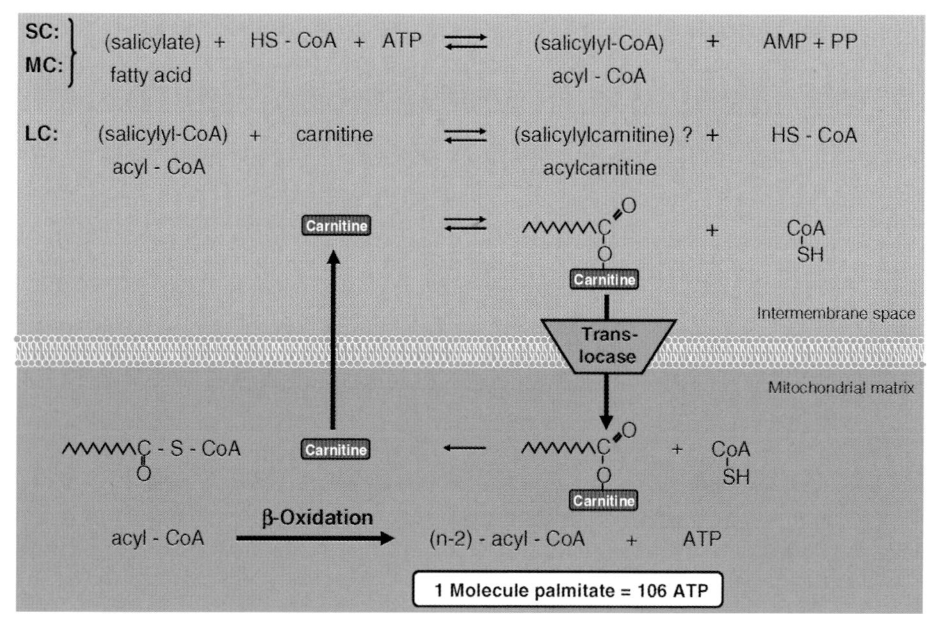

Figure 2.21 Mitochondrial β-oxidation of short (SC), medium (MC), and long-chain fatty acids and their modification by salicylate (for further explanation, see the text).

increased esterification in triglycerides. These data differ from animal studies with high-dose short-term aspirin treatment where increased triglycerides are a regular finding [199]. Of interest are also the marked reductions in phospholipids, possibly indicating an altered lipid signaling related to changes in membrane conductance. Unfortunately, apparently no further studies on this issue were conducted in men and probably will not be done in the future because high-dose long-term aspirin treatment is no longer the treatment of choice for these patients. Thus, it will probably never become elucidated whether aspirin-induced long-term changes in hepatic lipid metabolism are a general finding or superimposed to the altered immunologic status of rheumatic patients (Section 3.2.3).

Disturbed "Carnitine Shuttle" Like other fatty acids, salicylate is activated to salicylyl-CoA in mitochondria by a medium-chain fatty acid – CoA ligase [202]. This activation is a prerequisite for conjugation with glycine to form salicyluric acid [81] (Section 2.1.2). Generation of large amounts of salicylyl-CoA in the presence of high salicylate levels will deplete the cellular stores of coenzyme A and possibly carnitine (Figure 2.21). As a consequence, less carnitine and CoA are available for transport of long-chain fatty acids to the mitochondrial matrix and subsequent β-oxidation. In experimental studies, the reduced β-oxidation of long-chain fatty acids could be prevented by addition of carnitine and CoA to avoid exhaustion of these compounds [199]. Secondary events of disturbed β-oxidation are inhibition of gluconeogenesis and ureagenesis, though there is *in vitro* evidence that disturbed ureagenesis by salicylates can also be shown independent of its action on uncoupling of oxidative phosphorylation [203].

Appearance of Dicarboxylic Fatty Acids Another feature of impaired mitochondrial β-oxidation of long-chain fatty acids and their local accumulation is the appearance of long-chain dicarboxylic fatty acids as products of their omega-oxidation [204]. These acids are natural uncouplers of oxidative phosphorylation. Their physicochemical proper-

Table 2.10 Liver lipid composition in seven patients treated for years with 3.25–5.85 g aspirin daily as compared to seven age-matched controls.

	CON	Aspirin
Neutral lipids		
Total neutral lipids	49.5 ± 1.0	65.6 ± 0.7
Free fatty acids	12.6 ± 1.5	27.4 ± 2.4
Mono- and diacylglycerols	2.3 ± 0.1	6.4 ± 1.0
Triacylglycerols	11.9 ± 0.6	12.0 ± 3.2
Fatty acid esters	3.3 ± 0.2	4.8 ± 0.4
Cholesterol	8.0 ± 0.4	5.9 ± 0.7
Cholesteryl esters	6.8 ± 0.6	5.9 ± 0.6
Undetermined	5.3 ± 0.4	3.2 ± 0.4
Phospolipids		
Total phospholipids	50.5 ± 1.0	34.1 ± 0.6
Phosphatidylinositols	3.0 ± 0.1	2.3 ± 0.2
Phosphatidylethanolamines	13.5 ± 0.4	6.0 ± 1.5
Phosphatidyserines	4.6 ± 0.4	3.3 ± 0.3
Phosphatidylcholines	14.4 ± 1.1	6.0 ± 0.4
Lysophosphatidylcholines	1.0 ± 0.1	1.1 ± 0.1
Cardiolipins	0.6 ± 0.1	0.3 ± 0.0
Phosphatidic acids	9.2 ± 0.8	11.8 ± 1.2
Sphingomyelins	2.8 ± 0.3	2.3 ± 0.7
Undetermined	1.3 ± 0.4	1.1 ± 0.3

[a]All patients died from myocardial infarction and had no clinical liver pathology. All data are percentage of total lipids (modified after [201]).

ties [205] allow them to act as protonophores, whereas short and medium-chain fatty acids fail to do so [206] (see below). These abnormal fatty acids are not found in plasma in normal conditions but can be generated after high oral doses (600–700 mg/kg) or 1% of diet (for several days or weeks) of aspirin in animal experiments [197, 207]. Although the significance of this finding for the human is uncertain, it has been shown that the appearance of dicarboxylic acids in plasma may be associated with Reye-like symptoms [208–210].

2.2.3.2 Uncoupling of Oxidative Phosphorylation

Basic Mechanisms Energy coupling in the respiratory chain results in the generation of ATP from ADP and inorganic phosphate at the expense of energy. This energy is provided by the electron transport chain. The oxidative phosphorylation system is localized in the inner mitochondrial membrane. Uncoupling agents allow electron transport to oxygen to continue but prevent the phosphorylation of ADP to ATP, that is, they uncouple the energy-yielding from the energy-saving process. This results in increased mitochondrial oxygen uptake and reduced ATP levels despite an increased ATP-synthase activity.

The energy-yielding and energy-requiring processes are coupled by a high-energy intermediate state. An electrochemical gradient of H^+ ions across the mitochondrial inner membrane serves as means of coupling the energy flow from electron transport to the formation of ATP. An intact mitochondrial membrane that is impermeable to H^+ ions is essential for maintaining the proton gradient, necessary for oxidative phosphorylation. The electron transport chain pumps H^+ ions outward, and ATP formation is accompanied by an inward H^+ movement. Uncoupling agents, such as salicylate, allow protons to cross the otherwise impermeable membrane, thus destroying the proton gradient (Figure 2.22).

The system is devised as not to waste energy when it is not needed. When the utilization of ATP is low, there is little ADP in the mitochondrial matrix, little re-entry of protons through ATP synthase, and the high proton gradient slows down the activity of the respiratory chain by inhibition of ATP release from the ATP synthase. If ATP is consumed, the concentration of ADP increases, protons re-enter the matrix through ATP synthase and regenerate ATP. The electron transport through the respiratory chain causes H^+ to be pumped outward across the inner membrane of the mitochondrion, resulting in a gradient of H^+. This gradient is the energy-rich state to which electron transport energy is transformed and is the immediate driving force for the phosphorylation of ADP. The maintenance of this gradient, that is, the impermeability of the inner mitochondrial membrane for H^+, is essential for the functioning of this coupling process (Figure 2.22).

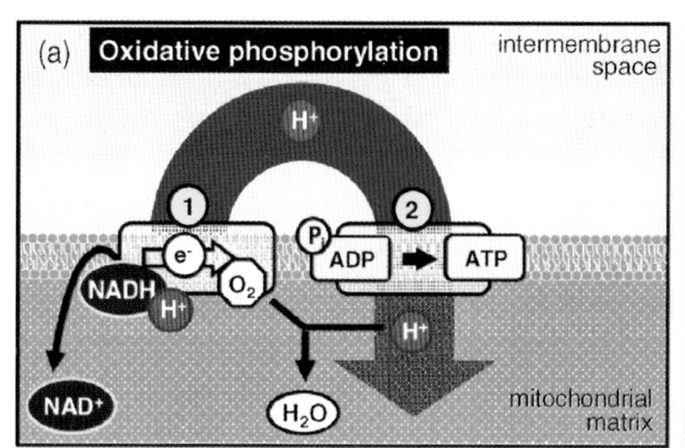

1. **Electron transfer chain:** Electrons are removed from NADH or $FADH_2$ and passed to the terminal electron acceptor O_2 via a series of redox reactions thereby forming of a proton gradient across the mitochondrial inner membrane.
2. **ATP synthase:** Since membranes are impermeable for protons the protons that reenter the matrix pass through special proton channel proteins called ATP synthase. The energy derived from the movement of these protons is used to synthesize ATP from ADP and phosphate.
3. **Salicylate** increases the number of proton channels available by competing for the proton (uncoupling) to form salicylic acid, leading to a decrease in protonmotive force and depletion of ATP.
4. **Influx of fluid** due to accumulation of sodium.

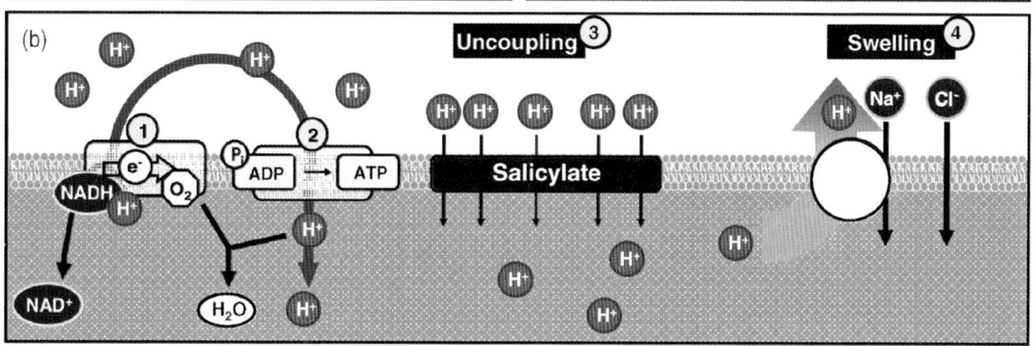

Figure 2.22 Oxidative phosphorylation in the absence (a) and presence (b) of salicylates. An electrochemical gradient of H^+ ions across the mitochondrial inner membrane couples the energy flow from electron transport to the generation of ATP. An intact mitochondrial membrane that is impermeable to H^+ ions is essential for maintaining the proton gradient. The electron transport chain pumps H^+ ions outward ① whereas ATP formation is accompanied by an inward H^+ movement ②. (a) Salicylate works as a protonophore and increases the number of proton channels. This results in a more than fourfold increase in membrane conductance at 1 mM ③, due to the weak anion current at neutral pH at which the compound is present as a lipid-soluble anion (the internal hydrogen bond delocalizes the negative charge). As a consequence, the proton gradient and membrane potential decrease and increased oxygen and substrate consumption will be required to maintain the protonmotive force. The increased proton accumulation inside the mitochondrion stimulates H^+/Na^+ exchange, resulting in mitochondrial swelling ④ (b).

Salicylates as Protonophores Uncoupling agents, such as 2,4-dinitrophenol or salicylate, increase the mitochondrial membrane proton conductance (Figure 2.22). This abolishes the energy-conserving proton gradient [211, 212] and results in bypassing the ATP synthase and release of protons into the matrix. In isolated mitochondria, the mitochondrial membrane proton conductance is increased more than fourfold at 1 mM salicylate [213]. Complete uncoupling occurs in model systems at 2–4 mM salicylate. The increasing H^+/Na^+ exchange causes swelling of mitochondria [214, 215]. Swelling of mitochondria and reduced urea generation were also found in "primarily living" intact rat hepatocytes [216]. This protonophoric effect reduces the mitochondrial membrane potential that normally will activate the respiratory chain. The energy is now wasted in the form of heat, instead of generating ATP. Interestingly, there are marked differences in this activity between different salicylate-related hydroxybenzoic acids [217].

At least two factors determine the activity of salicylate in uncoupling oxidative phosphorylation: the partition from an aqueous phase into a lipid-rich phase, allowing for penetration through the cell wall and access to the mitochondrion – the ultimate site of action. The second is a specific structural requirement to act as protonophore. Structure–activity comparisons for uncoupling oxidative phosphorylation in isolated mitochondria of 80 salicylate analogues showed that the essential pharmacophore for uncoupling activity is a salicylate with a negatively charged (carboxyl)group at the *o*-position, that is, *o*-hydroxybenzoate (salicylate) [218]. The *m*- and *p*-hydroxybenzoate analogues were inactive. This suggests that the *o*-position of the hydroxyl group is an essential steric requirement for this protonophoric action [219].

Consequences of Uncoupling of Oxidative Phosphorylation by Salicylates Administration of high-dose aspirin to man causes a marked and progressive increase in oxygen consumption [220] because of uncoupling of oxidative phosphorylation and becomes clinically evident as hyperventilation. This effect is dose dependent and typically occurs in initial stages of salicylate overdosing (Section 3.1.1) [221–224]. The uncoupling is not restricted to the liver but has also been found in isolated mitochondria of kidney, brain, and heart at higher salicylate concentrations (2–5 mM) [221, 225]. For thermodynamic reasons, glycogenolysis and glycolysis are enhanced to provide the necessary ATP for cell functions. Uncoupling is a generally reversible process that can be terminated by removal of the agent and is then followed by complete recovery [212].

In rats, orally treated for up to 1 week with sodium salicylate at doses causing toxic side effects (hyperventilation, body wastage), there was no disturbed oxidative phosphorylation by isolated mitochondria *ex vivo*. In contrast, complete inhibition of mitochondrial oxidative phosphorylation was seen after *in vitro* treatment of isolated mitochondria with 5 mM salicylate. The intracellular concentrations of salicylate in the liver of these rats after oral treatment were 0.8–4.0 mM, that is, close to the plasma levels of salicylate. Salicylate was taken up and washed out from liver mitochondria within minutes, even at 0 °C. The uncoupling, both *in vitro* and *ex vivo*, was reversible after washout of salicylate and did not cause irreversible tissue injury [212, 219].

2.2.3.3 Toxic Actions of Salicylates on the Liver

Reye's syndrome is a hepathoencephalopathy that has been brought into connection with aspirin-induced alterations in fatty acid metabolism, gluconeogenesis, and urea metabolism in the liver (Section 3.3.3). Serum of patients with clinical Reye-like symptoms stimulated oxygen consumption in isolated liver mitochondria, indicating uncoupling of oxidative phosphorylation [209]. In addition, there was the generation of dicarboxylic acids that corresponded directly to the reduction in ATP formation by Reye patients' serum, demonstrating that they are central to the general disturbance of mitochondrial function [209]. These findings were taken as evidence by some authors for a causal relationship between salicylate (overdosing) and Reye's syndrome. There are, however, serious doubts in this hypothesis (Section 3.3.3). This chapter compares metabolic alterations in liver metabolism by salicylate intoxication with symptoms of Reye-associated hepatic failure.

Impaired β-Oxidation In both situations, the free fatty acid content in liver and plasma is markedly increased, in one study up to about 10-fold above normal [209]. However, similar to *in vitro* experiments with high-concentration salicylate exposure, the mitochondrial lesions in general were transient and fully reversible [226]. Morphologically, severe salicylate intoxication causes microvesicular steatosis of the liver that is also seen in Reye's syndrome. However, histopathology and ultrastructural pathology of liver biopsy specimens in Reye patients were different from those in salicylate intoxication [227, 228]. Acute high-dose aspirin treatment in mice causes only mild microvesicular steatosis [199]. Moreover, microvesicular steatosis of the liver is not a unique etiologic entity and is seen in different forms of mitochondrial injury [229]. Inborn errors of ureagenesis may present with Reye-like microvesicular steatosis but with a morphology that is different from that observed in Reye's syndrome [230].

Medium-chain acyl coenzyme A dehydrogenase deficiency, an inherited defect of mitochondrial β-oxidation of fatty acids, was found to be associated with Reye-like symptoms [231] as was an inborn defect in the carnitine shuttle [232]. There is also fatty infiltration of the liver in these impairments of mitochondrial β-oxidation [198]. However, impaired β-oxidation of short or medium-chain fatty acids is not a typical feature of salicylate-induced liver toxicity [199]. Thus, salicylates will impair oxidative phosphorylation in the liver at high toxic concentrations [233, 234] (Section 3.2.3). These changes, however, are different from those obtained in patients with Reye's symptoms.

Dicarboxylic Acids A considerable percentage, at least 55% of total serum free fatty acids in Reye's syndrome, are dicarboxylic acids, the vast majority of them, 85–90%, being long chain [208–210]. Generation of these abnormal fatty acids is also seen at high salicylate levels. However, dicarboxylic acids are also found in inborn errors of metabolism in mitochondria or peroxisomes, such as Zellweger syndrome or neonatal adrenodystrophy [235]. In general, long-chain dicarboxylic acids may be formed as a secondary event in all forms of severely disturbed β-oxidation and are by no means Reye specific.

Plasma Salicylate Levels Exhaustion of the carnitine shuttle with subsequent generation and accumulation of dicarboxylic acid require high toxic concentrations of salicylate. In most reported cases of Reye's syndrome, serum salicylate levels, if measured at all, were not in the toxic range, and there are considerable doubts regarding the reliability of measurement of circulating salicylates in the clinical conditions of Reye-like disease by methods like the Trinder assay (Section 3.3.3).

Thus, there are not only some similarities between symptoms of Reye-like diseases and aspirin-related liver toxicity but also significant differences. In particular, there is definitely more than one reason for "Reye's syndrome" in the clinics – the diagnosis being one of exclusion – and the finding that salicylates at selected experimental conditions may cause liver injury, specifically at high toxic *in vitro* concentrations maintained for many hours in largely protein-free media. This does not provide sufficient evidence or even causality between aspirin and Reye. Finally, there is no animal model for Reye's disease allowing studying the unique two-step process of this hepatic injury, initiated by certain virus infections.

Summary

Salicylates exert a number of effects on energy metabolism that become most prominent in the liver. There is impaired β-oxidation, in particular, of long-chain fatty acids, generation of dicarboxylic acids, and uncoupling of oxidative phosphorylation. These effects are concentration-dependent and most prominent at high toxic concentrations of salicylates maintained *in vitro* over longer periods of time.

Mechanistically, salicylate becomes activated by CoA to be able to penetrate the mitochondrial membranes. This might result in exhaustion of carnitine that is necessary for this carnitine "shuttle" of long-chain fatty acids. In addition, salicylates act as a proton carrier, thus bypassing the "normal" way of energy generation and preventing the buildup of an H^+ gradient across the mitochondrial membrane, which is essential for energy storage in the form of ATP.

These actions of salicylates share some similarities with the liver pathology in Reye's syndrome. However, there are many differences too and several inherited disorders of fatty acid metabolism in the liver exhibit similar laboratory and clinical features. Until now, no causality between Reye's syndrome and salicylate-induced liver pathology has been shown.

References

193 Segar, W.E. and Holliday, M.A. (1958) Physiologic abnormalities of salicylate intoxication. *The New England Journal of Medicine*, **259**, 1191–1198.

194 Weinbach, E.C. and Garbus, J. (1964) Protein as the mitochondrial site for action of uncoupling phenols. *Science*, **145**, 824–826.

195 Tolman, K.G., Peterson, P., Gray, P. *et al.* (1978) Hepatotoxicity of salicylates in monolayer cell culture. *Gastroenterology*, **74**, 205–208.

196 Raucy, J.L., Lasker, J.M., Lieber, C.S. *et al.* (1989) Acetaminophen activation by human liver cytochromes P450IIE1 and P450IA2. *Archives of Biochemistry and Biophysics*, **271**, 270–283.

197 Okita, R. (1986) Effect of acetylsalicylic acid on fatty acid ω-hydroxylation in rat liver. *Pediatric Research*, **20**, 1221–1224.

198 Fromenty, B. and Pessayre, D. (1995) Inhibition of mitochondrial beta-oxidation as a mechanism of hepatotoxicity. *Pharmacology & Therapeutics*, **67**, 101–154.

199 Deschamps, D., Fisch, C., Fromenty, B. *et al.* (1991) Inhibition by salicylic acid of the activation and thus oxidation of long chain fatty acids. Possible role in the development of Reye's syndrome. *The Journal of Pharmacology and Experimental Therapeutics*, **259**, 894–904.

200 Rognstad, R. (1991) Effects of salicylate on hepatocyte lactate metabolism. *Biomedica Biochimica Acta*, **50**, 921–930.

201 Rabinowitz, J.L., Baker, D.G., Villanueva, T.G. *et al.* (1992) Liver lipid profiles of adults taking therapeutic doses of aspirin. *Lipids*, **27**, 311–314.

202 Killenberg, P.G., Davidson, E.D. and Webster, L.T., Jr (1971) Evidence for a medium-chain fatty acid: coenzyme A ligase (adenosine monophosphate) that activates salicylate. *Molecular Pharmacology*, **7**, 260–268.

203 Kay, J.D.S., (1988) Inhibition by salicylates of urea synthesis by isolated rat hepatocytes and citrullin synthesis by isolated rat mitochondria: an effect independent of uncoupling. *Hormone and Metabolic Research*, **20**, 333–335.

204 Mortenson, P.B. (1981) C_6–C_{10} dicarboxylic aciduria in starved, fat-fed and diabetic rats receiving decanoic acid or medium-chain diacylglycerol. *Biochimica et Biophysica Acta*, **664**, 349–355.

205 Spector, A.A., John, K. and Fletcher, J.E. (1969) Binding of long-chain fatty acids to bovine serum albumin. *Journal of Lipid Research*, **10**, 56–67.

206 Wojtczak, L. and Schönfeld, P. (1993) Effect of fatty acids on energy coupling processes in mitochondria. *Biochimica et Biophysica Acta*, **1183**, 41–57.

207 Kundu, R.K., Tonsgard, J.H. and Getz, G.S. (1991) Induction of omega-oxidation of monocarboxylic acids in rats by acetylsalicylic acid. *The Journal of Clinical Investigation*, **88**, 1865–1872.

208 Ng, K.J., Andresen, B.D., Hilty, M.D. *et al.* (1983) Identification of long-chain dicarboxylic acids in the serum of two patients with Reye's syndrome. *Journal of Chromatography*, **276**, 1–10.

209 Tonsgard, J.H. and Getz, G.S. (1985) Effects of Reye syndrome serum on isolated chinchilla liver mitochondria. *The Journal of Clinical Investigation*, **76**, 816–825.

210 Tonsgard, J.H. (1986) Serum dicarboxylic acids in Reye's syndrome. *The Journal of Pediatrics*, **109**, 440–445.

211 Charnock, J.S. and Opit, L.J. (1962) The effect of salicylate on adenosine-triphosphatase activity in rat liver mitochondria. *The Biochemical Journal*, **83**, 596–602.

212 Charnock, J.S., Opit, L.J. and Hetzel, B.S. (1962) An evaluation of salicylate on oxidative phosphorylation in rat-liver mitochondria. *The Biochemical Journal*, **83**, 602–606.

213 Haas, R., Parker, W.D., Jr, Stumpf, D. *et al.* (1985) Salicylate-induced loose coupling: protonmotive force measurements. *Biochemical Pharmacology*, **34**, 900–902.

214 Gutknecht, J. (1990) Salicylate and proton transport through lipid bilayer membranes: a model for salicylate-induced uncoupling and swelling in mitochondria. *Journal of Membrane Biology*, **115**, 253–260.

215 Gutknecht, J. (1992) Aspirin, acetaminophen and proton transport through phospholipid bilayers and mitochondrial membranes. *Molecular and Cellular Biochemistry*, **114**, 3–8.

216 Venerando, R., Miotto, G., Pizzo, P. *et al.* (1996) Mitochondrial alterations induced by aspirin in rat hepatocytes expressing mitochondrial targeted green fluorescent protein (mtGFP). *FEBS Letters*, **382**, 256–260.

217 Thompkins, L. and Lee, K.H. (1969) Studies on the mechanism of action of salicylates. IV. Effect of salicylates on oxidative phosphorylation. *Journal of Pharmaceutical Sciences*, **58**, 102–105.

218 Whitehouse, M.W. (1964) Biochemical properties of antiinflammatory drugs. III. Uncoupling of oxidative phosphorylation in a connective tissue (cartilage) and liver mitochondria by salicylate analogues: relationship of structure to activity. *Biochemical Pharmacology*, **13**, 319–336.

219 You, K. (1983) Salicylate and mitochondrial injury in Reye's syndrome. *Science*, **221**, 163–165.

220 Cochran, J.B. (1952) The respiratory effect of salicylate. *British Medical Journal*, **2**, 964–967.

221 Brody, T.M. (1956) Action of sodium salicylate and related compounds on tissue metabolism *in vitro*. *The Journal of Pharmacology and Experimental Therapeutics*, **117**, 39–51.

222 Bosund, I. (1957) The effect of salicylic acid, benzoic acid and some of their derivatives on oxidative phosphorylation. *Acta Chemica Scandinavica*, **11**, 541–544.

223 Miyahara, J.T. and Karler, R. (1965) Effect of salicylate on oxidative phosphorylation and respiration of mitochondrial fragments. *The Biochemical Journal*, **97**, 194–198.

224 Petrescu, I. and Tarba, C. (1997) Uncoupling effects of diclofenac and aspirin in the perfused liver and isolated hepatic mitochondria of rat. *Biochimica et Biophysica Acta*, **1318**, 385–394.

225 Nulton-Persson, A.C., Szweda, L.I. and Sadek, H.A. (2004) Inhibition of cardiac mitochondrial respiration by salicylic acid and acetylsalicylate. *Journal of Cardiovascular Pharmacology*, **44**, 591–595.

226 Segalman, T.Y. and Lee, C.P. (1982) Reye's syndrome: plasma-induced alterations in mitochondrial structure and function. *Archives of Biochemistry and Biophysics*, **214**, 522–530.

227 Daugherty, C.C., McAdams, A.J. and Partin, J.S. (1983) Aspirin and Reye's syndrome. *The Lancet*, **2**, 104.

228 Partin, J.S., Daugherty, C.C., McAdams, A.J. *et al.* (1984) A comparison of liver ultrastructure in salicylate intoxication and Reye's syndrome. *Hepatology*, **4**, 687–690.

229 Hautekeete, M.L., Degott, C. and Benhamou, J.P. (1990) Microvesicular steatosis of the liver. *Acta Clinica Belgica*, **45**, 311–326.

230 Heubi, J.E., Partin, J.C., Partin, J.S. *et al.* (1987) Reye's syndrome: current concepts. *Hepatology*, **7**, 155–164.

231 Coates, P.M., Hale, D.E., Stanley, C.A. *et al.* (1985) Genetic deficiency of medium-chain acyl coenzyme A dehydrogenase: studies in cultured skin fibroblasts and peripheral mononuclear leukocytes. *Pediatric Research*, **19**, 671–676.

232 Scaglia, F., Scheuerle, A.E., Towbin, J.A. *et al.* (2002) Neonatal presentation of ventricular tachycardia and a Reye-like syndrome episode associated with disturbed mitochondrial energy metabolism. *BMC Pediatrics*, **2**, 12.

233 Troll, M.M. and Menten, M.L. (1945) Salicylate poisoning. Report of four cases. *American Journal of Diseases of Children*, **69**, 37–43.

234 Starko, K.M. and Mullick, F.G. (1983) Hepatic and cerebral pathology findings in children with fatal salicylate intoxication: further evidence for a causal relation between salicylate and Reye's syndrome. *The Lancet*, **1**, 326–329.

235 Rocchiccioli, F., Aubourg, P. and Bougneres, P.F. (1986) Medium and long-chain dicarboxylic aciduria in patients with Zellweger syndrome and neonatal adrenoleukodystrophy. *Pediatric Research*, **20**, 62–66.

2.3
Actions on Organs and Tissues

Section 2.2 has described the cellular and subcellular targets of aspirin with the intention to cover the spectrum of its pharmacological activities as complete as possible. In this context, the experimental conditions, including the selection of doses (concentrations) and duration of action (incubation), were of secondary interest. This view changes as the consequences of these cellular activities come into the focus of pharmacological actions on multicellular integrated functions at the tissue and organ levels, respectively. Here, different cell types with mutually different sensitivities to aspirin and salicylates and different cellular responses are incorporated in one and the same functional unit. The local concentration of the active compound, that is, aspirin or salicylate, depends on blood supply and (hepatic) metabolism. Finally, the dosing – a nonissue in conventional cell culture studies – becomes an important variable because of possible side effects to not only the organ or tissue of pharmacological interest but also other tissues without direct relation to the primary target.

Three major areas of pharmacological aspirin actions at the tissue and organ level are of therapeutic interest, but differ by the different dosing that is necessary to elicit the desired pharmacological effect: 0.1–0.3 g/day for antiplatelet actions, the principal mechanism for antithrombotic activities of aspirin (Section 2.3.1) and 1–2 g, mostly single-dose or short-term treatment, for analgesic/antipyretic effects (Section 2.3.2). Higher anti-inflammatory doses, that is, 3–4 g/day and more, were previously used for long-term treatment of rheumatic diseases but are now restricted to initial anti-inflammatory treatment of children with Kawasaki's disease (30–60 mg/kg) to prevent late complications such as coronary aneurysms (Section 4.2.3). An overview on therapeutic and toxic actions of aspirin in relation to the total plasma level of salicylate is shown in Figure 2.23.

An interesting new issue of clinical aspirin research is its use for cancer prevention, in particular, prevention of colorectal carcinoma. This probably involves both prostaglandin-dependent and independent mechanisms (Section 2.3.3). According to available clinical data, the required (minimum) dosing is in a medium range, about 0.3 g/day, in the antiplatelet rather than in the anti-inflammatory range. However, the required duration of treatment is at least 10 years (Section 4.3.1).

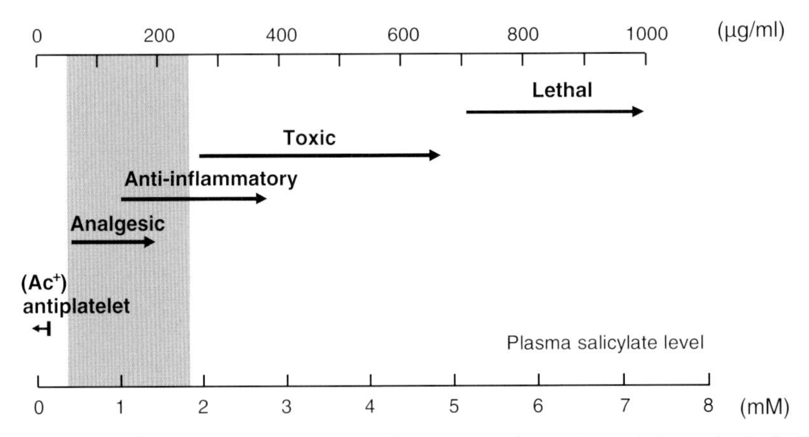

Figure 2.23 Therapeutic and toxic actions of aspirin in relation to the total plasma level of salicylate. The shaded area represents the plasma salicylate level at analgesic/anti-inflammatory doses. Note that the free salicylate is only about 1% at analgesic doses and increases to 10–30% at anti-inflammatory and toxic doses, respectively. The plasma level of salicylate does not affect the antiplatelet actions of aspirin because they are entirely due to acetylation (Ac) of target proteins (COX-1, prothrombin?).

2.3.1
Hemostasis and Thrombosis

Hemostasis The rapid cessation of bleeding after vessel injury is a vital function of the organism. To reach this goal, a variety of chemical factors have been developed that form together the functional unit of the clotting cascade. This system becomes activated within seconds after tissue injury to avoid life-threatening blood loss. In physiological conditions, hemostasis is well controlled and carefully balanced by a variety of procoagulant and anticoagulant factors. In arteries, clotting starts with the targeted adhesion of platelets to the subendothelium in an area of endothelial injury. There is activation of the arachidonic acid cascade with subsequent thromboxane formation, platelet aggregation, and secretion of vasoconstrictor, inflammatory, and mitogenic factors. The activation of the coagulation cascade, culminating in thrombin formation, occurs at the surface of activated platelets that not only provide clotting factors but also act as a matrix to localize thrombin formation to the side where it is needed. The result is an occluding thrombus that stops bleeding mechanically by "plugging" the site of vessel injury while local vasoconstriction prevents the thrombus washout. Generation of thromboxane A_2 by activated platelets acts as amplifying factor for platelet aggregation and vessel constriction (Figure 2.24).

At about the same time, platelet inhibitory, antithrombotic mechanisms (prostacyclin, NO, and endothelial nucleotidases) become activated in the noninjured endothelium in the vicinity of the thrombus. These anticoagulatory factors are also activated by thrombus formation and limit thrombus growth to the site of vessel injury. Activation of the fibrinolytic system (tissue plasminogen activator (tPA)) and subsequent clot lysis allow recanalization of the thrombus and restitution of blood flow and initiate the healing phase of vessel wall injury.

Aspirin can principally modify all three components of the hemostatic system, that is, platelet function, plasmatic coagulation, and fibrinolysis,

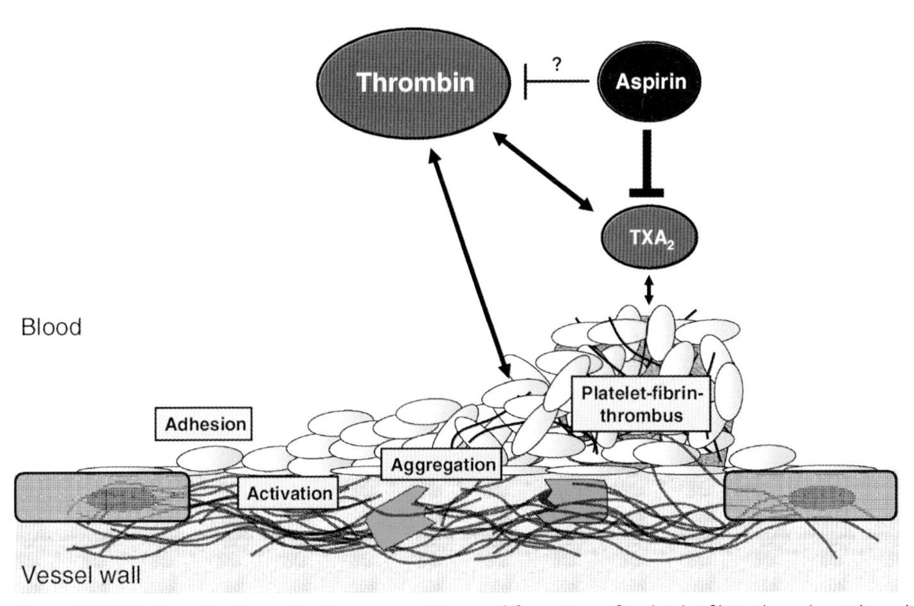

Figure 2.24 Platelet adhesion, activation, aggregation, and formation of a platelet-fibrin thrombus. Thrombin generation occurs at the surface of activated platelets that also synthesize thromboxane A_2 (TXA_2). Both compounds are released from the thrombus and stimulate thrombus growth and stability. Aspirin not only inhibits thromboxane formation but may also reduce thrombin generation.

though inhibition of platelet function is clearly the most significant – and most intensively studied – component. Platelet activation is also a trigger event with at least two components sensitive to aspirin: thromboxane biosynthesis and thrombin generation. The actions of aspirin on platelets, plasmatic coagulation, and fibrinolysis are discussed separately for formal reasons. However, *in vivo* they form a functional unit and a separate alteration of only one component without changing the others does not usually occur.

Thrombosis This well-balanced dynamic equilibrium between hemostatic factors derived from the vessel wall, circulating in plasma or generated by blood cells, is disturbed in atherosclerosis, the most frequent cause of atherothrombosis. Atherosclerosis is a chronic inflammatory disease that is associated with "endothelial dysfunction," that is, loss of the antithrombotic properties of the endothelium and its conversion into a prothrombotic surface that expresses adhesion molecules and becomes a target for inflammatory cytokines and growth factors [236]. Platelet adhesion and activation, the initial processes of arterial thrombosis, occur and are facilitated by platelet hyperreactivity. Acute thrombotic events in atherosclerotic vessels, clinically appearing as acute coronary syndromes, transient ischemic attacks, or atherothrombotic stroke are initiated by erosions or rupture of an atherosclerotic plaque [237]. This exposes thrombogenic material from inside the plaque to the flowing blood. Plaque material contains procoagulant tissue factor, generated by macrophages and smooth muscle cells [238]. The availability of TF starts the extrinsic pathway of coagulation, eventually resulting in thrombin formation at the surface of activated platelets. Inhibition of platelet function by aspirin will, therefore, block not only thromboxane formation but also thrombin generation at the platelet surface [239, 240].

Arteries Versus Veins The mechanisms causing thrombus formation in the arterial and venous circulations are basically the same, that is, local disturbances of hemostasis caused by a pathological interaction between blood constituents and the vessel wall. However, the pathomechanisms of thrombus formation differ: in arterial thrombosis, the platelets and their adhesion to the vessel wall under high shear stress conditions initiate thrombus formation in endothelium-denuded or dysfunctioning areas. In the low-pressure venous system, it is stasis and fibrin formation facilitated by the accumulation of nondegraded activated clotting factors that cause thrombosis. This is the reason for different pharmacological approaches in the treatment of arterial and venous thrombosis and the different efficacy of aspirin in the prevention of arterial (Sections 4.1.1, 4.1.2 and 4.1.3) but much less if any in venous (Section 4.1.4) thrombotic events.

2.3.1.1 Platelets

Inhibition of Platelet Functions by Aspirin Aspirin selectively inhibits prostaglandin production in human platelets [241]. This effect is now more precisely referred to as the inhibition of thromboxane A_2 formation and the generally accepted mode of antiplatelet action of aspirin (Section 2.2.1). Aspirin does not modify ADP-induced platelet aggregation, that is, the contraction of the platelet cytoskeleton, granule secretion, and thromboxane formation, in native blood [242]. Nor does it change the number of thromboxane receptors [243]. Inhibition of (stimulated) thromboxane formation, for example, in Ca^{2+}-depleted media, such as citrated plasma, will provide more substrates for biosynthesis of 12-hydro(pero)xyarachidonic acid (12-H(P) ETE) that can also modify platelet function by interaction with cyclooxygenases (Section 2.2.1). In addition to thromboxane inhibition, aspirin inhibits thrombin generation at least at analgesic doses (0.5–1 g) [239, 240, 244]. However, since thienopyridines also inhibit platelet-dependent thrombin formation [245, 246], it is likely that antiplatelet effects of these compounds are more important for their antithrombotic efficacy than their specific actions on clotting factors (Section 3.1.2). Aspirin might increase platelet-dependent NO production at high intravenous doses (800 mg) [247]. Whether this effect is clinically significant, considering also the activation of endothelial NO formation by aspirin (Section 2.2.2), is unknown.

Time-Dependent Inhibition of Platelet Function by Aspirin The antiplatelet action of aspirin is maximum *in vitro* after about 15 min [248] and irreversible because of the absence of sufficient synthesis of new enzyme proteins in platelets. *In vivo*, significant inhibition of thromboxane formation and platelet aggregation in the absence of an aspirin loading-dose requires about 1 h of continuous i.v. infusion of aspirin at effective plasma levels of 2 or 4 μM [249]. After oral administration, at least two daily doses (e.g., 75 mg) of standard aspirin are necessary to obtain sufficient inhibition of serum thromboxane. Oral single doses of 325, 162 (plain), and 75 mg (slow-release) aspirin result in peak plasma aspirin levels of 10.7, 6.8, and 0.3 μM [23]. Intravenous low-dose aspirin (50 mg) requires about 1 h for maximum inhibition of thromboxane formation [250] whereas nearly complete inhibition of thromboxane formation and arachidonic acid-induced platelet aggregation was seen within 5 min after 250 and 500 mg i.v. soluble aspirin (Figure 2.25). For oral dosing of 500 mg, 20 min were required for a maximum effect [251]. This is the reason for a "loading" dose if immediate inhibition of platelet function is required, for example, as an emergency first-line treatment in acute coronary syndromes (Section 4.1.1). Once sufficient acetylation is obtained, only a maintenance dose of aspirin per day is necessary.

Pretreatment with (nonselective) reversible COX-1 inhibitors, such as ibuprofen or indomethacin, will antagonize the antiplatelet effects of aspirin because of interference with aspirin (salicylate) binding within the substrate channel of COX-1 (Section 2.2.1). Theoretically and under certain experimental conditions with a high salicylate/aspirin ratio (≥20 : 1), this type of interaction can also be shown for salicylate [252, 253]. However, there is no evidence for a clinically relevant antagonistic antiplatelet interaction of aspirin and salicylate at

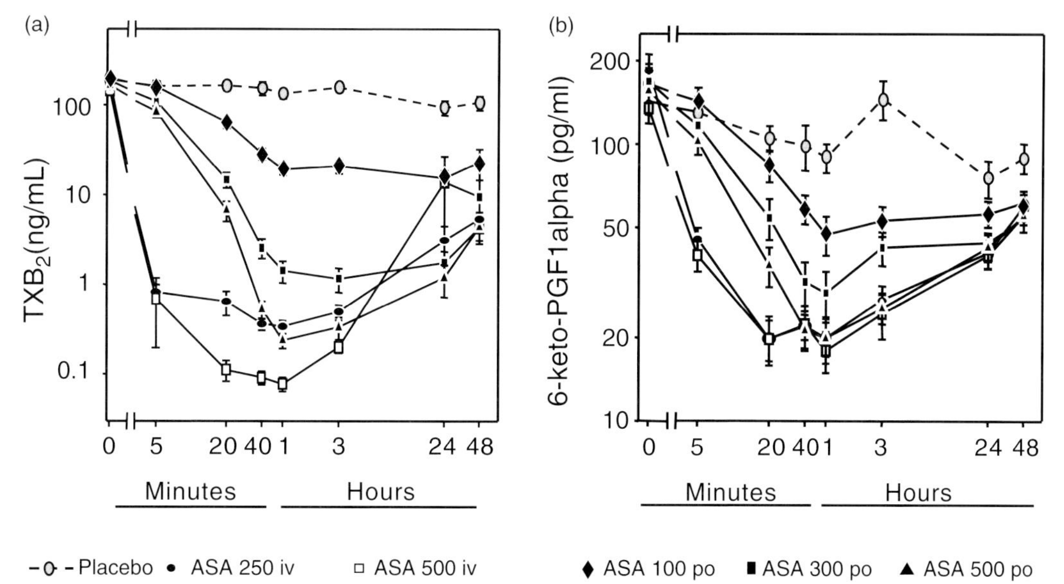

Figure 2.25 Time- and dose-dependent inhibition of arachidonic acid-induced formation of TXB_2 (a) and prostacyclin (6-keto-$PGF_{1\alpha}$) (b) in platelet-rich plasma in healthy volunteers *ex vivo* after i.v. or oral aspirin (note the different scaling at the ordinate for TXB_2 and 6-keto-$PGF_{1\alpha}$) (modified after Hohlfeld and Nagelschmitz, in preparation).

conventional antiplatelet doses *in vivo*. Even pretreatment with 1200 mg/day sodium salicylate for 3 days did not antagonize the inhibition of platelet function and thromboxane formation after 350 mg i.v. single-dose aspirin [254].

The inhibition of platelet COX-1 and platelet function by aspirin is functionally antagonized by the 10–15% fresh platelets that enter the circulation every day from the bone marrow. Thus, some reduction of the antiplatelet activity after cessation of aspirin is seen at 1–2 days, but 4–5 days are required to fully restore normal platelet function in healthy volunteers [255].

Dose-Dependent Inhibition of Platelet Function by Aspirin No other issue in aspirin research has been discussed more intensively than the question of the optimal antithrombotic dose – frequently though not necessarily correctly put on a level with the antiplatelet dose. There is general agreement that daily doses of 75–100 mg aspirin are sufficient to inhibit platelet-dependent thromboxane formation. Somewhat higher doses may be required if enteric-coated preparations are used, possibly because of a lower bioavailability of aspirin [256]. In terms of platelet-dependent thromboxane formation in serum – a surrogate estimate for the platelets' thromboxane-forming capacity – more than 95% have to be eliminated for a clinically relevant inhibition of platelet function [257–259]. However, this thromboxane-synthesizing capacity is an *in vitro* artifact useful to determine the pharmacological potency of aspirin to inhibit thromboxane formation but without any physiological correlate to circulating thromboxane levels *in vivo*, which are significantly lower (Section 3.1.2).

A direct comparison between single-dose and repeated-dose administration of aspirin to men shows two parallel dose–response curves differing in IC_{50} values by a factor of 8 that is equivalent to the platelet turnover rate and suggests that a daily maintenance dose that compensates for the entry of new platelets into the circulation is sufficient [255, 260] (Figure 2.26). However, this is also a pharmacological assay and most if not all doses of

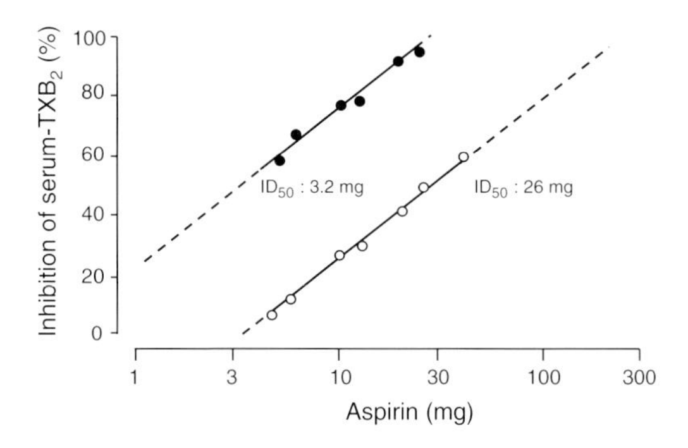

Figure 2.26 Dose-dependent inhibition of thromboxane (B_2) (TXB_2) formation in serum after single (○) and repeated (5 days, ●) oral administration of aspirin to healthy volunteers. Inhibition of thromboxane formation in serum by 50% (ID_{50}) requires 26 mg aspirin as a single dose. For the same effect, only 3.2 mg is necessary at repeated administration. The about eightfold lower ID_{50} at repeated administration corresponds to the maintenance dose of aspirin necessary to acetylate fresh platelets that enter the circulation every day from the bone marrow. (modified after [260]).

aspirin used here inhibit thromboxane formation by less than 95%, that is, they do not inhibit platelet function and are, therefore, clinically ineffective.

The subcellular targets of thromboxane in platelets and their interactions are still not completely understood. At least theoretically, it is likely that thromboxane per se rather acts as a platelet-derived amplification factor for further platelet recruitment and activation than being a direct stimulus for platelet secretion [261]. There is also evidence for thromboxane-dependent thrombin activation that then might connect platelet activation with the clotting process.

Platelet stimulation by thromboxane results in G_q-$G_{12/13}$ coupled phospholipase C (PLC) activation and subsequent activation of protein kinase C (PKC). Similar activation pathways are used by thrombin. Since the receptor–G-protein–effector systems are complex networks, synergistic signaling between TXA_2 and other platelet-activating agents, acting via the common G proteins (G_q), such as thrombin and thromboxane A_2, occurs [262]. In addition, thrombin transcriptionally upregulates thromboxane receptors in acute myocardial infarction [263–265]. Interruption of this positive feedback loop by aspirin prevents thromboxane formation and its synergism with thrombin. This synergism is highly desirable and effective, for example, in treatment of myocardial infarction. Similar considerations may apply to other situations with an overproduction of thromboxane or enhanced thromboxane receptor numbers, such as preeclampsia (Section 4.1.5), erythromelalgia, exposition to testosterone, or cigarette smoking [266].

The high sensitivity of platelet COX-1 against aspirin, the fact that platelet-dependent thromboxane formation occurs largely (≥99%) via COX-1, and the functional irreversibility of this effect for the platelet were the main pharmacological reasons to search for the lowest dose of aspirin, in particular, for long-term use in cardiovascular prevention. However, prophylactic treatment of arterial thrombosis is frequently performed in patients at enhanced vascular risk, that is, patients at more or less advanced stages of atherosclerosis. In these (inflammatory) conditions, nonplatelet sources of prostaglandin endoperoxides, the immediate precursors of thromboxane A_2, become increasingly important, for example, an upregulated COX-2 in endothelial cells or in monocytes/macrophages. This allows for transcellular precursor exchange and for platelet COX-1-independent thromboxane formation that is less sensitive to aspirin [267]. Alternatively, endothelial cells may use platelet-derived prostaglandin endoperoxides to increase prostacyclin production [268]. Moreover, studies in knockout mice have shown that prostacyclin inhibits thromboxane formation and intimal thickening via the prostacyclin receptor [269], pointing to the significance of (COX-2)-derived prostacyclin production for control of thromboxane formation. Thus, aspirin dosage recommendations, which are based solely on clotting blood assays or platelet-rich plasma of healthy volunteers *ex vivo*, may not apply directly to patients with atherosclerotic diseases and hyperreactive platelets *in vivo*.

Red Cells and Platelet "Recruitment" Red cells can markedly enhance platelet reactivity *in vivo* and *in vitro*. This becomes evident by increased platelet thromboxane formation, release of intracellular platelet granule components, and recruitment of additional platelets from the microenvironment into the developing thrombus [270]. In consequence, antiplatelet effects of aspirin may also be modulated by red cells, in particular, at low aspirin doses. In one study, collagen-induced platelet activation in the presence of red cells was only partially inhibited by 50 mg aspirin/day in healthy volunteers and 200 or 300 mg aspirin/day in patients with ischemic heart disease, despite a nearly complete (>94%) inhibition of serum thromboxane formation. It was concluded that inhibition of platelet function *in vitro* in platelet-rich plasma might not reflect sufficiently the therapeutic efficacy of aspirin *in vivo* because of the absence of red cells [271]. Interestingly, a loading dose of 500 mg aspirin was found to be sufficient to inhibit the proaggregatory activity of red cells on platelets [272]. This agrees with the clinical experience on aspirin "loading" doses mentioned above.

Aspirin Versus Other Antiplatelet Drugs Aspirin is the only antiplatelet agent in clinical use that acts by blocking thromboxane formation. Therefore, aspirin will act synergistically with other antiplatelet compounds, such as ADP-receptor antagonists or antithrombins, that separately interact with other pathways of platelet activation. All currently used platelet antagonists eventually inhibit clustering and activation of the platelet GPIIb/IIIa receptor, that is, prevent binding of fibrinogen and von Willebrand factor and aggregation by interplatelet bridging (Table 2.11). The clinical potency of aspirin as an antiplatelet drug depends entirely on the significance of the thromboxane pathway for platelet activation and secretion. This explains why aspirin is more effective in some clinical situations of platelet hyperreactivity than in others. An overview on the mode of antiplatelet action of aspirin as compared with other antiplatelet agents is shown in Figure 2.27.

Aspirin also differs from other antiplatelet agents with respect to target selectivity. Aspirin

Table 2.11 Some pharmacological properties of antiplatelet drugs.

Parameter	Aspirin	Clopidogrel	Eptifibatie/Tirofiban
Mechanism of antiplatelet action	Inhibition of TX generation	Blockade of $P2Y_{12}$-ADP receptor	Blockade of GPIIb/IIIa receptors
Action platelet specific	No	(Yes)	Yes
Action reversible	No	No	Yes
Oral administration possible	Yes	Yes	No
Additional effects within the circulation	(Yes)[b]	(Yes)[a]	No

[a]Clopidogrel might indirectly modify cell function via inhibition of release of platelet-derived products (CD40, PDGF, and serotonin, etc.).
[b]Aspirin at doses ≥300 mg not only has additional endothelial protective and anti-inflammatory properties but also largely prevents vascular prostacyclin production.

exhibits a broad spectrum of biological activities that are not restricted to or even specific for the platelet, whereas both thienopyridines and GPIIb/IIIa antagonists are largely platelet specific. This is because of the fact that their molecular targets, the $P2Y_{12}$ and GPIIb/IIIa receptors, are specifically expressed on platelets. This does not exclude pleiotropic actions of aspirin and thienopyridines via the modification of generation and release of platelet-derived mediators and their action on nonplatelet targets, for example, P-selectin receptors (PSGL) or CD40L.

Thromboxane-Independent Pathways of Platelet Activation The antiplatelet potency of aspirin depends on the stimulus. At antiplatelet doses, aspirin will not inhibit platelet stimulation by thrombin, the most potent platelet-activating agent. Aspirin will also not antagonize platelet aggregation [273] or secretion [274] induced by ADP, shear stress [273, 275, 276], or high-dose collagen. Aspirin does only partially [277], if at all [278], antagonize the platelet activation by norepinephrine. Aspirin will also not antagonize the platelet stimulatory actions of isoprostanes and nonenzymatic products of lipid

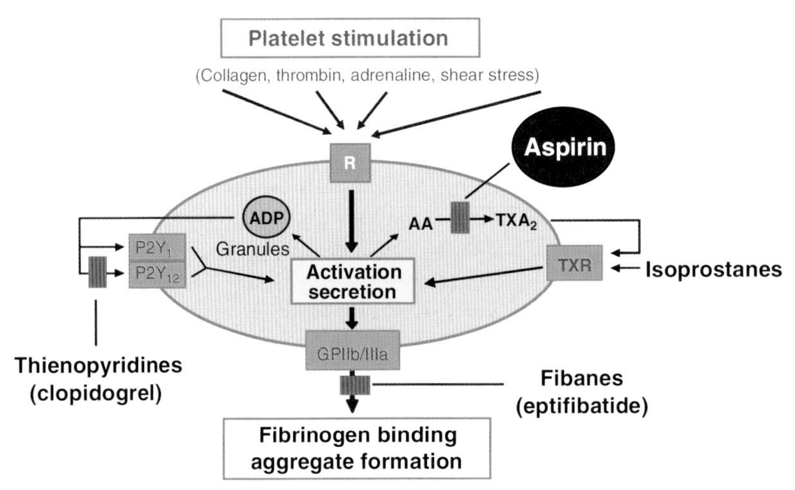

Figure 2.27 Sites of action of antiplatelet drugs.

peroxidation that bind to the platelet thromboxane receptor [279] and might act synergistically with enzymatically generated TXA_2 (Figure 2.27). Isoprostane formation is, therefore, considered as one explanation for aspirin "resistance" (Section 4.1.6). Finally, aspirin will also not inhibit platelet activation induced by psychical stress [280, 281].

These variable actions of aspirin on platelet function in dependency of the activation mode explain its variable antiplatelet activity in clinical use. *In vivo*, not one but several different stimuli act together simultaneously and determine platelet reactivity and aggregation. This offers a simple explanation for the variable treatment efficacy or even treatment failure with the drug *in vivo*. Thus, treatment failures are not necessarily related to a pharmacological inability of aspirin to work, that is, to prevent platelet-/COX-1-derived thromboxane biosynthesis.

2.3.1.2 Endothelial Cells

Healthy Endothelium Neither spontaneous platelet aggregation nor clot formation occurs in the intact circulation because of the antithrombotic properties of healthy endothelium. Consequently, endothelial injury results in a loss of its antiplatelet/antithrombotic properties and allows for local platelet adherence and thrombus formation that is sensitive to aspirin [282]. Aspirin also reduces prostacyclin (PGI_2) production in vascular endothelial cells. In contrast to platelets, this is functionally antagonized by *de novo* protein synthesis. In cell culture, this requires about 36 h for full recovery [283, 284].

The inhibitory effect of aspirin on (bradykinin)-stimulated endothelial prostacyclin production *in vivo* disappears faster, within 6 h in healthy subjects [285]. Repeated administration of 500 mg aspirin twice daily reduced the excretion of prostacyclin for only about 3 h [286]. In any case, inhibition of vascular prostacyclin generation is long lasting but incomplete [287, 257]. One explanation for this phenomenon is that vascular (endothelial) cells of healthy volunteers not only have a significant protein turnover rate (in contrast to the platelet) but also generate significant amounts of PGI2 (more than 50%) via COX-2 [288] that has a rather short half-life [289] and is less sensitive to aspirin at antiplatelet doses (Section 2.2.1). Thus, antiplatelet doses of aspirin will cause only a minor and transient inhibition of endothelium-dependent PGI_2, the most important natural inhibitor of platelet activation.

Endothelium in Atherosclerosis A different situation exists in atherosclerotic vessels. Because of the inflammatory nature of the disease [236], there are marked alterations of endothelial function, including the generation and release of antithrombotic/vasodilatory/fibrinolytic agents. It is known for a long time that vascular prostacyclin production is enhanced in atherosclerosis [290, 291] whereas the generation of other anticoagulants and profibrinolytics is reduced. Overall, this results in a disturbed hemostatic balance between the vessel wall and platelets and subsequent platelet hyperreactivity toward about every kind of platelet stimulus. Another consequence is upregulation of endothelial COX-2, providing more PGI_2 and PGE_2 for regulation of hemostasis and vessel tone (Figure 2.28). Furthermore, there is an enhanced release of proinflammatory cytokines in advanced atherosclerosis that is reduced by aspirin [292]. Aspirin also reduces the number of macrophages in atherosclerotic lesions and increases the number of smooth muscle cells and interstitial collagen in an animal model of atherosclerosis [293]. These plaque-stabilizing and endothelial-protective actions of aspirin [294] might result in reduced progression of atherosclerosis and reduced likelihood of plaque rupture, the starting event of arterial thrombus formation. Cotreatment with selective COX-2 inhibitors appears not to have any significant effect on endothelial dysfunction or vascular inflammation in atherosclerotic patients concomitantly treated with low-dose aspirin [295], whereas selective COX-2 inhibitors may increase the risk of myocardial infarctions in patients at advanced stages of atherosclerosis. However, nonselective NSAIDs might

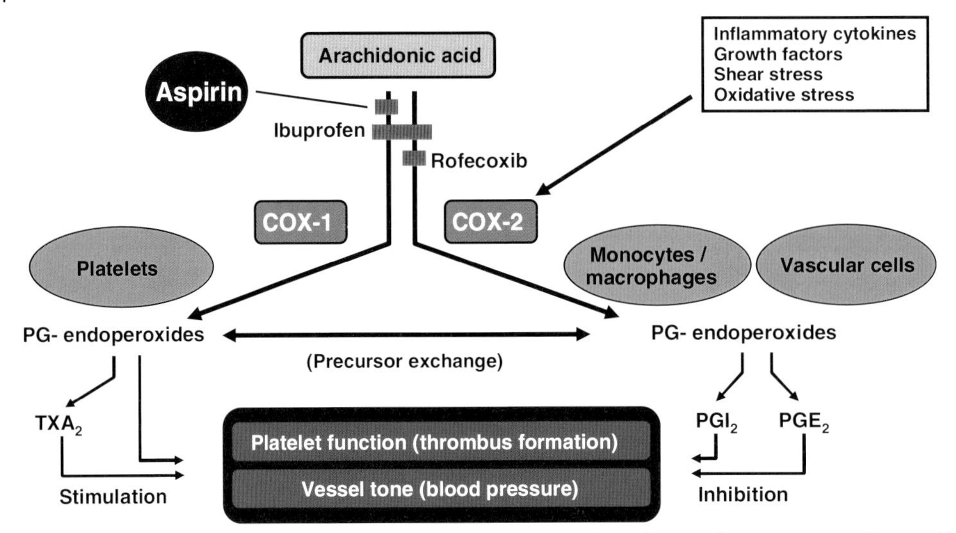

Figure 2.28 Arachidonic acid metabolism via COX-1 and COX-2 in the cardiovascular system. Note the possible exchange of PG-endoperoxide precursors between platelets, vessel wall, and inflammatory cells (for further explanation, see the text).

have a similar effect (Section 4.1.1). The clinical significance of an interaction between aspirin and reversible COX inhibitors is not proven. However, the possible interaction of aspirin with vascular PGI_2 and platelet TXA_2 is of considerable interest.

Weksler and colleagues have studied the inhibition of vascular PGI_2 in vessel segments *ex vivo* and platelet TXA_2 formation in serum in patients with angiographically documented coronary artery disease, undergoing elective aortocoronary bypass surgery.

Aspirin caused a dose-dependent inhibition of thromboxane formation in serum and PGI_2 generation in specimens of the aorta and saphenous vein. A single dose of 325 mg completely prevented thromboxane generation but reduced PGI_2 formation for significantly less amounts. No significant reduction of prostacyclin generation was seen at lower doses. There was no difference in perioperative blood loss at 325 mg aspirin as compared to untreated controls.

The conclusion was that low-dose aspirin can largely inhibit platelet aggregation and thromboxane formation but is much less effective on vascular prostacyclin production in arterial and venous endothelium in these patients (Figure 2.29) [296, 297].

At this point, it should be noted that inhibition of PGI_2 generation is not necessarily equivalent to inhibition of PGI_2 function. PGI_2 acts via specific G-protein-coupled receptors at the cell membrane that are subject to agonist-induced (down)regulation in number and affinity. Unfortunately, there are only very few studies addressing the issue in prostacyclin receptors. Bioassay data suggest that inhibition of tissue-derived prostaglandin synthesis is generally associated with an enhanced sensitivity of the target tissue. This has also been shown for inhibition of platelet aggregation by prostacyclin after aspirin treatment [298]. The opposite is seen in situations of extensive endogenous prostacyclin production, for example, ischemia-induced prostacyclin formation in acute myocardial infarction. This is associated with a marked reduction in prostacyclin receptor number and sensitivity [299, 300].

Aspirin and Other Endothelium-Derived Antiplatelet Factors Prostacyclin is not the only endothelium-derived product that inhibits platelet function. Two others are the endothelial cell ADPase (CD39) [301] and nitric oxide. Although the cleavage of ADP by the ADPase activity of the 5′-nucleotidase is not changed or only modestly reduced [302], the

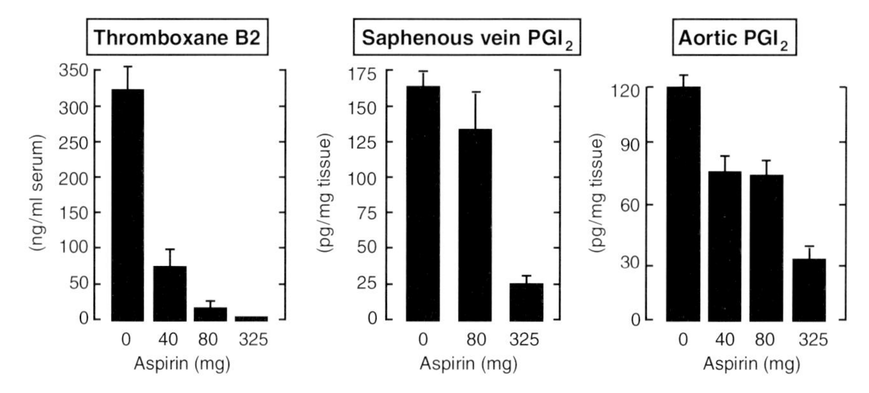

Figure 2.29 Dose-dependent inhibition of thromboxane and PGI_2 formation in serum and vessel specimens of patients with ischemic heart disease undergoing elective aortocoronary bypass surgery. All patients received single-dose aspirin at the dose indicated. There was no significant inhibition of PGI_2 formation by up to 80 mg aspirin and only about 75% reduction at 325 mg but a nearly complete prevention of thromboxane formation at these doses (modified after [296]).

endothelial generation and action of NO might be significantly enhanced by aspirin. Taubert *et al.* [188] have shown that aspirin acetylates eNOS protein, an effect independent of COX inhibition or inhibition of superoxide-mediated NO degradation (Section 2.2.2). Aspirin treatment of patients at advanced stages of atherosclerosis or hypercholesterolemia improves the reduced endothelium-dependent relaxation. No such effect was seen in healthy subjects [303, 304] (Figure 2.30).

Interactions Between Aspirin and COX Inhibitors The thrombotic side effects of coxibs, eventually resulting in myocardial infarction and stroke, are currently of considerable concern [305–308] (Section 4.1.1). Interestingly, patients on coxibs (pare-

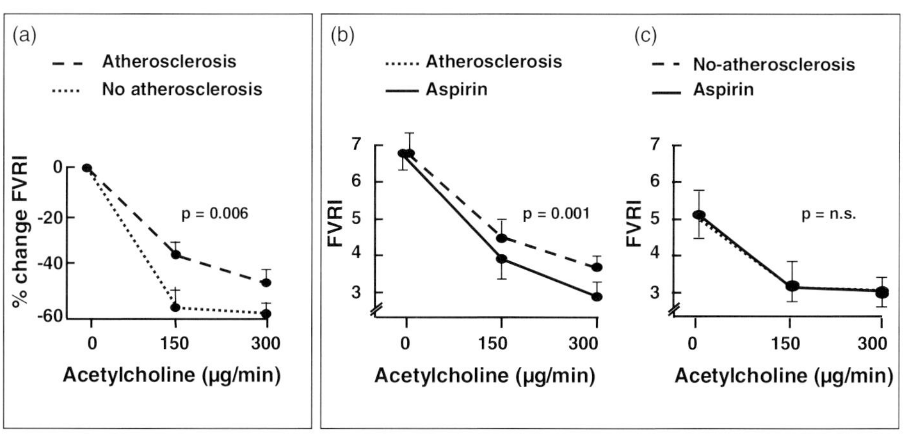

Figure 2.30 Reduced endothelium-dependent relaxation of femoral arteries in patients with advanced atherosclerosis (a). Improvement by intravenous soluble aspirin (1 g) in patients with atherosclerosis (b) but not in nonatherosclerotic controls (c). (modified after [303]).

coxib/valdecoxib) subjected to coronary artery bypass surgery exhibited a significant increase in cardiovascular events within a few days – though being on aspirin treatment [309]. The reason for this therapeutic failure of aspirin is unknown. Possibly, the hemostatic equilibrium in these patients depends on the continuous formation of antiplatelet, vasodilatory prostaglandins via an up-regulated COX-2 in the vessel wall. A more detailed discussion of the clinical aspects of these drug interactions is found in Section 4.1.1.

2.3.1.3 Plasmatic Coagulation

Aspirin does not directly alter the plasmatic coagulation at antithrombotic doses ($\leq$300 mg) in healthy individuals [310, 311]. However, there is evidence that aspirin might indirectly interfere with the plasmatic clotting system via inhibition of thrombin formation [239]. Whether this activity is primarily a thromboxane-related or a thromboxane-independent event, for example, due to acetylation of prothrombin [239] and/or fibrinogen [312, 313] remains to be determined. It is also unknown to what extent inhibition of thrombin formation contributes to the antithrombotic action of aspirin *in vivo* [239]. Interestingly, the inhibition of thrombin formation by aspirin is abolished in patients after acute myocardial infarction and thus might be clinically relevant [314].

Inhibition of plasmatic coagulation by aspirin at higher doses ($\geq$3–4 g) is known for a long time and is caused by inhibition of hepatic synthesis of vitamin K-dependent zymogens of clotting factors. This includes prothrombin (factor II) as well as factors VII, IX, and X. The subcellular mode of action is unknown.

Link and colleagues originally detected the anticoagulatory action of coumarins in the 1940s and also described a fall in plasmatic prothrombin levels by aspirin and sodium salicylate. Salicylic acid is formed quantitatively from these compounds during metabolism by the liver. These findings, a bleeding tendency after intake of about 10 g of salicylates and the fact that coumarins had no direct anticoagulant effect *in vitro*, prompted him to speculate that the antithrombotic action of coumarins was due to intermediate generation of salicylate as the active metabolite. He also thought that this was the reason for the slow onset of the anticoagulatory action of coumarins after oral intake.

It is now known that this is not true. However, it is interesting from a medical–historical point of view that sometimes even formally absolutely logical concepts, apparently verified by well-done experiments, providing the correct, expected result, may lead to wrong conclusions [315].

According to current knowledge, the marked prolongation of bleeding time at high doses of aspirin results from the combined inhibition of platelet function and reduced generation of the zymogens of plasmatic clotting factors, in particular, prothrombin (Section 3.1.2).

2.3.1.4 Fibrinolysis

Fibrinolysis, that is, reopening of an occluded vessel by dissolution of the thrombus, marks the beginning of repair mechanisms, eventually restoring the original perfusion conditions. Aspirin affects fibrinolysis, both directly and indirectly, at different levels and acts via different mechanisms. The two most important components are the platelets, facilitating the generation of procoagulant and antifibrinolytic factors and vascular endothelium, generating anticoagulant and profibrinolytic factors. The net effect depends on the actual conditions.

Fibrinolysis and Endothelium Aspirin does not change plasma levels of endothelium-derived tPA or plasminogen activator inhibitor-1 (PAI-1) [316]. Aspirin also does not affect enhanced fibrinolysis after vessel injury [317] or physical exercise [318] or does so only at toxic doses in animal experiments [319]. However, several studies in healthy volunteers have demonstrated an inhibition of ischemia-induced fibrinolysis by aspirin. The suggested mechanism of action was inhibition of endothelial tPA release at unchanged PAI-1 activity [318, 320, 321]. Consequently, inhibition of fibrinolysis by aspirin was prevented by prostacyclin [322, 319]. These data suggest that ischemia-related

Table 2.12 Inhibition of ischemia-induced stimulation of fibrinolysis by aspirin *ex vivo* and antagonism of this effect by iloprost.

	Euglobulin lysis (mm^2)			
Treatment	**Before ischemia**	**After ischemia**	**Before/after**	***p***
Placebo	67 ± 13	232 ± 56	4.4	<0.01
Aspirin	92 ± 15	197 ± 42	2.2	n.s.
Iloprost	79 ± 19	255 ± 93	4.5	<0.01
Aspirin + iloprost	76 ± 12	285 ± 37	4.3	<0.01

[a]Healthy volunteers received 2×650 mg aspirin and/or 1 ng/kg/min i.v. iloprost for 1 h prior to 10 min of upper arm ischemia by venous stasis, mean $\pm$ SEM. (after [322]).

stimulation of fibrinolysis is at least partially caused by prostacyclin release that is aspirin sensitive (Table 2.12). At high doses (650 mg bid), aspirin might additionally enhance fibrinolysis by acetylation of fibrinogen, independent of tPA levels [310]. The fibrinolytic response to urokinase is enhanced by aspirin pretreatment [323].

Fibrinolysis and Platelets Lysis of an arterial thrombus releases a variety of platelet-activating agents, including active thrombin and thromboxane A_2 from still functioning (over hours) platelet cyclooxygenases and other serine proteases, such as factor Xa. Streptokinase and fibrin degradation products enhance platelet reactivity. Coronary thrombolysis with streptokinase in patients with acute myocardial infarction markedly stimulates platelet function and thromboxane formation [324]. Plasmin and tPA show the opposite effect and inhibit platelet function [325]. Thus, streptokinase appears to directly stimulate platelet function whereas other fibrinolytics, such as tPA or urokinase do not. Aspirin does not directly affect generation of antiplasmin by platelets [320] nor does it stimulate plasmin activity at antithrombotic concentrations [326].

As a net effect for the clinics, there is a tendency toward synergistic actions between aspirin and fibrinolytics [327] (Section 4.1.3). Thus, antiplatelet treatment should be performed prior to any fibrinolytic treatment in order to antagonize the release of platelet activating product after lysis of an occluding thrombus. The synergistic effect of thrombolysis by streptokinase plus aspirin treatment has been convincingly demonstrated in the ISIS-II trial (Section 4.1.1).

Summary

The main effect of aspirin on hemostasis and thrombosis is the inhibition of platelet function via inhibition of thromboxane formation. At least in experimental settings, aspirin also antagonizes generation of thrombin, the most potent procoagulatory factor, and modulates fibrinolysis. Thus, aspirin affects all three components of the hemostatic system in a direction toward prevention of thrombotic events and facilitation of fibrinolysis.

The potency of aspirin as inhibitor of platelet function is dose dependent and also determined by the stimulating factor. Aspirin does not directly inhibit platelet function *ex vivo* after stimulation by ADP, high-dose thrombin or collagen, and shear stress at physiological Ca^{2+} levels. "Loading" doses of 250–500 mg soluble aspirin will result in a rapid and complete inhibition of platelet-dependent thromboxane formation whereas regular intake of lower doses, that is, about 100 mg/day, is sufficient for clinically relevant inhibition of platelet function in cardiovascular prevention (Section 4.1.1). This effect of aspirin is irreversible for each acetylated platelet and can only be functionally overcome by new platelets that enter the circulation from the bone marrow.

The antiplatelet action of aspirin is unique and differs from that of other antiplatelet drugs. This is the rationale for combined treatment, for example, with ADP-receptor antagonists (clopidogrel) or GPIIb/IIIa blockers (abciximab or fibanes). In contrast to clopidogrel and fibanes, the pharmacological actions of aspirin are not only restricted to the platelet but may also involve actions on the vessel wall and endothelium. In advanced atherosclerosis, this might result in reduced expression of adhesion receptors and circulating inflammatory cytokines as well as stimulation of endothelial NO production. Whether these effects are clinically significant and, possibly, might become more prominent at higher doses of aspirin, remains to be determined.

Aspirin might also modify plasmatic coagulation, possibly via inhibition of thrombin generation at the surface of activated platelets. Consequently, there might be synergism with coumarin-type anticoagulants.

Fibrinolysis results in the release of platelet-activating prothrombotic factors that can be antagonized by aspirin treatment. Aspirin also inhibits the release of profibrinolytic factors from the endothelium. As a net result, there is a trend toward synergistic actions between aspirin and fibrinolytics.

References

236 Ross, R. (1999) Atherosclerosis – an inflammatory disease. *The New England Journal of Medicine*, **340**, 115–126.

237 Davies, M.J. and Thomas, A.C. (1985) Plaque fissuring: the cause of acute myocardial infarction, sudden ischaemic death and crescendo angina. *British Heart Journal*, **53**, 363–373.

238 Taubman, M.B., Fallon, J.T., Schecter, A.D. *et al.* (1997) Tissue factor in the pathogenesis of atherosclerosis. *Thrombosis and Haemostasis*, **78**, 200–204.

239 Szczeklik, A., Krzanowski, M., Góra, P. *et al.* (1992) Antiplatelet drugs and generation of thrombin in clotting blood. *Blood*, **80**, 2006–2011.

240 Kessels, H., Béguin, S., Andree, H. *et al.* (1994) Measurement of thrombin generation in whole blood – the effect of heparin and aspirin. *Thrombosis and Haemostasis*, **72**, 78–83.

241 Smith, J.B. and Willis, A.L. (1971) Aspirin selectively inhibits prostaglandin production in human platelets. *Nature: New Biology*, **231**, 235–237.

242 Pengo, V., Boschello, M., Marzari, R. *et al.* (1985) ADP-induced α-granules release from platelets of native whole blood is not inhibited by the intake of aspirin in healthy volunteers. *Thrombosis and Haemostasis*, **54**, 183.

243 Covatto, R.H. and Niewiarowski, S. (1990) Platelet thromboxane A_2/prostaglandin H_2 receptors in human volunteers on low doses of aspirin. *Biochemical Pharmacology*, **40**, 1559–1561.

244 Undas, A., Brummel, K., Musial, J. *et al.* (2001) Blood coagulation at the site of microvascular injury: effects of low-dose aspirin. *Blood*, **98**, 2423–2431.

245 Herault, J.-P., Dol, F., Gaich, C. *et al.* (1999) Effect of clopidogrel on thrombin generation in platelet-rich plasma in the rat. *Thrombosis and Haemostasis*, **81**, 957–960.

246 Gurbel, P.A., Bliden, K.P., Guyer, K. *et al.* (2007) Delayed thrombin-induced platelet fibrin clot generation by clopidogrel: a new dose-related effect demonstrated by thrombelastography in patients undergoing coronary artery stenting. *Thrombosis Research*, **119**, 563–570.

247 O'Kane, P.D., Queen, L.R., Ji, Y. *et al.* (2003) Aspirin modifies nitric oxide synthase activity in platelets: effects of acute vs. chronic aspirin treatment. *Cardiovascular Research*, **59**, 152–159.

248 Zucker, M.B. and Peterson, J. (1970) Effect of acetylsalicylic acid, other nonsteroidal anti-inflammatory agents and dipyridamole on human blood platelets. *The Journal of Laboratory and Clinical Medicine*, **76**, 66–75.

249 Wilson, K.M., Siebert, D.M., Duncan, E.M. *et al.* (1990) Effect of aspirin infusions on platelet function in humans. *Clinical Science*, **79**, 37–42.

250 Böger, R.H., Bode-Böger, S.M., Gutzki, F.M. *et al.* (1993) Rapid and selective inhibition of platelet aggregation and thromboxane formation by intravenous low dose aspirin. *Clinical Science*, **84**, 517–524.

251 Nagelschmitz, J., Krätschmar, J., Voith, B. *et al.* (2007) Inhibition of platelet aggregation and thromboxane synthesis after intravenous and oral acetylsalicylic acid administration and pharmacokinetics of ASA and SA. Abstracts of the Meeting of the EACPT, Amsterdam.

252 De Gaetano, G., Cerletti, C., Dejana, E. *et al.* (1985) Pharmacology of platelet inhibition in humans: implications of the salicylate–aspirin interaction. *Circulation*, **72**, 1185–1193.

253 Philp, R. and Paul, M. (1986) Salicylate antagonism of acetylsalicylic acid inhibition of platelet aggregation in male and female subjects: influence of citrate concentration. *Haemostasis*, **16**, 369–377.

254 Rosenkranz, B., Fischer, C., Meese, C.O. *et al.* (1986) Effects of salicylic and acetylsalicylic acid alone and in combination on platelet aggregation and prostanoid synthesis in man. *British Journal of Clinical Pharmacology*, **21**, 309–317.

255 Patrignani, P., Filabozzi, P. and Patrono, C. (1982) Selective cumulative inhibition of platelet thromboxane production by low-dose aspirin in healthy subjects. *The Journal of Clinical Investigation*, **69**, 1366–1372.

256 Cox, D., Maree, A.O., Dooley, M. *et al.* (2006) Effect of enteric coating on antiplatelet activity of low-dose aspirin in healthy volunteers. *Stroke*, **37**, 2153–2158.

257 FitzGerald, G.A., Oates, J.A., Hawiger, J.A. *et al.* (1983) Endogenous biosynthesis of prostacyclin and thromboxane and platelet function during chronic administration of aspirin in man. *The Journal of Clinical Investigation*, **71**, 676–688.

258 Reilly, I.A.G. and FitzGerald, G.A. (1987) Inhibition of thromboxane formation *in vivo* and *ex vivo*: implication for therapy with platelet inhibitory drugs. *Blood*, **69**, 180–186.

259 Patrono, C., Coller, B., Dalen, J.E. *et al.* (2001) Platelet-active drugs. The relationship among dose, effectiveness, and side effects. *Chest*, **119**, 39S–63.

260 Patrono, C., Ciabattoni, G., Patrignani, P. *et al.* (1985) Clinical pharmacology of platelet cyclooxygenase inhibition. *Circulation*, **72**, 1177–1184.

261 Paul, B.Z.S., Jin, J. and Kunapuli, S.P. (1999) Molecular mechanism of thromboxane A_2-induced platelet aggregation. *The Journal of Biological Chemistry*, **274**, 29108–29114.

262 Djellas, Y., Antonakis, K. and Le Breton, G.C. (1998) A molecular mechanism for signaling between seven-transmembrane receptors: evidence for a redistribution of G-proteins. *Proceedings of the National Academy of Sciences of the United States of America*, **95**, 10944–10948.

263 Dorn, G.W., II, Liel, N., Trask, J.L. *et al.* (1990) Increased platelet thromboxane A_2/prostaglandin H receptors in patients with acute myocardial infarction. *Circulation*, **81**, 212–218.

264 D'Angelo, D.D., Davis, M.G., Houser, W.A. *et al.* (1995) Characterization of 5′ end of human thromboxane receptor gene. Organizational analysis and mapping of protein kinase C-responsive elements regulating expression in platelets. *Circulation Research*, **77**, 466–474.

265 Modesti, P.A., Colella, A., Cecioni, I. *et al.* (1995) Increased number of thromboxane A2-prostaglandin H2 platelet receptors in active unstable angina and causative role of enhanced thrombin formation. *American Heart Journal*, **129**, 873–879.

266 Halushka, P.V. (2000) Thromboxane A_2 receptors: where have you gone? *Prostaglandins & Other Lipid Mediators*, **60**, 175–189.

267 Weber, A.-A. (2004) Aspirin and activated platelets, in *The Eicosanoids* (ed. P. Curtis-Prior), John Wiley & Sons, Ltd, London, pp. 373–385.

268 Marcus, A.J., Weksler, B.B., Jaffe, E.A. *et al.* (1980) Synthesis of prostacyclin from platelet-derived endoperoxides by cultured human endothelial cells. *The Journal of Clinical Investigation*, **66**, 979–986.

269 Cheng, Y., Austin, S.C., Rocca, B. *et al.* (2002) Role of prostacyclin in the cardiovascular response to thromboxane A2. *Science*, **296**, 539–541.

270 Valles, J., Santos, M.T., Aznar, J. *et al.* (1991) Erythrocytes metabolically enhance collagen-induced platelet responsiveness via increased thromboxane production. ADP release and recruitment. *Blood*, **78**, 154–162.

271 Valles, J., Santos, T., Aznar, J. *et al.* (1998) Erythrocyte promotion of platelet reactivity decreases the effectiveness of aspirin as an antithrombotic therapeutic modality. The effect of low-dose aspirin is less than optimal in patients with vascular disease due to prothrombotic effects of erythrocytes on platelet reactivity. *Circulation*, **97**, 350–355.

272 Santos, M.T., Valles, J., Aznar, J. *et al.* (1997) Prothrombotic effects of erythrocytes on platelet reactivity. Reduction by aspirin. *Circulation*, **95**, 63–68.

273 Moake, J.L., Turner, N.A., Stathopoulos, N.A. *et al.* (1988) Shear-induced platelet aggregation can be mediated by VWF released from platelets, as well as by exogenous large or unusually large VWF multimers, requires adenosine-diphosphate, and is resistant to aspirin. *Blood*, **71**, 1366–1374.

274 Pengo, V., Boschello, M., Marzari, R. *et al.* (1986) Adenosine diphosphate (ADP)-induced α-granules release from platelets of native whole blood is reduced by ticlopidine but not by aspirin or dipyridamole. *Thrombosis and Haemostasis*, **56**, 147–150.

275 Rajagopalan, S., McIntire, L.V., Hall, E.R. *et al.* (1988) The stimulation of arachidonic acid metabolism in human platelets by hydrodynamic stress. *Biochimica et Biophysica Acta*, **958**, 108–115.

276 Maalej, N. and Folts, J.D. (1996) Increased shear stress overcomes the antithrombotic platelet inhibitory effect of aspirin in stenosed dog coronary arteries. *Circulation*, **93**, 1201–1205.

277 Lanza, F., Beretz, A., Stierle, A. *et al.* (1988) Epinephrine potentiates human platelet activation but is not an aggregating agent. *The American Journal of Physiology*, **255**, H1276–H1288.

278 Larsson, P.T., Wallén, N.H. and Hjemdahl, P. (1994) Norepinephrine-induced human platelet activation *in vivo* is only partly counteracted by aspirin. *Circulation*, **89**, 1951–1957.

279 Audoly, L.P., Rocca, B., Fabre, J.E. *et al.* (2000) Cardiovascular responses to the isoprostanes iPF (2alpha)-III and iPE(2)-III are mediated via the thromboxane A(2) receptor *in vivo*. *Circulation*, **101**, 2833–2840.

280 Gordon, J.L., Bower, D.E., Evans, D.E. *et al.* (1973) Human platelet reactivity during stress in diagnostic procedures. *Journal of Clinical Pathology*, **26**, 958–962.

281 Haft, J.I. and Fani, K. (1973) Intravascular platelet aggregation in the heart induced by stress. *Circulation*, **47**, 353–358.

282 Folts, J.D., Crowell, E.D. and Rowe, G.G. (1976) Platelet aggregation in partially obstructed vessels and its elimination with aspirin. *Circulation*, **54**, 365–370.

283 Czervionke, R.L., Smith, J.B., Fry, G.L. *et al.* (1979) Inhibition of prostacyclin by treatment of endothelium with aspirin. *The Journal of Clinical Investigation*, **63**, 1089–1092.

284 Jaffe, E.A. and Weksler, B.B. (1979) Recovery of endothelial prostacyclin production after inhibition by low-dose aspirin. *The Journal of Clinical Investigation*, **63**, 532–535.

285 Heavey, D.J., Barrow, S.E., Hickling, N.E. *et al.* (1985) Aspirin causes short-lived inhibition of bradykinin-stimulated prostacyclin production in man. *Nature*, **318**, 186–188.

286 Vesterqvist, O. (1986) Rapid recovery of *in vivo* prostacyclin formation after inhibition by aspirin. *European Journal of Clinical Pharmacology*, **30**, 69–73.

287 Hanley, S.P., Bevan, J., Cockbill, S.R. *et al.* (1981) Differential inhibition by low-dose aspirin of human venous prostacyclin synthesis and platelet thromboxane synthesis. *The Lancet*, **8227**, 969–971.

288 McAdam, B.F., Catella-Lawson, F., Mardini, I.A. *et al.* (1999) Systemic biosynthesis of prostacyclin by cyclooxygenase (COX)-2 – the human pharmacology of a selective inhibitor of COX-2. *Proceedings of the National Academy of Sciences of the United States of America*, **96**, 272–277.

289 Rimarachin, J.A., Jacobson, J.A., Szabo, P. *et al.* (1994) Regulation of cyclooxygenase-2 expression in aortic smooth muscle cells. *Arteriosclerosis, Thrombosis, and Vascular Biology*, **14**, 1021–1031.

290 FitzGerald, G.A., Smith, B., Pedersen, A.K. *et al.* (1984) Prostacyclin biosynthesis is increased in patients with severe atherosclerosis and platelet activation. *The New England Journal of Medicine*, **310**, 1065–1068.

291 Belton, O., Byrne, D., Kearney, D. *et al.* (2000) Cyclooxygenase-1 and -2-dependent prostacyclin formation in patients with atherosclerosis. *Circulation*, **102**, 840–845.

292 Ikonomidis, I., Andreotti, F., Economou, E. *et al.* (1999) Increased proinflammatory cytokines in patients with chronic stable angina and their reduction by aspirin. *Circulation*, **100**, 793–798.

293 Cyrus, T., Sung, S., Zhao, L. *et al.* (2002) Effect of low-dose aspirin on vascular inflammation, plaque stability, and atherogenesis in low-density lipoprotein receptor-deficient mice. *Circulation*, **106**, 1282–1287.

294 Kharbanda, R.K., Walton, B., Allen, M. *et al.* (2002) Prevention of inflammation-induced endothelial dysfunction. A novel vasculo-protective action of aspirin. *Circulation*, **105**, 2600–2604.

295 Title, L.M., Giddens, K., McInerney, M.M. *et al.* (2003) Effect of cyclooxygenase-2 inhibition with rofecoxib on endothelial dysfunction and inflammatory markers in patients with coronary artery disease. *Journal of the American College of Cardiology*, **42**, 1747–1753.

296 Weksler, B.B., Pett, S.B., Alonso, D. *et al.* (1983) Differential inhibition by aspirin of vascular and platelet prostaglandin synthesis in atherosclerotic patients. *The New England Journal of Medicine*, **308**, 800–805.

297 Weksler, B.B., Tack-Goldman, K., Subramanian, V.A. *et al.* (1985) Cumulative inhibitory effect of low-dose aspirin on vascular prostacyclin and platelet thromboxane production in patients with atherosclerosis. *Circulation*, **71**, 332–340.

298 Philp, R. and Paul, M. (1983) Low-dose aspirin renders platelets more vulnerable to inhibition of aggregation by prostacyclin (PGI_2). *Prostaglandins, Leukotriene & Medicine*, **11**, 131–142.

299 Mueller, H.S., Rao, P.S., Greenberg, M.A. *et al.* (1985) Systemic and transcardiac platelet activity in acute myocardial infarction in man: resistance to prostacyclin. *Circulation*, **72**, 1336–1345.

300 Jaschonek, K., Karsch, K.R., Weisenberger, H. *et al.* (1986) Platelet prostacyclin binding in coronary artery disease. *Journal of the American College of Cardiology*, **8**, 259–266.

301 Marcus, A.J., Safier, L.B., Hajjar, K.A. *et al.* (1991) Inhibition of platelet function by an aspirin-insensitive endothelial cell ADPase. Thromboregulation by endothelial cells. *The Journal of Clinical Investigation*, **88**, 1690–1696.

302 Cheung, P.K., Visser, J. and Bakker, W.W. (1994) Upregulation of antithrombotic ectonucleotidases by aspirin in human endothelial cells *in vitro*. *The Journal of Pharmacy and Pharmacology*, **46**, 1032–1034.

303 Husain, S., Andrews, N.P., Mulcahy, D. *et al.* (1998) Aspirin improves endothelial dysfunction in atherosclerosis. *Circulation*, **97**, 716–720.

304 Noon, J.P., Walker, B.R., Hand, M.F. *et al.* (1998) Impairment of forearm vasodilatation to acetylcholine in hypercholesterolemia is reversed by aspirin. *Cardiovascular Research*, **38**, 480–484.

305 FitzGerald, G.A. (2003) COX-2 and beyond: approaches to prostaglandin inhibition in human disease. *Nature Reviews. Drug Discovery*, **2**, 879–890.

306 FitzGerald, G.A. (2004) Coxibs and cardiovascular disease. *The New England Journal of Medicine*, **351**, 1709–1711.

307 Jüni, P., Nartey, L., Reichenbach, S. *et al.* (2004) Risk of cardiovascular events and rofecoxib: cumulative meta-analysis. *The Lancet*, **364**, 4999–5007.

308 Schrör, K., Mehta, P. and Mehta, J.L. (2005) Cardiovascular risk of selective cyclooxygenase-2 inhibitors. *Journal of Cardiovascular Pharmacology and Therapeutics*, **10**, 95–101.

309 Nussmeier, N.A., Whelton, A.A., Brown, M.T. *et al.* (2005) Complications of the COX-2 inhibitors parecoxib and valdecoxib after cardiac surgery. *The New England Journal of Medicine*, **352**, 1081–1091.

310 Björnsson, T.D., Schneider, D.E. and Berger, H. (1989) Aspirin acetylates fibrinogen and enhances fibrinolysis – fibrinolytic effect is independent of changes in plasminogen-activator levels. *The Journal of Pharmacology and Experimental Therapeutics*, **250**, 154–161.

311 Fiore, L.D., Brophy, M.T., Lopez, A. *et al.* (1990) The bleeding time response to aspirin-identifying the hyperresponder. *American Journal of Clinical Pathology*, **94**, 292–296.

312 Ezratty, A., Freedman, J.E., Simon, D. *et al.* (1994) The antithrombotic effects of acetylation of fibrinogen by aspirin. *Journal of Vascular Medicine and Biology*, **5**, 152–159.

313 Upchurch, G.R., Ramdew, N., Walsh, M.T. *et al.* (1998) Prothrombotic consequences of the oxidation of fibrinogen and their inhibition by aspirin. *Journal of Thrombosis and Thrombolysis*, **5**, 9–14.

314 Szczeklik, A., Musial, J., Undas, A. *et al.* (1996) Inhibition of thrombin generation by aspirin is blunted in hypercholesterolemia. *Arteriosclerosis, Thrombosis, and Vascular Biology*, **16**, 948–954.

315 Link, K.P., Overman, R.S., Sullivan, W.R. *et al.* (1943) Studies on the hemorrhagic sweet clover disease. XI. Hypoprothrombinaemia in the rat induced by salicylic acid. *The Journal of Biological Chemistry*, **147**, 463–473.

316 Krishnamurti, C., Tang, D.B., Barr, C.F. *et al.* (1988) Plasminogen activator and plasminogen activator inhibitor activities in a reference population. *American Journal of Clinical Pathology*, **89**, 747–752.

317 Kyrle, P.A., Westwick, J., Scully, M.F. *et al.* (1987) Investigation of the interaction of blood platelets with the coagulation system at the site of plug formation *in vivo* in man – effect of low-dose aspirin. *Thrombosis and Haemostasis*, **57**, 62–66.

318 Keber, I., Jereb, M. and Keber, D. (1987) Aspirin decreases fibrinolytic potential during venous occlusion, but not during acute physical activity. *Thrombosis Research*, **46**, 205–212.

319 Iacoviello, L., De Curtis, A., Dádamo, M.C. *et al.* (1994) Prostacyclin is required for t-PA release after venous occlusion. *The American Journal of Physiology*, **266**, H429–H434.

320 Woods, A.I. and Lazzari, M.A. (1987) Aspirin effect on platelet antiplasmin release. *Thrombosis Research*, **47**, 269–277.

321 Levin, R.I., Harpel, P.C., Harpel, J.G. *et al.* (1989) Inhibition of tissue plasminogen activator activity by aspirin *in vivo* and its relationship to levels of tissue plasminogen activator antigen, plasminogen activator inhibitor, and their complexes. *Blood*, **74**, 1635–1643.

322 Bertelé, V., Mussoni, L., Pintucci, G. *et al.* (1989) The inhibitory effect of aspirin on fibrinolysis is reversed by iloprost, a prostacyclin analog. *Thrombosis and Haemostasis*, **61**, 286–288.

323 Terres, W., Beythien, C., Kupper, W. *et al.* (1989) Effects of aspirin and prostaglandin E_1 on *in vitro* thrombolysis with urokinase. Evidence for a possible role of inhibiting platelet activity in thrombolysis. *Circulation*, **79**, 1309–1314.

324 Fitzgerald, D.J., Catella, F., Roy, L. *et al.* (1988) Marked platelet activation *in vivo* after intravenous streptokinase in patients with acute myocardial infarction. *Circulation*, **77**, 142–150.

325 Heptinstall, S., Berridge, D.C. and Judge, H. (1990) Effects of streptokinase and recombinant tissue plasminogen activator on platelet aggregation in whole blood. *Platelets*, **1**, 177–188.

326 Mildwidsky, A., Finci-Yeheskel, Z. and Mayer, M. (1991) Stimulation of plasmin activity by aspirin. *Thrombosis and Haemostasis*, **65**, 389–393.

327 Basinski, A. and Naylor, C.D. (1991) Aspirin and fibrinolysis in acute myocardial infarction – metaanalytical evidence for synergy. *Journal of Clinical Epidemiology*, **44**, 1085–1096.

2.3.2
Inflammation, Pain, and Fever

Inflammation as a response to injury is an expression of the effective activation of cellular and humoral defense mechanisms. The local inflammatory syndrome is characterized by its five classical features: heat (calor), redness (rubor), pain (dolor), swelling (tumor), and tissue injury (functio laesa). These multiple symptoms of acute inflammation are caused by a complex interaction between inflammatory white cells, cell- and tissue-derived mediators, and the vessel wall endothelium [328]. Leukocyte-derived cytokines additionally cause an increase in core temperature (fever) that accelerates all of the numerous biochemical reactions designed to speed up the acute inflammatory response whereas (inflammatory) pain will act as an overall alarming signal, indicating tissue injury and its location. In pharmacological terms, this means that any effective anti-inflammatory treatment in turn will also inhibit the accompanying events fever and pain. Nevertheless, targeted removal of one of the mediator systems involved will never stop the whole process but rather ameliorate it. The inflammatory reaction as a whole is an essential life-preserving process, and in this aspect similar to the maintenance of hemostasis, which enables the organism to resist and to survive the exposure against a variety of noxious external stimuli [328].

The discovery by Sir John Vane that aspirin inhibits prostaglandin biosynthesis provided for the first time a unifying and simple explanation for the well-known anti-inflammatory, analgesic, and antipyretic actions of salicylates. This discovery was followed by the development of dozens of NSAIDs. These compounds were designed to reduce local prostaglandin levels via inhibition of injury-induced biosynthesis with prospective use as anti-inflammatory analgesics (Section 2.2.1). Second generation NSAIDs, the selective inhibitors of COX-2-derived prostanoids (coxibs), were even more specifically targeted to injury-induced upregulation of prostaglandin biosynthesis. Although the future of coxibs is still uncertain, not because of lack of potency but because of unwanted adverse effects (see below), traditional NSAIDs have meanwhile largely displaced aspirin from its use in long-term treatment of inflammatory diseases (Section 4.2.2). However, with increasing knowledge about the complexity of the inflammatory response, specifically the key role of cytokines, the elucidation of major immune regulatory mechanisms, and the identification of a plethora of chemicals that mediate these multiple responses and in addition interact with each other in a very complex manner, it became also clear that prostaglandins are neither the only nor the most important biomolecules that control the inflammatory process and its accompanying phenomen-pain and fever.

Inhibition of prostaglandin formation in an area of tissue injury is still the most widely accepted explanation for the analgesic/anti-inflammatory action of aspirin. However, the efficacy of aspirin as an anti-inflammatory analgesic can probably not solely be explained by inhibition of prostaglandin biosynthesis. For example, anti-inflammatory actions of aspirin are stronger at doses higher than those that cause maximum inhibition of prostaglandin formation. There are also no clear correlations between inhibition of prostaglandin biosynthesis and the analgesic efficacy of aspirin. Stimulation of the lipoxin pathway via 15-lipoxygenase of neutrophils might be one explanation, especially in inflammatory pain [329]. There might be a relation to the (endo) cannabinoid system [330] but clearly an interaction with pain transmission within the spinal cord and other parts of the central nervous system (Figure 2.31). Nevertheless, it is obvious that pain can result from many more reasons than just inflammation and the same is true for fever, for example, as a consequence of viral infections. This also means that the mode of action of aspirin to modify these processes might not be the same. Thus, the effects of aspirin on inflammation, fever, and pain, and the possible mode of action will be discussed separately.

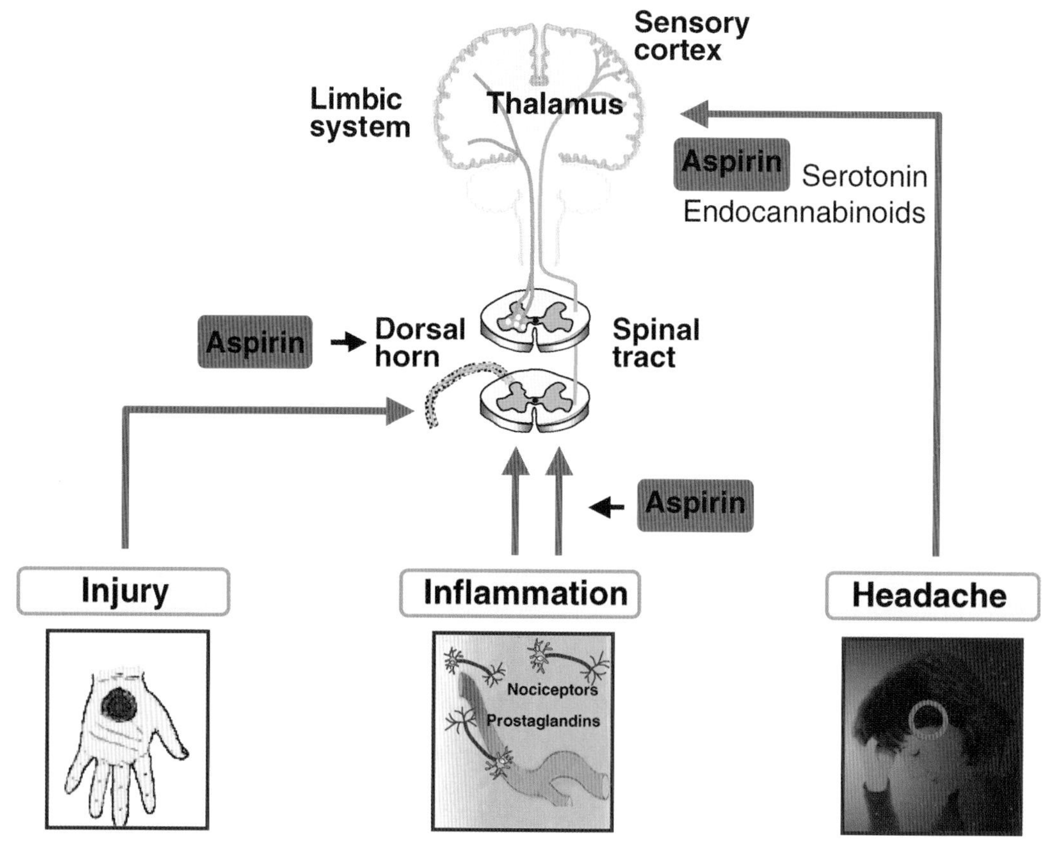

Figure 2.31 Release of mediators of inflammation and pain in acute tissue injury. Local tissue injury (inflammation, ischemia) is associated with generation and release of inflammatory mediators from local storage sites and invading white cells. This includes substance P (SP), bradykinin (BK), serotonin (5-HT), protons (acidic pH) and cytokines among others. Destruction of cell membranes results in release of arachidonic acid (AA), and generation of prostaglandins (PGs) and other eicosanoids. These eicosanoids amplify the local inflammatory reaction, including vessel dilatation, increase in vascular permeability, and sensitization of pain receptors.

2.3.2.1 Inflammation

Inflammation and Inflammation Mediators One of the first consequences upon tissue injury is the release of inflammatory mediators. The "inflammatory soup" [331] contains cytokines, kinins, growth factors, amines, protons, peptides, and a number of other chemicals, including arachidonic acid and prostaglandins, released from broken cells or generated locally. Prostaglandins are formed in the early vascular phase of the inflammatory process and modulate the inflammatory reaction in the injured area by enhancing the activity of other inflammatory mediators. This includes NO, generated by the iNOS and inflammatory cytokines such as interleukins 1 and 6 (IL-1 and IL-6) from leukocytes. Functionally, prostaglandins, acting synergistically with histamine and tryptase, cause local vasodilation (erythema and heat) and sensitize pain fibers to nociceptive stimuli. Tissue injury is enhanced – and cell debris removed – by invading white cells, such as neutrophils and monocyte/macrophages. They are attracted to the inflamed area by chemokines, such as leukotriene (LTB_4) or monocyte-chemoattractant protein-1 (MCP-1). In addition to their direct local effect, prostaglandins, produced via IL-6 upregulated COX-2, will also affect the speed and severity of inflammatory responses via resetting of central temperature control to higher levels (Section 2.3.2.3).

Mode of Anti-Inflammatory Actions of Aspirin Among the multiple pharmacological actions of aspirin, most of the attention has been focused on its analgesic/anti-inflammatory effects. This was also the reason why the substance was originally developed and clinically introduced (Section 1.1.3). Originally, it was thought that the anti-inflammatory action of aspirin (and salicylate) might result from its effects on cellular energy metabolism, that is, uncoupling of oxidative phosphorylation. This hypothesis, although attractive, was rejected [333] (Section 2.2.3). Future research was focused to more specific actions of salicylates on cells of interest. This included white cells, their numerous activation and secretion products, and the vascular endothelium. Aspirin appears to be a more potent inhibitor of COX-2-derived PGE_2 formation in cell culture *in vitro* than its metabolite salicylic acid while "activated" salicylic acid (salicylyl-CoA) was active [334]. Other investigators obtained different results with different models [97], though in vascular smooth cells, upregulated COX-2 was also less sensitive to salicylate than to aspirin (Figure 2.32). However, aspirin and salicylate are about equipotent anti-inflammatory compounds *in vivo* and reduce the prostaglandin content of inflammatory exudates to a comparable extent (Figure 2.32). Obviously, effects in addition to inhibition of enzymatic activity of COX-2 are involved (Section 2.2.2). These findings, together with the longer half-life of salicylate and its (intracellular) accumulation at an inflammatory site because of the acidic pH, led to the conclusion that salicylate rather than aspirin is the active anti-inflammatory component and aspirin is merely the prodrug of this active metabolite [335]. Similar conclusions were already reached by Dreser in his first description of pharmacological properties of aspirin in 1899 (Section 1.1.3).

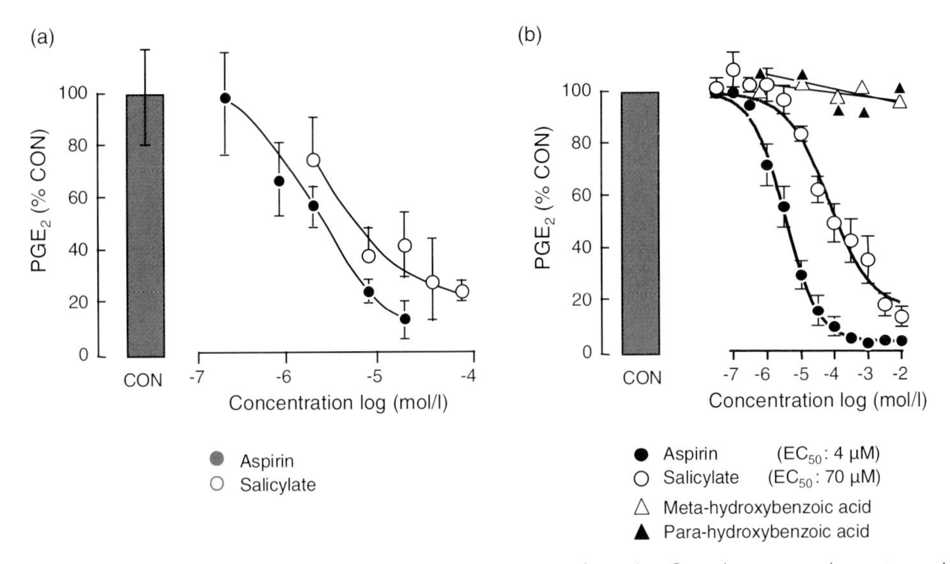

Figure 2.32 (a) Inhibition of PGE_2 generation in tissue explants of acutely inflamed rat synovia by aspirin and salicylate. Drugs were added to nonproliferating synovia cultures for 24 h. PGE_2 was assayed in the supernatant (modified after [335]). (b) Inhibition of COX-2-mediated PGE_2 generation in cultured human coronary artery smooth muscle cells by aspirin and salicylate (*o*-hydroxybenzoic acid) but no effect of *m*- and *p*-hydroxybenzoic acid. (Schrör and Kuger, unpublished).

Inhibition of Neutrophil Function and Recruitment Neutrophil granulocytes are the most abundant cell population in inflammatory exudates as well as in inflamed and ischemic tissues. Neutrophils cause tissue injury by two mechanisms: generation of free oxygen centered radicals and products thereof and release of tissue destructive enzymes, such as elastase and peroxidase. These reactions are amplified by LTB_4 release from activated neutrophils, which mediates leukocyte recruitment to the site of tissue injury. In this context, it should be noted that vasodilatory prostaglandins, most notably PGE_2 but not PGI_2, are potent inhibitors of neutrophil function [336] as might be aspirin via ATL formation (see below). Interestingly, there is evidence for a synergistic inhibitory effect of E-type prostaglandins and salicylate on neutrophil function [337]. An alternative explanation for the inhibition of leukocyte accumulation to an inflamed area by aspirin and salicylates is the local accumulation of adenosine in granulocytes after enhanced ATP breakdown [338] and/or preventing its rephosphorylation by inhibition of oxidative phosphorylation (Section 2.2.3). In this context, inhibition of leukocyte accumulation by salicylates was shown in one study to be independent of inhibition of prostaglandin synthesis and was suggested to be adenosine mediated [339].

Protection of Endothelial Function Inflammation and infections do profoundly affect the function of vascular endothelial cells. Endothelial dysfunction might be a common link between inflammation and the enhanced cardiovascular risk in patients suffering from chronic inflammatory diseases such as rheumatoid arthritis (Section 4.2.2). Inflammatory cytokines upregulate endothelial COX-2 and iNOS and enhance generation of reactive oxygen species (Section 2.2.2). Experimental systemic inflammation in healthy individuals, caused by *S. typhi* vaccination, was associated with endothelial dysfunction as seen from reduced endothelium-dependent relaxation that could be prevented by previous aspirin treatment (1.2 g). Aspirin treatment after vaccination had no effect, suggesting that these actions of aspirin were mediated by inhibition of the cytokine (IL-1) cascade [294]. In inflammatory conditions, treatment with aspirin might result in generation of 15-(*R*)-HETE that can be transformed to 15-epi-lipoxin A_4 or aspirin-triggered lipoxin (Section 2.2.1) by interaction with the 5- and 15-lipoxygenase of white cells, respectively (Figure 2.33). ATL, like other lipoxins, is a potent anti-inflammatory compound. It inhibits the neutrophil-induced increase in vascular permeability [329] and leukocyte recruitment to the inflammatory site. This last action might also involve

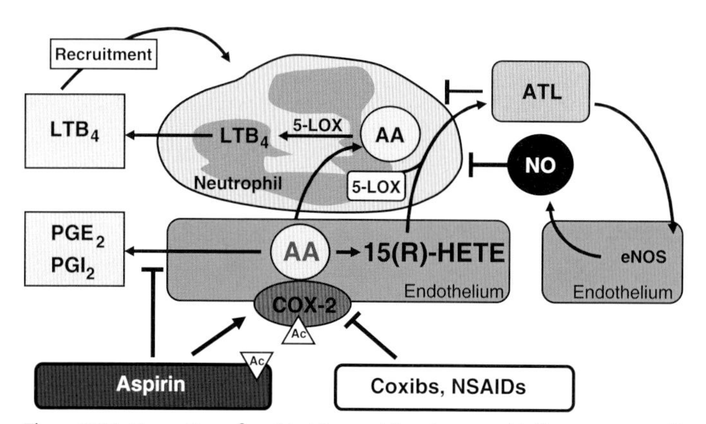

Figure 2.33 Generation of aspirin-triggered lipoxin, an anti-inflammatory mediator, by acetylation (Ac) of COX-2 in the presence of white-cell 5-lipoxygenase (5-LOX). Coxibs and traditional NSAIDs solely block COX-2 activity to a variant extent. ATL stimulates eNOS with subsequent NO formation and antagonizes the activation of neutrophils and their recruitment to the injured (inflamed) area.

NO formation via stimulation of the eNOS [340], possibly similar to the improved endothelium-dependent relaxation in patients with advanced systemic atherosclerosis (see Figure 2.30). Lipoxins will also improve endothelial oxygen defense by transcriptional upregulation of heme oxygenase-1 [191] (see Figure 2.20) and might be important as agents that terminate the "killing" phase of inflammation and initiate the healing process [328].

This mode of action of aspirin differs fundamentally from that of selective or nonselective COX inhibitors that rather tend to reduce ATL formation because of inhibition of COX-2 and cannot induce generation of the 15-(*R*)-HETE precursor because of their inability of target protein acetylation. In any case, the clinical significance of this attractive hypothesis of a central function of aspirin-generated ATL and other lipoxins for resolution of inflammation still needs to be confirmed in men.

Leukocyte Adhesion and Transmigration A prerequisite for tissue-destructing actions of leukocytes is their adhesion to and transmigration through the endothelial lining of blood vessels. This leukocyte traffic is controlled by signaling and adhesion molecules that also regulate leukocyte function. Salicylates inhibit cytokine-induced expression of adhesion molecules on endothelial cells [162, 341] and integrin-mediated neutrophil adhesion [342, 343]. No such effects were seen with indomethacin [162]. In healthy volunteers, inhibition of adhesion and transmigration of T lymphocytes to cultured cytokine-stimulated human endothelial cells were found *ex vivo* after one single i.v. injection of 500 mg soluble aspirin. This anti-inflammatory activity of salicylates was at least partially due to an interference with the integrin-mediated binding of resting T lymphocytes to "activated" endothelium with subsequent reduction of specific T-cell recruitment to the inflammatory site [342]. Finally, there is also evidence for inhibition of TNFα-induced monocyte adhesion to endothelial cells by aspirin. This action appears to be NFκB-mediated (Section 2.2.2) but requires salicylate concentrations of 5–10 mM *in vitro* for a significant effect [344].

Modulation of Cytokines Cytokines mediate many inflammation-associated immune reactions. However, they also mediate host defense, for example, by generation of antibodies or activation of lymphocytes [345, 346]. IL-1 and the related tumor necrosis factor (TNFα) from macrophages induce secretion of further cytokines and function as endogenous pyrogen. Aspirin has complex actions on cytokine-induced inflammatory and immunogenic responses in healthy man [347]. Interestingly, rebound phenomena on cytokine-induced cytokine synthesis have been described after short-term low-dose aspirin treatment and were possibly prostaglandin dependent since similar changes were seen after ibuprofen [348].

An interesting recent finding was the selective inhibition of IL-4 gene expression in human T cells by aspirin at therapeutic concentrations ($\leq$1 mM). IL-4 is the prototypic cytokine expressed in CD4$^+$ T cells and involved in several inflammatory diseases, such as juvenile rheumatoid arthritis and Kawasaki's disease. Salicylates inhibit IL-4 promoter activation *in vitro*. This action does not involve inhibition of prostaglandin biosynthesis or inhibition of NFκB activation [175].

Overall, the available data on the modification of cytokines by salicylates are complex and partly controversial. In some cases, this can be explained by the high concentrations of salicylates, required to obtain these actions in many *in vitro* studies. At higher levels, salicylates have multiple actions on C/CBP-β, NFκB, and other transcription factors in the promoter region of inducible genes, for example, those for cytokines (C/EBP-β), COX-2, and iNOS (Section 2.2.2). Another explanation for variable results is the different experimental settings, that is, different cell types, duration, and intensity of stimulation as well as complex interactions with other mediators such as immunomodulating prostaglandins. For example, PGE_2 generation might significantly alter cytokine release and activity *in vivo* but is virtually absent in many *in vitro* assays. Overall, available data show that aspirin and salicylates have multiple biochemical and pharmacological actions on important mediators of inflammation. This also includes prostaglandins. However, the full spectrum of putative anti-inflammatory actions of aspirin and salicylate might not solely be explained by inhibition of prostaglandin biosynthesis and is also not shared by traditional NSAIDs or coxibs, respectively.

2.3.2.2 Pain

Pain Mediators and Pain Perception Pain is an alarming signal indicating impending damage to an organ or tissue. Most forms of pain are caused by noxious stimuli, thus feeling of pain is essential for the survival of an organism in a potentially hostile environment [331]. Acute inflammatory or ischemic pain results from tissue injury and is caused by release of inflammatory mediators at a site of tissue injury as discussed above. Chronic pain is symptomatic of numerous pathological conditions, including not only chronic inflammation (rheumatoid arthritis, osteoarthritis, and low back pain) but also visceral diseases, cancer, and migraine. Aspirin and inhibitors of cyclooxygenases are standard compounds to treat inflammatory, that is, prostaglandin-related pain. However, because of the multiplicity of pharmacological actions of aspirin, the compound might also affect noninflammatory pain, most notably tension, headache, and migraine. Aspirin is rather ineffective in most forms of neuropathic pain, that is, pain due to lesions of pain signalling neurons [332].

Prostaglandins and Inflammatory Pain Among the numerous chemicals that are released at a site of inflammation, vasodilatory prostaglandins, such as PGE_2 and PGI_2, have an outstanding position in pain perception. This was originally described by *Sergio Ferreira* [349] and ultimately confirmed by the disturbed pain perception in animals with deleted prostaglandin receptors [350]. The local levels of prostaglandins in an injured area are probably too low to directly stimulate pain fibers [349, 351–353], though their effect may be enhanced by other mediators that enhance prostaglandin release, for example, bradykinin. Prostaglandins rather sensitize nociceptors in the affected skin and viscera to natural stimuli and other chemicals with the functional consequences of hyperalgesia and allodynia, that is, they shift the pain threshold to the left and increase pain perception (Figure 2.34). The actions of prostaglandins on peripheral nerves are receptor mediated, probably via the EP_1 and IP receptor

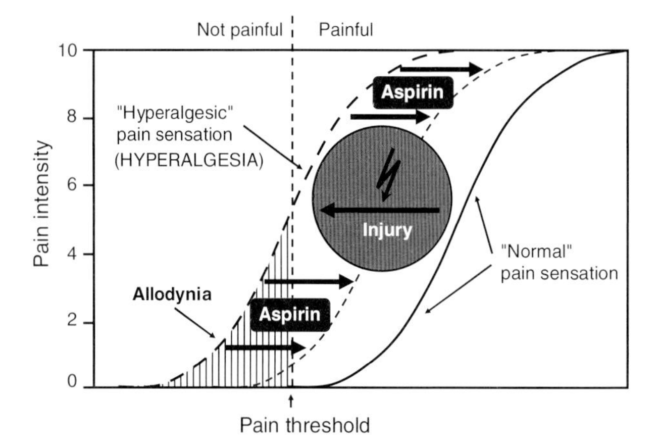

Figure 2.34 Changes in pain sensation induced by tissue injury: hyperalgesia and allodynia. Normal pain sensation occurs in response to pain stimuli. This dose–response curve is shifted to the left by the release of pain mediators from injured tissue. This causes *hyperalgesia*, that is, increased sensitivity to noxious stimuli and *allodynia*, that is, pain in response to stimuli that normally do not provoke pain. Drugs such as aspirin will shift the dose–response curve for pain backward to the right by peripheral and central modes of action. (modified after [358]).

subtypes. These in turn stimulate capsaicin receptors at unmyelinated C fibers that are considered to be critical for inflammatory pain [354]. The subcellular mode of pain sensitization by prostaglandins involves lowering of the activation threshold of voltage-sensitive, tetrodotoxin-resistant, nociceptor-specific sodium channels via increase in cAMP and cAMP-induced activation of protein kinase A. The net result of stimulation of capsaicin receptors and tetrodotoxin-resistant sodium channels is an increased excitability of the sensory neuron [355–357].

Peripheral Analgesic Actions of Aspirin There are several mechanisms that contribute to a peripheral analgesic action of aspirin. Probably best known is the inhibition of prostaglandin synthesis that explains actions of aspirin within the pain signal generating and processing (inflamed) site. Pricking of diluted emulsions of arachidonate, the precursor of prostaglandins, into the volar face of the human forearm causes pain after a latency period of 15–20 s, followed by erythema for 15–30 min [359]. Similar findings were obtained with intraarterial injection of bradykinin and additionally shown that its algetic effects could be antagonized by aspirin [360, 358] whereas algogenic actions of prostaglandinsPGE_2 remained unchanged [361, 362]. Furthermore, the potentiation of pain responses to threshold doses of other algogens (bradykinin, histamine) by intradermal injection of PGE_1 was also not reduced by aspirin [349]. These data suggest a peripheral prostaglandin-mediated antihyperalgesic (analgesic) action of aspirin in inflammation that is related to inhibition of prostaglandin biosynthesis at a site of injury.

More recent experimental and clinical data suggest that inhibition of peripheral prostaglandin synthesis is not the only mechanism of analgesic actions of aspirin [363]. For example, there is no clear correlation between the intensity of analgesia and inhibition of prostaglandin biosynthesis by aspirin. The short-lasting hyperalgetic actions of prostaglandins also suggest the contribution of additional factors for longer lasting responses, such as additional chemical mediators, released at an inflamed site [364, 365] or actions on afferent pain perception at the spinal or supraspinal levels [352, 366]. In this context, there might also be a role for neuronal COX-2.

Central Analgesic Actions of Aspirin Recent data indicate a cytokine-induced upregulation of "constitutive" COX-2 in spinal cord dorsal neurons and other regions of the CNS associated with hyperalgesia in response to inflammatory pain [367]. In addition to constitutive expression, COX-2 might also become induced in not only neuronal cells and white cells but also cerebral vascular cells after peripheral administration of inflammatory cytokines [368]. Thus, in addition to peripheral actions via inhibition of injury-induced prostaglandin release from an affected tissue, aspirin might also affect signal transduction in afferent nociceptive pathways [369], perhaps with some similarities to coxibs [367]. COX-1b, previously also known as "COX-3," might also be involved and inhibited by aspirin but not by acetaminophen [95]. The central antinociceptive actions of aspirin are not abolished by the opioid antagonist naloxon, suggesting that the endogenous opioid system of endorphins is not involved [370].

Another interesting class of compounds that are involved in pain control is endocannabinoids. They are monoacylglyerols, frequently containing arachidonic acid, anandamide being the prototypic compound. Appreciable concentrations of anandamide have been found in dorsal horn ganglia and spinal cord, suggesting a relation to pain transmission. Cannabinoid-induced analgesia has been shown to have components in brain, spinal cord, and peripheral sites and is mediated via specific G-protein-coupled receptors. Interestingly, desensitization of cannabinoid receptors by long-term treatment with cannabis (Δ^9-tetrahydrocannabinol) abolished the analgesic actions of aspirin and modified that of certain NSAIDs but not of acetaminophen in an animal model of visceral pain (Figure 2.35) [330]. Control experi-

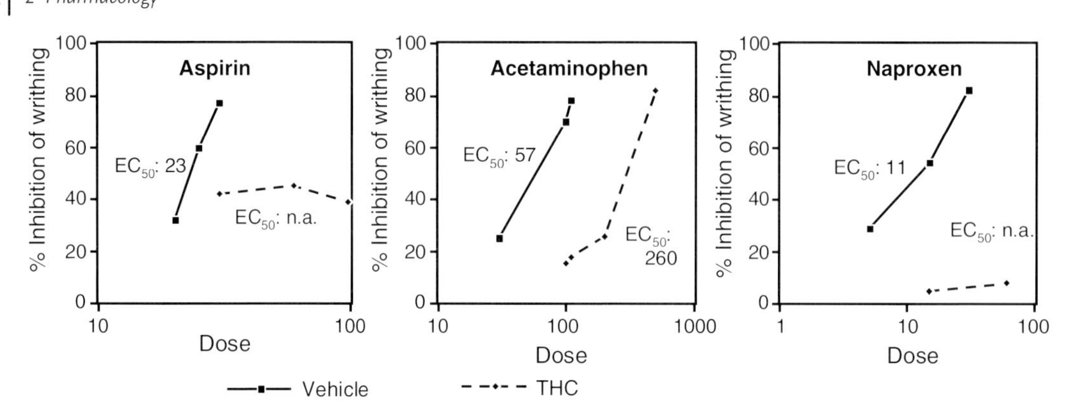

Figure 2.35 Dose-dependent analgesia by oral aspirin, acetaminophen, and naproxen in a mouse model of visceral pain before and after 1 week treatment with cannabis (Δ^9-tetrahydrocannabinol, THC). Note the parallel shift of the dose–response curve with acetaminophen and the absent analgesic effect of aspirin and naproxen after THC treatment. All doses are mg/kg; n.a.: nonapplicable (modified after [330]).

ments suggested a role for cannabinoid-induced prostaglandin release. In any case, this is the first evidence for the possible involvement of the cannabinoid system in the analgesic action of aspirin but less so for acetaminophen and deserves further studies.

Aspirin and Serotonin Serotonin – similar to norepinephrine – is a biogenic amine with particular relation to pain inside the brain, such as migraine and tension headache. An interaction between salicylates and serotonin has been postulated for a long time and is one of the major arguments for a central analgesic effect of aspirin [371, 372]. Several studies suggest a synergistic interaction between aspirin and serotonin in the central nervous system [370] that might be relevant not only to migraine (Section 4.2.1) but also to cognitive processing [373]. Mechanistically, interactions with serotonin might be due to displacement of the amino acid precursor tryptophan by salicylates from its binding to plasma proteins, eventually resulting in increased serotonin biosynthesis in the brain [374, 370]. Elevated brain serotonin was found in cortical and pontine areas of the rat brain, subsequent to parenteral aspirin treatment at anti-inflammatory doses. It was suggested that high brain serotonin levels will downregulate serotonin receptors [370, 375] and that this mechanism might be involved in the central antinociceptive actions of aspirin.

Direct Actions of Aspirin on Peripheral Pain Fibers and Central Pain Perception In noninflammatory conditions, nausea, vomiting, tinnitus, and dizziness are typical initial symptoms of aspirin overdosing (Sections 3.1.1 and 3.2.4). They are mediated by direct actions of aspirin on selected regions in the brain and suggest specific actions of aspirin on selected areas of the CNS that are independent of prostaglandins. Experimental studies have shown that intrathalamic injection of aspirin or salicylate depresses C-fiber-mediated nociceptive activity after nerve stimulation, suggesting a central (spinal cord) mechanism of the antinociceptive action of aspirin that is probably salicylate mediated [376]. More recent experimental studies have confirmed an analgesic action of oral aspirin at the spinal cord level and additionally demonstrated a nociceptive processing that was different from that of acetaminophen [377]. Further experimental evidence for a central analgesic action of aspirin in man was found by Bromm and colleagues [32, 378] in a human model of noninflammatory pain.

Thirty-two healthy volunteers were subjected to phased pain by electrical stimulation. Stimuli were applied via a specifically designed intracutaneous electrode that was inserted into the tip of the left middle finger in the immediate vicinity of the most superficial skin afferents, unmyelinated C fibers, and myelinated Aδ fibers, mostly belonging to the nociceptive system. The strength of electrical stimulation was adjusted to pain fibers and minimized concomitant excitation of receptors for other sensory modalities. All subjects were treated with aspirin (1 g) or a matching placebo in a double-blind crossover design. Measurements of subjective pain ratings and somatosensory evoked cerebral potentials were taken about 2 h later.

Electrical stimulation of the fingertip caused a well-localized pain response without overt tissue injury. Aspirin significantly reduced subjective pain sensation and somatosensory evoked pain-related cerebral potentials (SEPs). No effect of aspirin was observed on auditory evoked potentials, reaction time, and spontaneous electroencephalograms. Peak plasma levels of aspirin and salicylate at 2 h were 3 and 32 μg/ml, respectively, with large interindividual variation and no clear correlation with the analgesic effect.

The conclusion was that the analgesic action of aspirin in this model of nerve stimulation-induced noninflammatory pain in man is due to a central mode of action. No dose–response relationship could be established. Whether this was due to the large interindividual variations or just reflects an all-or-none response remains to be determined [32, 378] (Figure 2.36).

These data suggest that inhibition of (peripheral) prostaglandin biosynthesis by aspirin alone does not sufficiently explain the analgesic actions of the compound. Although it is clear that any anti-inflammatory action of aspirin will reduce prostaglandin formation and thereby remove a pain receptor sensitizing factor, it is not established whether a similar mechanism also explains the pain sensation, relating to neurogenic inflammation or direct electrical stimulation of nociceptors in the absence of tissue injury. Interactions with other mediator systems, for example, endocannabinoids or serotonin, are likely, as well as direct actions on pain transmission and perception in the central nervous system.

2.3.2.3 Fever

Fever and Mediators of the Febrile Response The febrile response is in most cases a part of a physiologic defense reaction and caused by cytokine-in-

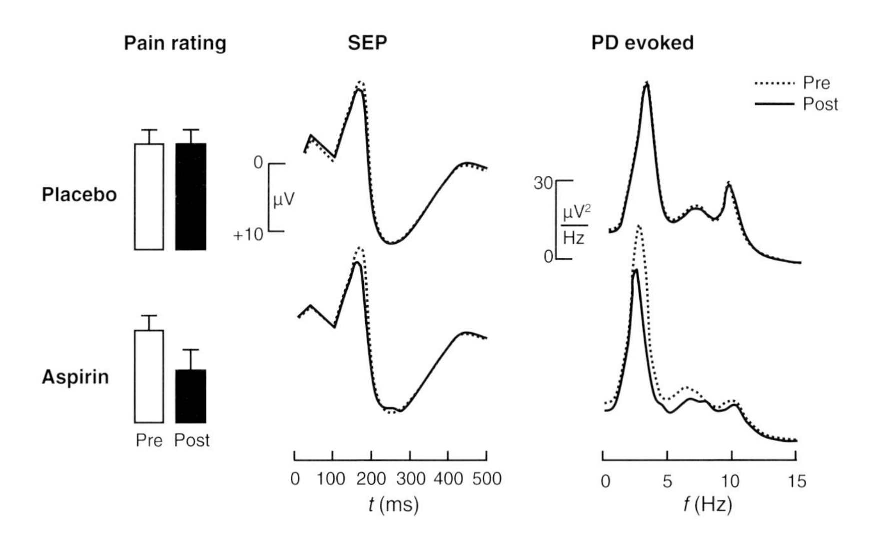

Figure 2.36 Effects of aspirin on pain rating, pain-related somatosensory evoked potentials (SEPs) and power density (PD) in response to pain-inducing stimuli in healthy volunteers. Encephalographic measurements (EEG) were taken before (pre) and 90 min after (post) oral aspirin (1 g) intake. Note the aspirin-induced alterations, suggesting reduced signal transmission. No changes were seen with a matching placebo (PLA) in a double-blind crossover design (modified after [32]).

duced rise in core temperature, generation of active phase reactants, and subsequent upregulation of peripheral defense systems [379, 380]. The febrile response starts with exposition of the organism to exogenous pyrogens, such as viruses, bacterial toxins, or other products of microbial origin. These enter the body and stimulate white cells to phagocytosis and generation of pyrogenic cytokines. These endogenous pyrogens IL-1, TNFα, interferon-γ and IL-6 have the capacity to raise the thermoregulatory center set point in the hypothalamus. They do so either by acting directly on thermosensitive neurons after crossing the blood–brain barrier and/or by release of other mediators, such as PGE_2, in circumventricular organs. This involves induction of COX-2 and results in subsequent increase of PGE_2 in the preoptic region of the hypothalamus. This is an area with high expression levels of prostaglandin EP_3 receptors [381], the putative sites of PGE action [382] (Figure 2.37).

Questions about risk–benefit ratio of fever have generated considerable controversies in recent years and are still not finally answered. Substantial data indicate both potentiating *and* inhibitory effects of the fever response to inflammatory reactions. The induction of heat-shock proteins by salicylates rather indicates stimulation of a protective system. The potential of the febrile response for harm is reflected in reports showing

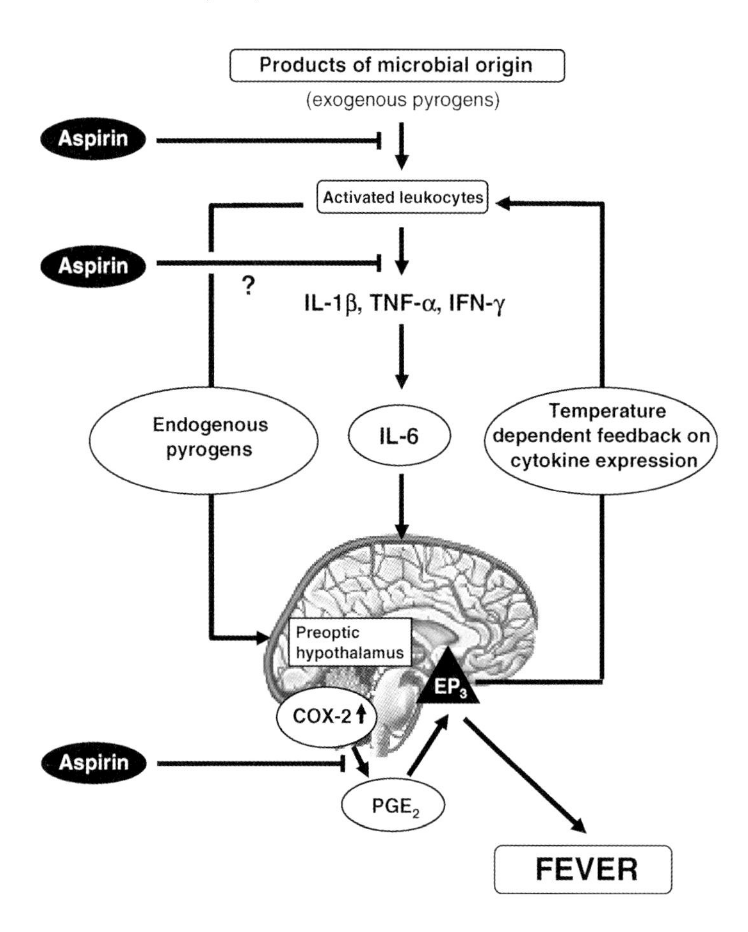

Figure 2.37 Hypothetical model for the febrile response and possible sites of action of aspirin and salicylates (modified after [380]).

that IL-1, TNFα, interferon-γ, and IL-6 mediate the physiological abnormalities of certain infections. Thus, treatment with antipyretic, anti-inflammatory analgesics, such as salicylates, will normalize the body temperature by fighting the generation and action of inflammatory pyrogenic cytokines [380].

Mechanisms of Antipyretic Actions of Aspirin Aspirin does not reduce normal body temperature, nor does it modify an elevated body temperature subsequent to physical exercise [384] or increased temperature in the environment [385]. Aspirin selectively reduces pyrogen-induced fever [386] by an interaction with the pyrogenic cytokines IL-1, TNFα, interferon-γ, and IL-6 (Figure 2.37). The antipyretic response to aspirin is probably salicylate mediated. It can be obtained after i.v. administration of sodium salicylate at antipyretic salicylate-plasma levels of about 1 mM (210–230 µg/ml). The antipyretic action of aspirin is typically associated with sweating, indicating extra "heat" production and "export" through the skin as a result of uncoupling of oxidative phosphorylation (Section 2.2.3) [383] (Figure 2.38). This might also explain the paradoxical "hyperpyrexia" seen in children with salicylate poisoning (Section 3.1.1). Thus, aspirin affects both sides of pyrogen-induced upregulation of temperature control: heat production and heat loss.

In addition to salicylate-mediated inhibition of endogenous pyrogens, salicylates also antagonize the actions of endogenous pyrogens on COX-2 expression and inhibit prostaglandin biosynthesis. This involves at least two mechanisms: inhibition of cytokine-induced expression of COX-2 protein [387] by salicylates and inhibition of enzymatic activity of COX-1 and COX-2. COX-2 appears to be the target cyclooxygenase since selective COX-2 inhibition in man reduces fever to the same extent as nonselective COX-1/COX-2 inhibitors [388]. Application of PGE_2 into the hypothalamus or the ventricles of the brain causes fever that, in contrast to that induced by IL-1 or TNFα, cannot be blocked by aspirin or salicylates. Thus, PGE_2 – generated via COX-2 – determines the severity of the febrile response and possibly acts via EP_3 receptors [382]. Aspirin inhibits the febrile response partly via inhibition of PGE_2 formation.

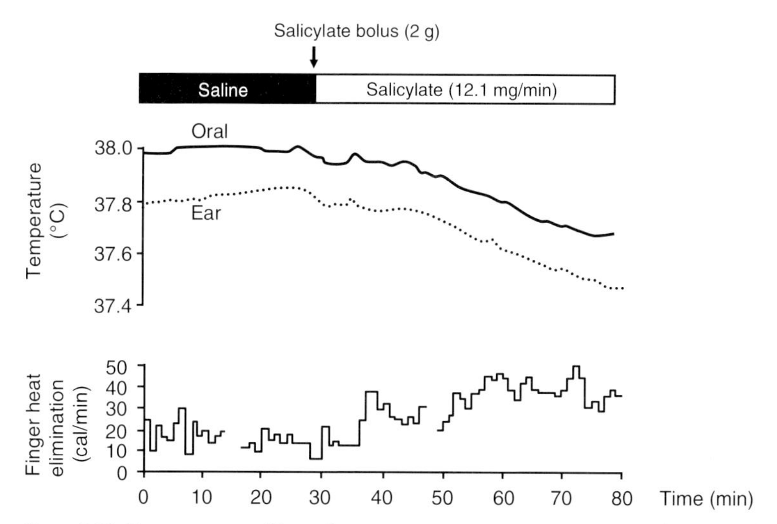

Figure 2.38 Temperature and heat elimination responses to intravenous sodium salicylate in a febrile patient. Oral and ear temperatures (°C) and finger heat elimination (cal/min) are shown during an intravenous infusion of saline (0–30 min), an intravenous bolus injection of 2 g, and subsequent intravenous infusion of sodium salicylate at 12.1 mg/min (30–80 min). Plasma salicylate concentrations at 20 and 42 min after start of salicylate infusion were 215 µg/ml and 229 µg/ml, respectively. Note the increased heat production that parallels the fall in body temperature, suggesting uncoupling of oxidative phosphorylation (Section 2.2.3) (modified after [383]).

A recently discovered additional mechanism of the antipyretic aspirin action might be the inhibition of "primary" pyrogen generation by inhibition of viral replication. This has been shown for influenza viruses not only *in vitro* but also in an animal experiment *in vivo*. Interestingly, this antiviral activity of aspirin was shared by salicylate but not by traditional NSAIDs, such as indomethacin [167].

Summary

Aspirin inhibits COX-2-dependent prostaglandin (PGE_2) formation *in vitro* whereas salicylic acid is considerably less potent. *In vivo*, both compounds are about equipotent inhibitors of inflammation, fever, and (inflammatory) pain. Inhibition of prostaglandin biosynthesis is a central component of the anti-inflammatory, antipyretic, and analgesic actions of aspirin but independently does not sufficiently explain all of the multiple biological activities. Therefore, NSAIDs that were designed to solely inhibit prostaglandin synthesis via modulation of COX-enzyme activity do not share the full spectrum of biological activities with salicylates.

The anti-inflammatory action of aspirin involves interactions with other mediator systems, allowing for the accumulation and activation of white cells and cytokine production. This might also include generation of 15-(*R*)-HETE via acetylation of COX-2 and subsequent generation of aspirin-triggered lipoxin, an anti-inflammatory mediator that stimulates resolution of inflammation. Further actions include transcriptional effects on regulation of gene expression, including inhibition of inflammatory cytokines and other mediators of inflammatory or immune responses, including COX-2.

The analgesic effect of aspirin involves peripheral and central sites of actions and also additional mediator systems. Both of them probably involve prostaglandins that are generated in both injured tissues and neuronal cells by an upregulated COX-2. Peripheral analgesic actions of aspirin include inhibition of prostaglandin release at a site of injury with subsequent reduced sensitization of nociceptive nerve terminals. Central effects involve changes in serotoninergic neurotransmission and interactions with antinociceptive hypothalamic neurons. There might be a relationship to the analgesic effects mediated via the endocannabioid system. In clinical conditions, different sites of action might be involved, dependent on the kind of injury and kind and intensity of the noxious stimulus. The functional consequences are hyperalgesia and allodynia that are both antagonized by aspirin.

The antipyretic action of aspirin is due to inhibition of PGE_2 formation in the central nervous system. In addition, aspirin and salicylate not only interfere with endogenous pyrogens and their induction of COX-2 expression and activity but will also reduce fever by peripheral anti-inflammatory effects. Whether aspirin in man directly interacts with viral replication, a possible additional explanation for an antipyretic action, remains to be determined.

References

328 Nathan, C. (2002) Points of control in inflammation. *Nature*, **420**, 846–852.

329 Takano, T., Clish, C.B., Gronert, K. *et al.* (1998) Neutrophil-mediated changes in vascular permeability are inhibited by topical application of aspirin-triggered 15-epi-lipoxin A_4 and novel lipoxin B_4 stable analogues. *The Journal of Clinical Investigation*, **101**, 819–826.

330 Anikwue, R., Huffman, J.W., Martin, Z.L. *et al.* (2002) Decrease in efficacy and potency of nonsteroidal antiinflammatory drugs by chronic Δ^9-tetrahydrocannabinol administration. *The Journal of*

Pharmacology and Experimental Therapeutics, **303**, 340–346.

331 Scholz, J. and Woolf, C.J. (2002) Can we conquer pain? *Nature Neuroscience (Supplement)*, **5**, 1062–1067.

332 Baron, R. (2006) Mechanisms of disease: neuropathic pain – a clinical perspective. *Nature Clinical Practice. Neurology*, **2**, 95–106.

333 Marks, V. and Smith, M.J.H. (1960) Anti-inflammatory activity of salicylates. *Nature*, **187**, 610.

334 Hinz, B., Kraus, V., Pahl, A. *et al.* (2000) Salicylate metabolites inhibit cyclooxygenase-2-dependent prostaglandin E_2 synthesis in murine macrophages. *Biochemical and Biophysical Research Communications*, **274**, 197–202.

335 Higgs, G.A., Salmon, J.A., Henderson, B. *et al.* (1987) Pharmacokinetics of aspirin and salicylate in relation to inhibition of arachidonate cyclooxygenase and antiinflammatory activity. *Proceedings of the National Academy of Sciences of the United States of America*, **84**, 1417–1420.

336 Hecker, G., Ney, P. and Schrör, K. (1990) Cytotoxic enzyme release and oxygen-centered radical formation in human neutrophils are selectively inhibited by E-type prostaglandins but not by PGI_2. *Naunyn-Schmiedebergs Archives of Pharmacology*, **341**, 308–315.

337 Kitsis, E.A., Weissmann, G. and Abramson, S.B. (1991) The prostaglandin paradox: additive inhibition of neutrophil function by aspirin-like drugs and the prostaglandin E_1 analog misoprostol. *The Journal of Rheumatology*, **18**, 1461–1465.

338 Newby, A.C., Holmquist, C.A., Illingworth, J. *et al.* (1983) The control of adenosine concentration in polymorphonuclear leucocytes, cultured heart cells and isolated perfused heart from the rat. *The Biochemical Journal*, **214**, 317–323.

339 Cronstein, B.N., Montosinos, M.C. and Weissman, G. (1999) Salicylates and sulfasalazine, but not glucocorticoids, inhibit leukocyte accumulation by an adenosine-dependent mechanism that is independent of inhibition of prostaglandin biosynthesis and p105 of NFκB. *Proceedings of the National Academy of Sciences of the United States of America*, **96**, 6377–6381.

340 Paul-Clark, M.J., van Cao, T., Moradi-Bidhendi, N. *et al.* (2004) 15-epi-Lipoxin A_4-mediated induction of nitric oxide explains how aspirin inhibits acute inflammation. *The Journal of Experimental Medicine*, **200**, 69–78.

341 Xia, L., Pan, J., Yao, L. *et al.* (1998) A proteasome inhibitor, an antioxidant or a salicylate but not a glucocorticoid blocks constitutive and cytokine-inducible expression of P-selectin in human endothelial cells. *Blood*, **91**, 1625–1632.

342 Gerli, R., Paolucci, C., Gresele, P. *et al.* (1998) Salicylates inhibit adhesion and transmigration of T-lymphocytes by preventing integrin activation induced by contact with endothelial cells. *Blood*, **92**, 2389–2398.

343 Pillinger, M.H., Capodici, C., Rosenthal, P. *et al.* (1998) Modes of action of aspirin-like drugs: salicylates inhibit Erk activation and integrin-dependent neutrophil adhesion. *Proceedings of the National Academy of Sciences of the United States of America*, **95**, 14540–14545.

344 Weber, C., Ertl, W., Pietsch, A. *et al.* (1995) Aspirin inhibits nuclear factor-κB mobilization and monocyte adhesion in stimulated human endothelial cells. *Circulation*, **91**, 1914–1917.

345 Libby, P., Warner, S.J.C. and Friedman, G.B. (1988) Interleukin 1: a mitogen for human vascular smooth muscle cells that induces the release of growth-inhibitory prostanoids. *The Journal of Clinical Investigation*, **81**, 487.

346 Dinarello, C.A. (1994) Interleukin-1. *Advances in Pharmacology*, **25**, 21–51.

347 Hsia, J. and Tang, T. (1992) Aspirin as a biological response modifier, in *Combination Therapies* (eds A.L. Goldstein and E. Garaci), Plenum Press, New York, pp. 131–137.

348 Endres, S., Whitaker, R.E., Ghorbani, R. *et al.* (1996) Oral aspirin and ibuprofen increase cytokine-induced synthesis of IL-1β and of tumor necrosis factor α *ex vivo*. *Immunology*, **87**, 264–270.

349 Ferreira, S.H. (1972) Prostaglandins, aspirin-like drugs and analgesia. *Nature: New Biology*, **240**, 200–203.

350 Murata, T., Ushikubi, F., Matsuola, T. *et al.* (1997) Altered pain perception and inflammatory response in mice lacking prostacyclin receptor. *Nature*, **388**, 678–682.

351 Arturson, G., Hamberg, M. and Jonsson, C.-E. (1973) Prostaglandins in human burn blister fluid. *Acta Physiologica Scandinavica*, **87**, 270.

352 Ferreira, S.H., Lorenzetti, B.B. and Correa, F.M.A. (1978) Central and peripheral antialgetic action of aspirin-like drugs. *European Journal of Pharmacology*, **53**, 39–48.

353 Brodie, M.J., Hensby, C.N., Parke, A. *et al.* (1980) Is prostacyclin the major proinflammatory prostanoid in joint fluid? *Life Sciences*, **27**, 603.

354 Moriyama, T., Higashi, T., Togashi, K. *et al.* (2005) Sensitization of TRPVI by EP_1 and IP reveals peripheral nociceptive mechanism of prostaglandins. *Molecular Pain*, **1** (3), doi:10.1186/1744-8069-1-3.

355 England, S., Bevan, S. and Docherty, R.J. (1996) PGE_2 modulates the tetrodotoxin-resistant sodium current in neonatal rat dorsal root ganglion neurones via the cyclic AMP protein kinase A cascade. *The Journal of Physiology*, **495**, 429–440.

356 Gold, M.S., Reichling, S.B., Shuster, M.J. *et al.* (1996) Hyperalgesic agents increase a tetrodotoxinresistant Na^+-current in nociceptors. *Proceedings of the National Academy of Sciences of the United States of America*, 1108–1112.

357 Brune, K. and Zeilhofer, H.U. (2006) Antipyretic analgesics: basic aspects, in *Wall and Melzack's Textbook of Pain*, 5th edn (eds A.B. McMahon and M. Koltzenburg), Elsevier, pp. 459–469.

358 Cervero, F. and Laird, J.M.A. (1996) Mechanisms of touch-evoked pain (allodynia): a new model. *Pain*, **68**, 13–23.

359 Jaques, R. (1959) Arachidonic acid, an unsaturated fatty acid which produces slow contractions of smooth muscle and causes pain. Pharmacological and biochemical characterisation of its mode of action. *Helvetica Physica Acta*, **17**, 255.

360 Coffman, J.D. (1966) The effect of aspirin on pain and hand blood flow responses to intra-arterial injection of bradykinin in man. *Clinical Pharmacology and Therapeutics*, **7**, 26–37.

361 Collier, H.O.J. and Schneider, C. (1972) Nociceptive response to prostaglandins and analgesic actions of aspirin and morphine. *Nature: New Biology*, **236**, 141–143.

362 Cunha, F.Q. and Ferreira, S.H. (2003) Peripheral hyperalgesic cytokines. *Advances in Experimental Medicine and Biology*, **521**, 22–39.

363 Brune, K., Beck, W.S., Geisslinger, G. *et al.* (1991) Aspirin-like drugs may block pain independently of prostaglandin synthesis inhibition. *Experientia*, **47**, 257.

364 Dray, A., Urban, L. and Dickenson, A. (1994) Pharmacology of chronic pain. *TiPS*, **15**, 190–197.

365 Dray, A. (1995) Inflammatory mediators of pain. *British Journal of Anaesthesia*, **75**, 125–131.

366 Besson, J.M. (1999) The neurobiology of pain. *The Lancet*, **353**, 1610–1615.

367 Samad, T.A., Moore, K.A., Sapirstein, A. *et al.* (2001) Interleukin-1β-mediated induction of COX-2 in the CNS contributes to inflammatory pain hypersensitivity. *Nature*, **410**, 471–475.

368 Engblom, D., Ek, M., Saha, S. *et al.* (2002) Prostaglandins as inflammatory messengers across the blood–brain barrier. *Journal of Molecular Medicine*, **80**, 5–15.

369 Bannwarth, B., Demotes-Mainard, F., Schaeverbeke, T. *et al.* (1995) Central analgesic effects of aspirin-like drugs. *Fundamental & Clinical Pharmacology*, **9**, 1–7.

370 Groppetti, A., Braga, C., Biella, G. *et al.* (1988) Effect of aspirin on serotonin and met-enkephalin in brain: correlation with the antinociceptive activity of the drug. *Neuropharmacology*, **27**, 499–505.

371 Shyu, K.W. and Lin, M.T. (1985) Hypothalamic monoaminergic mechanisms of aspirin-induced analgesia in monkeys. *Journal of Neural Transmission*, **62**, 285–293.

372 Pini, L.A., Sandrini, M. and Vitale, G. (1995) Involvement of brain serotonergic system in the antinociceptive action of acetylsalicylic acid in the rat. *Inflammation Research*, **44**, 30–35.

373 Austermann, M., Grotemeyer, K.-H., Evers, S. *et al.* (1998) The influence of acetylsalicylic acid on cognitive processing: an event-related potentials study. *Psychopharmacology*, **138**, 369–374.

374 Tagliamonte, A., Tagliamonte, P., Perez-Cruet, J. *et al.* (1973) Increase of brain tryptophan and stimulation of serotonin synthesis by salicylate. *Journal of Neurochemistry*, **20**, 909–912.

375 Pini, L.A., Vitale, G. and Sandrini, M. (1997) Serotonin and opiate involvement in the antinociceptive effect of acetylsalicylic acid. *Pharmacology*, **54**, 84–91.

376 Jurna, I., Spohrer, B. and Bock, R. (1992) Intrathecal injection of acetylsalicylic acid, salicylic acid and indomethacin depresses C-fibre-evoked activity in the rat thalamus and spinal cord. *Pain*, **95**, 289–290.

377 Choi, S.-S., Lee, J.-K. and Suh, H.-W. (2001) Antinociceptive profiles of aspirin and acetaminophen in formalin, substance P and glutamate pain models. *Brain Research*, **921**, 233–239.

378 Scharein, E. and Bromm, B. (1995) Comparative evaluation of analgesic efficacy of drugs. *Advances in Pain Research and Therapy*, **22**, 473–500.

379 Dinarello, C.A. (1996) Thermoregulation and the pathogenesis of fever. *Infectious Disease Clinics of North America*, **10**, 433–449.

tivity of malignant tumor cells to salicylates was markedly higher than that of nonmalignant adenoma cells. These and other data support the interesting hypothesis that aspirin and NSAIDs can possibly substitute for the physiological function of the (missing) APC tumor suppressor gene in colorectal cancer cells by inducing apoptosis.

Treatment with prostaglandins did not reverse the proapoptotic effects of aspirin and other NSAIDs, suggesting that they were independent of prostaglandin formation [423]. Moreover, structural analogues of NSAIDs that were non-COX inhibitors did induce the same anticarcinogenic changes in cell cycle and apoptosis [391]. Finally, aspirin also protected from tumor growth and stimulates apoptosis in COX-2 knockout mice [390, 424]. All of these data suggest that at least part of aspirin and NSAID-induced apoptosis in colon cancer cells is independent of prostaglandin formation. Alternatively, accumulation of arachidonic acid after COX inhibition will cause buildup of ceramide that sensitizes colon carcinoma cells to apoptosis [425] in addition to generating an apoptotic signal by its own [426, 393]. In addition, proapoptotic fatty acid peroxidation products might be generated, such as the 15-lipoxygenase products of linoleic acid [427].

Actions on DNA There are two different and separate mechanisms by which drugs can affect tumorigenesis via DNA modulations: (i) interactions with gene regulation, most notably transcription factors of tumor promoting or tumor suppressor genes and (ii) interaction with (disturbed) DNA-repair mechanisms. Defective DNA repair will allow for nonselected amplification of defective genes because of prevention of their degradation. There are many experimental data on this issue. However, they are often derived from immortal tumor cell lines *in vitro* and might not be directly transferable to malignancy *in vivo*. In many of these experiments, concentrations of aspirin in the medium-to-high millimolar range have been used, being toxic or even fatal also to nontumor cells.

Transcription Factors One of the primary actions exerted by oncogens is the modulation of gene transcription by interacting with transcription factors. The nuclear factor κB/RelA family of transcription factors (NFκB/RelA) is one of them and regulates the expression of numerous genes involved in the control of not only immune and inflammatory responses (Section 2.2.2) but also apoptosis and cell survival. NFκB acts either as a regulator of the apoptotic program for induction of apoptosis or, more commonly, as its inhibitor [159]. These disparate effects depend on the stimulus, cell type, and intracellular signaling pathways, eventually resulting in diverse target gene specificity and the more distal signaling pathways and their targets [428]. Aspirin causes activation of NFκB by signal-specific IκB degradation in colorectal carcinoma cell lines [429, 430]. This reaction was independent of the tumor suppressor gene p53 and other markers of genomic instability. However, as already seen with the role of NFκB in inflammation, salicylate treatment for 16 h with 1–10 mM aspirin *in vitro* is necessary to obtain this effect. These levels, specifically in terms of free salicylate, can probably not be obtained *in vivo* and also not maintained for such a long time period.

Another aspirin-sensitive transcription factor is activator protein-1 (AP-1) [431]. The inhibition by aspirin of this factor also required millimolar concentrations and was paralleled by a fall in intracellular pH. This strongly suggests a relationship of AP-1 inhibition to the protonophoric, that is, metabolic, actions of salicylates (Section 2.2.3).

The most interesting finding was that pretreatment with aspirin might downregulate the expression of the apoptosis gene Bcl-2 in human colorectal carcinoma cells via tumor necrosis factor-related apoptosis-inducing ligand (TRAIL). Pretreatment by aspirin (1 mM) of TRAIL-resistant cancer cells resulted in sensitization of these cells and augmented TRAIL-induced apoptotic death. This action required 12 h of pretreatment with aspirin and was related to down-regulation of Bcl-2 and decrease of the mitochondrial membrane potential.

The conclusion was that aspirin might enhance apoptosis in tumor cells by promotion of TRAIL cytotoxicity [432].

Thus, there are several tumor-promoting mechanisms at the level of gene transcription that could be modified by aspirin and might be involved in its tumor suppressor action in colorectal carcinoma. Upregulation of trefoil factor family (TFF) peptides, specifically TFF2, might be another mechanism involved in the chemopreventive action of aspirin in gastric adenocarcinoma cell lines [433].

DNA Injury and Repair Mismatch repair genes and proteins are important to correct for DNA instability by removing defective genes. Aspirin treatment of cultured colorectal cancer cells for 12 weeks markedly reduced DNA instability in cells genetically deficient for a subset of mismatch repair genes. It was suggested that aspirin induces a genetic selection for DNA stability that might be important for preventing hereditary nonpolyposis colorectal cancer [434]. However, the duration of treatment was long (over weeks) and the effective aspirin concentrations high (>1 mM). Others have shown that aspirin inhibits oxidative DNA strand breaks, mediated by reactive oxygen species [435]. Thus, aspirin might well modify the structure of DNA. Possibly these actions are related to the long known acetylation of DNA [101], the biological significance, however, of these alterations for carcinogenesis *in vivo* remains to be determined.

2.3.3.3 Nonspecific Actions of Salicylates

Independent of the rather specific actions on cellular signaling pathways, aspirin might also exert nonspecific actions on cell function, specifically at higher concentrations *in vitro*. This includes actions on cellular energy metabolism, that is, uncoupling of oxidative phosphorylation and disturbed cellular metabolism by impaired mitochondrial β-oxidation of long-chain fatty acids (Section 2.2.3). Both reactions are likely to be particularly effective in proliferating cells with increased energy requirements, that is, malignant tumor cells. It is surprising that their possible contribution to antitumor effects of aspirin has apparently not been studied in more detail so far. One possible explanation is that the basic research on aspirin-related changes in energy metabolism has been done – and completed – many years before an antitumor action of aspirin was described. Researchers also possibly found changes in gene transcription more attractive than nonspecific inhibition of kinases because of lack of ATP.

Summary

In addition to antithrombotic, anti-inflammatory, and analgesic effects, aspirin also modifies growth and proliferation of malignant tumors. An effective chemoprevention in malignant tumors has been most extensively studied – and shown – in colorectal carcinomas (Section 4.3.1). Possible mechanistic explanations are interactions with not only an upregulated COX-2, by oncogenes, and COX-2-dependent product formation but also COX-2-independent mechanisms.

There is significant transcriptional upregulation of COX-2 in colorectal cancer as well as adenomatous polyposis coli associated with a marked increase in prostaglandin (PGE_2) formation. These changes appear to be associated with a differential regulation of EP receptors and, as a net effect, result in elevated tissue cAMP levels. Aspirin not only reduces PGE_2 formation but also interacts with tumor-promoter-induced COX-2 expression and other COX-2 activities, specifically, the redox activation of (co)carcinogens.

In addition to interactions with COX-2, aspirin might also modify COX-independent pathways of cell signaling and survival. This includes modification of transcription factors (NFκB, AP-1) and interactions with mismatch repair genes. These actions are also seen in COX-2 knockout animals. The effects of aspirin are shared by salicylate. However, they require medium to high millimolar concentrations *in vitro*. At these concentrations, salicylates uncouple oxidative phosphorylation and might cause metabolic failure of cells, in particular, cells with increased turnover and energy metabolism, such as tumor cells.

Overall, the mechanisms of anticarcinogenic actions of aspirin are probably complex and in many important aspects poorly understood. However, the 50% reduced risk of colorectal cancer after long-term treatment with aspirin, according to clinical trials (Section 4.3.1), is an impressive finding and should stimulate basic research to improve the understanding of its biological background.

References

389 Lupulescu, A. (1978) Enhancement of carcinogenesis by prostaglandins. *Nature*, **270**, 634–636.

390 Marnett, L.J. (1992) Aspirin and the potential role of prostaglandins in colon cancer. *Cancer Research*, **52**, 5575–5589.

391 Levy, G.N. (1997) Prostaglandin H synthases, nonsteroidal antiinflammatory drugs, and colon cancer. *The FASEB Journal*, **11**, 234–247.

392 Courtney, E.D.J., Melville, D.M. and Leicester, R.J. (2004) Review article: chemoprevention of colorectal cancer. *Alimentary Pharmacology & Therapeutics*, **19**, 1–14.

393 Cao, Y., Pearman, T., Zimmerman, G.A. *et al.* (2000) Intracellular unesterified arachidonic acid signals apoptosis. *Proceedings of the National Academy of Sciences of the United States of America*, **97**, 11280–11285.

394 Rigas, B. and Shiff, S.J. (1999) Nonsteroidal anti-inflammatory drugs and the induction of apoptosis in colon cells: evidence for PHS-dependent and PHS-independent mechanisms. *Apoptosis*, **4**, 373–381.

395 Hardwick, J.C.H., van Santen, M., van den Brink, G.R. *et al.* (2004) DNA array analysis of the effects of aspirin on colon cancer cells: involvement of Rac1. *Carcinogenesis*, **25**, 1293–1298.

396 Eberhart, C.E., Coffey, R.J., Radhika, A. *et al.* (1994) Upregulation of cyclooxygenase-2 gene expression in human colorectal adenomas and adenocarcinomas. *Gastroenterology*, **107**, 1183–1188.

397 Kargman, S., O'Neill, G., Vickers, P. *et al.* (1995) Expression of prostaglandin G/H synthase-1 and -2 protein in human colon cancer. *Cancer Research*, **55**, 2556–2559.

398 Sano, H., Kawahito, Y., Wilder, R.L. *et al.* (1995) Expression of cyclooxygenase-1 and -2 in human colon cancer. *Cancer Research*, **55**, 3785–3789.

399 Kutchera, W., Jones, D.A., Matsunami, N. *et al.* (1996) Prostaglandin H synthase 2 is expressed abnormally in human colon cancer: evidence for a transcriptional effect. *Proceedings of the National Academy of Sciences of the United States of America*, **93**, 4816–4820.

400 Oshima, M., Dinchuk, J.E., Kargman, S.L. *et al.* (1996) Suppression of intestinal polyposis in $APC^{\Delta 716}$ knockout mice by inhibition of prostaglandin endoperoxide synthase-2 (COX-2). *Cell*, **87**, 803–809.

401 Tsuji, M., Kawano, S. and DuBois, R.N. (1997) COX-2 expression in human colon cancer cells increases metastatic potential. *Proceedings of the National Academy of Sciences of the United States of America*, **94**, 3336–3340.

402 Fujita, T., Matsui, M., Takaku, K. *et al.* (1998) Size- and invasion-dependent increase in cyclooxygenase-2 levels in human colorectal carcinomas. *Cancer Research*, **58**, 4823–4826.

403 Sheehan, K.M., Sheahan, K., O'Donoghue, D.P. *et al.* (1999) The relationship between cyclooxygenase-2 expression and colorectal cancer. *The Journal of the American Medical Association*, **282**, 1254–1257.

404 DuBois, R.N., Shao, J., Tsujii, M. *et al.* (1996) G1 delay in cells overexpressing prostaglandin endoperoxide synthase-2. *Cancer Research*, **56**, 733–737.

405 Tsuji, M. and DuBois, R.N. (1995) Alterations in cellular adhesion and apoptosis in epithelial cells, overexpressing prostaglandin endoperoxide synthase-2. *Cell*, **83**, 493–501.

406 Sheng, H., Shao, J., Kirkland, S.C. *et al.* (1997) Inhibition of human cancer cell growth by selective inhibition of cyclooxygenase-2. *The Journal of Clinical Investigation*, **99**, 2254–2259.

407 Kawamori, T., Uchiya, N., Sugimura, T. *et al.* (2003) Enhancement of colon carcinogenesis by prostaglandin E_2 administration. *Carcinogenesis*, **24**, 985–990.

408 Claria, J., Lee, M.H. and Serhan, C.N. (1996) Aspirin-triggered lipoxins (15-epi-LX) are generated by the human lung adenocarcinoma c ell line (A549)–neutrophil interactions and are potent inhibitors of cell proliferation. *Molecular Medicine*, **2**, 583–596.

409 Finley, P.R., Bogert, C.L., Alberts, D.S. *et al.* (1995) Measurement of prostaglandin E_2 in rectal mucosa in human subjects: a method study. *Cancer Epidemiology, Biomarkers & Prevention*, **4**, 239–244.

410 Bennett, A. and De Tacca, M. (1975) Proceedings: prostaglandins in human colonic carcinoma. *Gut*, **16**, 409.

411 Rigas, B., Goldman, I.S. and Levine, L. (1993) Altered eicosanoid levels in human colon cancer. *The Journal of Laboratory and Clinical Medicine*, **122**, 518–523.

412 Pugh, S. and Thomas, G.A. (1994) Patients with adenomatous polyps and carcinomas have increased colonic mucosal prostaglandin E_2. *Gut*, **35**, 675–678.

413 Earnest, D.L., Hixson, L.J. and Alberts, D.S. (1992) Piroxicam and other cyclooxygenase inhibitors: potential for cancer prevention. *Journal of Cellular Biochemistry*, **161**, 156–166.

414 Plescia, O.J., Smith, A.H. and Grinwich, K. (1975) Subversion of immune system by tumor cells and role of prostaglandins. *Proceedings of the National Academy of Sciences of the United States of America*, **72**, 1848–1851.

415 Baich, C.M., Doghert, P.A., Cloud, G.A. *et al.* (1984) Prostaglandin E_2-mediated suppression of cellular immunity in colon cancer patients. *Surgery*, **95**, 71–77.

416 Frommel, T.O., Dyavanapalli, M., Oldham, T. *et al.* (1997) Effect of aspirin on prostaglandin E_2 and leukotriene B_4 production in human colonic mucosa from cancer patients. *Clinical Cancer Research*, **3**, 209–213.

417 Eling, T.E., Thompson, D.C., Foureman, G.L. *et al.* (1990) Prostaglandin H synthase and xenobiotic oxidation. *Annual Review of Pharmacology and Toxicology*, **30**, 1–45.

418 Harris, R.M., Hawker, R.J., Langman, M.J.S. *et al.* (1998) Inhibition of phenolsulphotransferase by salicylic acid: a possible mechanism by which aspirin may reduce carcinogenesis. *Gut*, **42**, 272–275.

419 Craven, P.A. and DeRubertis, F.R. (1992) Effects of aspirin on 1,2-dimethylhydrazine-induced colonic carcinogenesis. *Carcinogenesis*, **13**, 541–546.

420 Shoji, Y., Takahashi, M., Kitamura, T. *et al.* (2004) Downregulation of prostaglandin E receptor subtype EP_3 during colon cancer development. *Gut*, **53**, 1151–1158.

421 Mutoh, M., Watanabe, K., Kitamura, T. *et al.* (2002) Involvement of prostaglandin E receptor subtype EP_4 in colon carcinogenesis. *Cancer Research*, **62**, 28–32.

422 Elder, D.J., Hague, A., Hicks, D.J. *et al.* (1996) Differential growth inhibition by the aspirin metabolite salicylate in human colorectal tumor cell lines: enhanced apoptosis in carcinoma and *in vitro*-transformed adenoma relative to adenoma cell lines. *Cancer Research*, **56**, 2273–2276.

423 Hanif, R., Pittas, A., Feng, Y. *et al.* (1996) Effects of non-steroidal antiinflammatory drugs on proliferation and on induction of apoptosis in colon cancer cells by a prostaglandin-independent pathway. *Biochemical Pharmacology*, **52**, 237–245.

424 Yu, H.G., Huang, J.A., Yang, Y.N. *et al.* (2002) The effects of acetylsalicylic acid on proliferation, apoptosis, and invasion of cyclooxygenase-2 negative colon cancer cells. *European Journal of Clinical Investigation*, **32**, 838–846.

425 Martin, S., Phillips, D.C., Szekely-Szucs, K. *et al.* (2005) Cyclooxygenase-2 inhibition sensitizes human colon carcinoma cells to TRAIL-induced apoptosis through clustering of DRS and concentrating death-inducing signalling complex components into ceramide-enriched caveolae. *Cancer Research*, **65**, 11447–11458.

426 Chan, T.A., Morin, P.J., Vogelstein, B. *et al.* (1998) Mechanisms underlying nonsteroidal antiinflammatory drug-mediated apoptosis. *Proceedings of the National Academy of Sciences of the United States of America*, **95**, 681–686.

427 Shureiqi, I., Chen, D., Lotan, R. *et al.* (2000) 15-Lipoxygenase-1 mediates non-steroidal anti-inflammatory drug-induced apoptosis independently of cyclooxygenase-2 in colon cancer cells. *Cancer Research*, **60**, 6846–6850.

428 Epinat, J.C. and Gilmore, T.D. (1999) Diverse agents act at multiple levels to inhibit the Rel/NFkappaB signal transduction pathway. *Oncogene*, **18**, 6896–6909.

429 Stark, L.A., Din, F.V.N., Zwacka, R.M. *et al.* (2001) Aspirin-induced activation of the NFκB signaling pathway: a novel mechanism for aspirin-mediated apoptosis in colon cancer cells. *The FASEB Journal*, **15**, 1273–1275.

430 Din, F.V.N. and Dunlop, M.G. (2005) Aspirin-induced nuclear translocation of NFκB and apoptosis in colorectal cancer is independent of p53 status and DNA mismatch repair proficiency. *British Journal of Cancer,* **92**, 1137–1143.

431 Dong, Z., Huang, C., Brown, R. *et al.* (1997) Inhibition of activator protein I activity and neoplastic transformation by aspirin. *The Journal of Biological Chemistry,* **272**, 9962–9970.

432 Kim, K.M., Song, J.J., An, J.Y. *et al.* (2005) Pretreatment of acetylsalicylic acid promotes tumor necrosis factor-related apoptosis-inducing ligand-induced apoptosis by down-regulating BCL-2 gene expression. *The Journal of Biological Chemistry,* **280**, 41047–41056.

433 Azarschab, P., Al-Azzeh, E., Kornberger, W. *et al.* (2001) Aspirin promotes TFF2 gene activation in human gastric cancer cell lines. *FEBS Letters,* **488**, 206–210.

434 Rüschoff, J., Wallinger, S., Dietmaier, E. *et al.* (1998) Aspirin suppresses the mutator phenotype associated with hereditary nonpolyposis colorectal cancer by genetic selection. *Proceedings of the National Academy of Sciences of the United States of America,* **95**, 11301–11306.

435 Hsu, C.S. and Li, Y. (2002) Aspirin potently inhibits oxidative DNA strand breaks: implications for cancer chemoprevention. *Biochemical and Biophysical Research Communications,* **293**, 705–709.

436 Shiff, S.J., Koutsos, M.I., Qiao, L. *et al.* (1996) Nonsteroidal antiinflammatory drugs inhibit the proliferation of colon adenocarcinoma cells: effects on cell cycle and apoptosis. *Experimental Cell Research,* **222**, 179–188.

3 Toxicity and Drug Safety

3.1 Systemic Side Effects
- 3.1.1 Acute and Chronic Toxicity
- 3.1.2 Bleeding Disorders
- 3.1.3 Safety Pharmacology in Particular Life Situations

3.2 Organ Toxicity
- 3.2.1 Gastrointestinal Tract
- 3.2.2 Kidney
- 3.2.3 Liver
- 3.2.4 Audiovestibular System

3.3 Non-Dose-Related (Pseudo)allergic Actions of Aspirin
- 3.3.1 Aspirin Hypersensitivity (Widal's Syndrome)
- 3.3.2 Urticaria/Angioedema, Stevens–Johnson and Lyell Syndromes
- 3.3.3 Reye's Syndrome

Acetylsalicylic Acid. Karsten Schrör

ISBN: 978-3-527-32109-4

3
Toxicity and Drug Safety

Drug safety is a key issue and subject of detailed and sophisticated legal regulations. Safety is particularly important for over-the-counter (OTC) medications that are used without prescription by a doctor and where dosing and indications are subject to consumers' discretion. The diagnosis is also made by the consumer, and side effects are usually only considered to the extent to which they determine the (subjective) tolerability. Antipyretic analgesics, including aspirin, acetaminophen, and ibuprofen, belong to this category of drugs. In addition, salicylates are ingredients of a large variety of nonprescriptional drug combinations. In many cases, it is not immediately apparent from the "fantasy" names of these products, for example, Soma Compound, Norgesic, Darvon, Percodan, and others [1], that these preparations contain aspirin as an active constituent. Addition of vitamins, caffeine, codeine, and other potentially habit-forming components is another problem. These combinations became "popular" in connection with chronic misuse of analgesics (analgesic nephropathy) that, fortunately, has now largely disappeared after the removal of phenacetin in analgesic mixtures (Section 3.2.3).

These days, millions of daily doses of antipyretic analgesics are used worldwide for OTC treatment of acute and chronic pain. Nevertheless, not all users are aware of the fact that the desired actions of drug treatment, such as disappearance of headache, might also be associated with unwanted side effects that may not necessarily cause subjective symptoms, for example, gastrointestinal (GI) bleedings after aspirin and liver toxicity with paracetamol. This makes safety aspects an important issue, specifically, for the medical lay.

For formal reasons, side effects of aspirin might be divided into three categories: first are the systemic effects due to acute and chronic overdosing or intoxication. To this category also belongs the bleeding tendency and toxic effects, in particular life situations, such as pregnancy or older age (Section 3.1). In addition to systemic effects, some organs, in particular, those with a preexisting injury or increased sensitivity, might be affected even by therapeutic doses of aspirin in the absence of systemic side effects. This involves the GI tract, liver, kidney, and the audiovestibular system (Section 3.2). Finally, there are no dose-related side effects that are due to a particular predisposition of the patient. These "hypersensitivities" might be inherent or acquired. This involves the Widal triad (aspirin-sensitive asthma), allergic reactions at the skin or mucosa (urticaria), and Reye's syndrome with a still unclear relationship with aspirin (Section 3.3).

3.1
Systemic Side Effects

Systemic side effects of aspirin result from acute or chronic overdosing and become detectable at plasma salicylate levels of about 300–400 μg/ml (≥2 mM). These toxic effects can affect almost every organ and tissue in the body because of the diversity of phar-

Acetylsalicylic Acid. Karsten Schrör

ISBN: 978-3-527-32109-4

macological actions of aspirin and salicylates, respectively. The ubiquitous distribution of salicylates within the body and the accumulation within cells at high doses facilitate tissue toxicity. Inhibition of platelet cyclooxygenase is associated with a subsequent bleeding disorder but not with irreversible bleeding. All of the other toxic side effects of aspirin are mainly caused by salicylate (Section 3.1.1).

Prolongation of bleeding time by aspirin is a frequent – and to some extent desired – event in long-term use but is neither life threatening nor significant in single or short-term use. Nevertheless, aspirin-induced bleeding may become a clinically relevant side effect in particular life situations. This includes urgent operations on aspirin-treated patients such as appendicitis or surgical treatment of traumatic accidents. Bleeding might also become a compliance-limiting factor in long-term cardiocoronary prevention because of, for example, epistaxis and gingival bleeding (Section 3.1.2). Prolonged bleeding after aspirin intake is also important in late pregnancy and older age. In patients at older age, bleeding problems may also arise from cotreatment with other drugs because of complex morbidities. For example, sulfonylureas or warfarin may interact with the pharmacokinetics and pharmacodynamics of aspirin, respectively, eventually resulting in overt bleeding in patients who take aspirin for long-term prevention of atherothrombotic events (Section 3.1.3).

In spite of legitimate concerns about the consequences of uncontrolled use, aspirin is a remarkably safe drug when used circumspectly [2]. The rate of side effects at single or short-term use is low and may be further reduced by use of appropriate galenic formulations. Unwanted side effects frequently result from unnecessary or careless use of the compound. *All* effective drugs have side effects, and aspirin is no exception from this rule.

3.1.1 Acute and Chronic Toxicity

Systemic intoxications with salicylates today are relatively seldom. However, they are facilitated by the easy access to the compounds as well as the broad and largely uncontrolled use. In addition, in public understanding, salicylates are often considered harmless household remedies because of their apparently unrestricted use by both health professionals and medical layers ("take an aspirin"). This is clearly an underestimation of the pharmacological potential of these agents at both sides [2].

3.1.1.1 Occurrence and Symptoms

Occurrence Acute aspirin intoxication results from suicide attempts, accidental (toddlers!), or iatrogenic overdosing. Iatrogenic overdosing, that is, acquired intoxication during therapeutic use, occurs predominantly in the elderly because of overloading the body's clearing capacity (salicylism) in long-term treatment. Most fatal cases of chronic salicylate poisoning occurred in toddlers or the elderly, the two patient populations in which mental deterioration is most difficult to identify [3].

Aspirin used to be a popular remedy for suicide attempts in teenagers and young adults. However, in the meantime, it has been replaced for this "indication," at least in the United Kingdom, by acetaminophen [4]. In the older literature, there is an interesting Hungarian study that gives also an explanation for the preference of aspirin in suicide attempts in earlier days.

Balazs reported 792 cases of acute aspirin poisoning in Budapest (Hungary) during an observational period of 7 years (1923–1929). The vast majority of overdosing (590) was suicide attempts in adults but only 4 cases of them, that is, less than 1%, were terminated fatally.

This number of overdosing corresponded to about one case every third day. Balazs explained this unusually high figure by the fact that aspirin was used as an *ultimum refugium* in partnership problems. For this purpose, aspirin was very popular because the clinical picture of salicylate poisoning was quite impressive to lay persons and at the same time not associated with a too high risk really to die [5].

Thus, oral aspirin appears not to be an "effective" suicide drug, also because nausea and/or vomiting belong to the first clinical symptoms after acute

Table 3.1 Clinical symptoms of aspirin intoxication as seen 6 h after intake of the drug as an oral standard formulation in relation to plasma salicylate levels (modified after [12, 20]).

Severity of intoxication	Plasma level of salicylate (μg/ml)	(mM)	Symptoms
Mild/early	300–600 (adults)	2.2–4.3	Nausea/vomiting, abdominal pain, tinnitus
	200–450 (children/elderly)	1.4–3.2	Dizziness, lethargy
Moderate	600–800 (adults)	4.3–5.8	All of the above plus tachypnea, sweating, hyperpyrexia,
	450–700 (children/elderly)	3.2–5.0	Dehydration, loss of coordination, restlessness
Severe	>800 (adults)	>5.8	All of the above plus hypotension
	>700 (children/elderly)	>5.0	Severe metabolic acidosis (after rehydration), bleeding tendency, purpuric lesions. CNS symptoms: hallucinations, stupor, coma renal insufficiency: oliguria, uremia pulmonary edema

overdosing (Table 3.1). It is known from self-experiments and patient studies that even 20 g aspirin/day (!) can be taken over a longer period of time without significant toxicity (cited after [6]). There is one case of a suicide attempt with aspirin after self-administration of approximately 700 (!) aspirin tablets dissolved in water as an enema form. The patient survived with chronic hypoxic encephalopathy after severe acidosis and transient cardiac arrest [7]. However, occasionally aspirin is "successfully" used for suicide attempts even in recent times [8]. More frequent is accidental salicylate overdosing due to (erroneous) ingestion of salicylate-containing products, directed to topical or external use. This includes salicylic acid (SA) (keratolytic) and wintergreen oil or methyl salicylate (one teaspoon = 5 ml of wintergreen oil contains 7000 mg salicylate, which is equivalent to 22 standard aspirin (325 mg) tablets (!)) [1, 9–11]. Methyl salicylate appears to be the most toxic salicylate, possibly due to the rapid uptake and tissue distribution [6]. Interestingly, the historical first cases of salicylate poisoning were reported with wintergreen oil or methyl salicylate, respectively.

Dose Dependency of Symptoms Clinical symptoms of acute aspirin intoxication become detectable in most individuals at serum salicylate levels above 300–400 μg/ml (≥2 mM) [1, 13–15]. Gastrointestinal symptoms (nausea and vomiting) and tinnitus are frequent initial symptoms. Vomiting occurs in about 50% of patients when salicylate plasma levels exceed 300 μg/ml (Table 3.1). Life-threatening intoxications after acute ingestion of salicylates (mostly aspirin) in adults start with doses above 12–15 g but with 3 g in children [16]. There is a large interindividual variability. Patients have reportedly died at plasma salicylate levels of less than 150 μg/ml whereas others have been relatively asymptomatic with levels of 500–600 μg/ml [17] or died at these tissue levels of salicylate [8].

Thisted and colleagues studied the clinical course of 177 patients who were treated in an intensive care unit during a period of 15 years because of severe salicylate poisoning, mostly due to suicide attempts. On admission, cerebral depression (lethargy) was seen in 61% of patients, respiratory failure in 47%, acidosis in 37%, and cardiovascular dysfunctions in 14%. The mortality rate was 15% (27 patients) and proportionally higher in patients aged above 40 years and in patients with delayed diagnosis. Disturbed acid–base balance was found in 50% of cases and pulmonary complications in 43%. Interestingly, coagulation problems were only seen in 38% of cases. Fever occurred in 20% and hypotension in 14%.

An autopsy was performed in 26 out of the 27 patients who died. The main findings were as follows: ulcers of the gastrointestinal tract in 46%, pulmonary edema in 46%, cerebral edema in 31%, and cerebral hemorrhage in 23% [18].

The percentage of salicylate poisoning, according to an epidemiological survey in Canada, amounted to about 20% of all lethal drug intoxications with salicylate plasma levels in the range of 6–8 mmol/l (830–1100 μg/ml) [19]. According to a US survey, no ingested amounts equivalent to less than 0.125 g/kg are harmless, a moderate risk exists at 0.15–0.30 g/kg, a severe and prolonged risk at 0.30–0.50 g/kg whereas doses greater than 0.50 g/kg are considered potentially lethal [12]. The mortality in individuals with clinical features of severe salicylate poisoning amounts to 5% (see below) but can increase to 15% if treatment is started (too) late, frequently because of delayed diagnosis [20].

The severity of intoxication is determined by the accumulation of salicylate and its prolonged action at the cell and tissue level as metabolic acidosis progresses. Because the major excretion pathway via salicyluric acid by glycine conjugation is saturated (Section 2.1.2), salicylate accumulates in plasma as also seen from the increasing percentage of unmetabolized salicylate in urine (Table 3.2). This is associated with a drastic prolongation of salicylate plasma half-life from about 2–3 h at therapeutic doses to 20–30 h and more at massive overdosing (Section 2.1.2). The symptoms of intoxication (see below) are further enhanced by the generalized metabolic acidosis, eventually resulting in a higher percentage of nonprotonized salicylate that penetrates the cell membrane and accumulates inside the cells, eventually resulting in severe metabolic disturbances owing to the uncoupling of oxidative phosphorylation (Section 2.2.3).

Clinical Symptoms of Overdosing Burning pain in the upper GI tract because of local irritation of the mucosa, eventually associated with or followed by nausea, and vomiting are early clinical symptoms of acute salicylate overdosing. Tinnitus and hyperventilation are further early signs of intoxication (Table 3.1). Hyperpnoea is seen at salicylate plasma levels of 300 μg/ml and becomes significant at about 500 μg/ml. The reason for hyperpnoea is direct stimulation of the respiratory center in the medulla oblongata. This effect is amplified at higher salicylate levels by dose-dependent disturbances of cellular energy metabolism as a result of uncoupling of oxidative phosphorylation [22, 23]. This is associated with increased tissue oxygen demand and CO_2 production. Elevated CO_2 levels in plasma stimulate the respiratory center with subsequently enhanced exhalation of CO_2 (hypercapnia). The pCO_2 in plasma remains unchanged because of the simultaneously enhanced renal bicarbonate excretion.

Aspirin-induced hyperventilation was clinically tested for the treatment of sleep apnea. Daily administration of 9–11 g (!) aspirin to nine patients with clinically relevant obstructive sleep apnea doubled the ventilation from 8 to 15 l/min and at the same time reduced the occurrence

Table 3.2 Aspirin metabolites in urine (percentage of total metabolites) after oral intake of 600 mg aspirin by 45 healthy volunteers as compared to 37 patients with salicylate (SA) intoxication.

Metabolite	Healthy volunteers (600 mg aspirin)	Intoxication plasma SA (240–600 μg/ml)	Intoxication plasma SA (715–870 μg/ml)
Salicylic acid	9 ± 1	32 ± 4	65 ± 4
Salicyluric acid	75 ± 1	47 ± 34	22 ± 4
Salicylic acid phenol glucuronide	11 ± 1	23 ± 2	15 ± 4
Gentisic acid	5 ± 1	10 ± 2	7 ± 2
Total salicylates (mg salicylic acid equivalent)	246 ± 8	2999 ± 374	8092 ± 1470

Note the increased % excretion of nonmetabolized salicylic acid and the reduced % excretion of salicyluric acid with increasing severity of poisoning, indicating the saturation of the salicyluric acid pathway (glycine conjugation) [21].

of breath standstills from 42 to 28 apneas/h. This was accompanied by a decrease in pCO_2 and an increase in pO_2. However, because of possible side effects, this high-dose treatment was not recommended for general clinical use [24].

Salicylate-induced noncardiogenic pulmonary edema occurs in both severe acute aspirin intoxication and long-term overdosing of the substance [25]. The edema occurs only at advanced stages of intoxication and may be lethal. The incidence according to recent surveys was 7% (29) in 397 patients with salicylate intoxication. At the same time, there might be proteinuria, indicating a generally increased vascular permeability [26–29]. Renal failure is rare and usually restricted to patients with preexisting renal diseases, specifically elderly persons with hypoalbuminemia (Section 3.2.2).

Toxic symptoms of the central nervous system dominate the clinical picture with increasing severity of poisoning. The initial cerebral excitation is converted into an increased cerebral depression. Finally, there is stupor, coma with cardiovascular failure, and death from respiratory arrest.

Laboratory Findings Laboratory findings are mainly the consequence of uncoupling of oxidative phosphorylation and inhibition of β-oxidation of (long-chain) fatty acids by high-level salicylates (Section 2.2.3). In this situation, metabolic CO_2 production exceeds its respiratory elimination. This effect is further enhanced by the depressive action of high salicylate levels on the respiratory center. With increasing inhibition of oxidative phosphorylation, there is also increasing accumulation of acids from the disturbed energy metabolism (lactate, pyruvate, and others) with further aggravation of acidosis and dehydration. Eventually, this results in anion-gap acidosis [30]. Salicylate itself contributes only minimally to the anion gap: about 3 mval/l at serum levels of 500 μg/ml [1]. There is an increased renal excretion of bicarbonate (followed by K^+ and Na^+) and impaired kidney function, possibly also related to disturbed energy metabolism within the tubular cells. Water and electrolyte imbalance as well as heat production (sweating!) may cause dehydration. This and the decreased renal blood flow, if untreated, will result in oliguria and finally renal failure. Disturbances in the acid–-base equilibrium are most prominent in babies and toddlers [15].

Another typical symptom of salicylate overdosing is the changes in blood glucose levels, mostly hypoglycemia [31]. This is due to enhanced insulin secretion as result of activation of the NFκB pathway in the pancreas by high-dose salicylate (Section 2.2.2). This might result in low glucose levels in the liquor, which require glucose substitution. However, hyperglycemia has also been reported [32].

Salicylate Intoxication in Children In relation to the discussion on the relationship between aspirin toxicity and Reye's syndrome, the signs and symptoms of salicylate intoxication are of particular interest in children. Similar to adults, toxic salicylate effects in children are also determined by the extreme prolongation of salicylate half-life, the markedly increased plasma levels, and its enhanced distribution into cells and tissues with decreasing pH.

Aspirin overdosing in children becomes clinically evident as a consequence of uncoupling of oxidative phosphorylation. Conventional light microscopic examination of liver biopsies of children with severe salicylate intoxication showed intrahepatocytic microvesicular steatosis without inflammation or necrosis but with depletion of glycogen stores. Findings of this kind in 10 out of 12 children who died from assumed salicylate intoxication and had cerebral edema (according to a review of records) suggested a causal relationship between aspirin intoxication and the occurrence of Reye's syndrome in a frequently cited paper [16]. However, ultrastructural microscopy of liver biopsies (not done in this particular study) of children with Reye's syndrome is different from that of children with salicylate intoxication [33], and a causal relationship between aspirin and Reye's syndrome in children has never been established [34] (Section 3.3.3).

Uncoupling of oxidative phosphorylation is compensated by an increase in metabolic turnover and associated with increased oxygen consumption, depletion in liver glycogen, and increased production of heat. This increased production of heat is responsible for the dangerous hyperpyrexia that is a prominent symptom of salicylate poisoning in infants [35].

Hyperpyrexia in salicylate poisoning is somehow difficult to understand because salicylates were frequently and effectively used as antipyretic analgesics in children until the Reye's syndrome discussion was started. The possible explanation is that salicylates "reset" the disturbed temperature regulation in the hypothalamus via their interaction with endogenous pyrogens (Section 2.3.2) but are unable to block the production of "extra" heat as a consequence of uncoupling of oxidative phosphorylation in peripheral organs (Section 2.2.3).

Physical temperature control functions by the production of large quantities of sweat as long as enough fluid for sweat production is available. When this mechanism becomes exhausted because of large water losses and dehydration, unbalanced hyperpyrexia develops because the "upregulation" of temperature control by endogenous pyrogens is still in function. Dehydration is also associated with convulsions, in particular, at severe intoxication of children below 2 years of age [35].

These metabolic disturbances, including respiratory alkalosis and metabolic acidosis, are the most important life-threatening effects of salicylates. In children, metabolic acidosis usually predominates over respiratory alkalosis whereas the opposite is seen in adults [3, 12, 36].

Chronic Overdosing Chronic salicylate intoxication in adults (salicylism) usually results from iatrogenic overdosing during long-term aspirin treatment and is frequently overlooked because of the absence of specific symptoms [26]. Symptoms of chronic overdosing are tinnitus, multiple neurological deficits, including headache (!), confusion and central excitation, sweating, hyperventilation, gastrointestinal bleeding, and ulcers. GI side effects appear to dominate in younger age whereas tinnitus and other audiovestibular toxicities (Section 3.2.4) are more frequent in the elderly. Reversible hepatic injury as another feature of long-term aspirin overdosing was, in principle, only seen in patients with a pathologic immune status, for example, rheumatoid arthritis, where the patients required long-term aspirin treatment at high doses [37] (Section 3.2.3). However, because of available therapeutic alternatives in this indication, this finding is now more of historical interest.

3.1.1.2 Treatment

Severe salicylate poisoning is an acute life-threatening, though rarely fatal, medical emergency situation. The treatment is entirely symptomatic because no specific antidote is available. As with other systemic intoxications, there are two basic therapeutic principles: reduction or prevention of absorption and stimulation of excretion of salicylates. Both measures are supported by symptomatic treatment of functional and metabolic disturbances. An actual evidence-based consensus guideline for out-of-hospital treatment of salicylate poisoning [38] as well as recommendations for emergency department management [20, 39] is available. However, obviously, not all recommendations at least in the past were identical.

A survey on recommendations by health professionals for optimum treatment of acute aspirin poisoning by retarded-release formulations provided interesting results. Seventy-six poison-control centers of North America were asked for their recommendations to treat a hypothetical case of an adult male patient presented 1 h after ingestion of 500 mg/kg enteric-coated aspirin with normal vital signs.

There was a considerable variability between the center-based recommendations. Some of the 36 (!) different recommended courses of action were considered potentially harmful [40].

Inhibition of Absorption Treatment of the "conventional" oral intoxication starts with an interruption of further salicylate uptake from the GI tract since intestinal absorption in the presence of toxic doses still continues for several hours and might be even longer in the case of enteric-coated preparations because of their retarded absorption [41]. Usual procedures for preventing further absorption include gastric lavage and administration of activated charcoal [42]. As to be expected, the earlier they can be started the more effective these procedures are. An optimum time frame would be within 1 h after ingestion.

In one prospective trial, 13 healthy adults received 1.9 g aspirin (24 tablets, each containing 81 mg aspirin). In a randomized crossover design, each subject received additionally 50 g charcoal as a single dose, two times or three times in a 4 h interval. Urinary salicylate excretion was measured and each protocol repeated in weakly intervals.

In the absence of charcoal, 91 ± 6% (mean ± SD) of total salicylate was recovered in urine. This amount was reduced to 68 ± 12, 66 ± 13, and 49 ± 12% after one, two, and three single 50 g doses of charcoal, respectively. All of these changes were significant and document for this model a 30–50% reduction in salicylate absorption by charcoal [43].

This study and most of the other pharmacokinetic studies on the modification of aspirin kinetics used aspirin doses between 1.5 and 3.0 g. These doses are far below those where toxic effects had to be expected and, therefore, may not sufficiently reflect the reality of salicylate poisoning. Nevertheless, the data indicate that charcoal treatment will reduce the aspirin absorption if it is started early and charcoal is given at sufficiently high doses. In addition, charcoal may recoat the surface of aspirin concretions within the stomach [44] and thus reduce ongoing absorption. Administration of repeated doses of charcoal may be particularly useful in patients who have ingested overdoses of enteric-coated or other slow-release formulations [41, 45]. Thus, repeated charcoal doses (4 × 50 g in 1 h intervals to adults or 1 g/kg body weight to children) are recommended until the plasma salicylate reaches peak levels [20]. In the postabsorption phase, there is no accelerated clearance of plasma salicylate by charcoal [46, 47] and no change in the prolonged half-life [48].

Stimulation of Elimination Determinations of salicylate plasma levels by any suitable method (Section 1.2.2) and of the acid–base equilibrium to detect ionic gaps are helpful because most of the clinical symptoms of salicylate poisoning are well correlated with these parameters. Measurements of salicylate plasma levels should be done initially and should be repeated at appropriate time intervals until peak plasma levels are obtained. The combined metabolic/respiratory acidosis should be corrected by appropriate treatment with sodium bicarbonate together with 20–40 mM K^+ i.v. over 3 h under control of kidney function and rehydration [20]. This procedure works with several mechanisms: inhibition of reabsorption of salicylate in the kidney by alkaline diuresis and improvement of the acid–base equilibrium in blood with normalization of plasma pH. This facilitates the rediffusion of (acetyl) salicylic acid from tissues in the blood. Particularly important is rediffusion from the central nervous system. An urinary pH of 7.5 or higher is suggested whereas the pH of blood should not exceed 7.55. Renal salicylate clearance is stimulated about 20-fold when the urinary pH increases from 6.1 to 8.1 [49], indicating that renal clearance of salicylates depends much more on urinary pH than on the renal flow rate [20]. An additional approach to stimulate salicylate clearance is conversion into salicyluric acid by substitution of glycine [21].

Severe acute poisoning, that is, plasma salicylate levels above 1200 or 1000 μg/ml 6 h after ingestion, refractory acidosis or other symptoms of severe intoxication (Table 3.1), volume overload, and renal failure are indications for hemodialysis. In chronic overdose, hemodialysis may be considered in symptomatic patients with serum salicylate levels above 600 μg/ml [1]. Hemodialysis has been shown to reduce both morbidity and mortality of salicylate poisoning [50]. An actual flowchart on algorithms for the treatment of acute salicylate poisoning was published by Dargan *et al.* [20] (Figure 3.1). Further details and dose recommendations can be found in the original publication.

Further Measures Hyperthermia and dehydration require immediate treatment, that is, cooling and fluid uptake. Ketoacidosis and hypoglycemia additionally require the administration of glucose

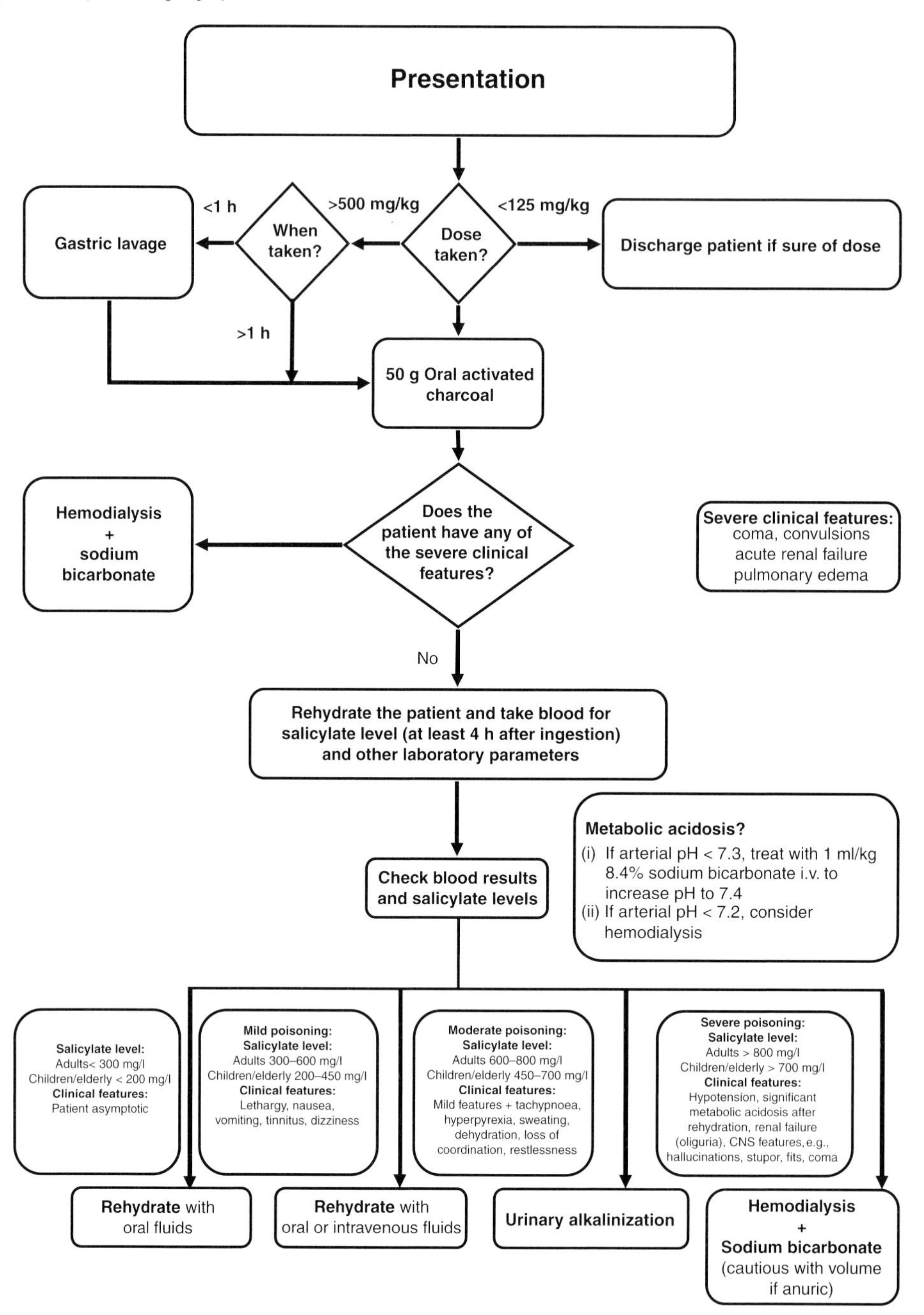

Figure 3.1 Flow chart with algorithms for the treatment of acute aspirin (salicylate) poisoning after [20]. For more details and practical recommendations, see the full publication of Dargan *et al.* [20].

(see [15] for details). Administration of dextrose may help to avoid low cerebrospinal glucose level [1]. Pulmonary edema usually resolves quickly with standard supportive therapy, though it might also be lethal (see above) [29].

3.1.1.3 Habituation

The many tons of aspirin consumed every year worldwide have occasionally led to the opinion that the drug is habit forming. However, it is generally accepted that antipyretic analgesics, such as aspirin or acetaminophen in contrast to morphine-type analgesics, do not cause physical dependence. This is also confirmed by the scarcity of reports on "addiction" or "habituation" to salicylates. There might be some psychological desire for drug intake, for example, regular use for pain relief, but only to the extent that frequent use of any substances that gives relief, real or imaginary, from pain, is a habit [51].

In contrast, there might be abuse of aspirin if used in (fixed) combinations with other constituents of analgesic mixtures, most notably, caffeine and codeine. A separate issue is the few reports on abuse of aspirin at high doses when toxic effects, such as salicylism with exaltation and deafness, were used for "therapeutic" purposes.

A 59-year-old man took about 100 tablets of aspirin within 2 weeks for "encouragement." A 30-year-old man with eplilepsy and alcoholic problems took 20–30 tablets of aspirin within 1 h for the same purpose and a 58-year-old female with alcoholic problems took up to 100 aspirin tablets against crapulousness and also because she was unable to tolerate the noise at her working place (after [52]).

Taken together, there is no evidence that aspirin as a single preparation has a habit-forming potential.

Summary

Acute life-threatening salicylate intoxication in adults occurs at doses of about 12–15 g and above and 3 g and above in children. This is equivalent to plasma levels of $\geq$300 μg/ml or $\geq$2 mM. Initial symptoms are nausea and vomiting, tinnitus, and tachypnea with respiratory alkalosis and central excitation, eventually resulting in combined respiratory/metabolic acidosis. At severe intoxications, there are increasing central nervous dysfunctions (hallucinations, stupor, and coma), renal failure, and, finally, death from respiratory arrest. All of these symptoms are caused by salicylate accumulation in organs and tissues, most notably, in the central nervous system, and probably due to salicylate actions on cellular metabolism. Despite large interindividual variability, the salicylate plasma levels correlate well with clinical symptoms.

Treatment of salicylate poisoning is symptomatic. Reduced absorption by repeated administration of activated charcoal is an effective measure in early stages of intoxication, that is, as long as absorption from the GI tract is not completed. With ingestion of enteric-coated aspirin, this interval is longer than that with plain compounds. Renal salicylate excretion can be considerably enhanced by correction of acidosis and appropriate alkalinization of urine by sodium bicarbonate. The final outcome is largely determined by an early diagnosis, that is, beginning of the treatment. Under optimum conditions, mortality of severe intoxications amounts to $\leq$5% but may increase to 15–20% if the beginning of treatment is delayed.

There is no evidence for addiction or habituation with salicylates, even at long-term use. The risk of persistent injuries of liver or kidney function, the main sites of salicylate metabolism and excretion, respectively, is very small and clinically only relevant at preexisting diseases, for example, high-dose long-term treatment of inflammatory disorders with an altered immune status, such as rheumatoid arthritis. However, this is currently no indication of aspirin use.

References

1 Mokhlesi, B., Leikin, J.B., Murray, P. *et al.* (2003) Adult toxicity in critical care. Part II. Specific poisonings. *Chest*, **123**, 897–922.

2 Mills, J.A. (1991) Aspirin, the ageless remedy? *The New England Journal of Medicine*, **325**, 1303–1304.

3 Krause, D.S., Wolf, B.A. and Shaw, L.M. (1992) Acute aspirin overdose: mechanisms of toxicity. *Therapeutic Drug Monitoring*, **14**, 441–451.

4 Hawton, K., Ware, C., Mistry, H. *et al.* (1995) Why patients choose paracetamol for self poisoning and their knowledge of its dangers. *British Medical Journal*, **310**, 164.

5 Bekemeier, H. (1960) Über die toxikologische Bedeutung der Salizylate. *Zeitschrift für ärztliche Fortbildung*, **15**, 859–867.

6 Bekemeier, H. (1962) Salicylamid- und Salicylsäurevergiftung bei der Katze im Vergleich zu anderen Tieren. 2. Mitteilung. *Archives Internationales de Pharmacodynamie et de Therapie*, **137**, 212–217.

7 Watson, J.-E. and Tagupa, E.-T. (1994) Suicide attempt by means of aspirin enema. *Annals of Pharmacotherapy*, **28**, 467–469.

8 Wollersen, H., Preuss, J., Thierauf, A. *et al.* (2007) Suicide with acetylsalicylic acid [in German]. *Archives Kriminology*, **219**, 115–123.

9 Brubacher, J.R. and Hoffman, J.S. (1996) Salicylism from topical salicylates: review of the literature. *Journal of Toxicology – Clinical Toxicology*, **34**, 431–436.

10 Chan, T.Y. (1996a) Potential dangers from topical preparations containing methyl salicylate. *Human & Experimental Toxicology*, **15**, 747–750.

11 Chan, T.Y. (1996b) The risk of severe salicylate poisoning following the ingestion of topical medicaments or aspirin. *Postgraduate Medical Journal*. **72**, 109–112.

12 Temple, A.R. (1981) Acute and chronic effects of aspirin toxicity and their treatment. *Archives of Internal Medicine*, **1451**, 364–369.

13 Gross, M. and Greenberg, L.A. (1948a) Salicylate poisoning, in *The Salicylates. A Critical Bibliographic Review*, Hillhouse Press, New Haven, CT, pp. 152–190.

14 Hill, J.B. (1973) Salicylate intoxication. *The New England Journal of Medicine*, **288**, 1110–1113.

15 Insel, P.A. (1996) Analgesic-antipyretics and antiinflammatory agents and drugs employed in the treatment of gout, in *Goodman & Gilman's The Pharmacological Basis of Therapeutics*, 9th edn (eds J.G. Hardman, L.E. Limbird, P.B. Molinoff, R.W. Ruddon and A.G. Gilman), McGraw-Hill, New York, pp. 617–657.

16 Starko, K.M. and Mullick, F.G. (1983) Hepatic and cerebral pathology findings in children with fatal salicylate intoxication: further evidence for a causal relation between salicylate and Reye's syndrome. *The Lancet*, **1**, 326–329.

17 Done, A.K. (1960) Salicylate intoxication: significance of measurements of salicylate in blood in cases of acute ingestion. *Pediatrics*, **26**, 800–807.

18 Thisted, B., Krantz, T., Strom, J. *et al.* (1987) Acute salicylate self-poisoning in 177 consecutive patients treated in ICU. *Acta Anaesthesiologica Scandinavica*, **31**, 312–316.

19 McGuigan, M.A. (1987) A two-year review of salicylate deaths in Ontario. *Archives of Internal Medicine*, **147**, 510–512.

20 Dargan, P.I., Wallace, C.I. and Jones, A.L. (2002) An evidence based flowchart to guide the management of acute salicylate (aspirin) overdose. *Emergency Medicine Journal*, **19**, 206–209.

21 Patel, D.K., Ogunbona, A., Notarianni, L.J. *et al.* (1990) Depletion of plasma glycine and effect of glycine by mouth on salicylate metabolism during aspirin overdose. *Human & Experimental Toxicology*, **9**, 389–395.

22 Tenney, S.M. and Miller, R.M. (1965) The respiratory and circulatory actions of salicylate. *The American Journal of Medicine*, **198**, 498–508.

23 Cameron, I.R. and Semple, S.J.R. (1968) The central respiratory stimulation action of salicylates. *Clinical Science*, **35**, 391–401.

24 Oliven, A., Pilar, G. and Bassan, H. (1990) Improvement in sleep apnea by salicylate-induced hyperventilation. *The American Review of Respiratory Disease*, **141**, A194.

25 Pei, Y.P. and Thompson, D.A. (1987) Severe salicylate intoxication mimicking septic shock. *The American Journal of Medicine*, **82**, 381–382.

26 Anderson, R.J., Potts, D.E., Grabow, P.A. *et al.* (1976) Unrecognized adult salicylate intoxication. *Annals of Internal Medicine*, **85**, 745–748.

27 Hormaechea, E., Carlson, R.W., Rogove, H.,*et al.* (1979) Hypovolemia, pulmonary edema and protein changes

in severe salicylate poisoning. *The American Journal of Medicine*, **66**, 1046–1050.

28 Heffner, J.E. and Sahn, S.A. (1981) Salicylate-induced pulmonary edema. *Annals of Internal Medicine*, **95**, 405–409.

29 Reed, C.R. and Glauser, F.L. (1991) Drug induced noncardiogenic pulmonary edema. *Chest*, **100**, 1120–1124.

30 Smith, M.J.H. and Dawkins, P.D. (1971) Salicylate and enzymes. *The Journal of Pharmacy and Pharmacology*, **23**, 729–744.

31 Cotton, E. and Fahlberg, V. (1964) Hypoglycemia with salicylate poisoning. *American Journal of Diseases of Children*, **108**, 171–173.

32 Schadt, D.C. and Purnell, D.C. (1958) Salicylate intoxication in an adult. *Archives of Internal Medicine*, **102**, 213–216.

33 Partin, J.S., Daugherty, C.C., McAdams, A.J. *et al.* (1984) A comparison of liver ultrastructure in salicylate intoxication and Reye's syndrome. *Hepatology*, **4**, 687–690.

34 Schrör, K. (2007) Aspirin and Reye syndrome – a review of the evidence. *Pediatrics Drugs*, **9**, 195–204.

35 Segar, W.E. and Holliday, M.A. (1958) Physiologic abnormalities of salicylate intoxication. *The New England Journal of Medicine*, **259**, 1191–1198.

36 Gabow, P.A., Anderson, R.J., Potts, D.E. *et al.* (1978) Acid–base disturbances in the salicylate-intoxicated adult. *Archives of Internal Medicine*, **138**, 1481–1484.

37 Zimmermann, H.J. (1981) Effects of aspirin and acetaminophen on the liver. *Archives of Internal Medicine*, **141**, 333–342.

38 Chyka, P.A., Erdman, A.R., Christianson, G. *et al.* (2007) Salicylate poisoning: an evidence-based consensus guideline for out-of-hospital management. *Clinical Toxicology*, **45**, 95–131.

39 O'Malley, G.F. (2007) Emergency department management of the salicylate-poisoned patient. *Emergency Medicine Clinics of North America*, **25**, 333–346.

40 Juurlink, D.N. and McGuigan, M.A. (2000) Gastrointestinal decontamination for enteric-coated aspirin overdose: what to do depends on whom you ask. *Journal of Toxicology – Clinical Toxicology*, **38**, 465–470.

41 Pierce, R.P., Gazewood, J. and Blake, R.L. (1991) Salicylate poisoning from enteric-coated aspirin: delayed absorption may complicate management. *Postgraduate Medicine*, **89**, 61–64.

42 Danel, V., Henry, J.A. and Glucksman, E. (1988) Activated charcoal, emesis, and gastric lavage in aspirin overdose. *British Medical Journal*, **296**, 1507.

43 Barone, J.A., Raia, J.J. and Huang, Y.C. (1988) Evaluation of the effects of multiple-dose activated charcoal on the absorption of orally administered salicylate in a simulated toxic ingestion model. *Annals of Emergency Medicine*, **17**, 34–37.

44 Yip, L., Dart, R.C. and Gabow, P.A. (1994) Concepts and controversies in salicylate toxicity. *Emergency Medicine Clinics of North America*, **12**, 351–364.

45 Wortzman, D.J. and Grunfeld, A. (1987) Delayed absorption following enteric-coated aspirin overdose. *Annals of Emergency Medicine*, **16**, 434–436.

46 Kirshenbaum, L.A., Mathews, S.C., Sitar, D.S. *et al.* (1990) Does multiple-dose charcoal therapy enhance salicylate excretion? *Archives of Internal Medicine*, **150**, 1281–1283.

47 Mayer, A.L., Sitár, D.S. and Tenenbein, M. (1992) Multiple-dose charcoal and whole bowel irrigation do not increase clearance of absorbed salicylate. *Archives of Internal Medicine*, **152**, 393–396.

48 Ho, J.L., Tierney, M.G. and Dickinson, G.E. (1989) An evaluation of the effect of repeated doses of oral activated charcoal on salicylate elimination. *Journal of Clinical Pharmacology*, **29**, 366–369.

49 Prescott, L.F., Balali-Mood, M., Critchley, J.A. *et al.* (1982) Diuresis or urinary alkalinisation for salicylate poisoning? *British Medical Journal*, **285**, 1383–1386.

50 Chapman, B.J. and Proudfoot, A.T. (1989) Adult salicylate poisoning: deaths and outcome in patients with high plasma salicylate concentrations. *Quarterly Journal of Medicine*, **72**, 699–707.

51 Gross, M. and Greenberg, L.A. (1948b) The question of addiction or habituation, in *The Salicylates. A Critical Bibliographic Review*, Hillhouse Press, New Haven, CT, pp. 191–193.

52 Bressel, R. (1973) Zur Toxikologie der Salizylsäurederivate. Inauguraldissertation, Erlangen-Nürnberg.

3.1.2
Bleeding Disorders

It is known for more than half a century that aspirin at therapeutic doses can cause bleeding problems (Section 1.1.6). A bleeding risk caused by aspirin is, for the most part, an issue of concern. However, these concerns might not be justified in all cases because the incidence of bleeding during single or short-term aspirin treatment, if any, is very low. In long-term use, aspirin-related bleeding risks have to be balanced against the risk to facilitate atherothrombotic events after aspirin withdrawal, for example, in patients who use aspirin for cardiocoronary prevention and have to undergo a surgical intervention. Interestingly, abnormal bleeding was seen only in a minority (38%) of patients hospitalized because of severe life-threatening salicylate poisoning [53]. In real life, aspirin-induced bleeding is of low clinical significance [54] and was even considered to be one of the "myths" of minor surgical interventions such as tooth extractions [55].

3.1.2.1 Prolongation of Bleeding Time by Aspirin

Dose Dependency *John Quick* [56] published one of the first mechanistic studies on the effect of salicylates on bleeding time. He found that single-dose aspirin in healthy subjects caused prolonged bleeding only when given at high doses (1300 mg) but not at medium doses (650 mg). Only aspirin prolonged bleeding but not salicylate, and there was a large interindividual variability (Figure 3.2). There is only one study demonstrating inhibition of platelet aggregation by salicylate *in vitro*. However, the salicylate concentrations (3 mM) were too high to be clinically relevant [57]. These data suggest an aspirin-specific acetylation of proteins of interest, including clotting factors in the hemostatic system. The dose dependency of this effect made it unlikely that inhibition of platelet-dependent thromboxane formation was the explanation – this inhibition should be complete at 650 mg (Section 2.3.1).

A dose-dependent (80–1300 mg) increase in bleeding time after single-dose aspirin was confirmed by later investigators [58, 59]. These authors also confirmed the observation of Quick that the bleeding time was highly variable and was not prolonged in all individuals, only about 60% being responders. This variability was seen not only in healthy volunteers but also in patients at advanced stages of atherosclerosis. Conversely, 30% of patients with a history of aspirin-related gastrointes-

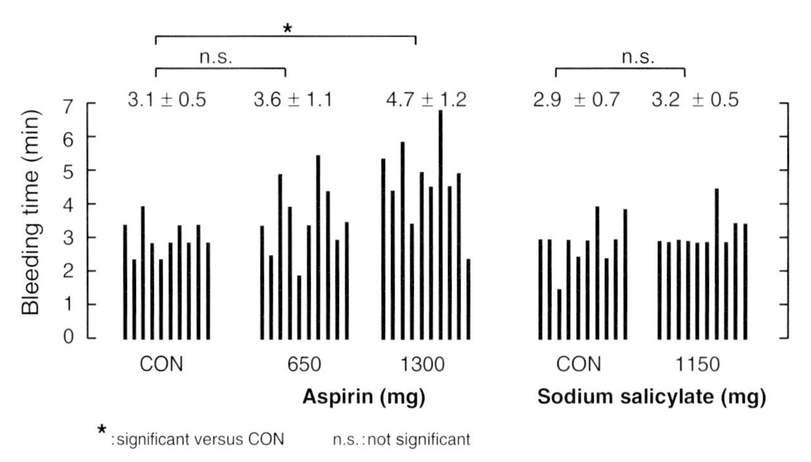

Figure 3.2 Dose-dependent increase in bleeding time in 10 healthy subjects after single oral ingestion of 650 or 1300 mg aspirin. No prolonged bleeding time after ingestion of sodium salicylate. No changes in clotting time, prothrombin time, clot retraction time, and prothrombin consumption. The numbers above the columns are means ± SD (modified after [56]).

tinal bleeding had an exaggerated prolongation of bleeding time in response to single-dose (375 mg) aspirin compared with 10% of controls [60]. However, an absence of prolongation of bleeding does not indicate an absence of risk of bleeding. Clinically relevant bleedings, for example, during surgical interventions or ulcer bleeding in the GI tract may not necessarily be correlated with a prolonged bleeding time [61–63].

Bleeding Time and Platelet Function The maximum prolongation of bleeding time after single-dose aspirin is seen at about 2–3 h. Depending on the method used, the absolute time of prolongation differs but usually is about twice the absolute time before aspirin [58, 64]. This time frame is similar to aspirin-induced inhibition of platelet function (Section 2.3.1). Thus, the first explanation for prolonged bleeding by aspirin after tissue injury was inhibition of platelet function. However, the mechanisms involved in blood–vessel wall interaction subsequent to vascular lesions are more complex than inhibition of platelet function [65]. Consequently, the prolongation of bleeding time and antithrombotic actions of aspirin are no parallel events [66], and recovery of bleeding time and platelet function have a different time course [67]. This is consistent with the hypothesis that inhibition of platelet function by aspirin is only *one* factor for aspirin-induced prolongation of bleeding.

The amounts of thromboxane produced in bleeding-time blood are considerably lower than those produced by the same volume of blood during clotting *in vitro*. This is not surprising since clotting in a glass vial *in vitro* is a highly artificial procedure and determines the thromboxane-forming capacity of platelets when time (h) is no limiting factor. This is different from the *in vivo* situation where thrombus formation starts within seconds and is complete within minutes. At this time, only about 5% of the thromboxane-forming capacity are generated. Thus, the capacity of clotting blood to form thromboxane *in vitro* appears not to be a suitable parameter to describe *in vivo* hemostasis and its alteration by aspirin treatment [68].

Determinants of Aspirin-Induced Prolongation of Bleeding Time The aspirin-induced abnormality in platelet function does not directly translate into disturbed hemostasis of injured capillaries. Factors that might be involved in addition to inhibition of platelet function are coagulation factors, changes in capillary permeability, and an altered endothelial function. Considering this variety in variables, it is not surprising that even standardized measurements of bleeding time under well-defined conditions show a high interindividual variability and until now, there has been no generally accepted definition of normal values for bleeding time in man.

Mechanistically, aspirin could prolong bleeding by acetylation of target proteins. These transacetylations are nonspecific and not restricted to cyclooxygenases. Acetylation of endothelial NO synthase (eNOS), followed by an increased formation of nitric oxide (NO) [69], a vasodilator and inhibitor of platelet function, could also be involved. In addition, aspirin acetylates prothrombin, fibrinogen, and other zymogens of clotting factors as well as proteins of the platelet membrane. It has been shown that aspirin at a single dose of 500 mg significantly retards thrombin formation *ex vivo*. This effect is not shared by indomethacin (IND), ticlopidine (TIC), or a thromboxane synthase inhibitor (TSI), though all compounds caused a marked prolongation of bleeding time. Thus, the mechanisms of prolongation of bleeding time by aspirin are probably different from inhibition of thromboxane formation and possibly also different from that of other antiplatelet agents [70] (Figure 3.3).

The sequence of events in tissue factor-dependent activation of the clotting system and its modifications by aspirin were studied *ex vivo* in capillary blood of healthy volunteers subsequent to vessel injury by skin incision. Measurements were performed 1 week before and after daily intake of 75 mg aspirin. Blood was collected in 30 s intervals. Activation of clotting factors was determined by quantitative immunoassays.

Vascular injury was followed by an immediate, continuous fall of prothrombin levels, approaching less than 10% of the initial value at the end of bleeding, that is,

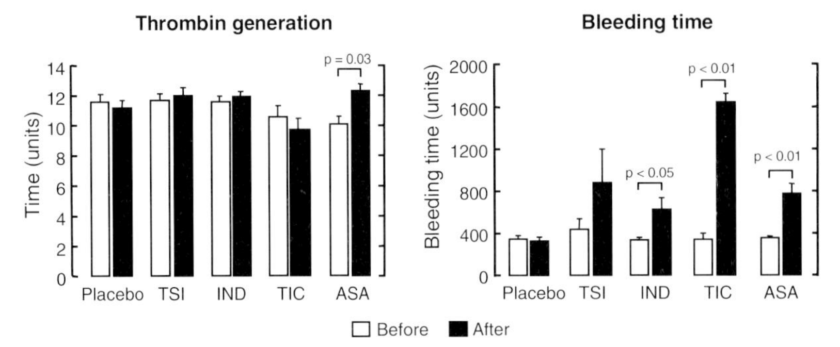

Figure 3.3 Thrombinogenesis *in vitro* (time to peak thrombin formation) and skin bleeding time in healthy volunteers before and after treatment with placebo, ticlopidine, thromboxane synthase inhibitor, indomethacin or aspirin (ASA, 500 mg). At the doses selected, all substances prolonged the bleeding time but only aspirin retarded thrombin formation (modified after [70]).

thrombus formation. This prothrombin was converted to thrombin, reaching peak values of 38 nM (!). These amounts of thrombin were much higher than those that were required for maximum platelet activation. Fibrinogen level fell and became undetectable at about 3 min of bleeding, indicating maximum fibrin formation. (Thrombin-induced) activation of clotting factor V to factor Va (amplification mechanism for thrombin formation) was detected after thrombin generation had started and was later followed by the inactivation of FVa by activated protein C. This indicates thrombin-induced stimulation of anticoagulant factors.

Aspirin treatment markedly reduced all of these activation markers, on average by about 30%. This was associated with a significant downregulation of the clotting process.

It was concluded that aspirin at antiplatelet doses not only inhibits thromboxane formation but also impairs thrombin generation and all follow-up reactions catalyzed by thrombin at the site of tissue injury. This also involves thrombin-induced vasoconstriction by contraction of vascular smooth muscle cells after endothelial defects [71].

These findings suggest that activation of the tissue-factor-induced "extrinsic" pathway of blood coagulation is partially antagonized by aspirin and that this action of aspirin finally causes inhibition of thrombin formation. (Activated) platelets contribute to thrombus formation, that is, cessation of bleeding, by providing a matrix for activation of clotting factors, such as factor Va. The exact role of thromboxane for these processes beyond stimulating platelet recruitment and, in particular, its "threshold" level for activation of the clotting processes is unknown.

3.1.2.2 Aspirin-Related Bleeding Risk in Surgical Interventions

Because of a possible bleeding risk, it is generally recommended to stop the intake of aspirin (or other antiplatelet agents), such as clopidogrel, prior to a surgical intervention. However, inhibition of platelet function is the therapeutic goal for the clinical use of aspirin in the prevention of atherothrombotic events. Problems may arise if patients at increased atherothrombotic risk have to undergo surgical interventions. It is a clinically highly relevant question whether withdrawal of aspirin in these patients, specifically in disease-related operations such as coronary artery bypass surgery or carotid endarterectomy, improves the clinical outcome by reducing perioperative blood loss or increases the risk of atherothrombotic events [72] (Section 4.1.1). A similar though less dramatic issue is the question whether minor surgical interventions, such as tooth extractions, can be performed in aspirin-treated patients without the risk of major bleeding.

Aspirin and Bleeding Risk After Minor Surgical Interventions Because of the risk of uncontrolled periprocedural bleeding, patients receiving long-term aspirin treatment are frequently asked to discontinue the use of the drug for 5–7 days before surgery. However, no double-blind randomized studies support this practice in oral surgery. Aspirin use at antiplatelet doses (100 mg/day) did cause significant, though minor, increases in bleeding time after oral surgery. However, local hemostatis is sufficient to control bleeding [73]. Similar results were obtained on patients at the same aspirin dosage when no increase in secondary bleedings was found after tooth extraction [74]. The conclusion of these studies and one earlier critical review on this issue [55] were that aspirin, taken at antiplatelet doses for atherothrombotic prophylaxis, needs not be discontinued for minor dentoalveolar surgery.

Aspirin and Atherothrombotic Risk After Major Surgical Interventions Ischemic events (stroke, (re) infarctions) were repeatedly described in patients with previous myocardial infarction [75] or stroke (Section 4.1.2), when aspirin prophylaxis was discontinued because of an elective surgical intervention. A recently published overview of retrospective trials came to the conclusion that discontinuation of aspirin in these high-risk patients may cause acute thromboembolic vessel occlusion in up to 10% of patients. This suggests a considerable clinical benefit of continuation of aspirin prophylaxis, in particular, in larger surgical interventions that are associated with a significant activation of the clotting system [76]. Importantly, atherothrombotic events usually do not occur immediately but only some days after the surgical intervention and may not be seen by the surgeon [77].

The impact of aspirin use on perioperative blood loss in cardiovascular risk patients was studied prospectively in a series of 317 consecutive patients undergoing elective reoperative coronary artery bypass surgery with extracorporeal circulation. A total of 215 patients received aspirin or aspirin-containing medications within the 7 preoperative days. These patients were compared with 102 cases who had not taken aspirin within this time interval (control). Neither autotransfusions of mediastinal blood or platelet rich plasma nor treatment with aprotinin or desmopressin was used in either group.

No significant differences were observed between aspirin-treated patients and controls with respect to postoperative hematocrit, mediastinal drainage, or blood substitution. There was also no significant interaction with preoperative heparin therapy.

The conclusion was that preoperative aspirin is not an important determinant of mediastinal drainage or allogeneic transfusions, even in repeated bypass surgery. Surgical and patient characteristics are more important predictors of clinical outcome [78].

A 50% increased risk of periprocedural blood loss is to be expected if aspirin treatment is not discontinued prior to surgery. However, this blood loss is usually moderate and is not associated with an increase in the severity of bleeding complications or even bleeding-related death [77–79]. With the exception of intracranial surgery and transurethral prostatectomy – fetal aspirin bleedings have been described for both – the decision to discontinue aspirin before surgical interventions has to consider the markedly elevated risk of atherothrombotic vessel occlusions, in particular, in cardiovascular risk patients. This risk is further enhanced by the proinflammatory and prothrombotic conditions of major surgeries [72] (Section 4.1.1). Platelet function tests prior to surgery may be valuable for making decisions [80]; however, they are not standardized and there are no generally accepted "normal" values.

An interesting analysis on the effect of perioperative aspirin on the clinical outcome was published by Neilipovitz *et al.* [81].

A statistical model based on 138 study protocols indicated that continued aspirin use reduced the total perioperative mortality rates from 2.8 to 2.0% and the number of myocardial infarctions from 4.6 to 2.7%. This was associated with a slight increase in life expectancy. However, aspirin increased the number of hemorrhagic complications by 2.5%, primarily because of an increased risk of non-life-threatening bleedings.

It was concluded that the beneficial reductions in mortality from thrombotic problems by aspirin account for the increase in life expectancy and overweigh the increased bleeding tendency. In the absence of randomized prospective trials, this raises doubts whether the practice of routinely stopping aspirin before surgery and during the early postoperative period is justified [81].

A recent review on the effect of preoperative aspirin on the bleeding risk of patients undergoing coronary artery bypass surgery came to the conclusion that bleeding risk exists but may be largely avoided by the use of aspirin at low doses (<325 mg) [82].

3.1.2.3 Aspirin, Other Drugs, and Alcohol

Aspirin and NSAIDs In contrast to aspirin, the increased bleeding tendency with reversible COX inhibitors, such as traditional nonsteroidal anti-inflammatory drugs (NSAIDs) [83] can be prevented if the compounds are withdrawn sufficiently early before surgery. Five half lives are recommended [84]. For most of the compounds (with the exception of naproxen), this is equivalent to about 1 day. Since these drugs have not been shown to protect from ischemic events in atherothrombosis, considerations regarding the loss of putative beneficial effects are unnecessary.

Aspirin and Alcohol Ingestion of aspirin together with moderate amounts of alcohol (50 g) causes a significant prolongation of bleeding time. Similar effects were seen with indomethacin and ibuprofen but not with choline salicylate [83]. An explanation is the inhibition of gastric mucosal alcohol dehydrogenase. This effect is seen only in men and not in women, possibly because of the low or absent first-pass metabolism of alcohol in the female stomach (Section 2.1.1). Thus, alcohol will enhance the effect of aspirin on hemostasis and probably also contribute to the beneficial effects of low-to-moderate drinking on prevention of myocardial infarctions – with and without aspirin (Section 4.1.1).

3.1.2.4 Prevention and Treatment of Bleedings

Aspirin-related bleedings with few but important exceptions, including GI bleedings (Section 3.2.1) and hemorrhagic cerebral infarctions (Section 4.2.1), are not life threatening and usually do not require particular therapeutic measures. Any increased bleeding tendency, however, is inconvenient to the consumer (nose bleed, gingival, and shaving bleeding) and might result in a poor compliance in long-term use. Bleeding may also become a clinical problem in case of nonelective or unexpected surgery in otherwise healthy individuals, including accidents or urgent operations, for example, appendicitis. The clinical problem of bleeding in aspirinized patients is aggravated by the absence of a specific antidote and the irreversibility of inhibition of platelet function. Two drugs have been used as functional antagonists of aspirin-induced bleeding: desmopressin (1-deamino-8-D-arginine-vasopressin) (DDAVP) [85] and aprotinin [86].

DDAVP DDAVP improves platelet adhesion to the vessel wall by generation of abnormally large factor VIII/von Willebrand factor multimers. These multimers bind platelets are particularly effective for subendothelial collagen. The action is specific for platelets but independent of the kind of antiplatelet agent and has been shown to be effective also after treatment with (platelet-selective) GPIIb/IIIa antagonists [87]. DDAVP is well tolerated and the treatment of choice for von Willebrand's disease and mild hemophilia [88]. The effect on bleeding time is maximum 1–2 h after i.v. administration (0.3 μg/kg) and lasts for about 4 h [89, 90].

DDAVP has also been found, however, to stimulate platelet function and to promote thrombus growth [91] without changing platelet-dependent thromboxane formation [85, 92]. This is clearly not a desired activity in surgical interventions, particularly in patients undergoing cardiovascular surgery (Section 4.1.1). Thus, the use of DDAVP appears to be of limited value in patients at elevated atherothrombotic risk who need continuous aspirin treatment as a result of thrombosis prophylaxis.

Aprotinin Aprotinin is a nonselective inhibitor of serine proteases, including procoagulatory (factor XIIa, thrombin) and fibrinolytic (plasmin) enzymes. Aprotinin reduces bleeding after surgery at maintained platelet hemostatic function and inhibition of thromboxane formation by aspirin. Several well-controlled prospective double-blind studies have shown that aprotinin reduces blood loss in cardiac surgery [93–96].

Aprotinin has been withdrawn from the market by the producer in November 2007 because of the pending final results from the Canadian Blood Conservation Using Antifibrinolytics Trial (BART). At that time, the available data suggested an increased mortality in the aprotinin arm that "almost reached conventional statistical significance." This prompted the data and safety monitoring board to stop the trial. Nevertheless, there are still a number of unsolved issues regarding the risk/benefit ratio of aprotinin [97] and a recent though nonrandomized clinical trial in patients undergoing coronary artery bypass grafting found aprotinin both efficacious and safe [98].

Taken together, the available evidence suggests that aspirin-associated bleeding during surgical interventions in most indications is not a critical issue and there is no increased mortality but rather an improved clinical outcome in patients at elevated cardiovascular risk. If necessary, DDAVP is available though not free from an increased thrombotic risk. Platelet infusions are not the treatment of choice in high-risk patients and may even increase the risk for adverse outcome [99].

Summary

A prolongation of bleeding time, about twofold, is frequently seen after (regular) aspirin intake. This action is dose dependent and the mechanism multifactorial, involving both vascular and circulating blood components of primary hemostasis. There is a considerable interindividual variability.

Aspirin-induced bleeding cannot be solely explained by inhibition of thromboxane formation and does not correlate with the antithrombotic efficacy of the compound in cardiocoronary prophylaxis. Mechanistically, acetylation of enzymes in addition to cyclooxygenases might be involved because salicylate does not show comparable effects. Possible candidates are inhibition of thrombin formation (acetylation of prothrombin?) after stimulation of tissue-factor-dependent coagulation and enhanced endothelial NO formation (acetylation of eNOS) among others.

Increased aspirin-related bleedings are particularly relevant in surgery. Preoperative aspirin within 5–7 days prior to the intervention increases the risk of periprocedural bleedings by about 50%. Available data from retrospective trials suggest that this is not a critical issue if the aspirin doses are low (<325 mg) and there is no increased mortality. Any withdrawal of aspirin prior to surgery has also to be balanced against an enhanced thrombotic risk due to the proinflammatory and prothrombotic conditions of surgical interventions. This is particularly relevant to patients at elevated cardiovascular risk (Section 4.1.1).

There is no specific antidote to antagonize the aspirin-induced bleeding tendency. Therefore, treatment or prevention of aspirin-induced bleedings is symptomatic, if necessary at all. Aprotinin used to be the drug of choice to inhibit excessive bleedings but is no longer available for this indication because of a thrombotic tendency. DDAVP is an alternative, though not free from increased thrombotic risk. There is no place for platelet infusions.

References

53 Thisted, B., Krantz, T., Strom, J. *et al.* (1987) Acute salicylate self-poisoning in 177 patients treated in ICU. *Acta Anaesthesiologica Scandinavica*, **31**, 312–316.

54 Schafer, A.I. (1995) Effects of nonsteroidal antiinflammatory drugs on platelet function and systemic hemostasis. *Journal of Clinical Pharmacology*, **35**, 209–219.

55 Alexander, R.E. (1998) Eleven myths of dentoalveolar surgery. *Journal of the American Dental Association*, **129**, 1271–1279.

56 Quick, A.J. (1966) Salicylates and bleeding: the aspirin tolerance test. *The American Journal of the Medical Sciences*, **252**, 265–269.

57 Davies, D.T.P., Hughes, A. and Tonks, R.S. (1969) The influence of salicylate on platelets and whole blood adenine nucleotides. *British Journal of Pharmacology*, **36**, 437–447.

58 Dybdahl, J.H., Daae, L.N.W., Eika, C. *et al.* (1981) Acetylsalicylic acid-induced prolongation of bleeding time in healthy men. *Scandinavian Journal of Haematology*, **26**, 50–56.

59 Buchanan, M.R. and Brister, S.J. (1995) Individual variation in the effects of ASA on platelet function: implications for the use of aspirin clinically. *The Canadian Journal of Cardiology*, **11**, 221–227.

60 Lanas, A.I., Arroyo, M.T., Esteva, F. *et al.* (1996) Aspirin related gastrointestinal bleeders have an exaggerated bleeding time response due to aspirin use. *Gut*, **39**, 654–660.

61 O'Laughlin, J.C., Hoftiezer, J.W., Mahoney, J.P. *et al.* (1981) Does aspirin prolong bleeding from gastric biopsies in man? *Gastrointestinal Endoscopy*, **27**, 1–5.

62 Rodgers, R.P.C. and Levin, J. (1990) A critical reappraisal of the bleeding time. *Seminars in Thrombosis and Hemostasis*, **16**, 1–20.

63 Lind, S.E. (1991) The bleeding time does not predict surgical bleeding. *Blood*, **77**, 2547–2552.

64 Mielke, C.H., Jr, Kaneshiro, M.M., Maher, I.A. *et al.* (1969) The standardized normal bleeding time and its prolongation by aspirin. *Blood*, **34**, 204–215.

65 Preston, F.E., Whipps, S., Jackson, C.A. *et al.* (1981) Inhibition of prostacyclin and platelet thromboxane A_2 after low-dose aspirin. *The New England Journal of Medicine*, **304**, 76–79.

66 Frith, P.A. and Warlow, C.P. (1988) A study of bleeding time in 120 long-term aspirin trial patients. *Thrombosis Research*, **49**, 463–470.

67 Velo, G.P. and Milanino, R. (1990) Nongastrointestinal adverse reactions to NSAID. *The Journal of Rheumatology*, **17** (Suppl. 20), 42–45.

68 Thorngren, M., Shafi, S. and Born, G.V.R. (1983) Thromboxane A_2 in skin bleeding-time blood and clotted venous blood before and after administration of acetylsalicylic acid. *The Lancet*, **1**, 1075–1078.

69 Taubert, D., Berkels, R., Grosser, N. *et al.* (2004) Aspirin induces nitric oxide release from vascular endothelium: a novel mechanism of action. *British Journal of Pharmacology*, **143**, 159–165.

70 Szczeklik, A., Krzanowski, M., Góra, P. *et al.* (1992) Antiplatelet drugs and generation of thrombin in clotting blood. *Blood*, **80**, 2006–2011.

71 Undas, A., Brummel, K., Musial, J. *et al.* (2001) Blood coagulation at the site of microvascular injury: effects of low-dose aspirin. *Blood*, **98**, 2423–2431.

72 Ferraris, V.A., Ferraris, S.P., Joseph, O. *et al.* (2002) Aspirin and postoperative bleeding after coronary artery bypass grafting. *Annals of Surgery*, **235**, 820–827.

73 Ardekian, L., Gaspar, R., Peled, M. *et al.* (2000) Does low-dose aspirin therapy complicate oral surgical procedures? *JADA*, **131**, 331–335.

74 Hemelik, M., Wahl, G. and Kessler, B. (2006) Zahnextraktionen unter Medikation mit Acetylsalicylsäure (ASS). *Mund, Kiefer und Gesichtschirurgie*, **10**, 3–6.

75 Collet, J.P., Himbet, F. and Steg, P.G. (2000) Myocardial infarction after aspirin cessation in stable coronary artery disease patients. *International Journal of Cardiology*, **76**, 719–748.

76 Merritt, J.C. and Bhatt, D.L. (2004) The efficacy and safety of perioperative antiplatelet therapy. *Journal of Thrombosis and Thrombolysis*, **17**, 21–27.

77 Burger, W., Chemnitius, J.-M., Kneissl, G.D. *et al.* (2005) Low-dose aspirin for secondary prevention – cardiovascular risks after its perioperative withdrawal versus bleeding risks with its continuation – review and meta-analysis. *Journal of Internal Medicine*, **257**, 399–414.

78 Tuman, K.J., Mc Carthy, R.J., O'Connor, C. *et al.* (1996) Aspirin does not increase allogeneic blood transfusion in reoperative coronary artery surgery. *Anesthesia and Analgesia*, **83**, 1178–1184.

79 Reich, D.L., Patel, G.C., Vela-Cantos, F. *et al.* (1994) Aspirin does not increase homologous blood requirements in elective coronary bypass surgery. *Anesthesia and Analgesia*, **79**, 4–8.

80 Bracey, A.W., Grigore, A.M. and Nussmeier, N.A. (2006) Impact of platelet testing on presurgical screening and implications for cardiac and noncardiac surgical procedures. *The American Journal of Cardiology*, **98** (Suppl.), 25N–32N.

81 Neilipovitz, D.T., Bryson, G.L. and Nichol, G. (2001) The effect of perioperative aspirin therapy in peripheral vascular surgery: a decision analysis. *Anesthesia and Analgesia*, **93**, 573–580.

82 Sun, J.C.J., Whitlock, R., Cheng, J. *et al.* (2008) The effect of pre-operative aspirin on bleeding, transfusion, myocardial infarction, and mortality in coronary artery bypass surgery: a systematic review of randomized and observational studies. *European Heart Journal*, **29**, 1057–1071.

83 Deykin, D., Janson, P. and McMahon, L. (1982) Ethanol potentiation of aspirin-induced prolongation of the bleeding time. *The New England Journal of Medicine*, **306**, 852–854.

84 Connelly, C.S. and Panush, R.S. (1991) Should nonsteroidal anti-inflammatory drugs be stopped before elective surgery? *Archives of Internal Medicine*, **151**, 1963–1966.

85 Mannucci, P.M. (1988) Desmopressin: a nontransfusional form of treatment for congenital and acquired bleeding disorders. *Blood*, **72**, 1449–1455.

86 Van Oeveren, W., Eijsman, L., Roozendaal, K.J. *et al.* (1988) Platelet preservation by aprotinin during cardiopulmonary bypass. *The Lancet*, **1**, 644.

87 Reiter, R.A., Mayr, F., Blazicek, H. *et al.* (2003) Desmopressin antagonizes the *in vitro* platelet dysfunction induced by GPIIb/IIIa inhibitors and aspirin. *Blood*, **102**, 4594–4599.

88 Franchini, M. (2007) The use of desmopressin as a hemostatic agent: a concise review. *American Journal of Hematology*, **82**, 731–735.

89 Lethagen, S. and Rugarn, P. (1992) The effect of DDAVP and placebo on platelet function and prolonged bleeding time induced by oral acetyl salicylic acid intake in healthy volunteers. *Thrombosis and Haemostasis*, **67**, 185–186.

90 Flordal, P.A. and Salin, S. (1993) Use of desmopressin to prevent bleeding complications in patients treated with aspirin. *The British Journal of Surgery*, **80**, 723–724.

91 Peter, F.W., Benkovic, C., Muehlberger, T. *et al.* (2002) Effects of desmopressin on thrombogenesis in aspirin-induced platelet dysfunction. *British Journal of Haematology*, **117**, 658–663.

92 Schulman, S. (1991) DDAVP – the multipotent drug in patients with coagulopathies. *Transfusion Medicine Reviews*, **5**, 132–144.

93 Tabuchi, N., Huet, R.C.G., Sturk, A. *et al.* (1994) Aprotinin preserves hemostasis in aspirin-treated patients undergoing cardiopulmonary bypass. *The Annals of Thoracic Surgery*, **58**, 1036–1039.

94 Murkin, J.M., Lux, J. and Shannon, N.A. (1994) Aprotinin significantly decreases bleeding and transfusion requirements in patients receiving aspirin and undergoing cardiac operations. *The Journal of Thoracic and Cardiovascular Surgery*, **107**, 554–561.

95 Bidstrup, B.P., Hunt, B.J., Sheikh, S. *et al.* (2000) Amelioration of the bleeding tendency of preoperative aspirin after aortocoronary bypass grafting. *The Annals of Thoracic Surgery*, **69**, 541–547.

96 Alvarez, J.M., Jackson, L.R., Chatwin, C. *et al.* (2001) Low-dose postoperative aprotinin reduces mediastinal drainage and blood product use in patients undergoing primary coronary artery bypass grafting who are taking aspirin: a prospective, randomized, double-blind, placebo-controlled trial. *The Journal of Thoracic and Cardiovascular Surgery*, **122**, 457–463.

97 McEvoy, M.D., Reeves, S.T., Reves, J.G. *et al.* (2007) Aprotinin in cardiac surgery: a review of conventional and novel mechanisms of action. *Anesthesia and Analgesia*, **105**, 949–962.

98 Bittner, H.B., Lemke, J., Lange, M. *et al.* (2008) The impact of aprotinin on blood loss and blood transfusion in off-pump coronary artery bypass grafting. *The Annals of Thoracic Surgery*, **85**, 1662–1668.

99 Spiess, B.D., Royston, D., Levy, J.H. *et al.* (2004) Platelet transfusions during coronary artery bypass graft surgery are associated with serious adverse outcomes. *Transfusion*, **44**, 1143–1148.

3.1.3 Safety Pharmacology in Particular Life Situations

Data on pharmacokinetic and pharmacodynamic properties of aspirin are usually obtained from healthy middle-aged individual. However, these data may not apply to every single patient or consumer, respectively, and may also vary in different ethnic populations and particular life situations. Pregnancy and older age are two examples for particular life situations in which pharmacological actions of aspirin may differ in a clinically relevant manner from those in "normal" situations and, therefore, require particular attention.

In early pregnancy, there is the possible interference of COX inhibition with fertility and embryogenesis, and in late pregnancy, the transplacental transfer of aspirin into the fetal circulation with possible consequences for fetal blood flow and hemostasis. However, aspirin is also a valuable therapeutic for the prevention of preeclampsia (Section 4.1.5).

A quite different, though not less important, issue is the use of aspirin in older age. Older patients are frequently long-term users of aspirin, for example, because of prophylaxis of ischemic stroke (Section 4.1.2). In the elderly, biotransformations and renal excretion of salicylates may be reduced. This enhances the bioavailability of the active compound(s), that is, salicylate, eventually resulting in too high plasma levels and toxic side effects, mainly tinnitus or other disturbances of the audiovestibular system (Section 3.2.4). Additionally, drug interactions have to be considered because of frequent comorbidities and cotreatments in these patients.

3.1.3.1 Pregnancy and Fetal Development

Low-dose aspirin is generally considered to be relatively safe during early pregnancy [100]. Nevertheless, no drug should be used in this particular life situation without a clear therapeutic need that outweighs any possible risk.

Prostaglandin-Dependent Events of Ovulation Prostaglandins are involved in all stages of pregnancy and fetal development. This starts with ovulation, that is, the release of the fertile oocyte from the ovarian follicle subsequent to rupture of the luteinized follicle membrane. This is followed by ovum transport through the Fallopian tube and nidation in the endometrium of the uterus. Prostaglandins, in particular, those derived from COX-2, appear to be crucial to these processes since COX-2-deficient female mice are infertile and have deficits in all components of early pregnancy.

NSAIDs, Aspirin, and Fertility Normal implantation requires the release of the mature oocyte from the luteinized follicle. If the first step of this process, that is, rupture of the follicular wall, does not occur, the cycle remains unovulatory, eventually resulting in the luteinized unruptured follicle (LUF) syndrome. Mechanistically, rupture of the follicular wall is mediated by proteases that in turn are activated by prostaglandins. Consequently, inhibition of prostaglandin synthesis will prevent follicle rupture.

NSAIDs, such as indomethacin and diclofenac, have been shown to inhibit ovulation and to delay follicle rupture, that is, to induce a luteinized but unruptured follicle. This might be associated with female infertility [101–103]. Delaying or prevention of follicle wall rupture has potential in the form of nonhormonal oral contraception since steroid genesis is not affected by inhibition of prostaglandin synthesis [104]. There are no definite reports that aspirin interferes with these processes in oocyte maturation. This might be because of the short half-life of aspirin or a different pharmacological profile of action, that is, the preferential inhibition of COX-1 and not COX-2, the more important COX isoform involved in these processes. Alternatively, there might be pharmacokinetic reasons, that is, a low lipophilicity as opposed to NSAIDs or coxibs.

Teratogeneity Several animal studies have shown that aspirin may increase the risk of congenital abnormalities when given at high toxic doses. It is, however, questionable, whether these animal data can be transferred to human. No increased risk of malformations was found in a large cohort study, the "Collaborative Perinatal Project," evaluating more than 44 000 women [105]. In 2000 children with congenital heart malformations, there was no increased risk for cardiac malformations if the mothers had taken aspirin in early pregnancy [106]. No increased risk of malformations by aspirin was seen in two randomized trials [107, 108] and confirmed in a large meta-analysis on aspirin consumption in early pregnancy and the risk of congenital abnormalities [109]. The only possible malformation with a relation to aspirin was gastroschisis. However, the significance of this finding is unknown. Overall, fetal exposure to aspirin taken during early pregnancy (first trimester) at therapeutic doses by the mother appears not to be associated with a higher risk of congenital abnormalities in otherwise healthy individuals [110–113].

Risk of Miscarriage Two large population-based cohort studies in Denmark and California addressed the question of a relationship between the use of NSAIDs and the risk of miscarriage. In the Danish study, prenatal use of NSAIDs was associated with an enhanced risk of miscarriage but appeared not to increase the risk of adverse birth outcome. The use of aspirin was not specifically addressed [114]. In the Californian Study, the association of prenatal aspirin with the risk of miscarriage was weaker than that of other NSAIDs and the estimates unstable because of the small number of aspirin users [115].

There are only few randomized trials studying the role of aspirin in miscarriage. These failed to show a relationship between aspirin use and miscarriage [116, 117]. On the opposite, aspirin has even been used to improve the rate of *in vitro* fertilization and to reduce the rate of miscarriages and intrauterine growth retardation in a meta-analysis [118]. This is the reason for the consideration of aspirin (alone or in combination with heparin) for improving fertilization rates in women with recurrent fetal loss (Section 4.1.5). Thus, aspirin appears not to increase the risk of miscarriages in normal pregnancies.

Aspirin and Pregnancies at Increased Risk Similar considerations apply to pregnancies at increased risk, specifically preeclampsia where aspirin is used as a therapeutic agent (Section 4.1.5). According to a recent large meta-analysis, the risk of miscarriages and the rate of perinatal death were not different for aspirin and placebo. There was also no significant difference in perinatal mortality or the rate of "small-for-gestational-age" infants. However, women, who took aspirin, had a significantly lower risk of preterm deliveries, possibly because of inhibition of prostaglandin synthesis (see below) [113]. Taken together, there is no convincing evidence that the use of aspirin – possibly opposed to other NSAIDs – is a particular risk factor for mother or fetus in early pregnancy and appears to be safe also in women with moderate and high-risk pregnancies.

Late Pregnancy Intake of low-dose aspirin (100 mg/day) in the last trimester of pregnancy will neither reduce the placental weight nor retard fetal growth and differentiation [119]. Similar results were obtained in a cohort study of more than 40 000 pregnant women. No relation was found between aspirin intake (high, medium, and low) and the rates of stillbirth, perinatal mortality, and mean birth weights [120]. The interesting hypothesis of a negative correlation between intake of aspirin during pregnancy and the IQ at the age of 4 years, suggested from data in a small study [121], was not confirmed in a large prospective trial [122].

Ductus Arteriosus Botalli Vasodilatory prostaglandins are involved in the low vascular resistance of the fetal circulation [123] and probably also contribute to the low incidence of thrombosis in the placental circulation. In that respect, vasodilatory prostaglandins have also a key role for the maintenance of blood flow through the ductus. Consequently, NSAIDs, such as indomethacin and ibuprofen, are currently used for pharmacological closure of the ductus. The question arises whether aspirin has a similar effect and might also cause functionally relevant reductions in ductal blood flow.

In a placebo-controlled double-blind trial, 43 pregnant women at risk of preeclampsia or intrauterine growth retardation were allocated to 100 mg/day aspirin or placebo. Doppler measurements of the uterine artery, several fetal arteries, and the ductus arteriosus Botalli were performed in 2-week intervals from 18th gestational week until delivery.

No differences between aspirin and placebo were seen in any of the parameters measured.

The conclusion was that low-dose aspirin during the second and third trimesters of pregnancy does not alter uteroplacental or fetoplacental hemodynamics and does not cause detectable constriction of the ductus arteriosus Botalli [124].

Reversible, transient reductions of ductal blood flow have been reported in early studies, usually at much higher doses than used today [125, 126]. According to later studies and other available data, it is not likely that aspirin will cause a functionally relevant constriction of the ductus with consequences for the fetus.

Bleeding Aspirin rapidly passes the placenta and enters the fetal circulation, approaching about 50% of the plasma level at steady-state conditions [127, 128]. Thus, aspirin may interfere with hemostasis not only in the maternal but also in the fetal circulation and the newborn. Regular doses of 100 mg are sufficient for antiplatelet effects in the fetus and neonate [129, 130]. The antiplatelet effect is less pronounced or absent at lower doses of aspirin, such as 60 mg/day [131], which have been frequently used in the treatment of preeclampsia (Section 4.1.5).

Platelets of neonates have abnormalities regarding adhesion, aggregation, and granule secretion in response to standard stimuli *in vitro* [130]. Whether platelets from newborn are also more susceptible to aspirin than those of the mother is unclear [130–132]. There is, however, no doubt of a bleeding risk in the newborn if the mother had taken aspirin shortly, that is, within 1 week prior to labor and this risk might be greater in premature newborns with immature drug clearance systems.

Ingestion of aspirin during the last week of pregnancy by the mother was associated with a remarkable risk of severe bleeding disorders in premature newborns ($\leq$34 weeks of gestation or body weight 1500 g or less) associated with intracranial hemorrhage. Computer tomography showed that 53 out of 108 infants (49%) exhibited intracranial hemorrhage within the first week postpartum. The incidence of hemorrhage in the infants whose mothers had ingested aspirin was significantly greater than that in infants whose mother did not take either aspirin or paracetamol.

The conclusion was that the use of aspirin is associated with increased risk of intracranial hemorrhage in premature newborn. Therefore, aspirin should not be taken within the past 3 months of pregnancy [133].

Despite formal criticisms, mainly regarding the uncertainty about aspirin doses and its duration of use, similar results were also seen in a prospective case–control trial on full-term pregnancies [134]. Thus, intake of aspirin should be avoided during late pregnancy, specifically shortly before delivery.

Maternal Aspirin Intoxication and Consequences for the Fetus Because of the slow metabolic and renal clearance of aspirin in the fetal circulation as opposed to the maternal circulation, effective salicylate levels are maintained in the fetal circulation for a longer period of time [110]. This will result in an increased fetal sensitivity to salicylate overdosing of the mother [135], a factor to be particularly considered in suicide attempts of young females.

A mature female newborn (3.5 kg) was admitted to the hospital at 20 h of age with a history of increasing hyperpnoea since 3 h of age. The mother had taken 15–18 g aspirin 27 h before delivery. Her salicylate levels 20 h before birth were 380 μg/ml. The serum salicylate of the patient on admission was 350 μg/ml whereas the mother's level, drawn at 19 h after delivery, had already declined to 220 μg/ml. Four hours after starting intravenous fluid infusion, serum salicylate in the patient was still 330 μg/ml. An exchange transfusion was performed without incident. The postexchange salicylate level was 220 μg/ml, suggesting that a total of 135 mg salicylate(s) was removed by this procedure.

The data suggest that in the case of moderate intoxication of a newborn, serum levels of salicylate in the fetus 20 h before birth are the same as the mother's and, in the absence of sufficient urine formation and vomiting, remain nearly unchanged during the first 20 h after birth, probably because of renal immaturity [136].

Retardation of Labor In addition to the maintenance of blood flow through the ductus arteriosus Botalli, prostaglandins are also critical factors in control of uterine tone, in particular, in late pregnancy and during delivery. NSAIDs can retard labor in late pregnancy or *ante partum* because of inhibition of prostaglandin synthesis. Ingestion of high-dose aspirin (>3.2 g/day) over a longer period of time, that is, more than 6 months, will significantly prolong the duration of pregnancy [137]. Because of this, aspirin was used in early days as tokolytic agent in incipient abortion. Today, better tolerable and more effective tokolytics, such as β_2-sympathomimetics, are available.

3.1.3.2 The Elderly Patient

The elderly patient, that is, individuals at the age of ≥75 years, is frequently multimorbid and takes several medications simultaneously. This is the background for drug interactions, for example, in long-term prophylaxis of stroke (Section 4.1.2) and other atherothrombotic diseases. In some cases, chronic inflammatory diseases, such as rheumatoid arthritis, may exist simultaneously (Section 4.2.2). In addition, the elderly is more prone to compliance problems and age-related changes in pharmacokinetics. This requires particular attention, also, because the consequences are more severe in the elderly than in medium-age individuals.

Drug Metabolism in the Elderly Pharmacokinetics and, in some cases, pharmacodynamics of drugs are changed in the elderly. Unwanted pharmacokinetic effects of long-lasting administration of aspirin in the elderly may be due to restricted renal function and changes in hepatic clearance (Section 2.1.2).

Typical early symptoms of aspirin or salicylate overdosing in the elderly are those from the CNS, such as bilateral hearing disorders or tinnitus (Section 4.3.4). Further disturbances involve dizziness, loss of motoric speech control, hallucinations, and changes in mood rather than GI toxicity [138]. Nevertheless, (micro) bleedings in the GI tract though less deteriorating, because not painful, might become problematic if they exceed a certain limit (Section 3.2.1).

The use of antiplatelet agents, including aspirin in elderly patients also increases the risk of traumatic intracranial hemorrhage subsequent to traumatic head injury. In a retrospective analysis, the use of antiplatelet drugs in the elderly was correlated with increased mortality [139]. Although this finding is not totally surprising, the elderly patients should be made aware of this increased risk if they are taking the compound for atherothrombotic prevention.

Probably one of the largest population-based prospective trials on side effects of long-term OTC use of aspirin in the elderly was published by Paganini-Hill *et al.* [140].

A total of 13 987 older citizens of a settlement close to Los Angeles (medium age: 73 years) were asked for aspirin use and afterward studied for 6.5 years or until the occurrence of an acute event. Primary end points were death or hospitalization. The study population included 5106 males and 8881 females. Controls were those individuals who denied aspirin use.

Daily intake of aspirin resulted in a significant increase of renal carcinomas in man, and a nonsignificant increase in women. There was a gender-independent increase of colon carcinomas. There was a tendency for reduced risk of myocardial infarctions in man whereas the risk for stroke tended to be increased. Interestingly, exclusion of study participants with previous myocardial infarction, stroke, or a history of angina pectoris resulted in a considerable (doubling) of risk of ischemic heart disease in both sexes. These changes were seen only in study participants, who, according to self-reports, took daily or "several fold" daily aspirin but not in those who took aspirin less than once daily.

The conclusion was that regular intake of aspirin increases the risk of renal carcinomas and myocardial infarctions [140].

These data were unexpected, in particular, with respect to cardiocoronary prevention (Section 4.1.1) and prevention of colorectal carcinomas (Section 4.3.1) and the study itself became subject of several criticisms. Most importantly, there was no information about dose and duration of aspirin intake. No differentiation was made between different aspirin formulations and there was no objective validation of the anamnestic data or compliance control, for example, by pill count. Interestingly, no differences in any of the parameters were seen, if the study participants regularly took aspirin but less than once (or more than once) daily. Finally, all information about aspirin intake was based on one single interview at the beginning of the trial. Another interview, done 5 years later, indicated that about one-third of study participants had changed their aspirin intake habits in the meantime [141].

Despite these limitations and opposite results in essentially all other published epidemiological trials – specifically in chemoprevention of colorectal carcinoma (Section 4.3.1) – this study nevertheless is of epidemiological interest for geriatrics and another argument against nonprescriptional long-term use of drugs, including aspirin, in the elderly when the individual risk–benefit ratio is unknown.

Summary

Pregnancy and older age are two particular life situations with peculiarities regarding aspirin's efficacy and safety. In pregnancy, possible actions of the drug on pregnancy development and birth have to be considered along with possible effects on the fetus and neonate, respectively. In older age, there are general alterations in pharmacokinetics, possibly associated with drug interactions because of frequent comorbidities. In addition, compliance problems might exist because of the possible daily intake of many different drugs.

There is no reproduction without (COX-2-derived) prostaglandins because of female infertility as seen in COX-2 knockout mice. Most available clinical data suggest that aspirin, in contrast to traditional NSAIDs or coxibs, does not significantly interfere with ovulation and early stages of pregnancy. Although aspirin passes the placental barrier and enters the fetal circulation, there is no conclusive evidence that aspirin ingestion causes malformations or growth retardation in the fetus. There is also no evidence from prospective trials that aspirin increases the rate of miscarriages. Low-dose aspirin used therapeutically for the prevention of preeclampsia in high-risk pregnancies (Section 4.1.5). Aspirin may also be used for *in vitro* fertilization and will increase the efficacy of this procedure, both as a single medication and in combination with heparin.

Intake of aspirin shortly (within 1 week) before delivery is associated with a significant bleeding risk for not only the mother but also the fetus and newborn and should be avoided. There is no conclusive evidence that aspirin in late pregnancy causes functionally relevant constriction of the ductus arteriosus Botalli. Nevertheless, because of bleeding and general safety reasons, aspirin should not be taken in the third trimester of pregnancy.

Aspirin is frequently used in the elderly as single medication or in combination with other drugs. In the elderly, drug clearance may be retarded, eventually resulting in toxic symptoms, specifically in long-term use. At conventional antiplatelet doses, this risk of overdosing is low. At analgesic or anti-inflammatory doses, there might be a higher risk of side effects, mostly symptoms in the CNS, such as tinnitus. These actions are transient and can be corrected by adjustment of doses if further aspirin use is required for preventive reasons.

be quantified by measuring the gastric transmucosal potential difference. Changes of this gradient by salicylates are transient, concentration-dependent, and probably caused by the same physicochemical mechanisms as salicylate-induced depletions of the membrane potential of liver mitochondria (Section 2.2.3) or outer hair cells in the cochlea (Section 3.2.4). After ingestion of one 600 mg single dose of predissolved aspirin, maximum changes (decreases) in gastric transmucosal potential are obtained at about 10 min, recovery starts within 30–60 min, and is complete within 6 h [160].

Repeated Injury and Restitution of Mucosal Surface Epithelium Longer lasting or more intense irritations of the stomach mucosa will result in morphologically detectable defects of the mucosal epithelium that require repair. This restitution of the surface epithelium starts with migration of epithelial cells from gastric pits that start to cover the area of damage, later followed by cell division for complete coverage. There is no correlation between the severity of mucosal injury and the reduction of the potential difference, that is, the total amount of rediffused protons [144, 158]. More recent experimental data suggest that aspirin-induced changes in transmucosal potential difference probably reflect damage to structures in the oxyntic glands and not the breaking of the surface and pit cell mucosal barrier. Thus, not all cells in the stomach mucosa might be affected by aspirin to the same extent [161].

Adaption of the Gastric Mucosa Gastric mucosa has the unique property to become tolerant against noxious stimuli of any kind after repeated or continuous challenging. This long-known phenomenon of "adaption" has been ascribed to morphological alterations in the gastric mucosal epithelium, eventually resulting in the emergence of a new cell population with an increased rate of cell turnover and replacement or greater resistance to noxious stimuli, including aspirin [144, 162]. Thus, restitution of gastric mucosal epithelium means covering of an epithelium-denuded area by a new cell lining. These processes appear not to require the presence of prostaglandins [163]. However, mucosal adaption can be considerably accelerated by supply with exogenous prostaglandins [164]. This suggests that inhibition of endogenous prostaglandin synthesis is not the cause of gastric mucosal injury but rather an aggravating factor for any preexisting functional impairment or injury, respectively.

In man, adaption, that is, resolution of mucosal injury, starts within 1 week of continuous aspirin exposure and presents clinically as (chronic) gastritis. Biochemically, there is enhanced generation of growth factors, such as TGF-β, and increased cell regeneration, eventually resulting in healing of the injury [165, 166]. The last effect is accelerated by prostaglandins. Endoscopically, there is a reduction of petechiae and clearance of the superficial erosions, being completed within a few weeks. The stomach mucosa appears then morphologically normal [165, 167]. The development of a chronic ulcer subsequent to aspirin and other noxious stimuli thus reflects rather a focal failure of adaption, that is, disturbed healing process, than a direct effect of the noxious stimulus on tissue integrity.

The molecular biology of gastric adaption to repeated or long-term aspirin exposure is still only partially understood. In animal experiments, gastric adaption to high doses of aspirin was associated with the expression of particular proteins, including regenerating protein (RegI) in areas with increased cell proliferation [168]. This suggests that protein expression by gastric epithelial cells may generate mediators of resolution (healing) of mucosal injury, characteristic of gastric adaption to aspirin. This process may additionally be enhanced by aspirin-related suppression of apoptosis [169] and transcriptional upregulation of COX-2 with enhanced generation of prostaglandins or even "aspirin-triggered lipoxin" (ATL) as part of the resolution process (see below).

GI Bleeding Another complication of a disturbed mucosal function is microbleeding in the stomach and upper duodenum. The average spontaneous GI blood loss in healthy individuals is low, amounting to less than 1 ml/day [170]. Aspirin increases this blood loss dose dependently, on average, to 2–6 ml/day [144, 171, 172]. However, marked interindividual variations exist and even a small regular daily blood loss may become clinically relevant if it persists over a certain period of time. High-dose long-term use of aspirin may even cause iron deficiency anemia [171]. The inhibition of platelet function by aspirin and the associated disturbances in hemostasis will additionally accelerate bleeding from any preexisting lesion anywhere in the GI tract. Consequently, the GI blood loss by sodium salicylate or other nonacetylated salicylic acid derivatives is much smaller, if at all present [171]. Interestingly, there is no direct correlation between aspirin-induced gastrointestinal microbleeding and the prolongation of skin bleeding time [156] (Section 3.1.2) and also no correlation between gastric mucosal microbleeding and gastric mucosal injury [173]. Consequently, histamine H_2-receptor antagonists, such as ranitidine, protect from aspirin-induced gastric microbleedings but not from ulcer formation. Aspirin appears also to cause significantly less GI bleeding after parenteral (i.v.) administration [174]. This suggests that aspirin-induced gastric microbleedings typically are local events, requiring both the physical presence of aspirin and its acetylation potential but without direct relation to bleeding time.

In contrast to the established risk of microbleedings in the gastrointestinal mucosa after aspirin intake, there are mixed data regarding whether aspirin intake also causes overt, severe GI bleeding [175]. However, there is clear evidence for increased (GI) bleeding in the large prospective randomized trials on aspirin in primary and secondary prevention of cardiovascular events, including severe, that is, life-threatening GI or cerebral bleeding (Sections 4.1.1 and 4.1.2).

Pharmacokinetic Aspects The cellular actions of aspirin on stomach mucosal cells are determined by the lipophilic proportion of the drug that can penetrate the cell membrane following its concentration gradient by passive diffusion from the lumen into mucosal cells (Section 2.1.1). At a luminal pH of 3.5 and a pK_a of 3.5 of aspirin, 50% of the dissolved compound is nonionized and can enter – and leave – the cells from the lumen. This percentage increases with decreasing pH and decreases when the pH becomes more alkaline [176]. Thus, the total amount of aspirin absorption into the stomach mucosal cells is determined by the local concentration of the dissolved, nonionized proportion, that is, the dose and pH of the gastric juice (Section 2.1.2).

In practical terms, this means that about 10% of a 250 mg dose (1.25 mg/ml) of soluble aspirin is absorbed within the stomach from an acid solution at a pH of 1–2, but much less if the compound is administered in a highly buffered solution, that is, transferred into the ionized, nonpermeable form. However, there is "ionic trapping," that is, accumulation of aspirin and salicylate, respectively, inside the gastric mucosal cells (pH 7.4) because of the pH-dependent ionization. Similar considerations apply to the upper duodenum with a pH between 2 and 4.

3.2.1.2 Prostaglandins and Gastric Mucosal Protection

The role of prostaglandins in gastric mucosal defense has been extensively studied after first reports of a "cytoprotective" action of these compounds on the stomach [177, 178]. In rats, it was shown that certain prostaglandins, in particular, PGE_2, the main prostaglandin synthesized by stomach mucosal cells, improve the tolerance of the stomach wall against any kind of noxious stimuli. At higher doses, PGE_2 additionally inhibits acid secretion and stimulates bicarbonate and mucus secretion. This suggests that NSAIDs and aspirin might reduce mucosal resistance by depleting the stomach mucosa from cytoprotective and antisecretory prostaglandins. In addition, aspirin might modulate mucosal blood flow via inhibition of prostaglandin synthesis and NO formation [162].

Originally, it was thought that COX-1 is the only COX isoform that generates prostaglandins in the stomach mucosa. Later studies have challenged this concept and have shown that in addition to COX-1, COX-2 mRNA is also expressed in human stomach mucosal tissue [179, 180] and will con-

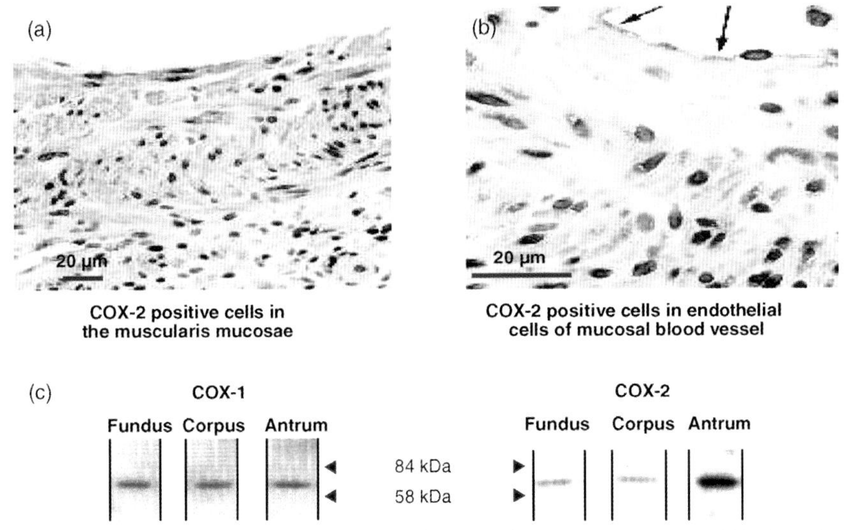

Figure 3.5 Immunohistochemical localization of COX-2 in human gastric mucosa. COX-2 positive cells were detected in the mucosal layer, muscularis mucosae (a), and endothelium (b) of mucosal blood vessels (arrow). COX-1 and COX-2 proteins were identified by Western blotting in stomach mucosa homogenates (c) with the highest expression of COX-2 in the antrum (modified after [181]).

stitutively generate COX-2 protein even under unstimulated "resting" conditions [181]. A different location of COX-2 and COX-1 in stomach mucosal cells additionally suggests a different function of these two isoenzymes [182]. In man, COX-2 immunoreactivity is mainly localized in the gastric mucosal layer and in endothelial cells of mucosal blood vessels (Figure 3.5). This suggests that COX-2 expression is permanently regulated by physiological, that is, mechanical, thermal, osmotical, or chemical stimuli, all of them present during food intake [183]. Both COX-1- and COX-2-derived prostaglandins may contribute to gastric protection as also seen from the incomplete inhibition of gastric PGE_2 production versus COX-1-dependent platelet thromboxane formation by aspirin (Figure 3.6). Consequently, inhibition of both COX isoforms is probably related to gastric damage by inhibition of prostaglandin synthesis [184].

Prostaglandin production in the stomach is age dependent, and is reduced by more than 50% in the elderly (52–72 years) [186]. This reduced prostaglandin formation is probably related to the doubling in basal acid output [187] and might also contribute to the higher sensitivity of stomach mucosa to injury in this population.

3.2.1.3 Mode of Aspirin Action

In the early days of prostaglandin research, inhibition of prostaglandin formation by aspirin was considered to offer a simple and logical explanation to understand the gastric side effects of the compound. After detection of a second COX isoform, COX-2, this hypothesis was modified to a more or less selective inhibition of COX-1 by aspirin that then causes GI side effects, in particular, a GI bleeding tendency, even at low doses. However, pharmacological studies in wild-type and COX-knockout animals have now clearly shown that it may be too simple to refer the complex actions of aspirin – and other NSAIDs and gastric irritants – solely to inhibition of prostaglandin synthesis by inhibition of COX-1. In contrast, (topic) mucosal irritation by aspirin and other salicylates will rather stimulate local prostaglandin synthesis that renders the mucosa less susceptible to injury by increasing mucosal resistance. Thus, actions independent of COX-1 inhibition will contribute to aspirin-induced GI injury [188] (Section 2.2.2).

Mucosal Tissue Injury by Physical Contact Administration of plain aspirin results in endoscopically visible acute gastric mucosal injury, evident as mucosal and submucosal hemorrhages (petechiae). These changes are seen within 1 h after drug ingestion. After oral administration of aspirin tablets, the most frequent and most severe injury is seen in the antrum [144]. Initially, small gastric erosions appear, typically in areas with previous petechiae. If the ulcerative process breaks the muscularis mucosae, larger erosions and, finally, gastric ulcers may develop. This local injury is visible wherever salicylates have contacted the gastric mucosa. One study in rats compared parenteral with intragastric application of aspirin. Both applications inhibited gastric prostaglandin biosynthesis by more than 95% but only the intragastric application caused mucosal injury [189]. Probably, salicylate is the main determinant of gastric injury by aspirin. One reason is its long half-life (17 h!) as a result of the poor solubility in gastric juice (Section 2.1.1) [190], the other being its ability to accumulate in cell membranes and to destabilize the proton gradient and energy metabolism (Section 2.2.3). Consequently, aspirin-related gastric injury is more pronounced at acidic pH and apparently nonexistent or largely reduced if the compound is given in alkaline solution [191] or predissolved in a *sufficiently* buffered medium [192]. These data suggest that the undissociated aspirin and/or salicylate cause mucosal injury by direct contact and subsequent breaking of the mucosal barrier.

Inhibition of Prostaglandin Synthesis Inhibition of prostaglandin synthesis with subsequently reduced defense of stomach mucosal cells, impaired cell renewal, and reduced mucosal blood flow enhances tissue injury. The main prostaglandin in the stomach mucosa is PGE_2. Clinical studies have unequivocally shown only an incomplete inhibition of PGE_2 generation by aspirin, even at the high dose of 2.6 g, whereas inhibition of platelet-dependent thromboxane formation was already complete at regular daily intake of 30 mg (Figure 3.6) [147, 185]. Recovery of gastric prostaglandin (PGE_2) biosynthesis with daily aspirin at antithrombotic doses (81 mg/day) is linear and apparently complete after 5 days. Fifty percent recovery is seen after 2 days. At this time, platelet-dependent thromboxane formation is still completely blocked [193]. This suggests that inhibition of gastric PG(E) synthesis by aspirin, as opposed to inhibition of thromboxane formation by platelets, cannot solely be explained by inhibition of gastric mucosal COX-1. The data provide strong though speculative evidence for a contribution of COX-2 and/or a rapid COX-protein turnover for gastric mucosal prostaglandin biosynthesis.

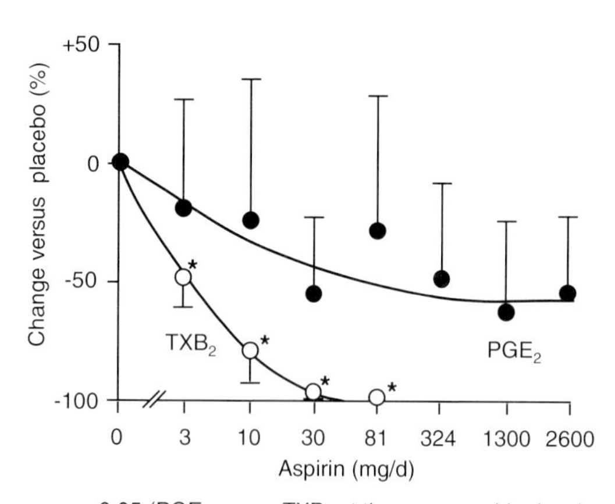

Figure 3.6 Dose-related inhibition of serum thromboxane (TXB_2) and gastric juice PGE_2 formation by oral aspirin. The ingestion of aspirin at the doses indicated or placebo was done in a double-blind manner daily for 8 days with each dose, followed by a 2-week washout interval. The ID_{50} for inhibition of thromboxane formation by aspirin was 3 mg/day and was complete at 30 mg/day. The ID_{50} for inhibition of PGE_2 was 30 mg/day. The inhibition was not stronger at higher doses (modified after [185]).

Aspirin and COX-2 As mentioned before, there is evidence for constitutive COX-2 expression in the stomach mucosa (Figure 3.5). COX-2 may become upregulated not only in response to physiological stimuli but also to inflammatory cytokines and tumor promoters. In the presence of aspirin, COX-2 generates 15-(*R*)-HETE as a major product in the human stomach [181]. 15-(*R*)-HETE could then be utilized by neutrophil-derived 5-lipoxygenase – neutrophils playing a key role in the pathogenesis of aspirin-induced gastric damage – [194] to synthesize aspirin-triggered lipoxin that has protective actions on the injured stomach mucosa [195] (Section 2.3.2). Thus, aspirin may trigger the generation of gastroprotective eicosanoids via the interaction of (acetylated) COX-2 and 5-lipoxygenase [196]. Gastric endothelial cells [197] are a rich source of COX-2 in the human stomach mucosa [181] and may be one site where these reactions occur. These data suggest a complex interaction between aspirin, (inducible) COX-2, (constitutive) COX-1, (acetylated) COX-2 plus 5-lipoxygenases of inflammatory cells (ATL), and gastric mucosal defense.

An elegant technology to evaluate the relative contributions of COX-1 and COX-2 to mucosal cell function is the use of gene-manipulated animals. Deletion of the COX-1 gene in mice did not cause spontaneous gastric ulcerations [198]. However, in COX-1 deleted animals, the gastric mucosa became more severely injured by HCl (0.6 M), both as single application and in the presence of a high concentration (20 mM) intragastric aspirin. No such changes were seen in wild-type and COX-2 knockout mice, respectively [199].

The mucosal toxicity of aspirin in COX-1 knockout mice could be avoided by using phosphatidylcholine-associated aspirin instead of the plain compound [200], suggesting that appropriate "coating" of aspirin will reduce or avoid mucosal contacts of the compound and prevent subsequent mucosal injury.

The aspirin/HCl combination induced a four- to sixfold increase in gastric mucosal PGE_2 levels in COX-1 knockout animals as opposed to saline- or HCl-treated controls. This was explained by an aspirin-induced transcriptional upregulation of the COX-2 gene. In contrast, PGE_2 levels in wild-type and COX-2 knockout mice were reduced. The gastric lesion score appeared to be directly related to alterations in mucosal surface hydrophobicity by HCl but not to mucosal PGE_2 levels.

The conclusion was that aspirin causes gastric injury predominantly by a prostaglandin-independent mechanism, such as an attenuation of mucosal surface hydrophobicity. However, COX-1 may have a protective function in maintaining gastric mucosal barrier integrity via PGE_2 formation. Independent of this, aspirin can increase gastric mucosal PGE_2 levels even in the absence of COX-1, possibly by transcriptional upregulation of COX-2 [199] (Figure 3.7).

Although these are animal data that have to be interpreted with caution, considering also the

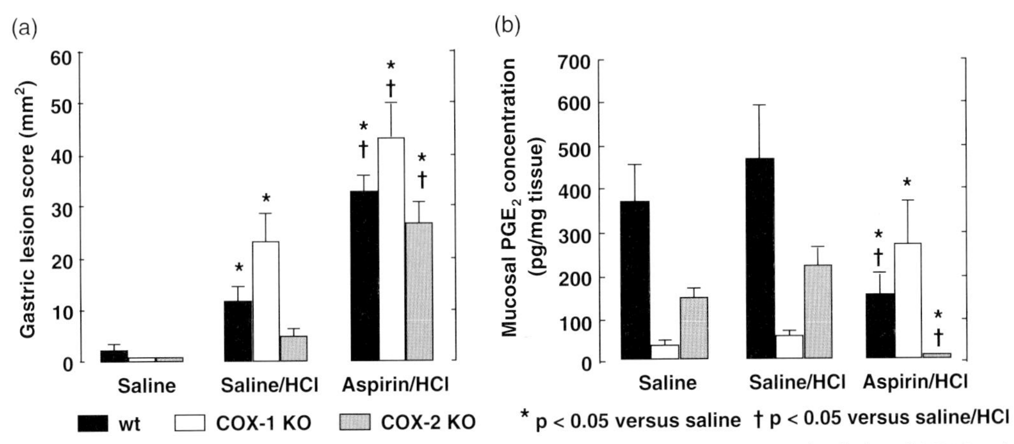

Figure 3.7 Gastric lesion score (a) and gastric mucosal PGE_2 (b) in wild-type (wt) as compared with that of COX-1 and COX-2 knockout mice. Animals were treated with saline, HCl, or aspirin (20 mM) + HCl (for further explanations, see the text) (modified after [199]).

well-known resistance of rodents, such as rats and mice, to the ulcerogenic actions of aspirin, they nevertheless allow several conclusions: the first is that the measurement of PGE_2 levels in gastric juice may not be a sufficient surrogate parameter to document a (causative) role of aspirin for gastric injury. Although it is generally appreciated that mucosal PGE_2 has a protective action on stomach mucosal barrier function, it could well be that COX-1- and COX-2-derived PGE_2 may serve different purposes. Upregulation of COX-2 and increased generation of COX-2-derived products including ATL occurs in response to injury and may be relevant for tissue (ulcer) healing. Conversely, removal of COX-2-derived products under these conditions might aggravate aspirin-induced GI injury [197], a hypothesis that is consistent with real life (see below). This suggests that upregulation of COX-2 mRNA by aspirin [201] could be a compensatory mechanism to generate more prostaglandins "on demand." Second, injury of the gastric mucosa by aspirin or other locally acting irritants is a prostaglandin-independent topic phenomenon and primarily due to local disruptions of the mucosal barrier and changes in its lipophilicity as a result of physicochemical properties of the compound [188, 202]. Generation of reactive oxygen species by aspirin may also be involved because antioxidants, such as vitamin C, protect from aspirin-induced mucosal injury in man [203]. Finally, the inhibition of (upregulated) COX-2 activity by selective COX-2 inhibitors in man is associated with exacerbation of aspirin-related gastric injury (see below). This suggests that simultaneous inhibition of COX-1 and COX-2 will damage the gastric mucosa at least to the same extent as nonselective NSAIDs [204]. This also means that the aspirin-related adaption of stomach mucosa to injury can be reversed by COX-2 inhibitors or vice versa [205], a hypothesis also supported by clinical experience (see below).

Aspirin and Gastric Blood Flow Another aspirin-sensitive variable that determines mucosal injury is the gastric mucosal blood flow. As in other organs, gastric mucosal blood perfusion is mainly regulated by nitric oxide, synthesized via the endothelial NO synthase. Gastric adaption to aspirin in man is associated with an enhanced expression of mucosal eNOS and a subsequent increase in mucosal blood flow [206]. If enhanced NO release was involved in enhanced mucosal blood flow, this might explain the absence of gastric injury by NO-releasing aspirin formulations as opposed to standard plain aspirin [204, 205]. Although the clinical benefits of "nitroaspirin" are not established, yet there is evidence that nitrovasodilators may independently reduce upper GI bleeding [207]. An overview of current concepts regarding GI effects of aspirin and the underlying mechanisms is shown in Figure 3.8.

3.2.1.4 Helicobacter pylori

A significant proportion of patients with gastrointestinal pathologies, including peptic ulcers, are infected with *H. pylori*. NSAIDs, aspirin, and *H. pylori* are considered independent risk factors for GI ulcer formation and bleeding [208, 209]. *H. pylori* eradication reduces aspirin-induced gastric injury and ulcer recurrence, probably by improving stomach adaption [210–212] (Figure 3.9). Eradication was also shown to be equieffective to comedication of proton-pump inhibitor in patients with previous GI bleeding who required long-term treatment with low-dose aspirin for cardiocoronary prevention [213]. Mechanistically, it is thought that *H. pylori* interacts with heat-shock protein 70 (HSP70) that is involved in adaptive reactions of stomach mucosa to aspirin [214].

As outlined above, GI intolerance and ulcer formation are frequent side effects of aspirin and NSAIDs. Interestingly, patients with rheumatoid arthritis who were supposed to take these compounds over years when no alternatives were available in earlier days are not more but less frequently infected with *H. pylori*. One possible explanation for this is a possible inhibitory effect of these compounds on the growth of *H. pylori*. In tissue culture studies, it was found that NSAIDs and salicylate exert a bacteriostatic or even bactericidal action on *H. pylori* and other bacteria at anti-inflammatory concentrations. This also includes salicylate at high concentrations [215].

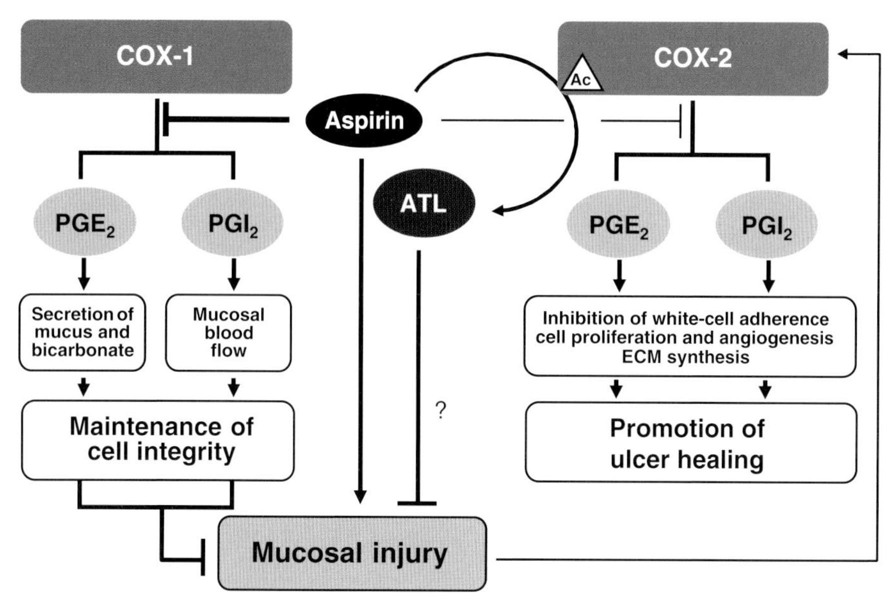

Figure 3.8 Aspirin, prostaglandins, and mechanisms of mucosal injury in human gastric mucosa.

Clearly, this inhibitory action of anti-inflammatory drugs on bacteria growth *in vitro* might be different from the *in vivo* situation, where drug kinetics, protein binding, and bacteria localization become additional variables for drug efficacy.

Another more important question regarding GI side effects of aspirin is whether there is an

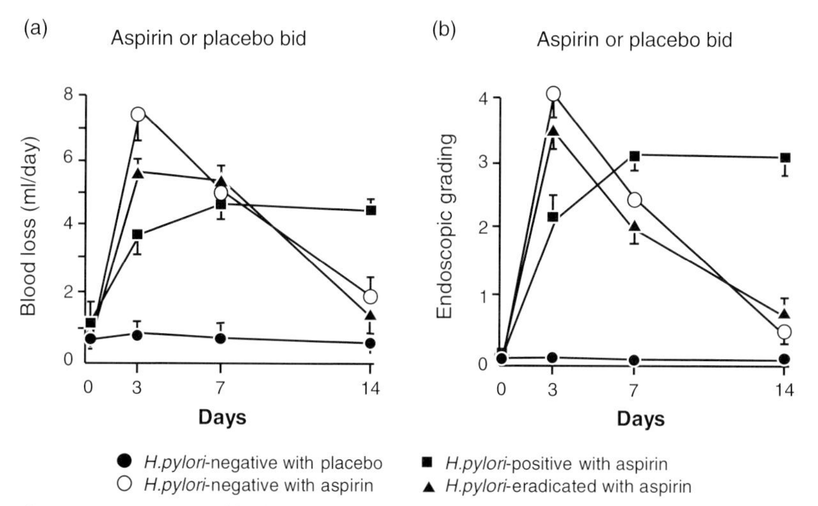

Figure 3.9 Gastric microbleeding (a) and endoscopic injury score (b) of human gastric mucosa before and after 14 days of aspirin (1 g/day) treatment in comparison with placebo. Aspirin causes significant gastric injury and enhances transiently microbleeding in *H. pylori*-negative and positive individuals. This effect is maintained at 14 days in *H. pylori*-positive individuals but has completely disappeared after *H. pylori* eradication (modified after [211]).

interaction between *H. pylori* infection, ulcer bleeding, and aspirin and NSAID use.

A recent meta-analysis has addressed the issue of the prevalence of peptic ulcer disease in adult NSAID takers or the prevalence of *H. pylori* infection and NSAID use in observational studies. Overall, the best 25 out of 463 citations were selected to quantify the relationship between *H. pylori* seropositivity, GI ulcers, and NSAIDs including aspirin and bleeding.

The relative risk of ulcer in *H. pylori* positive NSAID users versus *H. pylori* negative users was 61. *H. pylori* infection alone increased the risk for ulcer 18-fold, additional treatment with NSAIDs increased the risk 3.5-fold. This suggested that *H. pylori* is the main (not sole) cause of ulcers.

Given the presence of any ulcer, NSAIDs increased the risk of bleeding 4.8-fold, whereas *H. pylori* infection increased only 1.8-fold. Both factors together increased the risk of bleeding 6.1-fold. This suggested that NSAIDs are the main cause of (ulcer) bleeding.

The overall conclusion was that *H. pylori* infection and NSAIDs are independent risk factors for GI intolerance: *H. pylori* is the main cause of ulcers and NSAIDs are the main cause of bleeding in those patients with ulcers. There is synergism for the development of peptic ulcer and ulcer bleeding between *H. pylori* infection and NSAID use [216].

This raises the question whether the eradication of *H. pylori* is a useful preventive measure in all positive patients to improve gastric tolerance to aspirin (and NSAIDs), in particular, during long-term use. As outlined above, upregulation of COX-2 (but not of iNOS) may be involved in gastric adaption and subsequent mucosal protection from aspirin. This mechanism appears to be impaired in *H. pylori* infections [183]. Consequently, aspirin might enhance the inflammatory response to *H. pylori* infection [217]. The final answer is not yet available.

3.2.1.5 Clinical Studies

GI side effects of aspirin limit its clinical use, particularly in long-term treatment of elderly patients with age-related disturbances of upper gastrointestinal physiology. Peptic ulcers are common in the elderly and more than 50% are asymptomatic [212]. The elderly is also the group of patients who most frequently take aspirin for preventive purposes and might benefit the most from its use [218]. However, the elderly is also a high-risk group with respect to bleeding. According to an American Bleeding Registry, evaluating 1235 patients presenting with acute GI bleeding, aspirin was the drug most frequently associated with bleeding: 28% of upper GI bleeding and 22% of lower GI bleeding. Interestingly, this also included OTC use. Aspirin and NSAIDs were used significantly more often in the bleeding population (48%) than in controls (19%) [219]. Thus, an individual benefit/risk calculation is necessary, particularly in the elderly and other patients with increased bleeding risks, for example, with a history of GI bleeding or ulcers and comedications that interact with mucosal barrier integrity or the clotting system (antiplatelet agents, anticoagulants) [220], particularly with combined antithrombotic treatment [221].

Two Different Study Designs There are two different types of clinical studies targeted to investigate aspirin-related side effects in the GI tract: (i) studies that address the *risk* of aspirin-induced GI injury as a study end point in either observational or randomized trials and (ii) studies that investigate the *clinical benefits* of aspirin. Here, GI side effects of aspirin are generally expected – and accepted – but have to be balanced against the expected clinical benefit. The first type of trial is done in an experimental, prospective manner in predefined groups of (healthy) individuals. Observational investigations of the relationship between aspirin intake and GI bleeding also fall into this category. In contrast to randomized trials, the individual risk (end point) is largely determined by the composition of the study (and control!) groups and preexisting risk factors such as *H. pylori* infection, older age, or environmental factors such as alcohol (abuse), nicotine, or comedications. Thus, the information obtained about aspirin-related GI risk from observational and randomized trials may not be the same.

The second category of studies is done in patients with an intention to determine the clinical benefit of aspirin. The main groups of patients are those who take aspirin for long-term cardiovascular protection. This category also includes short-term OTC use of aspirin for pain relief and experimental clinical trials of aspirin in new clinical indications such as colorectal carcinoma. The frequency and severity of GI side effects are similar in all patient groups. The usefulness of aspirin in these patients, that is, the benefit/risk ratio is primarily determined not by the risk of GI bleeding but by the probability of preventing complicating events, such as myocardial infarction, stroke, or other end points of clinical efficacy. Importantly, in these studies, the site of GI bleeding is usually not determined.

GI Side Effects as Study End Points: Observational Studies Observational studies are – by definition – nonrandomized trials and have the major disadvantage of the absence of a randomized control group. Thus, these studies are more likely to overinterpret results in favor of treatment (investigator bias). However, they are closer to real life than preselected study populations of randomized trials and thus might be useful to generate hypotheses that then become subject to randomized studies [222]. Several large observational studies on GI side effects with aspirin are available and, as foreseeable, provided different results.

Weil *et al.* [148] conducted a case–control study in British subjects who were hospitalized for upper GI bleeding. The objective was to determine the risk of hospitalization for bleeding peptic ulcers in patients with current prophylactic aspirin use at antiplatelet doses of 300 mg/day or less. A total of 1121 patients with gastric or duodenal ulcer bleeding (cases) were included and matched with 1126 hospital and 989 community controls.

A total of 144 (12.8%) of the cases had been regular users of aspirin (taken at least 5 days a week for at least the previous month) as compared with 9.0 and 7.8% of the hospital and community controls, respectively. Odds ratios were raised for all doses of standard aspirin and amounted to 2.3 (95% CI: 1.2–4.4) at 75 mg, 3.2 (95% CI: 1.7–6.5) at 150 mg, and 3.9 (95% CI: 2.5–6.3) at 300 mg. Thus, 75 mg aspirin had a 40% lower risk than 300 mg. There was a clear dose-dependent increase in the risk of peptic ulcer bleeding from 75 to 300 mg plain aspirin (tablets or solutions) but no increased risk with enteric-coated formulations 1.1 (95% CI: 0.2–6.1). The risks seemed particularly high in patients who took non-aspirin NSAIDs concurrently with aspirin.

The conclusion was that no conventionally used prophylactic aspirin at antiplatelet doses seems free of the risk of peptic ulcer complications. However, the on average much lower incidence of ulcers with the enteric-coated preparation also indicated a better tolerability of this formulation [148].

Kelly *et al.* [223] performed a retrospective case–control study on drug-related gastrointestinal bleeding. Five hundred and fifty incident cases, admitted to Massachusetts hospitals because of acute upper gastrointestinal bleeding (confirmed by endoscopy), and 1202 controls identified from population census lists were asked about the use of aspirin, including the kind of aspirin formulation, and non-aspirin NSAIDs during the last week before the bleeding event (cases) or interview (controls).

The odds ratios for the risk of drug-related bleeding were similar, 2.6–3.1, between the different treatment groups and were also not different between different aspirin preparations, including enteric-coated aspirin.

The conclusion was that enteric-coated aspirin also carries a threefold increased risk in major upper GI bleeding and that this formulation is not less harmful than plain aspirin [223].

Sorensen *et al.* [224] also did a retrospective observational cohort study on the relationship between upper GI bleeding and aspirin intake in Denmark. The data of 27 694 users of low-dose aspirin (100–150 mg/day) were compared with the incidence rate in the general population in the same region. GI bleeding was 2.6-fold more frequent in aspirin users (95% CI: 2.2–2.9), and there was no difference between plain and enteric-coated preparations. However, the combined use of aspirin and traditional NSAIDs increased the incidence rate to 5.6 (95% CI: 4.4–7.0).

The conclusion was that regular low-dose aspirin is associated with an increased risk of upper GI bleeding that is about twice as much when combined with non-aspirin NSAIDs. Enteric coating of aspirin appears not to reduce the risk [224].

Thus, the results from nonrandomized observational trials unequivocally demonstrate an increased risk of GI intolerance by (regular) aspirin intake. Whether the different results are influenced by the different study populations and evaluation criteria, confounding biases by the open study design, or other non-aspirin-related factors remain to be determined. As mentioned above, epidemiological trials are open trials and suffer from the inherent difficulties of all observational studies, specifically the unknown aspirin dosage and exact duration of treatment. Further confounding factors are smoking, alcohol, *H. pylori* infections, or (additional) OTC use of high-dose (500 mg) aspirin tablets, possibly combined with other NSAIDs. All cited observational studies found an approximate doubling of the GI risk for the combined use of aspirin and traditional NSAIDs as opposed to aspirin alone. The same finding has also been recently published for the combination of aspirin and coxibs (see below). With respect to alcohol, it should be noted that the combination with aspirin results in a marked synergizing effect on overt gastric hemorrhage [225, 226]. Finally, endoscopic studies [227] may be more sensitive to detect aspirin-associated mucosal injury than cohort and case–control trials – though in the majority of cases without clinical significance.

A meta-analysis of 17 epidemiological studies published between 1990 and 2001 on the association between aspirin use and serious upper GI complications found an overall risk of serious upper GI complications (ulcer, bleeding, and perforations) of 2.2 (95% CI: 2.1–2.4) for cohort studies and nested case–control studies and 3.1 (95% CI: 2.8–3.3) for non-nested case–control studies. The overall risk was 2.6 (95% CI: 2.3–2.9) for plain, 5.3 (95% CI: 3.0–9.2) for buffered, and 2.4 (95% CI: 1.9–2.9) for enteric-coated aspirin formulations. The absence of less gastric but not duodenal injury by enteric-coated formulations, according to the authors, may partially be explained by the channeling of susceptible patients to these formulations [228].

GI Side Effects as Study End Points: Randomized Trials The GI bleeding due to aspirin in a small but significant proportion of patients in observational trials prompted the design of randomized studies including the supposedly safer aspirin formulations from which the enteric-coated form found most attention for repeated or long-term use. One of the first systematic investigations of the relationship between enteric-coated aspirin and GI mucosal injury in men was from Stubbé *et al.* [171].

The authors compared stomach-resistant enteric-coated preparations (made in the hospital-owned pharmacy) with a standard formulation, both preparations containing 500 mg aspirin. In all individuals studied, there was an increase of occult blood in stool with plain aspirin, but only in 4 out of 30 subjects with the enteric-coated formulation. This suggested for the first time that the increased GI blood loss subsequent to aspirin intake was mainly from the stomach and could be largely reduced or even avoided by appropriate coating of the preparation.

The conclusion was that enteric-coated formulations cause no or minor GI injury, in many cases being in the range of placebo [171].

Lanza [192] reviewed the gastric tolerance of different aspirin preparations in older trials and found a weak (<20% of standard aspirin) injury with the enteric-coated formulations. The first controlled prospective randomized trial on low-dose enteric-coated aspirin was conducted by Hawthorne *et al.* [227].

The authors compared the GI tolerance of different doses of plain and enteric-coated aspirin. Healthy subjects were treated for 5 days in a placebo-controlled, double-blind crossover design. Plain aspirin (300 mg) caused significant increases in gastric mucosal injury (Lanza-Score) and enhanced mucosal bleeding. Enteric coating apparently eliminated this type of gastric injury at the same dosage. Both formulations caused similar inhibition of gastric mucosal PGE_2 formation and suppressed equally well (>99%) the serum thromboxane generation, suggesting not only a comparable therapeutic effect but also a bleeding risk from platelet dysfunction.

The conclusion was that enteric-coated aspirin should be considered a useful preparation for long-term use, specifically in cardiovascular prophylaxis (Table 3.3) [227].

Although these data were confirmed by many other studies, usually small endoscopic studies in man [146, 229–235], they cannot rule out a bleeding risk that is (by definition) associated with the clinical use of aspirin in cardiocoronary prevention (see below).

Table 3.3 Effects of plain and enteric-coated (EC) aspirin as compared with placebo on gastric mucosal injury (gastric body) in 20 healthy volunteers.

Parameter	Placebo	Aspirin 300 plain	Aspirin 300 EC
Hemorrhage erosion score	0 (0–0.3)	2 (0–5)	0 (0–1.3)
Visual analogue injury score	0 (0–8)	5 (0–31)	0 (0–8)
Mucosal bleeding (μl/10 min)	0.9 (0.6–1.3)	2.8 (1.6–4.8)	1.0 (0.6–1.5)
Mucosal PGE_2 synthesis (pg/mg)	18 (1–51)	0.7 (0.4–11)	1.8 (0.5–9.2)
Mucosal TXB_2 synthesis (pg/mg)	19 (4.1–37)	1.4 (1.1–1.9)	2.5 (1.1–5.2)
Serum salicylate (μg/ml)	<5	16 (10–20)	7 (5–16)
Serum TXB_2 (% of placebo)	—	0.4 (0.2–1.1)	0.3 (0.2–0.9)

Treatment was for 5 days in a double-blind crossover design [227].Data are median and interquartile range. All differences between aspirin plain and aspirin EC were significant ($p < 0.05$–0.01) except serum thromboxane ($p > 0.05$).

Importantly, the inhibition of platelet-dependent thromboxane formation and platelet function does not differ between enteric-coated and plain aspirin at doses of 100 mg/day or above [173]. In addition, enteric-coated modified-release formulations cause less inhibition of prostacyclin generation in the systemic circulation because of the rapid breakdown of aspirin in the presystemic circulation [236]. Furthermore, gastroscopic (short-term) studies on enteric-coated preparations in controlled prospective trials show that the side effects of aspirin at antithrombotic doses (≤300 mg/day) on gastric mucosal injury but not bleeding [237] – which in many cases is from the duodenum – are largely prevented by appropriate enteric-coated preparations.

GI Side Effects in Cardiovascular Prevention Trials

Roderick *et al.* [238] have performed a meta-analysis on side effects of aspirin in randomized controlled trials, listed by the Antiplatelet Trialists Collaboration (ATC). In comparison with placebo, regular use of standard aspirin for at least 1 year at doses between 75 and 1500 mg/day was associated with a twofold elevated risk in GI bleeding and a 1.5-fold increase in the risk of peptic ulcers. Interestingly, there was no evidence that the incidence of GI bleeding was greater in the elderly [239]. Overall, these risks in randomized trials were lower than those found in observational studies.

The first double-blind, placebo-controlled long-term trial on thrombosis prevention in patients with cerebrovascular disease, the UK-TIA trial, also determined the dose dependency of GI side effects of aspirin. The patients received 300 mg/day aspirin, 1200 mg/day aspirin, or placebo for up to 7 years (Section 4.2.1). The rates for upper GI bleeding requiring hospitalization were 0.2% in the placebo group as opposed to 0.9% in the 300 mg and 1.2% in the 1200 mg aspirin groups. A dose-dependent increase in subjective GI side effects (nausea, vomiting, and indigestion) was also obtained [240, 241].

In the US American Physicians' Health study on primary prevention of cardiovascular events, the relative risk for bleeding requiring transfusion (site unspecified) was 1.71 (95% CI: 1.09–2.16) in the aspirin group (325 mg/second day) as compared with the placebo. Although this trial was not assigned to assess GI effects, there was a small though significant increase in gastrointestinal hemorrhage requiring transfusion: 0.5% in the aspirin versus 0.3% in the placebo group [242] and, interestingly, also an increased incidence of duodenal ulcers (Steering Committee of the Physicians' Health Study Research Group [243]). In this study, preexisting GI intolerance to aspirin was an exclusion criterion and, therefore, the real number of patients at risk might have been underestimated. However, a similar incidence of GI side effects was also noted in the open British Doctors' trial [244] and the randomized Dutch TIA trial, comparing 30 mg aspirin/day with 283 mg/day for 2 years [245].

The British Medical Research Council (MRC) primary prevention study (TPT) investigated 75 mg enteric-coated aspirin/day over 6 years in patients with high risk for atherothrombotic events. In comparison with placebo, there was an increase in minor bleeds but no significantly elevated risk of severe GI intolerance or bleedings (Section 4.1.1).

The Italian Primary Prevention project (PPP) also used low-dose (100 mg/day) enteric-coated aspirin in a randomized design and found significantly more total bleeding complications in the aspirin group as compared with controls. However, three out of the four deaths due to hemorrhage occurred in the placebo group [246, 247]. In the ISIS-2 study [248], a trial on secondary prevention in patients with acute myocardial infarction (Section 4.1.1), the gastric injury score with enteric-coated aspirin was not significantly different from that in the placebo group.

Overall, enteric-coated aspirin also appears to cause a bleeding tendency in these trials on cardiocoronary prevention. This is not surprising because of the (desired) inhibition of platelet function as the therapeutic goal in these studies. However, severe GI bleeding appears to be rare as was GI ulcer formation. None of these studies was designed to compare enteric-coated aspirin with standard plain aspirin formulations. Thus, although endoscopic studies in healthy volunteers have demonstrated a clear reduction of gastric mucosal injury by enteric-coated preparations, no clinical benefits in terms of reduction of GI bleeding have been shown in CABG patients [249].

The transfer of these data into clinical guidelines is still mixed. The latest available statement of the US Preventive Services Task Force [250] considers enteric-coated or buffered aspirin not to clearly reduce adverse gastrointestinal effects of aspirin whereas the American Diabetes Association in 2004 [251] favored enteric-coated aspirin more than the plain compound to prevent from thromboembolic events.

3.2.1.6 Aspirin and Other Drugs

Aspirin and Antiulcer Drugs *Sucralfate* did not affect aspirin-induced gastric mucosal injury but reduced aspirin-associated intragastric bleeding [150, 252]. Histamine H_2 antagonists, such as *ranitidine*, inhibit gastric acid secretion and microbleeding. They were not effective in the prevention of gastric ulcers but did prevent duodenal ulcers, the main source of GI bleeding [153, 253, 254]. *Misoprostol*, a PGE_1 analogue, did protect from ulcer formation and stimulated ulcer healing. However, the compound has several side effects, most notably diarrhea and dyspepsia and, as expected, is not an effective analgesic because of the sensitizing action of E-type prostaglandins on pain receptors (Section 2.3.2) [255–258].

The most convincing data for protection of the stomach from aspirin-induced mucosal injury are available for proton-pump inhibitors, such as *omeprazole* and its analogues [259–262]. These compounds – in contrast to misoprostol – can be given orally once daily and have been shown to markedly protect from aspirin-induced gastric ulcer formation by aspirin and traditional NSAIDs [207]. The risk of interference with the (desired) antiplatelet actions of aspirin is low. Antiplatelet effects of aspirin are largely determined by the amount of drug absorbed in the small intestine and appears not to be modified by comedication of antiulcer agents, including proton-pump inhibitors that act mainly on the stomach [263]. Interestingly, comedication of a proton-pump inhibitor to aspirin largely prevented recurrent GI ulcers in high-risk patients whereas clopidogrel alone did not [262, 264]. However, there was no effect of aspirin plus proton-pump inhibitor on the incidence of lower GI tract bleeding, suggesting that the protective action of the compound was restricted to the stomach.

Aspirin and COX-2 Inhibitors Selective inhibitors of COX-2, such as celecoxib and rofecoxib, were originally introduced because of less gastric injury, that is, endoscopically visible erosions and ulcers,

Table 3.4 Incidence of gastric erosions and ulcers at 12 week in patients with osteoarthritis.

Parameter	Placebo	Aspirin (81 mg EC)	Aspirin + rofecoxib (25 mg)	Ibuprofen (800 mg tid)
n	381	387	377	374
Patients with erosions (%)	0.17	0.85[a]	1.67[a]	1.91[a]
Patients with ulcers (%)	5.8	7.3	16.1[a]	17.1[a]

Double-blind, randomized placebo controlled trial [235].
[a]Significant versus placebo.

than conventional NSAIDs because of the absent inhibition of COX-1. Clinical data confirm a significantly reduced risk of GI bleeding and ulcus formation with these compounds. However, they also show that these benefits may be lost in patients, requiring additional aspirin cotreatment, for cardiocoronary prevention, for example, those suffering from rheumatoid arthritis. These patients are at a 32–55% higher risk for cardiocoronary events [265] than patients with osteoarthritis or controls [266]. This might provide an explanation why patients in the CLASS study, containing 28% patients with rheumatoid arthritis, who were allowed to take low-dose aspirin in addition to celecoxib, did not have evidence of improved GI safety with the selective COX-2 inhibitor [267]. They rather exhibited a higher frequency of upper GI ulcer complications (RR 4.5; $p = 0.01$) than patients receiving celecoxib alone [228]. This suggests that COX-2 inhibitors in patients taking aspirin offer no or little advantage over conventional NSAIDs. Furthermore, if aspirin has to be taken in combination with a COX-2 inhibitor, a low-dose enteric-coated formulation might be preferred [268]. According to a recent epidemiological survey in Northeastern United States, approximately 50% of long-term COX-2 users were also taking aspirin for cardioprotection. This rate is considerably higher than the 21% reported in the CLASS trial [269]. The reduced risk of GI side effects in the TARGET trial with lumiracoxib was nullified by aspirin cotreatment [270].

There was a fourfold increased GI bleeding in individuals treated with COX-2 inhibitors plus aspirin and a twofold increased GI ulcer formation in a randomized trial on osteoarthritis patients after combined treatment with aspirin and coxibs as compared with aspirin alone. Injury scores similar to the combined use of aspirin and coxibs (rofecoxib) were seen with an anti-inflammatory dose of ibuprofen (Table 3.4) [235].

Thus, the improved GI tolerance of COX-2 inhibitors appears to disappear if concomitant treatment with aspirin is necessary.

Aspirin and Traditional NSAIDs Traditional NSAIDs are the standard treatment for pain and inflammation but – in contrast to aspirin and with the possible exception of naproxen – appear not to reduce but rather increase the risk of myocardial infarctions and stroke (Sections 4.1.1 and 4.1.2). Traditional NSAIDs bear a similar risk of GI side effects as the standard aspirin. An important though unanswered question is whether concomitant low-dose aspirin use that occurs in more than 20% of patients will reduce the incidence of cardiovascular events, that is, maintain the preventive potency of aspirin. The concomitant aspirin will increase the risk of developing serious GI events in patients taking NSAIDs [271].

Summary

Gastric intolerance, although occurring only in a minority of patients, is the most common side effect of aspirin. GI intolerance usually presents with subjective symptoms such as dyspepsia, nausea, and heartburn. Clinically more important are gastrointestinal hemorrhages and ulcers. Subjective and objective symptoms of gastric intolerance appear not to be intercorrelated.

Mechanistically, aspirin has different and somehow antagonistic effects on the stomach mucosa. Although it may directly injure the mucosal lining by disturbing its barrier function, probably related to the lipophilicity of salicylate, it may also stimulate the more chronic (inflammatory) event of mucosal cell adaption. The role of prostaglandins in these processes is still a matter of debate. However, induced expression of COX-2 and possible generation of gastroprotective PGE2 and possibly lipoxins (ATL) are currently the issues of great scientific interest but yet of unknown clinical relevance.

In clinical trials, the incidence and severity of gastric injury by aspirin was found to be dose and time dependent and generally more likely to occur during repeated long-term use of the agent at higher doses. Randomized short-term trials show that the gastric injury apparently requires direct contact of the compound with the stomach mucosa and can be reduced by enteric-coated formulations. Whether these preparations are better tolerated than plain aspirin in long-term clinical trials in cardiocoronary prevention has not been shown. However, no head-to-head comparison has been performed yet.

Aspirin-related gastric injuries can be treated or prevented by eliminating gastric acid secretion, for example, by appropriate comedications such as proton-pump inhibitors. For OTC or short-term (analgesic) use, predissolved or soluble preparations are available. In this indication, aspirin is at least as well tolerable as conventional OTC analgesics such as acetaminophen and ibuprofen (Section 4.2.1). However, it also might cause GI bleeding, particularly in high-risk groups such as the elderly with frequent comedications and this is also valid for nonprescriptional use.

References

142 Kikendall, J.W., Friedman, A.C., Oyewole, M.A. *et al.* (1983) Pill-induced esophageal injury: case reports and review of the medical literature. *Digestive Diseases and Sciences*, **28**, 174–182.

143 McCarthy, D.M. (2007) Do drugs or bugs cause GERD? *Journal of Clinical Gastroenterology*, **41**, S59–S63.

144 Graham, D.Y. and Smith, J.L. (1986) Aspirin and the stomach. *Annals of Internal Medicine*, **104**, 390–398.

145 Hudson, N., Hawthorne, A.B., Cole, A.T. *et al.* (1992) Mechanisms of gastric and duodenal damage and protection. *Hepato-Gastroenterology*, **39** (Suppl. 1), 31–36.

146 Petroski, D. (1993) Endoscopic comparison of three aspirin preparations and placebo. *Clinical Therapeutics*, **15**, 314–320.

147 Cryer, B. and Feldman, M. (1999) Effects of very low dose daily, long-term aspirin therapy on gastric duodenal and rectal prostaglandin levels and on mucosal injury in healthy humans. *Gastroenterology*, **117**, 17–25.

148 Weil, J., Colin-Jones, D., Langman, M. *et al.* (1995) Prophylactic aspirin and risk of peptic ulcer bleeding. *British Medical Journal*, **310**, 827–830.

149 Derry, S. and Loke, Y.K. (2000) Risk of gastrointestinal haemorrhage with long-term use of aspirin: meta-analysis. *British Medical Journal*, **321**, 1183–1187.

150 Wolfe, M.M., Lichtenstein, D.R. and Singh, G. (1999) Gastrointestinal toxicity of nonsteroidal antiinflammatory drugs. *The New England Journal of Medicine*, **340**, 1888–1899.

151 Fries, J.F. and Bruce, B. (2003) Rates of serious gastrointestinal events from low dose use of acetylsalicylic acid, acetaminophen, and ibuprofen in patients with osteoarthritis and rheumatoid arthritis. *The Journal of Rheumatology*, **30**, 2226–2233.

152 Tarnawski, A.S. and Caves, T.C. (2004) Aspirin in the XXI century: its major clinical impact, novel mechanisms of action and safer formulations. *Gastroenterology*, **127**, 341–343.

153 Laine, L. (2001) Approaches to non-steroidal anti-inflammatory drug use in the high risk patient. *Gastroenterology*, **120**, 594–606.

154 Lanas, A., Perez-Aisa, M.A., Feu, F. *et al.* (2005) A nationwide study of mortality associated with hospital admission due to severe gastrointestinal events and those associated with nonsteroidal anti-inflammatory drug use. *The American Journal of Gastroenterology*, **100**, 1685–1693.

155 Voutilainen, M., Mantynen, T., Farkkila, M. *et al.* (2001) Impact of non-steroidal anti-inflammatory drug and aspirin use on the prevalence of dyspepsia and uncomplicated peptic ulcer disease. *Scandinavian Journal of Gastroenterology*, **36**, 817–821.

156 Wood, P.H.N., Harvey-Smith, E.A. and Dixon, A.J. (1962) Salicylates and gastrointestinal bleeding: acetylsalicylic acid and aspirin derivatives. *British Medical Journal*, **1**, 669–675.

157 Davenport, H.W. (1967) Salicylate damage to the gastric mucosal barrier. *The New England Journal of Medicine*, **267**, 1307–1312.

158 Kauffman, G.L. (1985) The gastric mucosal barrier. Component control. *Digestive Diseases and Sciences*, **30** (Suppl.) 69S–76S.

159 Kiviluoto, T., Mustonen, H. and Kivilaakso, E. (1989) Effect of barrier-breaking agents on intracellular pH and epithelial membrane resistance: studies in isolated *Necturus* antral mucosa exposed to luminal acid. *Gastroenterology*, **96**, 1410–1418.

160 Baskin, W.N., Ivey, K.J., Krause, W.J. *et al.* (1976) Aspirin-induced ultrastructural changes in human gastric mucosa: correlation with potential difference. *Annals of Internal Medicine*, **85**, 299–303.

161 Cook, G.A., Elliott, S.L., Skeljo, M.V. *et al.* (1996) Correlation between transmucosal potential difference and morphological damage during aspirin injury of gastric mucosa rats. *Journal of Gastroenterology and Hepatology*, **11**, 264–269.

162 Konturek, P.C. (1997) Physiological, immunohistochemical and molecular aspects of gastric adaptation to stress, aspirin and to *H. pylori*-derived gastrotoxins. *Journal of Physiology and Pharmacology*, **48**, 3–42.

163 Konturek, S.J., Brzozowski, T., Stachura, J. *et al.* (1994) Role of gastric blood flow, neutrophil infiltration, and mucosal cell proliferation in gastric adaption to aspirin in the rat. *Gut*, **35**, 1189–1196.

164 Olivero, J.J. and Graham, D.Y. (1992) Gastric adaptation to nonsteroidal anti-inflammatory drugs in man. *Scandinavian Journal of Gastroenterology*, **27** (Suppl.), 53–58.

165 Graham, D.Y., Smith, J.L., Spjut, H.J. *et al.* (1988) Gastric adaptation – studies in humans during continuous aspirin administration. *Gastroenterology*, **95**, 327–333.

166 Stachura, J., Konturek, J.W., Dembinski, A. *et al.* (1996) Growth markers in the human gastric mucosa during adaptation to continued aspirin administration. *Journal of Clinical Gastroenterology*, **22**, 282–287.

167 Graham, D.Y., Smith, J.L. and Dobbs, S.M. (1983) Gastric adaption occurs with aspirin administration in man. *Digestive Diseases and Sciences*, **28**, 1–6.

168 Alderman, B.M., Ulaganathan, M., Judd, L.M. *et al.* (2003) Insights into the mechanisms of gastric adaptation to aspirin-induced injury: a role for regenerating protein but not trefoil peptides. *Laboratory Investigation*, **83**, 1415–1425.

169 Alderman, B.M., Cook, G.A., Familiari, M. *et al.* (2000) Resistance to apoptosis is a mechanism of adaption of rat stomach to aspirin. *The American Journal of Physiology*, **278**, G839–G846.

170 Leonards, J.R., Levy, G. and Niemczura, R. (1973) Gastrointestinal blood loss during prolonged aspirin administration. *The New England Journal of Medicine*, **289**, 1020–1022.

171 Stubbé, L.Th.F.L., Pietersen, J.H., Van Heulen, C. (1962) Aspirin preparations and their noxious effect on the gastro-intestinal tract. *British Medical Journal* **1** 675–680.

172 Dybdahl, J.H., Daae, L.N., Larsen, S. *et al.* (1980) Acetylsalicylic acid-induced gastrointestinal bleeding determined by a ^{51}Cr method on a day-to-day basis. *Scandinavian Journal of Gastroenterology*, **15**, 887–895.

173 Hawkey, C.J., Hawthorne, A.B., Hudson, N. *et al.* (1991) Separation of the impairment of haemostasis by aspirin from mucosal injury in the human stomach. *Clinical Science*, **81**, 565–573.

174 Mielants, H., Veys, E.M., Verbruggen, G. *et al.* (1984) Salicylate-induced occult gastrointestinal blood loss:

comparison between different oral and parenteral forms of acetylsalicylates and salicylates. *Clinical Rheumatology*, **3**, 47–54.

175 Greenberg, P.D., Cello, J.P. and Rockey, D.C. (1999) Relationship of low-dose aspirin to GI injury and occult bleeding: a pilot study. *Gastrointestinal Endoscopy*, **50**, 618–622.

176 Dotevall, G. and Ekenved, G. (1976) The absorption of acetylsalicylic acid from the stomach in relation to intragastric pH. *Scandinavian Journal of Gastroenterology*, **11**, 801–805.

177 Robert, A., Nezamis, J.E., Lancaster, C. *et al.* (1979) Cytoprotection by prostaglandins in rats, prevention of gastric necrosis produced by alcohol, HCl, NaOH, hypertonic NaCl and thermal injury. *Gastroenterology*, **77**, 433–443.

178 Robert, A., Nezamis, J.E., Lancaster, C. *et al.* (1983) Mild irritants prevent gastric necrosis through "adaptive cytoprotection" mediated by prostaglandins. *The American Journal of Physiology*, **245**, 113–121.

179 O'Neill, G.P. and Ford-Hutchinson, A.W. (1993) Expression of mRNA for cyclooxygenase-1 and cyclooxygenase-2 in human tissues. *FEBS Letters*, **330**, 156–160.

180 Ristimäki, A., Hokanen, N., Jankala, H. *et al.* (1997) Expression of cyclooxygenase-2 in human gastric carcinoma. *Cancer Research*, **57**, 1276–1280.

181 Zimmermann, K.C., Sarbia, M., Schrör, K. *et al.* (1998) Constitutive cyclooxygenase-2-expression in healthy human and rabbit gastric mucosa. *Molecular Pharmacology*, **54**, 536–540.

182 Iseki, S. (1995) Immunocytochemical localization of cyclooxygenase-1 and cyclooxygenase-2 in the rat stomach. *The Histochemical Journal*, **27**, 323–328.

183 Fischer, H., Huber, V. and Boknik, P. (2001) Effect of *Helicobacter pylori* eradication on cyclooxygenase 2 (COX-2) and inducible nitric oxide synthase (iNOS) expression during gastric adaptation to aspirin (ASA) in humans. *Microscopy Research and Technique*, **53**, 336–342.

184 Wallace, J.L., McNight, W., Reuter, B.K. *et al.* (2000) NSAID-induced gastric damage in rats: requirement for inhibition of both cyclooxygenase 1 and 2. *Gastroenterology*, **119**, 706–714.

185 Lee, M., Cryer, B. and Feldman, M. (1994) Dose effects of aspirin on gastric prostaglandins and stomach mucosal injury. *Annals of Internal Medicine*, **120**, 184–189.

186 Goto, H., Sugiyama, S., Ohara, A. *et al.* (1992) Age-associated decreases in prostaglandin contents in human gastric mucosa. *Biochemical and Biophysical Research Communications*, **186**, 1443–1448.

187 Cryer, B., Redfern, J.S., Goldschmiedt, M. *et al.* (1992) Effect of aging on gastric and duodenal mucosal prostaglandin concentrations in humans. *Gastroenterology*, **102**, 1118–1123.

188 Lichtenberger, L.M. (2001) Where is the evidence that cyclooxygenase inhibition is the primary cause of non-steroidal-antiinflammatory drug (NSAID)-induced gastrointestinal injury? Topical injury revisited. *Biochemical Pharmacology*, **61**, 631–637.

189 Ligumsky, M., Golanska, E.M., Hansen, D.G. *et al.* (1983) Aspirin can inhibit gastric mucosal cyclo-oxygenase without causing lesions in the rat. *Gastroenterology*, **84**, 756–761.

190 Kauffman, G.L. (1989) Aspirin-induced gastric-mucosal injury – lessons learned from animal models. *Gastroenterology*, **96**, 606–614.

191 Thorsen, W.B., Jr, Western, D., Tanaka, Y. *et al.* (1968) Aspirin injury to the gastric mucosa: gastrocamera observations of the effect of pH. *Archives of Internal Medicine*, **121**, 499–506.

192 Lanza, F.L. (1984) Endoscopic studies of gastric and duodenal injury after the use of ibuprofen, aspirin, and other nonsteroidal anti-inflammatory agents. *The American Journal of Medicine*, **77**, 19–24.

193 Feldman, M., Shewmake, K. and Cryer, B. (2000) Time course inhibition of gastric and platelet COX activity by acetylsalicylic acid in humans. *The American Journal of Physiology*, **279**, G1113–G1120.

194 Wallace, J.L. (1997) Nonsteroidal anti-inflammatory drugs and gastroenteropathy: the second hundred years. *Gastroenterology*, **112**, 1000–1016.

195 Fiorucci, S., de Lima, O.M., Jr, Mencarelli, A. *et al.* (2002) Cyclooxygenase-2-derived lipoxin A4 increases gastric resistance to aspirin-induced damage. *Gastroenterology*, **123**, 1598–1606.

196 Wallace, J.L. and Fiorucci, S. (2003) A magic bullet for mucosal protection . . . and aspirin is the trigger! *Trends in Pharmacological Sciences*, **24**, 323–326.

197 Fiorucci, S., Distrutti, E., de Lima, O.M. *et al.* (2003a) Relative contribution of acetylated cyclo-oxygenase (COX)-2 and 5-lipoxygenase (LOX) in regulating gastric mucosal integrity and adaptation to aspirin. *The FASEB Journal*, **17**, 1171–1173.

198 Langenbach, R., Morham, S.G., Tiano, H.F. *et al.* (1995) Prostaglandin synthase 1 gene disruption in mice reduces arachidonic acid-induced inflammation and indomethacin-induced gastric ulceration. *Cell*, **83**, 483–492.

199 Darling, R.L., Romero, J.J., Dial, E.J. *et al.* (2004) The effects of aspirin on gastric mucosal integrity, surface hydrophobicity, and prostaglandin metabolism in cyclooxygenase knockout mice. *Gastroenterology*, **127**, 94–104.

200 Anand, B.S., Romero, J.J., Sanduja, S.K. *et al.* (1999) Phospholipid association reduces the gastric mucosal toxicity of aspirin in human subjects. *The American Journal of Gastroenterology*, **94**, 1818–1822.

201 Davies, N.M., Sharkey, K.A., Asfaha, S. *et al.* (1997) Aspirin causes rapid up-regulation of cyclo-oxygenase-2 expression in the stomach of rats. *Alimentary Pharmacology & Therapeutics*, **11**, 1101–1108.

202 Goddard, P.J. and Lichtenberger, L.M. (1987) Does aspirin damage the canine gastric mucosa by reducing its surface hydrophobicity? *The American Journal of Physiology*, **15**, G421–G430.

203 Pohle, T., Brzozowski, T., Becker, J.C. *et al.* (2001) Role of reactive oxygen metabolites in aspirin-induced gastric damage in humans: gastroprotection by vitamin C. *Alimentary Pharmacology & Therapeutics*, **15**, 677–687.

204 Fiorucci, S., Santucci, L., Wallace, J.L. *et al.* (2003b) Interaction of a selective cyclooxygenase-2 inhibitor with aspirin and NO-releasing aspirin in the human gastric mucosa. *Proceedings of the National Academy of Sciences of the United States of America*, **100**, 10937–10941.

205 Wallace, J.L., Zamuner, S.R., McKnight, W. *et al.* (2004) Aspirin, but not NO-releasing aspirin (NCX-4016), interacts with selective COX-2 inhibitors to aggravate gastric damage and inflammation. *American Journal of Physiology. Gastrointestinal and Liver Physiology*, **286**, G76–G81.

206 Fischer, H., Becker, J.C., Boknik, P. *et al.* (1999) Expression of endothelial cell-derived nitric oxide synthase (eNOS) is increased during gastric adaptation to chronic aspirin intake in humans. *Alimentary Pharmacology & Therapeutics*, **13**, 507–514.

207 Lanas, A., Bajador, E., Serrano, P. *et al.* (2000) Nitrovasodilators, low-dose aspirin, other nonsteroidal antiinflammatory drugs and the risk of upper gastrointestinal bleeding. *The New England Journal of Medicine*, **343**, 834–839.

208 Cullen, D.J.E., Hawkey, G.M., Greenwood, D.C. *et al.* (1997) Peptic ulcer bleeding in the elderly: relative roles of *Helicobacter pylori* and non-steroidal anti-inflammatory drugs. *Gut*, **41**, 459–462.

209 Feldman, M., Cryer, B., Mallat, D. *et al.* (2001) Role of *Helicobacter pylori* infection in gastroduodenal injury and gastric prostaglandin synthesis during long term/low dose aspirin therapy: a prospective placebo-controlled, double-blind randomized trial. *The American Journal of Gastroenterology*, **96**, 1751–1757.

210 Konturek, J.W., Dembinski, A., Konturek, S.J. *et al.* (1997) *Helicobacter pylori* and gastric adaptation to repeated aspirin administration in humans. *Journal of Physiology and Pharmacology*, **48**, 383–391.

211 Konturek, J.W., Dembinski, A., Konturek, S.J. *et al.* (1998) Infection of *Helicobacter pylori* in gastric adaptation to continued administration of aspirin in humans. *Gastroenterology*, **114**, 245–255.

212 McCarthy, D.M. (2001) *Helicobacter pylori* and NSAIDs – what interaction. *European Journal of Surgery*, **167**, 56–65.

213 Chan, F.K., Chung, F.C., Suen, B.Y. *et al.* (2001) Preventing recurrent upper gastrointestinal bleeding in patients with *Helicobacter pylori* infection who are taking low-dose aspirin or naproxen. *The New England Journal of Medicine*, **344**, 967–973.

214 Konturek, J.W., Fischer, H., Konturek, P.C. *et al.* (2001) Heat shock protein 70 (HSP70) in gastric adaptation to aspirin in *Helicobacter pylori* infection. *Journal of Physiology and Pharmacology*, **52**, 153–164.

215 Shirin, H., Moss, S.F., Kancherla, S. *et al.* (2006) Nonsteroidal anti-inflammatory drugs have bacteriostatic and bactericidal activity against *Helicobacter pylori*. *Journal of Gastroenterology and Hepatology*, **21**, 1388–1393.

216 Huang, J.-Q., Sridhar, S. and Hunt, R.H. (2002) Role of *Helicobacter pylori* infection and non-steroidal anti-inflammatory drugs in peptic-ulcer disease: a meta-analysis. *The Lancet*, **359**, 14–22.

217 Slomiany, B.L., Piotrowski, J. and Slomiany, A. (2001) Up-regulation of gastric mucosal inflammatory responses to *Helicobacter pylori* lipopolysaccharide by aspirin but not indomethacin. *Journal of Endotoxin Research*, **7**, 203–209.

218 Newton, J.L., Johns, C.E. and May, F.E. (2004) Review article: the ageing bowel and intolerance to aspirin. *Alimentary Pharmacology & Therapeutics*, **19**, 39–45.

219 Peura, D.A., Lanza, F.L., Gostout, C.J. *et al.* (1997) The American College of Gastroenterology Bleeding Registry: preliminary findings. *The American Journal of Gastroenterology*, **92**, 924–928.

220 Hernandez-Diaz, S. and Garcia-Rodriguez, L.A. (2006) Cardioprotective aspirin users and their excess risk of upper gastrointestinal complications. *BMC Medicine*, **4**, 22.

221 Hallas, J., Dall, M., Andries, A. *et al.* (2006) Use of single and combined antithrombotic therapy and risk of serious upper gastrointestinal bleeding: population-based case-control study. *British Medical Journal*, **333**, 726–729.

222 Wang, D. and Bakhai, A. (2006) *Clinical Trials. A Practical Guide to Design Analysis and Reporting*, Remedica, London.

223 Kelly, J.P., Kaufman, D.W., Jurgelon, J.M. *et al.* (1996) Risk of aspirin-associated major upper-gastrointestinal bleeding with enteric-coated or buffered product. *The Lancet*, **348**, 1413–1416.

224 Sorensen, H.T., Mellemkjaer, L., Blot, W.J. *et al.* (2000) Risk of upper gastrointestinal bleeding associated with use of low-dose aspirin. *The American Journal of Gastroenterology*, **95**, 2218–2224.

225 Needham, C.D., Kyle, J., Jones, P.F. *et al.* (1971) Aspirin and alcohol in gastrointestinal hemorrhage. *Gut*, **12**, 819–821.

226 Kaufman, D.W., Kelly, J.P., Wiholm, B.E. *et al.* (1999) The risk of acute major upper gastrointestinal bleeding among users of aspirin and ibuprofen at various levels of alcohol consumption. *The American Journal of Gastroenterology*, **94**, 3193–3196.

227 Hawthorne, A.B., Mahida, Y.R., Cole, A.T. *et al.* (1991) Aspirin-induced gastric mucosal damage: prevention by enteric-coating and relation to prostaglandin synthesis. *British Journal of Clinical Pharmacology*, **32**, 77–83.

228 Garcia-Rodriguez, L.A., Hernandez-Dias, S. and de Abajo, F.J. (2001) Association between aspirin and upper gastrointestinal complications: systematic review of epidemiologic studies. *British Journal of Clinical Pharmacology*, **52**, 563–571.

229 Hoftiezer, J.W., Silvoso, G.R., Burks, M. *et al.* (1980) Comparison of the effects of regular and enteric-coated aspirin on gastrointestinal mucosa of man. *The Lancet*, **2**, 609–612.

230 Lanza, F.L., Royer, G.L. and Nelson, R.S. (1980) Endoscopic evaluation of the effects of aspirin, buffered aspirin, and enteric coated aspirin on gastric and duodenal mucosa. *The New England Journal of Medicine*, **303**, 136–138.

231 Cole, A.T., Hudson, N., Liew, L.C. *et al.* (1999) Protection of human gastric mucosa against aspirin-enteric coating or dose reduction? *Alimentary Pharmacology & Therapeutics*, **13**, 187–193.

232 Dammann, H.G., Burkhardt, F. and Wolf, N. (1999) Enteric coating of aspirin significantly decreases gastroduodenal mucosal lesions. *Alimentary Pharmacology & Therapeutics*, **13**, 1109–1114.

233 Blondon, H., Barbier, J.P., Mahe, I. *et al.* (2000) Gastroduodenal tolerability of medium dose enteric-coated aspirin: a placebo controlled endoscopic study of a new enteric-coated formulation versus regular formulation in healthy volunteers. *Fundamental & Clinical Pharmacology*, **14**, 155–157.

234 Banoob, D.W., McCloskey, W.W. and Webster, W. (2002) Risk of gastric injury with enteric- versus nonenteric-coated aspirin. *Annals of Pharmacotherapy*, **36**, 163–166.

235 Laine, L., Maller, E.S., Yu, C. *et al.* (2004) Ulcer formation with low-dose enteric-coated aspirin and the effect of COX-2 selective inhibition: a double-blind trial. *Gastroenterology*, **127**, 395–402.

236 Clarke, R.J., Mayo, G., Price, P., *et al.* (1991) Suppression of thromboxane A2 but not of systemic prostacyclin by controlled-release aspirin. *The New England Journal of Medicine*, **325**, 1137–1141.

237 Kurata, J.H. and Abbey, D.E. (1990) The effect of chronic aspirin use on duodenal and gastric ulcer hospitalizations. *Journal of Clinical Gastroenterology*, **12**, 260–266.

238 Roderick, P.J., Wilkes, H.C. and Meade, T.W. (1993) The gastrointestinal toxicity of aspirin: an overview of randomized controlled trials. *British Journal of Clinical Pharmacology*, **35**, 219–226.

239 Pierson, R.N., Holt, P.R., Watson, R.M. *et al.* (1961) Aspirin and gastrointestinal bleeding: chromate^{51} blood loss studies. *The American Journal of Medicine*, **31**, 259–265.

240 UK TIA Study Group (1988) United Kingdom transient ischaemic attack aspirin trial: interim results. *British Medical Journal*, **296**, 316–320.

241 Slattery, J., Warlow, C.P., Shorrock, C.J. *et al.* (1995) Risks of gastrointestinal bleeding during secondary prevention of vascular events with aspirin. Analysis of gastrointestinal bleeding: the UK TIA-trial. *Gut*, **37**, 509–511.

242 Fuster, V., Chesebro, J.H., (1995) Aspirin for primary prevention of coronary disease. *European Heart Journal*, **16** (Suppl E), 16–20.

243 Steering Committee of the Physicians' Health Study Research Group (1989) Final report on the aspirin component of the ongoing Physicians' Health Study. *The New England Journal of Medicine*, **321**, 129–135.

244 Peto, R., Gray, R., Collins, R. *et al.* (1988) Randomized trial of prophylactic daily aspirin in British male doctors. *British Medical Journal*, **296**, 313–316.

245 Dutch TIA Trial Study Group (1991) A comparison of two doses of aspirin (30 mg vs. 283 mg a day) in patients after a transient ischemic attack or minor ischaemic stroke. *The New England Journal of Medicine*, **325**, 1261–1266.

246 De Gaetano, G. (1988) Primary prevention of vascular disease by aspirin. *The Lancet*, **14**, 1093–1094.

247 De Gaetano, G. on behalf of the Collaborative Group of the Primary Prevention Project (2001) Low-dose aspirin and vitamin E in people at cardiovascular risk: a randomized trial in general practice. *The Lancet*, **357**, 89–95.

248 ISIS-2 (Second International Study of Infarct Survival) Collaborative Group (1988) Randomised trial of intravenous streptokinase, oral aspirin, both, or neither among 17187 cases of suspected acute myocardial infarction: ISIS-2. *The Lancet*, **2**, 349–360.

249 Walker, J., Robinson, J., Stewart, J. *et al.* (2007) Does enteric-coated aspirin result in a lower incidence of gastrointestinal complications compared to normal aspirin? *Interactive Cardiovascular and Thoracic Surgery*, **6**, 519–522.

250 US Preventive Services Task Force (2002) Aspirin for the primary prevention of cardiovascular events: recommendations and rationale. *Annals of Internal Medicine*, **136**, 157–160.

251 American Diabetes Association (2004) Position statement: aspirin in diabetes. *Diabetes Care*, **27**, S72–S73.

252 Hudson, N., Murray, F.E., Cole, A.T. *et al.* (1997) Effect of sucralfate on aspirin induces mucosal injury and impaired haemostasis in humans. *Gut*, **41**, 19–23.

253 Berkowitz, J.M., Rogenes, P.R., Sharp, J.T. *et al.* (1987) Ranitidine protects against gastroduodenal mucosal damage associated with chronic aspirin therapy. *Archives of Internal Medicine*, **147**, 2137–2139.

254 Kitchingman, G.K., Prichard, P.J., Daneshmend, T.K. *et al.* (1989) Enhanced gastric-mucosal bleeding with doses of aspirin used for prophylaxis and its reduction by ranitidine. *British Journal of Clinical Pharmacology*, **28**, 581–585.

255 Lanza, F., Peace, K., Gustitus, L. *et al.* (1988) A blinded endoscopic comparative-study of misoprostol versus sucralfate and placebo in the prevention of aspirin-induced gastric and duodenal ulceration. *The American Journal of Gastroenterology*, **83**, 143–146.

256 Jiranek, G.C., Kimmey, M.B., Saunders, D.R. *et al.* (1989) Misoprostol reduces gastroduodenal injury from one week of aspirin – an endoscopic study. *Gastroenterology*, **96**, 656–661.

257 Roth, S., Agrawal, N., Mahowald, M. *et al.* (1989) Misoprostol heals gastroduodenal injury in patients with rheumatoid arthritis receiving aspirin. *Archives of Internal Medicine*, **149**, 775–779.

258 Donnelly, M.T., Goddard, A.F., Filipowicz, B. *et al.* (2000) Low-dose misoprostol for the prevention of low-dose aspirin-induced gastroduodenal injury. *Alimentary Pharmacology & Therapeutics*, **14**, 529–534.

259 Daneshmend, T.K., Stein, A.G., Bhaskar, N.K. *et al.* (1990) Abolition by omeprazole of aspirin-induced gastric-mucosal injury in man. *Gut*, **31**, 514–517.

260 Dent, J. (1998) Why proton pump inhibition should heal and protect against nonsteroidal anti-inflammatory drug ulcers. *The American Journal of Medicine*, **104** (3A), 52S–55.

261 Lai, K.C., Lam, S.K., Chu, K.M. *et al.* (2002) Lansoprazole for the prevention of recurrences of ulcer complications from long-term low aspirin use. *The New England Journal of Medicine*, **346**, 2033–2038.

262 Ng, F.H., Wong, B.C., Wong, S.Y. *et al.* (2004) Clopidogrel plus omeprazole compared with aspirin plus omeprazole for aspirin-induced symptomatic peptic ulcers/erosions with low to moderate bleeding/re-bleeding risk – a single blind, randomized controlled study. *Alimentary Pharmacology & Therapeutics*, **19**, 359–365.

263 Inarrea, P., Esteva, F., Cornudella, R. *et al.* (2000) Omeprazole does not interfere with the antiplatelet effect of low-dose aspirin in man. *Scandinavian Journal of Gastroenterology*, **35**, 242–246.

264 Chan, F.K.L., Ching, J.Y.L., Gung, L.C.T. *et al.* (2005) Clopidogrel versus aspirin and esomeprazole to prevent recurrent ulcer bleeding. *The New England Journal of Medicine*, **352**, 238–244.

265 Wallberg-Johnson, S., Ohmann, M.L. and Dahlqvist, S.R. (1997) Cardiovascular morbidity and mortality in patients with seropositive rheumatoid arthritis in Northern Sweden. *The Journal of Rheumatology*, **24**, 445–451.

266 FitzGerald, G.A. (2002) Cardiovascular pharmacology of nonselective nonsteroidal anti-inflammatory drugs and coxibs: clinical considerations. *The American Journal of Cardiology*, **89** (Suppl.) 26D–32D.

267 Silverstein, F.E., Faich, G., Goldstein, J.L. *et al.* (2000) Gastrointestinal toxicity with celecoxib vs. nonsteroidal antiinflammatory drugs for osteoarthritis and rheumatoid arthritis: the CLASS study – a randomized controlled trial. *The Journal of the American Medical Association*, **284**, 1247–1255.

268 Nathan, J.P., Castellanos, J.T., Rosenberg, J.M. *et al.* (2003) Impact of aspirin on the gastrointestinal-sparing effects of cyclooxygenase-2 inhibitors. *American Journal of Health-System Pharmacy*, **60**, 392–394.

269 Cox, E.R., Frisse, M., Behm, A. *et al.* (2004) Over the counter pain reliever and aspirin use within a sample of long-term cyclooxygenase 2 users. *Archives of Internal Medicine*, **164**, 1243–1245.

270 Farkouh, M.E., Kirshner, H., Harrington, R.A. *et al.* (2004) Comparison of lumiracoxib with naproxen and ibuprofen in the Therapeutic Arthritis Research and Gastrointestinal Event Trial (TARGET), cardiovascular outcomes: randomised controlled trial. *The Lancet*, **364**, 675–684.

271 Lanas, A. and Hunt, R. (2006) Prevention of antiinflammatory drug-induced gastrointestinal damage: benefits and risks of therapeutic strategies. *Annals of Medicine*, **38**, 415–428.

3.2.2
Kidney

The excretion of salicylates and their several phase-I and phase-II metabolites (Section 2.1.2) occurs exclusively through the kidney. Therefore, it is to be expected that the kidney is also a major target of salicylate toxicity. However, renal failure is no typical symptom of early acute salicylate poisoning but rather occurs in later or chronic stages of severe salicylate poisoning, mostly in predisposed individuals [272]. Moreover, treatment of salicylate intoxication with sodium bicarbonate and other alkalinizing agents rather requires a functioning kidney for appropriate salicylate "washout," that is, active drug clearance (Section 3.1.1).

Although kidney dysfunction with albuminuria was described in patients at high-dose salicylate as early as 1917 [273], any direct or even causal relationship to aspirin intake has not been established, neither in this nor in later trials. Today, repeated or long-term use of aspirin for cardiocoronary prophylaxis or pain control is generally considered safe in individuals with normal renal function [272, 274]. This low or absent renal toxicity of aspirin monotherapy differs fundamentally from renal toxicity caused by long-term (ab)use of analgesic mixtures. These may contain not only aspirin and acetaminophen together with other components, historically the most important phenacetin, but also compounds such as caffeine and codeine with habit-forming properties. In addition, subsets of individuals may exist who are more susceptible to renal toxicity of drugs than others. This includes patients with preexisting renal diseases including chronic kidney diseases and diabetes as well as elderly individuals with hypoalbuminemia and reduced kidney function [275, 276]. Finally, possible side effects on the kidney in patients taking long-term aspirin for cardiocoronary prevention are of interest.

In addition to potentially negative actions, there might also be beneficial effects of aspirin on the kidney function in renal diseases. This includes some types of glomerulonephritis, cyclosporine nephrotoxicity, and renal allograft function. Here, anti-inflammatory properties of aspirin as well as inhibition of thromboxane formation are probably therapeutically relevant [277–279]. These actions are still the subject of research and not discussed in greater detail here because their (positive) clinical relevance has still to be established in randomized controlled trials.

3.2.2.1 Analgesic Nephropathy

History A chronic interstitial nephritis after excessive long-term use of antipyretic analgesics was first described in Switzerland half a century ago [280]. About one-third of the 44 patients studied by the authors reported preceding chronic abuse of phenacetin-containing pain medications. The disease was associated with ultrastructural changes in the lower urinary tract [281]. Similar changes were also found after excessive long-term use of other analgesics containing phenacetin or acetaminophen as active ingredients whereas occasional reports' data on aspirin-related nephrotoxicity were less convincing and in most cases negative [282–286].

There was an interesting case report on analgesic abuse and renal function in 17 patients with rheumatoid arthritis. Each of them had consumed more than 5 kg (!) of aspirin. These patients only had some minor abnormalities in kidney function tests and none of them exhibited any clinically significant impairment of renal function [287]. In another case report, acute kidney failure was seen in an otherwise healthy 21-year-old man several hours after intake of 125 g (!) aspirin. There was massive polyuria that came back to normal within 1 week [288].

This "environmental disease of (Western) societies" [282], which is also known as "phenacetin nephritis," has raised serious concerns regarding the renal safety of NSAIDs, in particular, during long-term use of analgesic mixtures [274].

Incidence and Symptoms According to De Broe and Elseviers [289], the classic analgesic nephropathy is a slowly developing progressive disease resulting from the daily use for many (more than 5) years of mixtures containing at least two antipyretic analgesics and, usually, caffeine or codeine (or both). The nephropathy is characterized by renal papillary necrosis and chronic interstitial nephritis with an insidious progression to renal failure. Clinically, the disease presents initially with polyuria and renal colic, occasionally associated with acute renal failure due to bilateral obstruction of the ureters. Hematuria occurs with the elimination of necrotic papilla. With further progression of the disease, there appear the nonspecific symptoms of advanced renal failure. The patients may also suffer from chronic pain syndromes and several somatic conditions, including peptic ulcers and hypertension [289, 290]. The incidence of the disease is highly variable between different countries, possibly because of differences in the pattern of analgesic use. It was estimated to be about 2–30% [284] after one to two decades of abuse of analgesic antipyretics.

Mechanisms of Analgesic Nephropathy: The Role of Phenacetin and Acetaminophen Nephrotoxicity, due to abuse of analgesic mixtures containing phenacetin and acetaminophen, was established in several prospective case–control studies [291–294]. The risk for progression to kidney failure was increased six- to eightfold. As possible causes, phenacetin and its metabolite acetaminophen, generated during the hepatic metabolism of phenacetin, were the two compounds most intensively studied. In addition, their combined use with other analgesics, including salicylates, was another issue of concern.

Phenacetin undergoes first-pass metabolism in the liver, resulting in the generation of acetaminophen as the main metabolite. Acetaminophen is then taken up by the kidney and excreted. During excretion, acetaminophen becomes concentrated in the papillae. By oxidative metabolism, acetaminophen is converted to a reactive quinoneimine that is conjugated to glutathione. If glutathione is depleted, either by toxic doses of acetaminophen or by comedication with other drugs that also use glutathione for intermediate metabolism, such as salicylate (Section 2.1.2), depletion in cellular glutathione levels may occur, the reactive metabolite of acetaminophen then produces toxic products, ultimately resulting in necrosis of the papillae [294]. Experimental evidence also supports the notion that acetaminophen has a predominant role in analgesic nephropathy and that its actions may become potentiated by one or more of the other agents in mixed analgesic combinations, including aspirin [289, 294].

This concept of a key role of acetaminophen in analgesic nephropathy has recently been challenged by Mihatsch *et al.* [295]. In a series of about 600 adult autopsies at the Basle Institute of Pathology between 2000 and 2002, they found only 0.2% analgesic nephropathies (one case) as opposed to 3% in 1980. This coincided with the disappearance of phenacetin despite the fact that mixed analgesics containing acetaminophen continued to be widely used. The conclusion was that phenacetin was the "bad guy" and that the classic analgesic nephropathy apparently disappeared after phenacetin was banned from the market [295]. Unfortunately, there were no data in this study about drug use prior to death, including information about doses and duration of use. It can also not be excluded that the habits of analgesic use have changed during the past 20 years, including the availability of new OTC analgesics with a different mode of action but for similar indications like pain control, such as ibuprofen. With the currently available evidence, habitual and not occasional intake of acetaminophen should be discouraged [296].

3.2.2.2 Mode of Aspirin Action

The central biochemical process in the kidney that is affected by aspirin is prostaglandin biosynthesis. Prostaglandins in the kidney serve several functions: one is the regulation of renal blood flow and the other the contribution to sodium excretion [274]. Renal prostaglandin biosynthesis is low in healthy individuals. However, it becomes rapidly upregulated in situations of hemodynamic destabilization, associated with reduced circulating blood volume and subsequent activation of (intrarenal) pressure-control systems to avoid renal underperfusion. Another function of prostaglandins is the modulation of tubular function. In both situations, reactive stimulation of renal prostaglandin synthesis may become a critical factor for homeostasis, that is, control of blood pressure (BP) and fluid and electrolyte balance.

In contrast to traditional NSAIDs or coxibs, aspirin is only a weak inhibitor of renal prostaglandin biosynthesis. This might explain its relatively modest effects on kidney function both in "resting" and "stimulated" conditions.

3.2.2.3 Clinical Studies

Healthy Individuals The absence of renal side effects of aspirin, including regular long-term use over years, has recently been confirmed in a post hoc analysis of more than 11 000 participants of the American Physicians' Health Study (Section 4.1.1) [297]. Slightly different results were obtained in the Nurses' Health Study: lifetime analgesic consumption, including aspirin and non-aspirin NSAIDs, was not related to any dose-dependent decline in kidney function, whereas acetaminophen caused a dose-dependent increase in renal dysfunction that was nearly three times higher than that for aspirin at the same doses (Table 3.5) [298]. There is no evidence that long-term use of aspirin alone will result in renal dysfunction of otherwise healthy individuals. In a prospective radiographic study of 259 patients who had taken 1000–26 000 NSAID doses, papillary necrosis was found in 38 users, taking predominantly physician-prescribed NSAIDs. However, only 65% of these patients had renal functional impairment [299].

A recent population-based observational trial in the United States has studied habitual use of an analgesic, including aspirin, acetaminophen, and ibuprofen, as defined as ever intake for at least 1 month every day and its relation to decreased kidney function. The study was conducted in about 8000 adults and was also classified by the duration of use, that is, <1, 1–5, and >5 years. The outcome measure was albuminuria and reduced estimated glomerular filtration rate (GFR).

Twenty-four percent of the study population reported habitual analgesic use. The odds ratio for all compounds was 1.0 for GFR and 0.9 for albuminuria as compared with nonhabitual analgesic use. This was not significant. However, the reliability of self-reported analgesic use behavior could be assessed.

The conclusion was that habitual use of single or multiple analgesics was not associated with disturbed kidney function [300].

Table 3.5 Effect of lifetime analgesic consumption on decline in renal function in women (Nurses' Health Study) (modified after [298]).

Lifetime intake (g)	Participants ($n = 1697$)	Change (%)	Odds ratio (95% CI)
ASA			
<100	608	53 (9)	1.0 (reference)
100–499	176	16 (9)	0.7 (0.4–1.3)
500–2999	403	46 (11)	0.8 (0.5–1.3)
>3000	455	49 (11)	0.9 (0.6–1.4)
NSAIDs			
<100	790	67 (8)	1.0 (reference)
100–499	181	24 (13)	1.3 (0.8–2.2)
500–2999	376	41 (11)	1.1 (0.7–1.7)
>3000	292	31 (11)	1.1 (0.7–1.8)
Acetaminophen			
<100	819	56 (7)	1.0 (reference)
100–499	186	21 (11)	1.80 (1.0–3.2)
500–2999	288	40 (14)	2.23 (1.4–3.6)
>3000	352	45 (13)	2.04 (1.3–3.2)

Patients with Preexisting Kidney Diseases Some early nonrandomized studies in patients with preexisting kidney disease found reversible reductions in glomerular filtration by aspirin [301, 302]. Several epidemiologic studies have led to the hypothesis that habitual analgesic use contributes to the progression of chronic renal disease. This also includes aspirin as a single medication [284, 286, 303, 304].

The relationship between the intake of antipyretic analgesics and terminal kidney failure was studied in a case–control trial of 921 end-stage renal failure patients on hemodialysis as compared with 517 matched controls.

Regular long-term intake of a standardized minimum dose of more than 15 analgesic units per month for more than 1 year as monopreparations was not associated with an increased relative risk of terminal kidney failure. In contrast, there was a clear dose- and time-dependent relation between drug intake and terminal kidney failure for combined preparations with a particular high risk for combinations, containing caffeine; relative risk 52.6 as opposed to 4.0 with acetaminophen and 2.4 with aspirin.

The conclusion was that an increased risk of end-stage renal failure is related to both dose and exposure time of mixed analgesic compounds but not for the use of only single ingredient analgesics [303, 304].

Evidence for a low if any injury potential of aspirin – as opposed to acetaminophen and NSAIDs – came from several population-based case control studies in patients with chronic kidney diseases. The study included patients with early- [284] and end-stage [286] renal failure. Despite the different protocols, patient populations, and selection criteria, these studies found a significantly enhanced risk for kidney failure by acetaminophen but much less, if any, increased risk for kidney failure by aspirin. Another recent population-based case–control study [305] reported that regular analgesic intake (at least two tablets per week for at least 2 months) was associated with a 2.5-fold increased risk of chronic renal failure by aspirin and acetaminophen and concluded that both drugs may exacerbate chronic renal failure. This effect was more consistent with acetaminophen (Figure 3.10), and the odds ratio in 324 patients with chronic renal failure due to diabetes was about twice as much in regular acetaminophen users as compared with regular aspirin users. The possibility of bias due to the triggering of analgesic consumption by predisposing conditions was not excluded [305]. Thus, there appears to be evidence for an acetaminophen-related kidney dysfunction that is time and dose dependent but not for aspirin. In addition, available data strongly suggest avoiding analgesic mixtures containing centrally acting or dependence-producing drugs [289]. However, aspirin might prolong bleeding time in uremia by a mechanism, independent of inhibition of prostaglandin biosynthesis [306].

Patients with Preexisting Cardiovascular Diseases
Patients with established chronic kidney disease are also at elevated risk of cardiovascular diseases. A meta-analysis of 14 randomized trials of antiplatelet therapy, including more than 2600 hemodialysis patients, showed that antiplatelet therapy in these patients was associated with a significant 41% reduction in the risk of a severe vascular event. Chronic kidney disease might cause impaired hemostasis and, thus, there might be an excess risk in

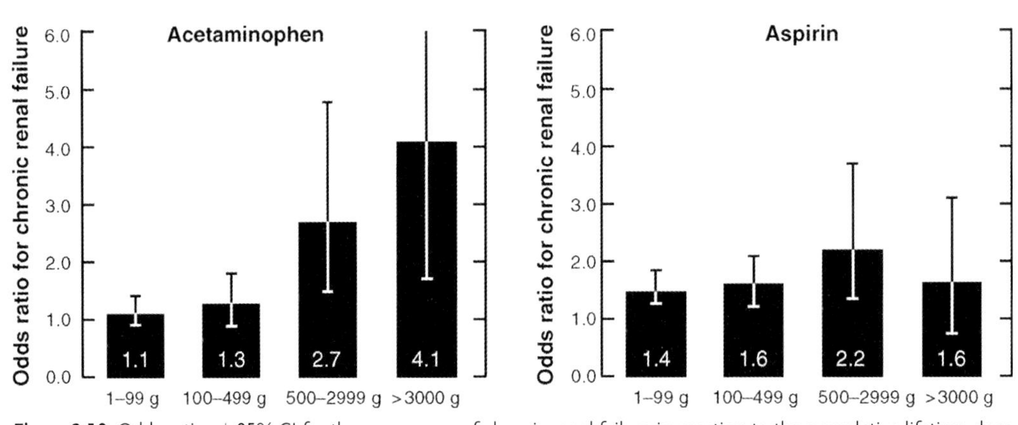

Figure 3.10 Odds ratios ± 95% CI for the occurrence of chronic renal failure in reaction to the cumulative lifetime dose of acetaminophen or aspirin. All patients had preexisting renal or systemic disease according to serum creatinine, >3.4 mg/dl in men and >2.8 mg/dl in women, respectively. The odds ratios are for comparisons with nonusers of both classes of analgesics until the time of interview [305].

major bleedings in association with antiplatelet therapy. However, only 46 major bleeds occurred in this meta-analysis of randomized trials [307]. This important issue was studied recently for the first time in a randomized prospective trial.

The First United Kingdom Heart and Renal Protection (UK-HARP-I) Study was a feasibility study to investigate the efficacy and safety of simvastatin and aspirin in a randomized 2 × 2 factorial design versus placebo in 448 patients with chronic kidney diseases. The aspirin group received 100 mg/day modified-release aspirin or a matching placebo for a median duration of 1 year.

Regarding safety, aspirin use was not associated with a significant excess of major bleed as compared with placebo. There was an about threefold increase in the risk in minor bleeds (RR 2.8, 95% CI: 1.5–5.3, $p = 0.001$) but no aspirin-related increase in kidney dysfunction nor increased urate levels or acute gout. Allocation to aspirin therapy was not associated with any significant difference in events, that is, decrease in renal function or time of initiation of dialysis therapy.

The conclusion was that aspirin at 100 mg/day is well tolerated in patients with chronic kidney disease but is associated with a threefold increase in minor bleeding. There was no increase in greater blood loss and no excess risk in renal dysfunction and/or its progression. However, a much larger trial is required to determine reliably whether low-dose aspirin has clinically significant effects on renal function in predialysis and dialysis patients [308].

Diabetes Diabetes, in particular, type 2, needs thrombosis prevention because of the significantly enhanced atherothrombotic risk. This is done with appropriate antiplatelet drugs, including aspirin (Section 4.1.1). Renal biosynthesis of prostaglandins is enhanced in both experimental and human diabetes [309], possibly because of upregulation of COX-2.

Microalbuminuria, associated with diabetic renopathy, can be reduced by high-dose aspirin in combination with dipyridamole (990 mg aspirin 225 mg dipyridamole/day) in type 1 diabetes, probably by a prostaglandin-related mechanism [310] whereas low-dose aspirin had no effect [311]. No effect on kidney function was seen with antithrombotic doses of aspirin (150 mg/day) in type 2 diabetics, suggesting that regular use of the drug in primary and secondary prevention of atherothrombosis in these patients does not have an impact on renal function [312] (Table 3.6).

Hypertension In essential hypertension renal generation of vasodilatory prostaglandins may become an important blood pressure regulating factor. In this case, inhibition of prostaglandin biosynthesis by aspirin may result in an amelioration of potency of antihypertensives, such as ACE inhibitors, β-blockers, and other compounds whose blood pressure lowering potency involves stimulation of renal vasodilator prostaglandin biosynthesis. Most available data do not suggest a clinically relevant interaction of aspirin at doses below 300 mg/day with antihypertensives, specifically ACE inhibitors (Section 4.1.1) at antithrombotic doses [313]. However, this issue is not finally answered yet.

Heart Failure The safety of aspirin in heart failure patients was studied in a community-based observational study in patients discharged from Canadian hospitals after a first hospitalization for heart failure. From more than 7000 patients, 44% had no coronary heart disease and 29% renal dysfunction.

Table 3.6 Effects of low-dose aspirin (150 mg/day) on GFR and blood pressure in type-2 diabetic patients with elevated urinary albumin secretion.

	Placebo	Aspirin	*p*
Urinary albumin excretion (mg/24 h)	205 (124–340)	201 (119–341)	0.78
GFR (ml/min/l/1.73 m^3)	102 (93–110)	103 (94–111)	0.58
Systolic BP (mmHg)	152 (146–158)	151 (143–158)	0.68
Diastolic BP (mmHg)	87 (82–91)	87 (83–91)	0.88

Double-blind crossover trial versus placebo in 31 patients. Data are mean and 95% CI (modified after [312]).

Compared with nonusers, aspirin users were not more likely to die or require heart failure readmission (RR: 1.02, 95% CI: 0.91–1.16), including patients without coronary heart disease and patients with renal dysfunction. Beneficial effects were seen with ACE inhibitors, and these were not reduced by aspirin cotreatment.

The conclusion was that aspirin in heart failure patients did not attenuate the beneficial effects of ACE inhibitors. This allays concerns about the safety of aspirin in patients with heart failure [314]. A similar conclusion was reached by Masoudi *et al.* [315] in patients with heart failure associated with coronary artery disease.

Summary

Repeated and even long-term use of aspirin in individuals with normal kidney function is not associated with an elevated risk of nephropathy or renal failure. There is also no established relationship between the progression of preexisting renal diseases and aspirin intake, including patients with diabetic nephropathy who regularly take aspirin for cardiovascular prevention.

Long-term (more than 5 years) abuse of analgesic mixtures (phenacetin nephritis) can cause papillary necroses and severe kidney dysfunction. Most likely, phenacetin in analgesic mixtures or a phenacetin metabolite is responsible for these alterations. There appears to be some nephrotoxic potential for acetaminophen, particularly in regular long-term use. However, generation of acetaminophen from phenacetin during intermediary metabolism will probably not explain the nephrotoxicity of phenacetin.

Inhibition of renal prostaglandin biosynthesis by aspirin, in general, is weak, but theoretically may interfere with antihypertensive drugs or other compounds where stimulation of renal prostaglandin formation is part of their clinical efficacy. There is no evidence that aspirin prophylaxis at conventional antithrombotic doses (about 150 mg/day) induces or aggravates preexisting kidney dysfunction, including high-risk groups, such as diabetics. Long-term use of aspirin will probably not aggravate cardiac failure because of ischemic or nonischemic conditions.

References

272 D'Agati, V. (1996) Does aspirin cause acute or chronic renal failure in experimental animals and in humans? *American Journal of Kidney Diseases*, **28** (Suppl. 1), S24–S29.

273 Hanzlik, P.J., Scott, R.W. and Thoburn, T.W. (1917) The salicylates. *Archives of Internal Medicine*, **19**, 1029–1041.

274 Bennett, W.M., Henrich, W.L. and Stoff, J.S. (1996) The renal effects of nonsteroidal anti-inflammatory drugs: summary and recommendations. *American Journal of Kidney Diseases*, **28** (Suppl. 1), S56–S62.

275 Kleinknecht, D. (1995) Intestinal nephritis, the nephritic syndrome and chronic renal failure secondary to nonsteroidal anti-inflammatory drugs. *Seminars in Nephrology*, **15**, 228–235.

276 Segal, R., Lubart, E., Leibovitz, A. *et al.* (2003) Early and late effects of low-dose aspirin on renal function in elderly patients. *The American Journal of Medicine*, **115**, 462–466.

277 Adams, K.E., Brown, P.A.J., Heys, S.D. *et al.* (1993) Alleviation of experimental cyclosporine A nephrotoxicity by low dose aspirin in the rat. *Biochemical Pharmacology*, **46**, 2104–2108.

278 Winchester, J.F. (1996) Therapeutic use of aspirin in renal diseases. *American Journal of Kidney Diseases*, **28** (Suppl. 1), S20–S23.

279 Grotz, W., Siebig, S., Olschewski, M. *et al.* (2004) Low-dose aspirin therapy is associated with improved allograft function and prolonged allograft survival after kidney transplantation. *Transplantation*, **77**, 1848–1853.

280 Spühler, O. and Zollinger, H.U. (1953) Die chronische interstitielle Nephritis. *Zeitschrift für Klinische Medizin*, **151**, 1–50.

281 Mihatsch, M.J., Torhorst, J., Amsler, B. *et al.* (1978) Capillarosclerosis of the lower urinary tract in analgesic (phenacetin) abuse. An electron-microscopic study. *Virchows Archives of Pathology and Anatomy*, **381**, 41–47.

282 Gsell, O. (1974) Von der Phenacetinniere zur Analgetikanephropathie 1953–1974. *Schweizerische Rundschau fur Medizin Praxis*, **63**, 1299–1302.

283 Segasothy, M., Suleiman, A.B., Puvaneswary, M. *et al.* (1988) Paracetamol: a cause for analgesic nephropathy and end-stage renal disease. *Nephron*, **50**, 50–54.

284 Sandler, D.L., Smith, J.C., Weinberg, C.R. *et al.* (1989) Analgesic use and chronic renal disease. *The New England Journal of Medicine*, **320**, 1238–1243.

285 Dubach, U.C., Rosner, B. and Sturmer, T. (1991) An epidemiologic study of abuse of analgesics. Effects of phenacetin and salicylate on mortality and cardiovascular disease. *The New England Journal of Medicine*, **324**, 155–160.

286 Perneger, T.V., Whelton, P.K. and Klag, M.J. (1994) Risk of kidney failure associated with the use of acetaminophen, aspirin and non steroidal antiinflammatory drugs. *The New England Journal of Medicine*, **331**, 1675–1679.

287 Macklon, A.F., Craft, A.W., Thompson, M.T. *et al.* (1974) Aspirin and analgesic nephropathy. *British Medical Journal*, **1**, 597–600.

288 Rupp, D.J., Seaton, R.D. and Wiegmann, T.B. (1983) Acute polyuric renal failure after aspirin intoxication. *Archives of Internal Medicine*, **143**, 1237–1238.

289 De Broe, M.E. and Elseviers, M.M. (1998) Analgesic nephropathy. *The New England Journal of Medicine*, **338**, 446–452.

290 Bennett, W.M. and De Broe, M.E. (1989) Analgesic nephropathy – a preventable renal disease. *The New England Journal of Medicine*, **320**, 1269–1271.

291 Dubach, U.C., Rosner, B. and Pfitzer, E. (1983) Epidemiologic study of abuse of analgesics containing phenacetin: renal morbidity and mortality (1968–1979). *The New England Journal of Medicine*, **308**, 357–362.

292 Elseviers, M.M. and de Broe, M.E. (1995) A long-term prospective controlled study of analgesic abuse in Belgium. *Kidney International*, **48**, 1912–1919.

293 Elseviers, M.M. and de Broe, M.E. (1996) Combination analgesic involvement in the pathogenesis of analgesic nephropathy: the European perspective. *American Journal of Kidney Diseases*, **28** (Suppl. 1), S48–S55.

294 Duggin, G.G. (1996) Combination analgesic-induced kidney disease: the Australian experience. *American Journal of Kidney Diseases*, **28** (Suppl. 1), S39–S47.

295 Mihatsch, M.J., Khanlari, B. and Brunner, F.P. (2006) Obituary to analgesic nephropathy – an autopsy study. *Nephrology Dialysis Transplantation*, **21**, 3139–3145.

296 Vadivel, N., Trikudanathan, S. and Singh, A.K. (2007) Analgesic nephropathy. *Kidney International*, **72**, 517–520.

297 Rexrode, K.M., Buring, J.E., Glynn, R.J. *et al.* (2001) Analgesic use and renal function in men. *The Journal of the American Medical Association*, **286**, 315–321.

298 Curhan, G.C., Knight, E.L., Rosner, B. *et al.* (2004) Lifetime nonnarcotic analgesic use and decline in renal function in women. *Archives of Internal Medicine*, **164**, 1519–1524.

299 Segasothy, M., Samad, S.A., Zulfigar, A. *et al.* (1994) Chronic renal disease and papillary necrosis with the long-term use of nonsteroidal antiinflammatory drugs as the sole or predominant analgesic. *American Journal of Kidney Diseases*, **24**, 17–24.

300 Agoda, L.Y., Francis, M.E. and Eggers, P.W. (2008) Association of analgesic use with prevalence of albuminuria and reduced GFR in US adults. *American Journal of Kidney Diseases*, **51**, 573–583.

301 Berg, K.J. (1977) Acute effects of acetylsalicylic acid in patients with chronic renal insufficiency. *European Journal of Clinical Pharmacology*, **11**, 111–116.

302 Kimberly, R.P. and Plotz, P.H. (1977) Aspirin-induced depression of renal function. *The New England Journal of Medicine*, **296**, 418–424.

303 Pommer, W., Bronder, E., Klimpel, A. *et al.* (1989) Regular intake of analgesic mixtures and risk of end-stage renal failure. *The Lancet*, **1**, 381.

304 Pommer, W., Bronder, E., Greiser, E. *et al.* (1989) Regular analgesic intake and the risk of end-stage renal failure. *American Journal of Nephrology*, **9**, 403–412.

305 Fored, C.M., Ejerblad, E., Lindblad, P. *et al.* (2001) Acetaminophen, aspirin and chronic renal failure. *The New England Journal of Medicine*, **345**, 1801–1808.

306 Gaspari, F., Vigano, G., Orisio, S. *et al.* (1987) Aspirin prolongs bleeding time in uremia by a mechanism distinct from platelet cyclooxygenase inhibition. *The Journal of Clinical Investigation*, **79**, 1788–1797.

307 Antithrombotic Trialists' Collaboration (2002) Collaborative meta-analysis of randomised trials of antiplatelet therapy for prevention of death, myocardial infarction, and stroke in high risk patients. *British Medical Journal*, **324**, 71–86.

308 Baigent, C., Landray, M., Leaper, C. *et al.* (2005) First United Kingdom Heart and Renal Protection (UK-HARP-I) study: biochemical efficacy and safety of simvastatin and safety of low-dose aspirin in chronic kidney disease. *American Journal of Kidney Diseases*, **45**, 473–484.

309 Mathiesen, E.R., Hommel, E., Olsen, U.B. *et al.* (1988) Elevated urinary prostaglandin excretion and the effect of indomethacin on renal function in incipient diabetic nephropathy. *Diabetic Medicine*, **5**, 145–149.

310 Hopper, A.H., Tindall, H. and Davies, J.A. (1989) Administration of aspirin-dipyridamole reduces proteinuria in diabetic nephropathy. *Nephrology Dialysis Transplantation*, **4**, 140–143.

311 Hansen, H.P., Gaede, P.H., Jensen, P.R. *et al.* (2000) Lack of impact of low-dose acetylsalicylic acid on kidney function in type 1 diabetic patients with miroalbuminuria. *Diabetes Care*, **23**, 1742–1745.

312 Gaede, P., Jansen, H.P., Parving, H.-H. *et al.* (2003) Impact of low-dose acetylsalicylic acid on kidney function in type 2 diabetic patients with elevated urinary albumin excretion rate. *Nephrology Dialysis Transplantation*, **18**, 539–542.

313 Teo, K.K., Yusuf, S., Pfeffer, M. *et al.* (2002) Effect of long-term treatment with angiotensin-converting enzyme inhibitors in the presence or absence of aspirin: a systematic review. *The Lancet*, **360**, 1037–1043.

314 McAlister, F.A., Ghali, W.A., Gong, Y. *et al.* (2006) Aspirin use and outcomes in a community-based cohort of 7532 patients discharged after first hospitalization for heart failure. *Circulation*, **113**, 2572–2578.

315 Masoudi, F.A., Wolfe, P., Havranek, P.E. *et al.* (2005) Aspirin use in older patients with heart failure and coronary artery disease: national prescription patterns and relationship with outcomes. *Journal of the American College of Cardiology*, **46**, 955–962.

3.2.3
Liver

The liver, located between the site of drug absorption in the GI tract and drug targets throughout the body after entering systemic circulation, is central to the metabolism of virtually every foreign substance. It is, therefore, not surprising that most drugs cause liver injury infrequently though drug-induced hepatic injury accounts for more than 50% of acute liver failure in the United States [316].

Toxic liver injury by aspirin may occur but is a very random event and not a typical symptom of acute salicylate overdosing. In this respect, aspirin differs qualitatively from acetaminophen and other antipyretic analgesics as well as traditional NSAIDs (e.g., diclofenac). From these compounds, toxic metabolites may be generated during hepatic biotransformation that injures hepatocytes in an irreversible fashion [316, 317]. A probably different etiology has the liver injury and subsequent encephalopathy of Reye's syndrome (Section 3.3.3) and possibly other viral infections [318].

3.2.3.1 Drug-Induced Liver Injury

Toxic liver injury may be caused by metabolites of xenobiotics that cause necrosis and apoptosis through direct toxicity. Acetaminophen and its reactive metabolites, respectively, are examples of direct toxic compounds [319]. Alternatively, drugs, such as aspirin, may cause mitochondrial dysfunction in energy production. In some patients, infections, cytokines, or inborn errors of metabolism (IEM) (β-oxidation) may favor drug-induced impairments in fatty acid metabolism, morphologically appearing as microvesicular steatosis [320] and clinically appearing as Reye-like syndrome (Section 3.3.3). In this case, there is disturbed metabolic function of the liver (and other organs) but no overt toxicity or tissue necrosis.

3.2.3.2 Mechanisms of Aspirin Action

Long-term use of aspirin at high anti-inflammatory doses (plasma salicylate levels of 100–350 μg/ml) may cause liver injury [317]. This is associated with increased serum transaminases but not jaundice. Whether prothrombin biosynthesis is also affected is uncertain [321, 322] but clinically less relevant since severe bleeding problems as a possible consequence of impaired coagulation are only seen in a minority of patients with severe salicylate poisoning (Section 3.1.1). These biochemical and functional disturbances, caused by the salicylate component of the drug, are fully reversible within a few days [323].

Detection of Hepatotoxicity of Salicylates First reports about a possible hepatoxic potential of aspirin appeared after more than half a century of extensive clinical use as an anti-inflammatory analgesic at anti-inflammatory doses of several grams per day. Possible explanations for this somewhat surprising late detection are the absence of typical clinical symptoms of liver toxicity, such as jaundice, and the inability to measure liver enzyme activities in routine laboratory settings. After these became available, several investigators reported elevated serum transaminases after repeated aspirin intake. For example, about one-third or more patients treated with salicylates because of different rheumatic diseases had elevated serum levels of liver enzymes [322, 324, 325]. In these and some other studies, the serum salicylate levels was related to the serum transaminase activity, suggesting a relationship between the two [317].

Incidence and Forms of Hepatic Injury Almost all of the reported patients on aspirin-induced hepatotoxicity were acute cases. The injury was reversible in nature, mild in severity, and biochemical parameters of hepatic injury, that is, increased serum transaminases, generally subsided after withdrawal of aspirin [317], sometimes even if the treatment was maintained [322].

In about 3% of reported cases, injury has been more severe. All of these patients received high anti-inflammatory doses of aspirin over longer periods. There are five reports about a relation between salicylate-induced hepatic injury and encephalopathy. Plasma levels of salicylate in these patients ranged between 270 and 540 μg/ml [317, 326]. There is only one report of a fatal case in a 17-year-old girl who received aspirin combined with acetaminophen. The death possibly related to liver necrosis [327]. There are also very few cases of chronic hepatic injury, histologically and clinically presenting as chronic active hepatitis. Similar to acute injury, the symptoms subsided on withdrawal of aspirin [317].

Determinants of Injury Major determinants of aspirin-induced hepatic injury are the dose, that is, the plasma levels of salicylate and the duration of treatment. Another determinant is the nature of the underlying disease, specifically, chronic rheumatoid diseases and preexisting hepatic or renal failure, allowing for the accumulation of the substance [317].

Signs of hepatotoxicity may develop at high salicylate plasma levels (200 μg/ml or 1 mM and more), though in only very few cases. All patients in whom hepatic injury developed had taken the drug for one to several weeks. A regular intake of high aspirin doses appears to be necessary for hepatotoxicity because even huge single overdoses of aspirin, which cause significant general toxicity (Section 3.1.1), do not produce any overt hepatic failure [328–330]. Regular intake of high doses will also result in significantly increased plasma levels of salicylate even at unchanged dosing because of the reduced clearance and marked prolongation of the salicylate half-life (Section 2.1.2).

Another factor, relevant to salicylate toxicity, is preexisting diseases with immunological background. Almost all reported cases occurred in patients with rheumatoid diseases who had taken the drug for a long time at high doses. Rheumatic diseases are known to be associated with the generation of inflammatory cytokines such as TNFα or IL-6 (Section 2.3.2). These cytokines might induce and maintain inflammatory processes in the liver. Consequently, serum transaminases after repeated high-dose aspirin were higher in patients with rheumatoid disease, when the patients were in an active stage of the disease [331]. The observation that hepatic toxicity of aspirin mainly appears in patients with active rheumatoid diseases suggests that immunological processes might be an aggravating factor. However, there is no positive evidence for this so far and possibly will not be since other anti-inflammatory agents, such as NSAID (Section 4.2.2), have replaced aspirin in this indication. Interestingly, at least some of these compounds, including diclofenac and sulindac, also bear a hepatotoxic potential by impairment of mitochondrial ATP synthesis and production of hepatotoxic reactive metabolites [332].

The finding that aspirin-related liver toxicity requires repeated administration of high doses suggests a relationship with salicylate-induced changes in hepatic energy metabolism, that is, impaired β-oxidation of fatty acids and uncoupling of oxidative phosphorylation (Section 2.2.3.). Both changes are reversible and only evident at large doses of salicylate. Elevated levels of serum transaminases, hepatic steatosis, impaired urea genesis and disturbed lipid metabolism also found in animals fed high aspirin doses [333–335]. Similar to man, these changes are generally reversible and do not cause necroses.

3.2.3.3 Aspirin and Other Drugs

Most drugs can cause liver injury infrequently [316]. However, in association with aspirin, the major compound of interest is acetaminophen, an alternative analgesic antipyretic as well as diclofenac.

Acetaminophen is an example of a drug with dose-related toxic effects after a certain threshold passed, determined by depletion of hepatic glutathione stores. There is rapid hepatocyte injury (centrilobular necrosis) due to increased and dose-dependent generation of reactive metabolites. Acetaminophen toxicity is the most common form of acute liver failure in the United States [316] and the United Kingdom [317]. A recent US-based multicenter prospective study has shown that the annual percentage of acetaminophen-related acute liver failure rose from 28% in 1998 to 51% in 2003. Without liver transplantation, 27% of patients died [336]. The toxicity of acetaminophen – in contrast to aspirin – results almost exclusively from large single overdose, usually suicidal, but there are also sporadic case reports that chronic overdosing of the compound may result in liver failure [337].

Summary

Aspirin, usually well tolerated by the liver, does not cause significant liver injury at single analgesic or repeated antithrombotic doses. Repeated administration of high anti-inflammatory doses might result in increased serum transaminases but no overt liver failure. These changes are transient, reversible after withdrawal of the drug, and salicylate mediated. They are of minor importance today because of the availability of alternative drugs, specifically for long-term treatment of inflammatory disorders.

Liver toxicity of aspirin is probably related to impaired hepatic β-oxidation of free fatty acids and uncoupling of oxidative phosphorylation, both occurring at higher doses of the compound (Section 2.2.3). These metabolic changes are also completely reversible. The possible relationship between aspirin intake and Reye's syndrome is discussed separately (Section 3.3.3). So far, no causal relationship with aspirin intake has been established.

Acetaminophen and other NSAIDs (diclofenac) are hepatotoxic agents. The mode of action is completely different from aspirin. In case of acetaminophen, generation of reactive metabolites occurs after exhaustion of hepatic metabolic pathways, eventually resulting in the formation of toxic metabolites that cause liver cell necrosis and even death from liver failure. On the contrary, the hepatotoxic effects of aspirin, subsequent to acute or chronic overdosing are usually completely reversible.

References

316 Lee, W.M. (2003) Drug-induced hepatotoxicity. *The New England Journal of Medicine*, **349**, 474–485.

317 Zimmermann, H.J. (1981) Effects of aspirin and acetaminophen on the liver. *Archives of Internal Medicine*, **141**, 333–342.

318 Cersosimo, R.J. and Matthews, S.J. (1987) Hepatotoxicity associated with choline magnesium trisalicylate: case report and review of salicylate-induced hepatotoxicity. *Drug Intelligence & Clinical Pharmacy*, **21**, 621–625.

319 Vermeulen, N.P.E., Bessems, J.G.M. and Van de Straat, R. (1992) Molecular aspects of paracetamol-induced hepatotoxicity and its mechanism-based prevention. *Drug Metabolism Reviews*, **24**, 367–407.

320 Pessayre, D., Mansouri, A., Haouzi, D. *et al.* (1999) Hepatotoxicity due to mitochondrial dysfunction. *Cell Biology and Toxicology*, **15**, 367–373.

321 Meyer, O.O. and Howard, B. (1943) Production of hypoprothrombinemia and hypocoagulability of the blood with salicylates. *Proceedings of the Society for Experimental Biology and Medicine*, **53**, 234–237.

322 Athreya, B.H., Moser, G., Cecil, H.S. *et al.* (1975) Aspirin induced hepatotoxicity in juvenile rheumatoid arthritis. *Arthritis and Rheumatism*, **18**, 347–352.

323 Kanada, S.A., Kolling, W.M. and Hindin, B.I. (1978) Aspirin hepatotoxicity. *American Journal of Hospital Pharmacy*, **35**, 330–336.

324 Manso, C., Taranta, A. and Nydick, I. (1956) Effect of aspirin administration on serum glutamic oxaloacetic and glutamic pyruvic transaminases in children. *Proceedings of the Society for Experimental Biology and Medicine*, **93**, 84–88.

325 Russell, A.S., Sturge, R.A. and Smith, M.A. (1971) Serum transaminases during salicylate therapy. *British Medical Journal*, **2**, 428–429.

326 Ulshen, M.H., Grand, R.J., Crain, J.D. *et al.* (1978) Hepatotoxicity with encephalopathy associated with aspirin therapy in rheumatoid arthritis. *The Journal of Pediatrics*, **93**, 1034–1037.

327 Koff, R.S. and Galdabini, J.J. (1977) Fever, myalgias and hepatic failure in a 17-year-old girl. *The New England Journal of Medicine*, **296**, 1337–1346.

328 Troll, M.M. and Menten, M.L. (1945) Salicylate poisoning: report of four cases. *American Journal of Diseases of Children*, **69**, 37–43.

329 Faivre, J., Faivre, M., Lery, N. *et al.* (1974) Hépatotoxicité de l'aspirine. *Journal de Medecine de Lyon*, **55**, 317–324.

330 Boss, G. (1978) Hepatotoxicity caused by acetaminophen or salicylates. *Western Journal of Medicine*, **129**, 50–51.

331 Miller, J.J. III and Weissman, D.B. (1976) Correlations between transaminase concentrations and serum salicylate concentration in juvenile rheumatoid arthritis. *Arthritis and Rheumatism*, **19**, 115–118.

332 O'Connor, N., Dargan, P.I. and Jones, A.L. (2003) Hepatocellular damage from non-steroidal anti-inflammatory drugs. *Quarterly Journal of Medicine*, **96**, 787–791.

333 Janota, I., Wincey, C.W., Sandiford, M. *et al.* (1960) Effect of salicylate on the activity of plasma enzymes in the rabbit. *Nature*, **185**, 935–936.

334 Glasgow, A.M. and Chase, H.P. (1977) Effect of salicylate on ureagenesis in rat liver. *Proceedings of the Society for Experimental Biology and Medicine*, **155**, 48–50.

335 Kundu, R.K., Tonsgard, J.H. and Getz, G.S. (1991) Induction of omega-oxidation of monocarboxylic acids in rats by acetylsalicylic acid. *The Journal of Clinical Investigation*, **88**, 1865–1872.

336 Larson, A.M., Polson, J., Fontana, R.J. *et al.* (2005) Acetaminophen-induced acute liver failure: results of a United States multicenter, prospective study. *Hepatology*, **42**, 1364–1372.

337 Cranswick, N. and Coghlan, D. (2000) Paracetamol efficacy and safety in children: the first 40 years. *American Journal of Therapeutics*, **7**, 135–141.

3.2.4 Audiovestibular System

Salicylate-related hearing loss, tinnitus, and vestibular dysfunctions are known since the clinical introduction of the substances more than hundred years ago [338]. Ototoxic side effects were also rather frequent because of the high doses of the compounds that were taken at the time for treatment of pain and inflammatory diseases. Tinnitus and deafness are also the typical early symptoms of acute salicylate overdosing (Section 3.1.1). Drug-induced hearing disturbances are neither specific nor even unique for salicylates but are also seen with many other compounds, including amino glycosides (Streptomycin) and quinine, though their appearance as well as the mechanisms behind might be quite different [339, 340].

Salicylate-related ototoxicity is typically bilateral symmetric, associated with a low-to-medium-degree-severity hearing loss and usually complete reversible within 1–3 days after salicylate withdrawal. Disturbed hearing functions are not associated with any morphologically detectable loss of hair cells. In addition to hearing disturbances, vestibular disturbances including nystagmus, vertigo, and imbalance are further manifestations of salicylate ototoxicity [341, 342].

As seen in other sections of this book (Sections 2.2.3 and 2.3.3), salicylate concentrations are also an issue in salicylate-related ototoxicity. In many experimental *in vitro* studies, concentrations in the range of 5–10 mM and above were used. Because salicylate levels in the perilymph are only about 30% of the plasma levels at steady-state conditions, this corresponds to *in vivo* plasma salicylate levels of about 15–30 mM that are presumably not only toxic for the ear but also fatal for the organism. Similarly, *in vivo* doses in animal experiments, frequently in the range of 300 mg/kg and more, are in the range of the LD_{50} even in aspirin-"resistant" animals, such as the rat. Thus, experimental data on the mode of ototoxic action of salicylates obtained at these disproportional high salicylate levels are of pharmacological interest but may not be transferable to any *in vivo* being.

3.2.4.1 Pathophysiology of Hearing and Equilibrium Disturbances

Inner Ear and Cochlea *Post mortem* studies of hearing bones and the inner ear (Corti's organ, cochlea, and hair cells) of patients with known regular aspirin consumption at high doses (5–10 g/day) over several months did not show any morphological abnormality. In guinea pigs also, sodium salicylate at high doses (375 mg/kg for 1 week) did not cause macroscopically or light microscopically detectable alterations of hair cells, the mechanotransducers of acoustic pressure. However, loss of the outer hair cells was seen by electron microscopy [343]. In cultured explants of the rat cochlea, degenerations of nerve cells were detected in the presence of high salicylate concentrations (3 mM and more for at least 2 days) but no cell loss [344]. At the subcellular level, alterations in peroxisomes were seen, which, however, were not considered pathognomonically for salicylate ototoxicity [345]. Thus, the majority of available morphological data does not support the concept of a salicylate-induced morphologically detectable loss of hair cells, neurons, or blood vessels in the cochlea. This agrees well with the principal reversibility of hearing dysfunctions by salicylates. However, the (few) available ultrastructural studies do not fully exclude functionally relevant morphological alterations in the cochlea by salicylates, specifically, loss of outer hair cells [346].

3.2.4.2 Mode of Aspirin Action

Hearing loss, tinnitus, and disturbed balance are the three clinical presentations of salicylate ototoxicity [342]. The site of these pathological alterations is the cochlea [347, 348] and the signals outgoing from it to the sensory cortex via the statoacoustic nerve.

Hearing Loss Salicylate-induced hearing loss presents with loss of absolute acoustic sensitivity and changes in sound perception. The reasons are functional disturbances in the cochlea, possibly amplified by a modification of signal transduction via the statoacoustic nerve. These effects of aspirin are caused by salicylate. Salicylate itself does not significantly change the endocochlear resting potential and also not the acoustic pressure, that is, mechanically induced action potentials. In osmotic experiments, it has been shown that salicylate cause hypotonic swelling of outer hair cells, eventually resulting in large volume increases and mechanic dysfunction [349]. The mechanical properties of the cochlea are disturbed after stimulation by sound waves in the presence of high salicylate concentrations (2–10 mM). This was associated with a disturbed conduction of sound waves, emitted from the cochlea after mechanical stimulation, a response that is normally mechanically amplified by the outer hair cells. These otoacoustic emissions are reduced or even abolished in man after high-dose aspirin (3.9 g for up to 4 days) [350–352]. Similar results were obtained in studies on isolated outer hair cells. Their mechanical properties were concentration dependent and reversibly lowered by salicylates (0.1–10 mM) *in vitro* [353].

The subcellular mechanism of salicylate-induced dysfunction of outer hair cells of the guinea pig cochlea has been studied in more detail in voltage-clamp experiments. Salicylates were found to directly affect the motility in the intracellular space in the charged, ionized form. Salicylate passes the cell membrane because of its lipophilic nature in the permeable, nondissociated form. According to a pK_a of 3.0, about 0.6 (M salicylic acid will exist in the cytosol as nondissociated form at an extracellular salicylate concentration of 10 mM. This dissociated form determines the voltage dependence shifts that were taken as a surrogate parameter for hair cell dysfunction [354]. The parallels to other toxic actions of salicylates on cell function, for example, energy metabolism (Section 2.2.3), are evident.

Thus, medium-to-high aspirin concentrations are sufficient to reduce motility and frequency selectivity of outer hair cells [355–358]. These data suggest disturbed mechanoelectrical sound transmission (otoacoustic emissions) by outer hair cells inside the cochlea as the site of salicylate-induced hearing disturbances. Reduction in their mechanosensory functions is particularly prominent at low sound pressure and even might result in complete disappearance of spontaneous emissions. The functional correlate of this salicylate-induced reduced hair cell motility are hearing losses after suprathreshold sound stimuli, associated with disturbed resolving power and localization.

Tinnitus Tinnitus (lat. “tinnere” = tinkle) is a subjective and, with respect to appearance, highly variable sound sensation (phantom sound), mostly 5–15 dB above the hearing threshold. Typical for tinnitus is the absence of any foreign sound source. There appears to be a direct relationship between the (subjective) occurrence of tinnitus and the (objective) hearing disturbances above a certain threshold concentration of salicylates [359]. Tinnitus is a typical side effect of aspirin at salicylate plasma levels of $\geq$200 μg/ml and is also an early symptom in salicylate poisoning (Section 3.1.1). Aspirin-induced tinnitus also originates at the level of outer hair cells [360]. Mechanistic studies on tinnitus are difficult because of its variability and highly subjective nature [361, 362]. Salicylates might activate cochlear NMDA receptors and cause tinnitus-like behavior responses that are antagonized by NMDA antagonists. Similar symptoms were seen with COX inhibitors such as mefenamate [363]. An interesting experimental study in rats, using an avoidance procedure to measure tinnitus, has suggested that salicylates cause tinnitus by activation of cochlear NMDA receptors that can be prevented by NMDA receptor blockade [363, 364].

Imbalance Imbalances (vertigo, dizziness, and disturbed balance) are another kind of ototoxic side effects of aspirin. Probably, disturbed balance is also related to functional disturbances in the inner ear (labyrinth) [365]. However, there are no detailed mechanistic studies available.

3.2.4.3 Aspirin, Arachidonic Acid, and Prostaglandins

Prostaglandins and Cochlear Blood Flow In addition to alterations in the mechanical behavior of outer hair cells, reduced cochlear blood flow is another side effect of aspirin that might reinforce hearing loss and tinnitus [342]. Both are related to changes in arachidonic acid metabolism. Intravenous administration of high-dose aspirin to guinea pigs (350 mg/kg) reduces prostaglandin formation by the cochlea within 30 min and prostaglandin levels in the perilymph after 3 h. Since the cochlear blood vessels are subject to hormonal and autonomous nerval rather than systemic nerval regulation, it has been suggested that the inhibition of prostaglandin biosynthesis by the cochlea will reduce cochlear blood flow with subsequent augmentation of functional disturbances [366]. Experimental studies in rabbits have shown a salicylate-induced reduction of cochlear blood flow by 30–40% [367]. Interestingly, intracochlear perfusion with aspirin or sodium salicylate caused comparable decreases in cochlear function and blood perfusion, whereas traditional NSAIDs (indomethacin) failed to do so [368]. Thus, disturbed cochlear function and reduced blood flow are associated with salicylate-induced ototoxicity in addition to inhibition of prostaglandin biosynthesis.

Leukotrienes COX inhibition by aspirin would also increase the synthesis of potentially cytotoxic leukotrienes such as LTC_4 [369]. Leukotrienes can reduce cochlear blood flow and cause hearing disturbances [367, 369]. As a pharmacological "proof of concept," it has been shown experimentally that salicylate-associated hearing loss can be prevented by treatment with a leukotriene antagonist [367, 370]. Although these data are interesting, they are currently incomplete and too inconsistent to demonstrate causality between salicylate ototoxicity and changes in arachidonic acid metabolism [342].

Arachidonic Acid It was suggested that COX inhibition might result in accumulation of arachidonic acid that has been shown previously to potentiate NMDA receptor currents [371]. However, any significant increase in cellular arachidonic acid levels might have actions of its own, for example, on apoptosis. Thus, the validity of this hypothesis is still unproven.

3.2.4.4 Clinical Trials

In subjects with normal hearing sensation, salicylate-induced hearing disturbances are bilaterally symmetric and seen at all frequencies. The maximum hearing loss is about 40–50 dB [342] and similar in individuals with normal hearing as compared to individuals with preexisting hearing disturbances. However, the intra- and interindividual variability is considerable. All salicylate-induced hearing changes are fully reversible and correlated with the plasma salicylate level. In 16 different studies involving more than 100 individuals, there was a remarkable linear correlation ($r = 0.7$) between absolute hearing loss and plasma salicylate levels (Figure 3.11) [342].

Hearing disturbances and, in particular, tinnitus are frequent side effects of high-dose aspirin treatment, particularly in the elderly. Hearing losses are apparent after suprathreshold sound stimuli and

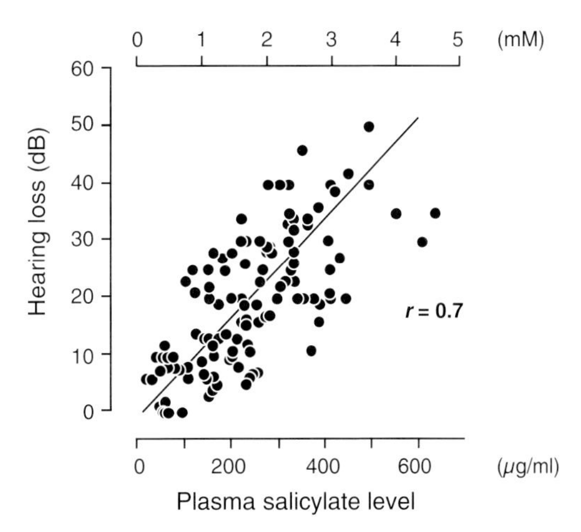

Figure 3.11 Absolute hearing loss in more than 100 subjects as determined in 16 different studies. There is an approximately linear correlation between plasma salicylate level and the loss of hearing sensitivity (modified after [342]).

are associated with a disturbed speech discrimination and sound localization. This has been shown in men after intake of aspirin at an anti-inflammatory dose (3.84 g/day for 3 days) in a double-blind randomized placebo-controlled trial [372]. However, most clinical studies on salicylate-related ototoxicity were done in rheumatics and patients with osteoarthritis, both being not a primary indication for aspirin use today. Consequently, ototoxic side effects of aspirin became considerably less frequent. However, tinnitus is still a side effect in cardiocoronary long-term prophylaxis even at low aspirin doses. Although it is no life-threatening complication, it is an important point for patients' adherence to drug medication.

Mongan and colleagues studied the occurrence of tinnitus after aspirin in a prospective trial in individuals with normal and hard of hearing individuals. All participants received aspirin in escalating doses until the appearance of tinnitus.

A total of 52 out of the 67 individuals developed tinnitus after aspirin treatment. Tinnitus occurred at a minimum salicylate plasma level of 200 μg/ml; interestingly, all individuals with normal hearing function developed tinnitus but only 7 out of the 22 hard of hearing individuals.

The conclusion was that tinnitus is tightly correlated with the plasma salicylate level and, therefore, might be considered a useful surrogate parameter for salicylate plasma levels in individuals with normal hearing [373].

Subsequent investigators did not agree with this conclusion [359], in particular, because symptomatic ototoxicity, as seen from tinnitus, is too unspecific [362]. Nevertheless, tinnitus is a safe index parameter for (too) high salicylate plasma levels in appropriately sensitive individuals.

Imbalances Imbalances (vertigo, dizziness) are another category of ototoxic side effects of aspirin in clinical trials. There is only one more systematic study on this issue. In this study, patients with rheumatoid arthritis were treated with high-dose 6–8 g/day aspirin. This resulted in disturbed vestibularis function and it was assumed that this was due to a functional disturbance in the inner ear (labyrinth) [365]. Again, this phenomenon is only relevant in acute or chronic overdosing.

Summary

Aspirin bears an ototoxic potential that is salicylate mediated. Clinical features are hearing loss, tinnitus, and imbalances. All of these disturbances are dose dependent and consequently appear predominantly at high-dose treatment or overdosing, are fully reversible, and usually disappear within 1–3 days after drug withdrawal.

The underlying mechanisms of salicylate-related ototoxicity are not completely understood but probably functional rather than morphological in nature. The site of action is the cochlea and here, in particular, the outer hair cells. Salicylates impair the mechanical properties of hair cells and the subsequent mechanoelectrical conversion of sound waves into electrical currents (otoacoustic emissions). Additional mechanisms include interactions with the arachidonic acid metabolism, eventually resulting in accumulation of free arachidonic acid and leukotriene formation as well as reduction of cochlear blood flow. The net response is a disturbed sound perception and tinnitus. Imbalances might also occur. The interesting hypothesis that tinnitus might be due to activation of cochlear NMDA receptors needs further experimental and clinical support.

The clinical significance of salicylate-related ototoxicity is steadily decreasing after the replacement of high-dose aspirin as an anti-inflammatory analgesic by NSAIDs. Nevertheless, tinnitus might still occur in long-term prevention even at low-dose aspirin and it is then the sign of individual overdosing that might influence patients' compliance.

References

338 Schwabach, D. (1884) Über bleibende Störungen im Gehörorgan nach Chinin- und Salicylsäuregebrauch. *Deutsche Medizinische Wochenschrift*, **10**, 163–166.

339 Falbe-Hansen, J. (1941) Clinical and experimental histological studies on effects of salicylate and quinine on the ear. *Acta Oto-Laryngologica*, **44** (Suppl.), 1–216.

340 Griffin, J.P. (1988) Drug-induced ototoxicity. *British Journal of Audiology*, **22**, 195–210.

341 Boettcher, F.A. and Salvi, R.J. (1991) Salicylate ototoxicity: review and synthesis. *American Journal of Otolaryngology*, **12**, 33–47.

342 Cazals, Y. (2000) Auditory sensory-neural alterations induced by salicylate. *Progress in Neurobiology*, **62**, 583–631.

343 Douek, E.E., Dodson, H.C. and Bannister, L.H. (1983) The effects of sodium salicylate on the cochlea of guinea pigs. *Journal of Laryngology and Otology*, **93**, 793–799.

344 Zheng, J.L. and Gao, W.Q. (1996) Differential damage to auditory neurons and hair cells by ototoxins and neuroprotection by specific neutrophils in rat cochlear organotypic cultures. *The European Journal of Neuroscience*, **8**, 1897–1905.

345 Dieler, R., Shehata-Dieler, W.E., Richter, C.P. *et al.* (1994) Effects of endolymphatic and perilymphatic application of salicylate in the pigeon. II. Fine structure of auditory hair cells. *Hearing Research*, **74**, 85–98.

346 Beveridge, H.A. and Brown, A.M. (1997) The effects of aspirin on frequency selectivity as measured psychophysically and from the periphery. *British Journal of Audiology*, **31**, 97–98.

347 McCabe, P.A. and Dey, F.L. (1965) The effect of aspirin upon auditory sensitivity. *The Annals of Otology, Rhinology, and Laryngology*, **74**, 312–325.

348 Stypulkowski, P.H. (1990) Mechanisms of salicylate ototoxicity. *Hearing Research*, **46**, 112–146.

349 Zhi, M., Ratnanather, J.T., Ceyhan, E. *et al.* (2007) Hypotonic swelling of salicylate-treated cochlear outer hair cells. *Hearing Research*, **228**, 95–104.

350 McFadden, D. and Plattsmier, H.S. (1984) Aspirin abolishes spontaneous otoacoustic emissions. *The Journal of the Acoustical Society of America*, **76**, 448–553.

351 Long, G.R. and Tubis, A. (1988) Modification of spontaneous and evoked otoacoustic emissions and associated psychoacoustic microstructure by aspirin consumption. *The Journal of the Acoustical Society of America*, **84**, 1343–1353.

352 Wier, C.C., Pasanen, E.G. and McFadden, D. (1988) Partial dissociation of spontaneous otoacoustic emissions and distortion products during aspirin use in human. *The Journal of the Acoustical Society of America*, **84**, 230–237.

353 Hallworth, R. (1997) Modulation of outer hair cell compliance and force by agents that affect hearing. *Hearing Research*, **114**, 204–212.

354 Kakehata, S. and Santos-Sacchi, J. (1996) Effects of salicylates and lanthanides on outer hair cell motility and associated gating charge. *The Journal of Neuroscience*, **16**, 4881–4889.

355 Shehata, W.E., Brownell, W.E. and Dieler, R. (1991) Effects of salicylate on shape, electromotility and membrane characteristics of isolated OHCs from guinea pig cochlea. *Acta Oto-Laryngologica*, **111**, 707–718.

356 Shehata-Dieler, W.E., Richter, C.P., Dieler, R. *et al.* (1994) Effects of endolymphatic and perilymphatic application of salicylate in the pigeon. I. Single fiber activity and cochlear potentials. *Hearing Research*, **74**, 77–84.

357 Russell, I.J. and Schauz, C. (1995) Salicylate ototoxicity effects: effects on the stiffness and electromotility of outer hair cells isolated from the guinea pig cochlea. *Auditory Neuroscience*, **1**, 309–319.

358 Tunstall, M.J., Gale, J.E. and Ashmore, J.F. (1995) Action of salicylate on membrane capacitance of outer hair cells from the guinea-pig cochlea. *The Journal of Physiology*, **485**, 739–752.

359 Day, R.O., Graham, G.G., Bieri, D. *et al.* (1989) Concentration–response relationships for salicylate-induced ototoxicity in normal volunteers. *British Journal of Clinical Pharmacology*, **28**, 695–702.

360 Janssen, T., Boege, P., Oestreicher, E. *et al.* (2000) Tinnitus and 2*f*? 1-*f* ?2 distortion product otoacoustic emissions following salicylate overdose. *The Journal of the Acoustical Society of America*, **107**, 1790–1792.

361 Halla, J.T. and Hardin, J.G. (1988) Salicylate ototoxicity in patients with rheumatoid arthritis: a controlled study. *Annals of the Rheumatic Diseases*, **47**, 134–137.

362 Halla, J.T., Atchison, S.L. and Hardin, J.G. (1991) Symptomatic salicylate ototoxicity: a useful indicator of serum salicylate concentration? *Annals of the Rheumatic Diseases*, **50**, 682–684.

363 Guitton, M.J., Caston, J., Ruel, J. *et al.* (2003) Salicylate induces tinnitus through activation of cochlear NMDA receptors. *The Journal of Neuroscience*, **23**, 3944–3952.

364 Puel, J.L. (2007) Cochlear NMDA receptor blockade prevents salicylate-induced tinnitus. *B-ENT*, **3** (Suppl. 7), 19–22.

365 Bernstein, J.M. and Weiss, A.D. (1967) Further observations on salicylate ototoxicity. *Journal of Laryngology and Otology*, **81**, 915–925.

366 Escoubet, P., Amsallem, P., Ferrary, E. *et al.* (1985) Prostaglandin synthesis by the cochlea of the guinea pig. Influence of aspirin, gentamicin, and acoustic stimulation. *Prostaglandins*, **29**, 589–599.

367 Jung, T.T., Hanf, A.L., Miller, S.K. *et al.* (1995) Effect of leukotriene inhibitor on cochlear blood flow in salicylate ototoxicity. *Acta Oto-Laryngologica*, **115**, 251–254.

368 Fitzgerald, J.J., Robertson, D. and Johnstone, B.M. (1993) Effects of intracochlear perfusion of salicylates on cochlear microphonic and other auditory responses in the guinea pig. *Hearing Research*, **67**, 147–156.

369 Park, Y.S., Jung, J.T., Choi, D.J. *et al.* (1994) Effect of corticosteroid treatment on salicylate ototoxicity. *The Annals of Otology, Rhinology, and Laryngology*, **103**, 896–900.

370 Arruda, J., Jung, T.T. and McGann, D.G. (1996) Effect of leukotriene inhibitor on otoacoustic emissions in salicylate ototoxicity. *American Journal of Otolaryngology*, **17**, 787–792.

371 Miller, B., Sarantis, M., Traynelis, S.F. *et al.* (1992) Potentiation of NMDA receptor currents by arachidonic acid. *Nature*, **355**, 722–725.

372 Beveridge, H.A. and Carlyon, R.P. (1996) Effects of aspirin on human psychophysical tuning curves in forward and simultaneous masking. *Hearing Research*, **99**, 110–118.

373 Mongan, E., Kelly, P., Nies, K. *et al.* (1973) Tinnitus as an indicator of therapeutic serum salicylate levels. *The Journal of the American Medical Association*, **226**, 142–145.

3.3 Non-Dose-Related (Pseudo)allergic Actions of Aspirin

Most of the unwanted effects of aspirin are dose related. This is valid not only for transacetylation reactions but also for the multiple actions of salicylate on cell function and energy metabolism. Systemic toxicity to salicylates is mainly caused by their metabolic effects and disturbed acid/base equilibrium (Section 3.1.1). A generalized intolerance to aspirin due to pathologic immune reactions is rare. This is somehow surprising since aspirin during hydrolytic degradation may transacetylate many proteins in the circulation with subsequent changes of their secondary and tertiary structure (Section 2.1.2). In addition, salicylates are tightly bound to plasma albumin. In this constellation, they might act as haptens, causing the generation of specific IgE-type antibodies. Finally, salicylate-mediated changes in eicosanoid metabolism, specifically, the increased generation of leukotrienes after blockade of the COX pathway might cause or enhance (pseudo)allergic reactions.

Pharmacologically, two different reasons of aspirin hypersensitivity have to be distinguished: pharmacological intolerance to aspirin and/or salicylates as chemicals or intolerance to aspirin-induced changes in cell metabolism, including an altered eicosanoid biosynthesis. The most frequent manifestation sites of hypersensitivity are the respiratory tract and the skin. The "Widal triad" in the respiratory tract (aspirin-sensitive asthma) (Section 3.3.1) is among the most intensively studied diseases related to aspirin intake in sensitive patients. Manifestations of aspirin intolerance at the skin are urticaria/angioedema. More severe anaphylactic reactions, such as acute toxic epidermolysis (Lyell's syndrome) or Stevens–Johnson syndrome, have been occasionally described but appear not to be causally related to aspirin intake (Section 3.3.2).

Another disease with a possible but unproven relationship with aspirin (and other chemicals), specifically in children, is Reye's syndrome (Section 3.3.3). Interestingly, both aspirin-induced asthma and Reye's syndrome are considered to develop at the background of a protracted preceding viral infection and thus might involve acquired change in the phenotype as a critical disease-modifying factor.

3.3.1 Aspirin Hypersensitivity (Widal's Syndrome)

The so-called "aspirin-sensitive asthma" belongs to the best known unwanted side effects of aspirin that occur in predisposed patients and are not critically dependent on the aspirin dose. Although the term "aspirin-induced" is imprecise and rather should be replaced by "analgesics' asthma" or named by the discoverer as "Widal triad" that also considers the fact that the disease is not restricted to the lung, it has been generally established and, therefore, is also used here.

3.3.1.1 Pathophysiology and Clinics

History and Epidemiology Widal *et al.* [374] were the first to describe a new clinical entity, consisting of aspirin intolerance, nasal polyposis, and bronchial asthma. This syndrome, subsequently named the "Aspirin triad," was later described in its clinical course by Samter and Beers [375]. The disease typically occurs in adult patients with asthma after ingestion of aspirin or other COX-1 inhibitors. According to recent epidemiological surveys in Europe and Australia, the prevalence amounts to about 10% of adult patients with asthma whereas the overall incidence in the general population is less than or equal to 1% [376]. Importantly, a significant percentage of affected people are unaware of their aspirin intolerance. In one study on 500 patients with asthma, 18% of the patients did not know that they were aspirin hypersensitive [377]. This suggests that aspirin hypersensitivity does occur in a measurable proportion of the population and is a relevant disease also from an epidemiological point of view with a comparable incidence worldwide [378–380]. The other important question is whether long-term aspirin use, for example, in cardiocoronary prevention will increase the occurrence of asthmatic diseases in otherwise healthy subjects. According to recent data from the Womens' Health Study, this appears not to be the case. There was rather a decreased risk of newly reported diagnosis of asthma [381].

Pathophysiology The reactions on aspirin are equivalent to an immediate allergic reaction. Although mean IgE levels may be elevated in some patients with aspirin hyperreactivity, no specific antibodies against aspirin or traditional NSAIDs are found in most cases. This and the cross-reactivity between aspirin and nonselective NSAIDs are arguments for pharmacological rather than immunological mechanisms as a causal attack precipitating factor [382]. A reduced aspirin-esterase activity has been described (Section 2.1.2), although at unchanged bioavailability of aspirin and salicylate [383]. Thus, the most likely explanation for the phenomenon is pharmacodynamic rather than pharmacokinetic in nature.

Role of Viruses The disease is typically preceded by a viral infection of upper airways. This led to the hypothesis that aspirin-sensitive asthma is of viral origin. Subsequent to a viral infection, but, frequently much later than the initial exposition, there is appearance of cytotoxic lymphocytes. These immunocompetent cells can generate and release a variety of inflammatory mediators. PGE_2 antagonizes these reactions. Consequently, inhibition of PGE_2 production by aspirin will prevent the PGE_2-mediated suppression of lymphocyte function [384]. Interestingly, selective COX-2 inhibitors appear not to precipitate asthmatic attacks, even in aspirin-sensitive individuals [385]. Although this might be taken as an argument against macrophages, that is, COX-2-dependent PGE_2 formation, there are other sources of PGE_2 production, for example, bronchial epithelial cells, which then could modify lymphocyte and macrophage function. It is also evident that changes in eicosanoid profiles of lymphocytes and macrophages occur after aspirin intake and have a tight relation to the occurrence and severity of asthmatic symptoms. This is also valid for generation and action of proinflammatory cytokines, most notably, IL-4. The efficacy of acyclovir in the treatment of patients with mild-to-moderate asthmatic attacks and the associated reduced urinary leukotriene excretion [386] support a role for viruses.

Mast Cells and Eosinophils Both mast cells and eosinophils are present in the nasal mucosa of patients with aspirin hypersensitivity. Mast cells are the natural source of not only PGD_2 and histamine but also other spasmogenic and permeability enhancing products such as cysteinyl leukotrienes. These eicosanoids are generated by both mast cells and eosinophils [387]. Aspirin challenge of sensitive individuals results in a marked increase in nasal symptoms. This is associated with mast cell activation and eosinophilia, as seen from increased release of (mast-cell-specific) tryptase and generation of cysteinyl leukotrienes. Both clinical symptoms and leukotriene release can be blocked by zileuton, a selective inhibitor of 5-lipoxygenase (5-LOX). This indicates that 5-LOX-derived products, that is, leukotrienes, are essential for the nasal response of these subjects to aspirin. There is also some evidence for an upregulation of leukotriene-related genes in these conditions [388]. Thus, although the reports on the number of mast cells in bronchi yielded different results, an activation of mast cells in the disease, both in the absence and presence of aspirin challenge, is a consistent finding and associated with increased plasma levels of stable PGD_2 metabolites [389–391].

Clinics The reactions on aspirin in intolerant individuals are those of an acute inflammatory immune response. They involve bronchospasm, profuse rhinorrhea, conjunctival injection, periorbital edema, and scarlet-like flushing of head and neck. The patients typically exhibit the first symptoms, such as perennial rhinitis in their third decade of life, subsequent to an upper respiratory tract viral infection. During the following 1–5 years, asthma and aspirin hypersensitivity follow [382]. Fifty percent of patients with aspirin intolerance already suffer from a chronic severe asthma requiring steroid treatment, 30% have moderate asthma [392]. In this context, it should be noted that aspirin treatment might be useful in "normal" asthma attacks in non(aspirin)-hypersensitive individuals because of its anti-inflammatory effects and has also been reported to reduce the occurrence of asthma in apparently healthy individuals [381].

A number of clinical studies have described beneficial effects of aspirin and salicylates in several allergic diseases including asthma [393–395]. These therapeutic actions of aspirin required higher doses than those necessary for inhibition of COX-1-dependent prostaglandin formation. In contrast to aspirin-induced asthma, similar beneficial effects were also seen with salicylate. Recent studies have shown that aspirin and salicylate inhibit IL-4- and IL-13-induced activation of STAT6, a transcription factor that is involved in the development of allergic diseases, including asthma [396] (Section 2.2.2). Thus, aspirin in addition to negative also might have beneficial effects on asthma. The direction of response depends on the etiology of the disease.

The clinical symptoms of hypersensitivity usually start about 1 h after aspirin intake. The attacks might even be life threatening – about 25% of asthmatic patients requiring emergency mechanical ventilation are intolerant against aspirin or other NSAIDs [397, 398].

As already noted in the original description by Samter and Beers [375], salicylate did not cause asthmatic symptoms in patients, reacting with asthmatic attacks to aspirin whereas nonselective NSAIDs will induce asthmatic symptoms in these patients [399]. These and other data suggested that the exacerbation of the disease is related to inhibition of COX-1-dependent prostaglandin production, specifically, an insufficient generation of bronchodilatory prostaglandins, that is, PGE_2.

3.3.1.2 Mode of Aspirin Action

Role of Prostaglandins *Andrew Szczeklik* and his group from Krakow (Poland) were the first to bring aspirin-induced asthma into connection with a pathology in eicosanoid metabolism. They detected that aspirin-sensitive asthma could be provoked not only by aspirin but also by a number of other structurally unrelated inhibitors of prostaglandin biosynthesis. This included indomethacin and phenylbutazone whereas paracetamol or salicylamide, two compounds that did not suppress prostaglandin formation at therapeutic doses, were ineffective. The conclusion was that precipitation of asthmatic attacks by aspirin in aspirin-intolerant patients involved inhibition of prostaglandin biosynthesis [399].

After leukotrienes were detected, this hypothesis was modified to an imbalance in eicosanoid formation. Aspirin-induced inhibition of prostaglandin synthesis will favor leukotriene generation that then in turn, causes the symptoms of aspirin-induced asthma. Today, it is generally appreciated that prostaglandins that inhibit white-cell function, such as PGE_2, can also inhibit white-cell-derived generation of inflammatory mediators, such as cysteinyl leukotrienes, which are generated via the 5-lipoxygenase pathway (Figure 3.12).

COX-1 versus COX-2 Both COX-1 and COX-2 isoforms are expressed for similar extent in normal respiratory epithelium. Neither isoform is upregulated in bronchi of aspirin-intolerant patients as compared with aspirin-tolerant individuals [389]. Interestingly, there is rather a downregulation of COX-2 in nasal polyps of aspirin-sensitive patients [400]. Compounds that preferentially (nimesulide) or selectively (rofecoxib) inhibit COX-2 do not induce hypersensitivity in aspirin-sensitive individuals at conventional therapeutic doses [401, 402]. These data confirm aspirin (and NSAID)-mediated inhibition of COX-1-dependent PGE_2 control of inflammatory cells in aspirin-induced asthma [385]. However, this does not answer the question that why inhibition of prostaglandin formation causes asthma in some individuals but not in others. In addition, COX-1 inhibitors should also block generation of PGD_2 by mast cells, which is a proinflammatory, asthma-promoting compound.

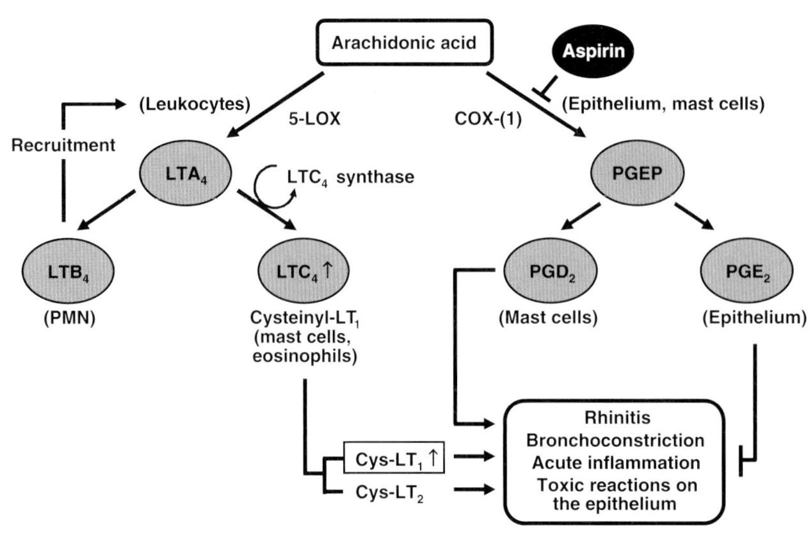

Figure 3.12 Arachidonic acid metabolism in the lung via 5-lipoxygenase and cyclooxygenase (COX-1) in "aspirin-induced asthma" – upregulation of LTC_4 synthase, as well as cysteinyl-LT_1 (Cys-LT1) receptors. Both mechanisms enhance LTC_4 production and action. Any reduction in PGE_2 formation via inhibition of COX-1 will act into the same direction.

The explanation is either an enhanced availability of arachidonic acid after COX inhibition for leukotriene formation or an inhibited action of leukotrienes on a (hyperreactive) bronchial epithelium in certain individuals.

Increase in LTC_4 Synthase Expression Cysteinyl leukotrienes (LTC_4, LTD_4, and LTE_4) induce bronchoconstriction, edema formation, and mucus production and are well-known mediators of asthma. They are synthesized via the intermediate LTA_4 by the LTC_4 synthase. This enzyme is rate limiting for synthesis of cysteinyl leukotrienes at an unrestricted availability of the arachidonic acid precursor (Figure 3.12). LTC_4 synthase but not 5-lipoxygenase is permanently overexpressed in patients with aspirin-induced asthma and accounts for enhanced leukotriene production in the airways of these patients [389, 403].

One explanation for enhanced LTC_4 synthase activity is gene polymorphisms. These have been described in the LTC_4 synthase promoter in some but not the majority of individuals with aspirin-induced asthma [390, 404]. It was suggested that overexpression of LTC_4 synthase in the bronchial wall may be the single most important determinant of acute respiratory reactions to aspirin in subjects with aspirin intolerance. In these subjects, removal of PGE_2 by aspirin and other COX-1 inhibitors will cause asthma via increased LTC_4 formation [392].

Cys-LT_1 and Cys-LT_2 Receptors In addition to increased production of cysteinyl leukotrienes, there is also increased responsiveness of the bronchi to these mediators in aspirin-sensitive patients with asthma. This suggests changes in cysteinyl-LT receptors that mediate these actions of leukotrienes. It is well known that bronchoconstriction in aspirin-induced asthma is associated with enhanced leukotriene E_4 excretion [405] and can be antagonized by LT-receptor antagonists [406]. Cys-LT_1 receptors are upregulated in aspirin-induced asthma. This upregulation can be antagonized by aspirin desensitization [407], a recommended treatment for aspirin hypersensitivity (see below). Cys-LT_1 receptors predominate on inflammatory leukocytes in aspirin-sensitive patients while the effects of cysteinyl leukotrienes on glands and epithelium are mediated predominantly through Cys-LT_2 [408]. Interestingly, gene polymorphism has also been described for the Cys-LT_2 receptor and was brought into connection with aspirin intolerance in patients with asthma [409]. Thus, combined inhibitors of both receptors might be more effective drugs to prevent aspirin-sensitive asthma than selective inhibitors of only one receptor subtype, that is, Cys-LT_1.

Lipoxins Lipoxins are anti-inflammatory eicosanoids that can be generated by an interaction of leukocytes with endothelial cells. Aspirin-triggered lipoxin can be formed from acetylated COX-2 and has also anti-inflammatory properties, specifically, with respect to white-cell recruitment (Section 2.2.1). A clinical study in a small group of patients with asthma who were aspirin tolerant and aspirin intolerant showed reduced lipoxin formation by the intolerant individuals [410]. Although this finding is interesting, its clinical significance has still to be established, and it should be noted that lipoxins will not directly interfere with cysteinyl leukotrienes, neither with their generation nor with their action on specific receptors.

3.3.1.3 Clinical Trials

Diagnosis Aspirin intolerance is an acquired disease that often occurs subsequent to (chronic) rhinitis, polyposis, and preexisting (severe) asthma [382]. The diagnosis of aspirin intolerance can only be validated by aspirin challenge. This can be done by oral ingestion, oral inhalation, or inhalation of aspirin via the nasal route. Oral tests are most common and are done with 30–150 mg of aspirin (average 60 mg) as a starting dose. In inhalation and nasal provocation tests, solute aspirin in the form of water-soluble lysine salt administered as aerosol is a reliable and comparably safe procedure for diagnosis of aspirin intolerance.

Treatment Since the primary mechanism of aspirin intolerance is rarely associated with drug-specific IgE production, desensitization to the drug, if clinically necessary, is the treatment of choice. Clinical experience shows that most, if not all, aspirin-sensitive patients can be desensitized by increasing doses of aspirin by continuous challenge over several days. The procedure should be performed by experts inside a specialized hospital.

The long-term clinical outcome of aspirin desensitization in aspirin-sensitive patients with rhinosinusitis-asthma was studied in a group of 65 patients. Desensitization to aspirin was performed with increasing doses of aspirin, until 650 mg were tolerated. Afterward, desensitized patients received aspirin therapy (median dose 1214 mg/day) for 1–6 years.

There was a significant reduction of numbers of sinus infections per year, hospitalizations for treatment of asthma per year, improvement in olfaction, and a marked reduction, on average from 10.2 to 2.5 mg/day, of doses of systemic corticosteroids. Emergency department visits and use of inhaled corticosteroids were unchanged. There was a dropout rate of 13–46% of patients. Fifteen to twenty percent of patients interrupted continuous aspirin treatment after desensitization, most of them because of gastric intolerance.

The conclusion was that aspirin desensitization followed by daily aspirin treatment can be considered in patients with uncontrolled respiratory symptoms or requiring repeated polypectomies or sinus surgeries, those with symptoms despite the use of corticosteroids, corticosteroids at unacceptably high doses and patients who needed aspirin for other diseases, most notably cardiocoronary prevention [394, 411].

After desensitization, a maintenance dose of 600 mg aspirin bid is recommended for at least 6–12 months or even lifelong. Desensitization followed by daily aspirin is effective by at least the first 6 months of treatment and continues to be effective for up to 5 years of follow-up [412]. The desensitized state is only maintained for 2–5 days after cessation of aspirin. If treatment is interrupted for more than 2 days, aspirin should not be restarted because of the risk of severe aspirin reaction – another graded dose desensitization procedure is necessary [382].

Aspirin desensitization might be particularly useful in patients who need prophylaxis from arterial thromboembolic events such as myocardial infarction and stroke [413]. However, there are no larger prospective trials with desensitization protocols in these patients [414]. Interestingly, a recent economic calculation indicated that ambulatory aspirin desensitization in these patients is less expensive than the use of clopidogrel as an alternative antiplatelet agent [415]. An overview on rapid desensitization protocols was recently published by Castells [416]. An actual guideline for the use of aspirin provocation tests for diagnosis of aspirin hypersensitivity is also available [417].

Alternative Drug Treatment and Drug Interactions Leukotriene modifying drugs do not sufficiently block aspirin-induced bronchial reactions in the respiratory system. In a group of 271 patients, those on leukotriene antagonists developed less bronchospasm but more upper airway and conjunctive reactions [418]. Whether this was due to different leukotriene receptor localization and function and may be overcome by more potent and selective compounds in the future remains to be shown. Importantly, comedication of leukotriene antagonists did not change the responses to aspirin treatment [412].

Selective COX-2 inhibitors can be safely used in aspirin-sensitive patients with asthma. Acetaminophen might also be an alternative at low doses since the compound is only a weak inhibitor of COX-1. It has been shown, however, that about one-third of patients with aspirin-sensitive asthma can cross-react with acetaminophen at conventional analgesic doses of 1–1.5 g [419]. Although the reactions were mild in most cases, a maximum dose of 1 g acetaminophen should not be exceeded. Alternatively, another type of analgesic drug, such as codeine, may be used [382].

Summary

A syndrome of aspirin intolerance (Widal triad, aspirin-sensitive asthma) can occur at an overall incidence of $\leq$1% in the general population as opposed to about 10% in patients with asthma. It involves bronchospasm, profuse rhinorrhea, conjunctive injection, periorbital edema, and scarlet-like flushing of head and neck. The symptoms start about 1 h after aspirin challenge in sensitive individuals. There is cross-reactivity with other nonsteroidal anti-inflammatory drugs that inhibit COX-1 but not with selective inhibitors of COX-2.

The disease is probably due to a pathology in eicosanoid metabolism that becomes manifest predominantly in the respiratory tract and is probably preceded by a viral infection of the upper airways. There is an upregulation of cysteinyl-leukotriene biosynthesis in mast cells and eosinophils, eventually resulting in enhanced production of LTC_4 and its metabolites. There is also an increased sensitivity of the upper airways against these mediators, probably related to an upregulation of Cys-LT_1 receptors. PGE_2 formation is unchanged or slightly reduced. Complete prevention of COX-1-mediated PGE_2 biosynthesis in these patients will evoke acute attacks that might be life threatening in sensitive individuals.

Treatment of choice is controlled desensitization by aspirin challenge. This is done by repeated incremental increase in (oral) dosing over several days, until about 600 mg are tolerated. A daily maintenance dose has to be given to maintain the tolerant state. Actual guideline recommendations are available and may be useful, in particular, in patients requiring regular long-term aspirin treatment because of cardiocoronary prevention.

References

374 Widal, M.F., Abrami, P. and Lenmoyez, J. (1922) Anaphylaxis et idiosyncrasies. *Presse Medicale*, **30**, 189–192.

375 Samter, M. and Beers, R.F. (1968) Intolerance to aspirin. Clinical studies and consideration of its pathogenesis. *Annals of Internal Medicine*, **68**, 975–983.

376 Jenneck, C., Juergens, U., Buecheler, M. *et al.* (2007) Pathogenesis, diagnosis and treatment of aspirin intolerance. *Annals of Allergy, Asthma & Immunology*, **99**, 13–21.

377 Szczeklik, A. and Nizankowska, E. (2000) Clinical features and diagnosis of aspirin-induced asthma. *Thorax*, **55** (Suppl. 2), S42–S44.

378 Hedman, J., Kaprio, J., Poussa, T. *et al.* (1999) Prevalence of asthma, aspirin intolerance, nasal polyposis and chronic obstructive pulmonary disease in a population-based study. *International Journal of Epidemiology*, **28**, 717–722.

379 Vally, H., Taylor, M.L. and Thompson, P.J. (2002) The prevalence of aspirin intolerant asthma (AIA) in Australian asthmatic patients. *Thorax*, **57**, 569–574.

380 Kasper, L., Sladek, K., Duplaga, M. *et al.* (2003) Prevalence of asthma with aspirin hypersensitivity in the adult population of Poland. *Allergy*, **58**, 1064–1066.

381 Kurth, T., Barr, R.G., Gaziano, J.M. *et al.* (2008) Randomised aspirin assignment and risk of adult-onset asthma in the Women's Health Study. *Thorax*, **63** (6), 514–518.

382 Morwood, K., Gillis, D., Smith, W. *et al.* (2005) Aspirin-sensitive asthma. *International Medicine Journal*, **35**, 240–246.

383 Dahlen, B., Boreus, L.O., Anderson, P. *et al.* (1994) Plasma acetylsalicylic acid and salicylic acid levels during aspirin provocation in aspirin-sensitive subjects. *Allergy*, **49**, 43–49.

384 Szczeklik, A. (1988) Aspirin-induced asthma as a viral disease. *Clinical Allergy*, **18**, 15–20.

385 Mastalerz, L., Sanak, M., Gawiewicz-Mroczka, A. *et al.* (2008) Prostaglandin E_2 systemic production in patients with asthma with and without aspirin hypersensitivity. *Thorax*, **63**, 27–34.

386 Yoshida, S., Sakamoto, H., Yamawaki, Y. *et al.* (1998) Effect of acyclovir on bronchoconstriction and urinary leukotriene E_4 excretion in aspirin-induced asthma. *The Journal of Allergy and Clinical Immunology*, **102**, 909–914.

387 Kowalski, M.L., Grzegorczyk, J., Wojciechowska, B. *et al.* (1996) Intranasal challenge with aspirin induces cell influx and activation of eosinophils and mast cells in nasal secretions of ASA-sensitive patients. *Clinical and Experimental Allergy*, **26**, 807–814.

388 Kim, S.H., Hur, G.Y., Choi, J.H. *et al.* (2008) Pharmacogenetics of aspirin-intolerant asthma. *Pharmacogenomics*, **9**, 85–91.

389 Cowburn, A.S., Sladek, K., Soja, J. *et al.* (1998) Overexpression of leukotriene C_4 synthase in bronchial biopsies from patients with aspirin-intolerant asthma. *The Journal of Clinical Investigation*, **101**, 834–846.

390 Szczeklik, A. and Stevenson, D.D. (2002) Aspirin-induced asthma: advances in pathogenesis, diagnosis and management. *The Journal of Allergy and Clinical Immunology*, **111**, 913–921.

391 Bochenek, G., Nagraba, K., Nizankowska, E. *et al.* (2003) A controlled study of $9\alpha 11\beta$-PGF_2 (a prostaglandin D_2 metabolite) in plasma and urine of patients with bronchial asthma and healthy controls after aspirin challenge. *The Journal of Allergy and Clinical Immunology*, **111**, 743–749.

392 Babu, K.S. and Salvi, S.S. (2000) Aspirin and asthma. *Chest*, **118**, 1470–1476.

393 Crimi, N., Polosa, R., Magri, G. *et al.* (1996) Inhaled lysine acetylsalicylate (L-ASA) attenuates histamine-induced bronchoconstriction in asthma. *Allergy*, **51**, 157–163.

394 Stevenson, D.D., Hankammer, M.A., Mathison, D.A. *et al.* (1996) Aspirin desensitization treatment of aspirin-sensitive patients with rhinosinusitis-asthma: long term outcomes. *The Journal of Allergy and Clinical Immunology*, **98**, 751–758.

395 Sestini, P., Refini, R.M., Pieroni, M.G. *et al.* (1999) Different effects of inhaled aspirin-like drugs on allergen-induced early and late asthmatic responses. *American Journal of Respiratory and Critical Care Medicine*, **159**, 1228–1233.

396 Perez-G, G.M., Melo, M., Keegan, A.D. *et al.* (2002) Aspirin and salicylates inhibit the IL-4- and IL-13-induced activation of STAT6. *Journal of Immunology*, **168**, 1428–1434.

397 Picado, C.I., Castillo, J.A., Monserrat, J.M. *et al.* (1989) Aspirin-intolerance as a precipitating factor of life-threatening attacks of asthma requiring mechanical ventilation. *The European Respiratory Journal*, **2**, 127–129.

398 Marquette, C.H., Saulnier, F., Leroy, O. *et al.* (1992) Long-term prognosis for near-fatal asthma. A 6-year follow-up study of 145 asthmatic patients who underwent mechanical ventilation for near-fatal attack of asthma. *The American Review of Respiratory Disease*, **146**, 76–81.

399 Szczeklik, A., Gryglewski, R.J. and Czerniawaska-Mysik, G. (1975) Relationship of inhibition of prostaglandin biosynthesis by analgesics to asthma attacks in aspirin-sensitive patients. *British Medical Journal*, **1**, 67–69.

400 Picado, C.I., Fernandez-Morata, J.C., Juan, M. *et al.* (1999) Cyclooxygenase-2 mRNA is downexpressed in nasal polyps from aspirin-sensitive asthmatics. *American Journal of Respiratory and Critical Care Medicine*, **160**, 291–296.

401 Stevenson, D.D. and Simon, R.A. (2001) Lack of cross-reactivity between rofecoxib and aspirin in aspirin-sensitive patients with asthma. *The Journal of Allergy and Clinical Immunology*, **108**, 47–51.

402 Jawien, J. (2002) A new insight into aspirin-induced asthma. *European Journal of Clinical Investigation*, **32**, 134–138.

403 Sampson, A.P., Cowburn, A.S., Sladek, K. *et al.* (1997) Profound overexpression of leukotriene C_4 synthase in bronchial biopsies from aspirin-intolerant asthmatic patients. *International Archives of Allergy and Immunology*, **13**, 355–357.

404 Sanak, M., Simon, H.U. and Szczeklik, A. (1997) Leukotriene C_4 synthase promoter polymorphism and risk of aspirin-induced asthma. *The Lancet*, **350**, 1599–1600.

405 Christie, P.E., Tagari, P., Ford-Hutchinson, A.W. *et al.* (1992) Urinary leukotriene E_4 after lysine-asparagine inhalation in asthmatic subjects. *The American Review of Respiratory Disease*, **146**, 1531–1534.

406 Christie, P.E., Smith, C.M. and Lee, T.H. (1991) The potent and selective sulfidopeptide leukotriene antagonist, SK&F 104353, inhibits aspirin-induced asthma. *The American Review of Respiratory Disease*, **144**, 957–958.

407 Sousa, A., Parik, A., Scadding, G. *et al.* (2002) Leukotriene receptor expression on nasal mucosal inflammatory cells in aspirin-sensitive rhinosinusitis. *The New England Journal of Medicine*, **347**, 1493–1498.

408 Corrigan, C., Mallett, K., Ying, S. *et al.* (2005) Expression of the cysteinyl leukotriene receptors $cysLT_1$ and $cysLT_2$ in aspirin-sensitive and aspirin-tolerant chronic rhinosinusitis. *The Journal of Allergy and Clinical Immunology*, **115**, 316–322.

409 Park, J.S., Chang, H.S., Park, C.S. *et al.* (2005) Association analysis of cysteinyl-leukotriene receptor 2 (CYSLTR2) polymorphism with aspirin intolerance in asthmatics. *Pharmacogenetics and Genomics*, **15**, 483–492.

410 Sanak, M., Levy, B.D., Clish, C.B. *et al.* (2000) Aspirin-tolerant asthmatics generate more lipoxins that aspirin-intolerant asthmatics. *The European Respiratory Journal*, **16**, 44–49.

411 Stevenson, D.D. and Zuraw, B.L. (2003) Pathogenesis of aspirin-exacerbated respiratory disease. *Clinical Reviews in Allergy & Immunology*, **24**, 169–188.

412 Berges-Gimeno, M.P., Simon, R.A. and Stevenson, M.D. (2003) Long-term treatment with aspirin desensitization in asthmatic patients with aspirin-exacerbated respiratory disease. *The Journal of Allergy and Clinical Immunology*, **111**, 180–186.

413 Schaefer, O.P. and Gore, J.M. (1999) Aspirin sensitivity: the role for aspirin challenge and desensitization in postmyocardial infarction patients. *Cardiology*, **91**, 8–13.

414 Gollapudi, R.R., Teirstein, P.S., Stevenson, D.D. *et al.* (2004a) Aspirin sensitivity. *The Journal of the American Medical Association*, **292**, 3017–3023.

415 Shaker, M., Lobb, A., Jenkins, P. *et al.* (2008) An economic analysis of aspirin desensitization in aspirin-exacerbated respiratory disease. *The Journal of Allergy and Clinical Immunology*, **121**, 81–87.

416 Castells, M. (2006) Desensitization for drug allergy. *Current Opinion in Allergy and Clinical Immunology*, **6**, 476–481.

417 Nizankowska-Mogilnicka, E., Bochenek, G., Mastalerz, L. *et al.* (2007) EAAC/GA2LEN guideline: aspirin provocation tests for diagnosis of aspirin hypersensitivity. *Allergy*, **62**, 1111–1118.

418 Berges-Gimeno, M.P., Simon, R.A. and Stevenson, D.D. (2002) The effect of leukotriene-modifier drugs on aspirin-induced asthma and rhinitis reactions. *Clinical and Experimental Allergy*, **32**, 1491–1496.

419 Settipane, R.A., Schrank, P.J., Simon, R.A. *et al.* (1995) Prevalence of cross sensitivity with acetaminophen in aspirin-sensitive asthmatic subjects. *The Journal of Allergy and Clinical Immunology*, **96**, 480–485.

3.3.2 Urticaria/Angioedema, Stevens–Johnson and Lyell Syndromes

In addition to the aspirin-exacerbated respiratory tract diseases (Section 3.3.1), the induction of aspirin-induced urticaria/angioedema is the other major aspirin-related allergic response. In contrast to asthma, aspirin-induced urticaria/angioedema is relatively rare and amounts to only 0.1–0.2% of the general population [420]. There are a few single-case reports on aspirin-induced skin hypersensitivity, for example, on the background of allergies against birch pollen [421]. Aspirin might also exaggerate food-related exercise-induced hypersensitivity reactions, due to wheat or gluten allergies, as detected by skin prick testing. These forms of aspirin sensitivity are rarely related to drug-specific IgE antibody production – although those have been detected in single cases [422]. They also extremely rarely lead to more severe anaphylactic reactions. However, aspirin might aggravate their symptoms in a certain proportion of patients subsequent to provocation tests [423]. The pathophysiology of these skin reactions is clearly more complex than aspirin-induced asthma. In any case, larger randomized trials to be able to discover any causative role for aspirin in any of those are missing.

3.3.2.1 Urticaria/Angioedema

Occurrence There are two different types of skin hypersensitivity toward aspirin: pharmacological reactions, due to salicylate-specific modifications of metabolic pathways, specifically, overproduction of or hyperreactivity to leukotrienes after inhibition of COX-1 (Section 3.3.1) and immunological reactions, based on an aspirin- or salicylate-specific IgE production. Both pharmacological and immunological reactions to aspirin can present clinically as urticaria/angioedema.

Patients with a History of Chronic Idiopathic Urticaria Similar to aspirin-induced asthma, patients with chronic idiopathic urticaria frequently react with an exacerbation of their symptoms after challenge with aspirin or nonselective NSAIDs [420]. The prevalence of NSAID-induced urticaria in these patients is 20–30% [424]. Similar to aspirin-induced asthma (Section 3.3.1), there appears to be an insufficient generation of COX-1-derived prostaglandins, such as PGE_2 and an excessive (over) production of leukotrienes. Leukotrienes increase vascular permeability. They cause subsequent urticaria [425] and in addition might amplify these responses by white-cell recruitment. Desensitization protocols with aspirin are usually not effective in these patients [426] and are also not recommended [420, 427, 428].

Patients Without a History of Chronic Idiopathic Urticaria Some patients without anamnestic hyperreactivity to aspirin may develop urticaria/angioedema after treatment with aspirin or NSAIDs that block COX-1. These patients can be effectively desensitized. In other subsets of patients, urticaria/angioedema may be caused by one single particular NSAID or aspirin. In that case, the drug will act as a hapten with subsequent production of drug-specific IgE antibodies against the specific compound [422]. On repeat exposure to the same NSAID, patients will experience an IgE-mediated immune reaction with histamine release [429]. Thus, the clinical symptoms of urticaria/angioedema are possibly IgE mediated [430]. These patients may also undergo desensitization treatment against the specific drug (antigen). However, no controlled randomized trials about the success rate are available [420].

Salicylate Protein Binding as a Possible Etiological Factor Salicylates, circulating in the blood, are bound exclusively to plasma albumin. The extent of binding is dose dependent and the bound fraction at anti-inflammatory concentrations (1 mM) was found to be about fivefold higher than the free concentration in steady-state conditions. This bound fraction of salicylate is markedly enhanced in several allergic diseases, including urticaria, allergic rhinitis, and bronchial asthma. Since the albumin concentrations are unchanged in these conditions, it was speculated that a change in binding properties of albumin to salicylates rather than a change in overall binding capacity of salicylates took place in these patients and might be involved in its pharmacological effects [431]. Unfortunately, no further reports about this interesting hypothesis have been published.

3.3.2.2 Stevens–Johnson Syndrome and Toxic Epidermal Necrolysis (Lyell Syndrome)

Occurrence Stevens–Johnson and Lyell syndromes are extremely rare but life-threatening immune reactions at the skin. These reactions can be caused by various medications, including nonsteroidal anti-inflammatory drugs [432]. A large epidemiological trial in the United States, including about 260 000 patients who were treated with several drugs including NSAIDs and aspirin over 16 years, did not establish any increased incidence of these diseases in relation to aspirin use [433]. Similar results were obtained in a retrospective epidemiological trial in France. There were 333 cases of Lyell syndrome, which were seen over several years. The incidence was to 1.2 cases per 1 million inhabitants and year. This confirms that Lyell syndrome is an extremely rare disease, although fatal in 30% of cases. Aspirin as a possible risk factor could be excluded. The relative risk for occurrence of the syndrome in aspirin-treated subjects was 1.1 as opposed to 1.9 for diclofenac, 4.0 for piroxicam, 13 for fenbufen, and 18 for oxyphenbutazone [434].

Similar results were obtained in a European Case–Control Study, conducted between 1989 and 1995 in Germany, France, Italy, and Portugal. This study searched specifically for Stevens–Johnson and Lyell syndromes and their possible relation with salicylates, including aspirin and salicylate combinations. Among 373 cases and 1720 controls, the multivariate relative risk estimate for any salicylate use was 1.3 and not different from controls (95% confidence interval: 0.8–2.2). This suggests that aspirin and other salicylates are not associated with a measurable increase in the risk of these severe anaphylactic events [435].

Summary

Aspirin-related urticaria/angioedema occurs in about 0.1–0.2% of the healthy population whereas about 20–30% of patients with chronic idiopathic urticaria might suffer an exacerbation of the disease when treated with aspirin or nonselective NSAIDs.

The pathophysiology of the disease is not uniform and involves pharmacological and immunological mechanisms. Pharmacologically, aspirin might inhibit COX-1-dependent prostaglandin formation, allowing for uncontrolled overproduction of leukotrienes, similar to aspirin-induced asthma (Section 3.3.1). Immunologically, aspirin (and salicylates) may act as haptens, allowing for antigen production and subsequent generation of drug-specific IgE antibodies. These antibodies might then cause pathologic immune reactions on the skin. Aspirin desensitization is probably useful in most cases, except chronic idiopathic urticaria [428].

Lyell syndrome and Stevens–Johnson syndrome are extremely rare but severe and life-threatening diseases with manifestations at the skin. The diseases might be caused by intolerance to several drugs, including nonselective NSAIDs. There is no epidemiological evidence that aspirin, either alone or in combination with other drugs, will cause these diseases.

References

420 Gollapudi, R.R., Teirstein, P.S., Stevenson, D.D. *et al.* (2004) Aspirin sensitivity. *The Journal of the American Medical Association*, **292**, 3017–3023.

421 Shelley, W.B. (1964) Birch pollen and aspirin psoriasis. A study in salicylate hypersensitivity. *The Journal of the American Medical Association*, **189**, 985–988.

422 Blanca, M., Perez, E., Garcia, J.J. *et al.* (1989) Angioedema and IgE antibodies to aspirin: a case report. *Annals of Allergy*, **62**, 295–298.

423 Aihara, M., Miyazawa, M., Osuna, H. *et al.* (2002) Food-dependent exercise-induced anaphylaxis: influence of concurrent aspirin administration on skin testing and provocation. *The British Journal of Dermatology*, **146**, 466–472.

424 Doeglas, H.M. (1975) Reactions to aspirin and food additives in patients with chronic urticaria, including the physical urticarias. *The British Journal of Dermatology*, **93**, 135–144.

425 Grattan, C.E. (2003) Aspirin sensitivity and urticaria. *Clinical and Experimental Dermatology*, **28**, 123–127.

426 Pleskow, W.W., Stevenson, D.D., Mathison, D.A. *et al.* (1982) Aspirin desensitization and characterization of the refractory period. *The Journal of Allergy and Clinical Immunology*, **69**, 11–19.

427 Simon, R.A. (2003) Prevention and treatment of reactions to NSAIDs. *Clinical Reviews in Allergy & Immunology*, **24**, 189–198.

428 Kong, J.S., Teuber, S.S. and Gershwin, M.E. (2007) Aspirin and nonsteroidal anti-inflammatory drug hypersensitivity. *Clinical Reviews in Allergy & Immunology*, **32**, 97–110.

429 Asad, S.I., Murdoch, R., Youlten, L.J. *et al.* (1987) Plasma level of histamine in aspirin-sensitive urticaria. *Annals of Allergy*, **59**, 219–222.

430 Stevenson, D.D. (2004) Aspirin and NSAID sensitivity. *Immunology and Allergy Clinics of North America*, **24**, 491–505.

431 Maehira, F., Nakada, F. and Hirayama, K. (1990) Alteration of salicylate binding to serum protein in allergic subjects. *Clinical Physiology and Biochemistry*, **8**, 322–332.

432 Roujeau, J.C. and Stern, R.S. (1994) Severe adverse cutaneous reactions to drugs. *The New England Journal of Medicine*, **331**, 1272–1285.

433 Chan, H.L., Stern, R.S., Arndt, K.A. *et al.* (1990) The incidence of erythema multiforme, Stevens-Johnson syndrome, and toxic epidermal necrolysis. A population-based study with particular reference to reactions caused by drugs among outpatients. *Archives of Dermatology*, **126**, 43–47.

434 Roujeau, J.C., Guillaume, J.C., Febre, J.P. *et al.* (1990) Toxic epidermal necrolysis (Lyell syndrome): incidence and drug etiology in France, 1981–1985. *Archives of Dermatology*, **126**, 37–42.

435 Kaufman, D.W. and Kelly, J.P. for the International Case-Control Study of Severe Cutaneous Adverse Reactions (2001) Acetylsalicylic acid and other salicylates in relation to Stevens-Johnson syndrome and toxic epidermal necrolysis. *British Journal of Clinical Pharmacology*, **51**, 174–176.

3.3.3 Reye's Syndrome

In 1963, Ralph Douglas Kenneth *Reye*, a pathologist from Sydney (Australia) and his colleagues Graeme *Morgan* and Jim *Baral* described a hitherto unrecognized disease in small children, morphologically presenting as noninflammatory encephalopathy associated with fatty degeneration of the liver. The disease was clinically preceded by an initial period of "malaise," mostly associated with upper airway infections. The 21 identified children presented to the hospital with hyperpnoea, severe protracted vomiting, hypoglycemia, and elevated liver enzyme levels. There were deteriorations in consciousness, including stupor or coma, sometimes followed by convulsions. Seventeen of the children died within the first 3 days after admission, exhibiting signs of severe encephalopathy, the surviving children recovered completely. Necropsy showed a fatty degeneration of the liver and other viscera as well as a noninflammatory cerebral edema with cell degeneration. Reye considered this disease, which was later named after him, a ... clinicopathological entity of unknown etiology ... but also stated that he was ... not convinced that the etiology is identical in every case ....

There was another case report on a similar disease a few months later in North Carolina (USA) during an outbreak of influenza B [436]. However, reports on patients with related diseases had already been published sporadically since 1929 [437]. Thus, a low number of patients suffering from this or a similar (hepato)encephalopathy obviously did already exist, and a relation to an antecedent (viral) infection as "primer" of the disease appears not unlikely. It had been speculated that the sudden appearance – and disappearance – of Reye's syndrome fits best with a precipitous mutation in a virus [438]. In this context, it is interesting to note that the virulence of a particular influenza strain is not a constant, but quite complex and involves host adaptation, transmissibility, tissue tropism, and replication efficacy. Genetically, different recombination may markedly change the virulence of viruses, eventually resulting in a marked increase in virulence with millions of victims, for example, in the 1918 "Spanish Flu" [439, 440] and very recently the "bird flu."

The cause for the transformation of a more or less trivial and frequent upper airway virus infection into a life-threatening follow-up disease and the fact that this occurs only in a very low proportion of cases were and are a matter of dispute and finally unexplained. The primary affection of (small) children might be related to their unprotected exposure to viruses. As already suggested by Reye, the syndrome may not be caused by one factor but rather initiated by different mechanisms with the (final) common feature of noninflammatory hepatoencephalopathy. This points to a decisive role of disease-modifying factors.

In addition to a possibly changed virulence of viruses mentioned above, toxins, environmental poisons, and drugs, presumably those that are frequently used for symptomatic treatment of febrile virus infections in children, are possible candidates. This includes aspirin and acetaminophen, both subject to hepatic metabolism and both being potentially hepatotoxic, though by quite different mechanisms (Section 3.2.2). At a first view, there may also be some parallels between Reye's syndrome and acute aspirin intoxication (Section 3.1.1), though there is no evidence for an increased use of aspirin in the 1970s and 1980s when several thousand cases of Reye's syndrome were identified in voluntary notification schemes [441]. In any case, the search for a possible association between Reye's syndrome and aspirin became an issue of epidemiological dimensions. A causal relationship has never been established and probably never will because of the current scarcity of the disease. However, there is also no evidence that aspirin is not related to Reye's syndrome. Interestingly, aspirin has been shown to be a potent anti-influenza agent not only in cell culture experiments [442] but also in an animal model *in vivo* [443]. This action was explained by inhibition of virus propagation via the inhibition of NFκB activation (Section 2.2.2). Nevertheless, aspirin use in children is still a matter of concern and each aspirin package throughout the world contains a warning label. The differing opinions of whether these warnings are (still) justified are discussed in greater detail in several overviews [438, 441, 444–447].

3.3.3.1 Clinics, Laboratory, and Morphological Findings

Clinics The syndrome is clinically dominated by severe protracted vomiting and an acute noninflammatory encephalopathy, associated with hepatopathy. It is typically preceded by a prodromal viral infection, most frequently, influenza B, A, or varicella (chicken pox), affecting the upper respiratory (more than 70%) or the gastrointestinal tract. The symptoms last for 3–5 days and are followed by a recovery phase for another 1–3 days. In a very low number of cases, there is abrupt onset of encephalopathy, dominated by pernicious vomiting associated with varying degrees of neurological impairment and cerebral edema. There are no focal neurological signs. Deterioration in consciousness is followed by delirium, alternating with stupor or lethargy and convulsions in 30% of cases. The patient either recovers or proceeds to coma. Death occurs in about 30–40% of cases as a result of brainstem dysfunction. Recovery may be complete; however, persistent neurological deficits can remain. The clinical picture is similar in children and adults [441, 448].

Laboratory Findings The laboratory findings typically indicate massive tissue breakdown with enormous losses of protein and nitrogen, associated with a severe liver pathology. In serum, alanine aminotransferase, aspartate aminotransferase, and ammonia levels are largely elevated. The prothrombin time is prolonged and increased bilirubin is found. Typical for the disease are marked elevations in plasma free fatty acids, the occurrence of long-chain dicarboxylic acids, and hypoglycemia with plasma glucose levels less than 40 mg/dl. There are no signs of inflammation in the cerebrospinal fluid.

The primary cause of these alterations is a severe disturbance of hepatic mitochondrial function with decreased mitochondrial enzymatic activity. A defect in mitochondrial β-oxidation [449] causes depletion of intramitochondrial ATP [448] with subsequently disturbed energy-dependent hepatic functions, including gluconeogenesis and urea synthesis [450]. Consequently, hyperammonemia and hypoglycemia are correlated with the severity of the disease. The uncoupling of oxidative phosphorylation and a general depletion of energy-providing products, such as ATP, are seen in all cells and tissues of the organism and markedly affect their function. The most dramatic changes are seen in neurons that are particularly dependent on sufficient energy supply. The symptoms might be aggravated by starvation and insufficient alimentary (glucose) uptake by food, possibly occurring during a febrile viral infection in children.

Morphological Findings The major morphological alterations were already in detail described by Reye *et al.* [451]. There is glycogen depletion and marked microvesicular steatosis (fat deposition) in the liver and other organs. On electron microscopy, there are enlarged and pleiomorphic mitochondria and proliferations of smooth endoplasmic reticulum and peroxisomes but no hepatocellular necroses. In enzymatic defects with Reye's syndrome-like manifestations, the mitochondria are dysfunctional but normal in size and appearance [452]. Similar changes are seen in the liver in salicylate intoxication [453].

3.3.3.2 Etiology

Heterogeneity Reye's syndrome is a descriptive term covering a group of etiologically heterogeneous disorders caused by infectious, metabolic, toxic, or drug-induced alterations [446]. A high index of suspicion is critical for diagnosis [454] and the "classical" Reye's syndrome must of necessity be a diagnosis of exclusion [455]. It is obvious that a primary genetic background, that is, inborn metabolic disorders as cause of Reye's syndrome, was not diagnosed or even considered in the 1960s and 1970s when the majority of Reye's syndromes were reported. This also means that the "changing clinical pattern of Reye's syndrome" [456] actually might reflect improved scientific knowledge and diagnostic possibilities to detect metabolic failure or a genetic abnormality. This is particularly true for inherent or acquired metabolic failure(s) of mitochondria, for example, enzyme defects of fatty acid β-oxidation or defects in the urea cycle, which are potentially rapidly fatal and may present clinically as Reye's syndrome. A hypothetical scheme on the possible etiology of Reye's syndrome is shown in Figure 3.13.

Infections and Immune Responses An antecedent viral infection appears to be an essential condition for the later development of Reye's syndrome in most cases [457] and might impair mitochondrial function and lipid metabolism in many organs. Activation of systemic host defense by inflammatory cytokines and other mediators subsequent to virus infection(s) results in altered gene expression of various induced and constitutive cytochrome P450 isoforms in the liver and many other organs and tissues throughout the body [458]. At least 19 different viruses have been brought into connection with the syndrome [441]. A number of these viruses disturb Kupffer cell function, eventually resulting in the release of inflammatory cytokines such as TNFα, a known mediator of metabolic toxicity [459]. Cytokines, such as TNFα and IL-1, cause metabolic alterations similar to those in Reye's syndrome. These cytokines are found in significant amounts after viral and bacterial infections in relation to elevated levels of endotoxin. High levels of an endotoxin-like activity were found by bioassay in serum of patients with Reye's syndrome [460].

Acquired Hepatic Metabolic Failure Viral infections or associated mediators, generated and released as a result of these infections, might sensitize tissues such as the liver for subsequent injury by exogenous factors (Figure 3.13) [447]. Many exogenous noxious factors have been described.

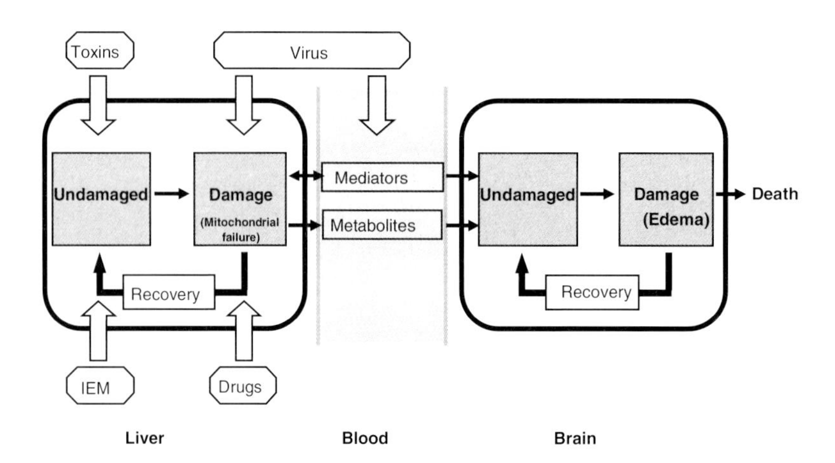

Figure 3.13 Hypothetical etiology of Reye's syndrome. An altered immune response, possibly related to pathogenic viruses, causes the release of inflammatory cytokines and other mediators, eventually resulting in mitochondrial injury. In the liver, this results in metabolic failure, reduced gluconeogenesis, and release of toxic metabolites (e.g., ammonia and branched fatty acids). This reaction is aggravated by environmental factors (toxins, pesticides, and certain drugs), or there may be preexisting genetic abnormalities in the mitochondria, such as inborn errors of metabolism. Hypoglycemia, hyperammonemia, and perhaps additional noxious factors cause neurological deficits and cerebral edema. Complete or partial recovery may follow or the disease progresses to death from brainstem dysfunction (after [440]).

These include aflatoxin, insecticides, solvents, and several drugs, including salicylates, phenothiazines, zidovudine, valproic acid, metoclopramide, and others [446]. As a consequence, mitochondrial failure can result with subsequent disturbances in mitochondrial metabolism. This is not exclusive for the liver, the major site of energy-driven drug metabolism, but also occurs in brain, kidney, and skeletal muscle.

Salicylates at high toxic concentrations can cause metabolic alterations in liver mitochondria, specifically when given over a longer period of time under conditions of an impaired immune response, such as rheumatoid arthritis [461]. Giles [462] originally developed the concept that some children suffering from Reye's disease may have carbohydrate metabolizing enzymes that are hypersensitive to salicylates. Disturbed mitochondrial fatty acid oxidation by salicylate was shown in animal experiments [463] and is one of the major features of salicylate toxicity in the liver (Section 3.1.2). The mitochondrial oxidation of long-chain fatty acids (palmitate) in skin fibroblasts from children who had recovered from Reye's syndrome was more sensitive to inhibition by salicylate (1–5 mM) than in healthy controls [459]. In addition, energy-dependent hepatic salicylate conjugation to glycine, resulting in salicyluric acid, might be disturbed, eventually yielding toxic levels of salicylates that not only further injure (liver) mitochondria but may also contribute to the encephalopathy.

The possible relationship between serum salicylate and Reye's syndrome (confirmed by liver biopsy) was studied in 218 children, diagnosed between 1963 and 1980 in Cincinnati. Mean salicylate serum levels in 27 children who died or survived with neurological deficits were 150 µg/ml (range: 0–460 µg/ml) but only 100 µg/ml (range: 0–480 µg/ml) ($p = 0.01$) in the 103 patients who recovered without neurological deficit. In contrast, serum salicylate in a group of 27 age-matched controls was less than 20 µg/ml.

It was concluded that increased salicylate concentrations at admission in Reye patients could result from excessive dosage because of a greater severity of the prodromal illness or to diminished salicylate clearance in case of impaired hepatic metabolism. But it was also found impossible to determine from these data whether salicylates are involved in the etiology or in determining the outcome of the disease [464].

Unfortunately, no information was provided in this study about aspirin dosage and the interval between the last administration and the time when the sample was taken. Thus, the real salicylate levels *in vivo* remained unknown.

Methods of Salicylate Determination Although not outlined in detail, an automated variant of the Trinder method [465] has probably been used to determine salicylate levels in the study mentioned above [464]. This method is nonspecific for salicylate in the presence of many other chemicals. As many as 63 (!) organic acids and amines, several of them elevated in Reye's disease, have been found to interfere with the assay, eventually resulting in wrong positive results [466]. A comparison of the Trinder assay with more sensitive and selective HPLC methods showed that salicylate levels in liquor and serum of Reye children not only were much lower – less than 1% of the Trinder assay – but also did not correlate with the severity of the disease [467]. Another study found an impaired oxidative metabolism of salicylates (measured by HPLC), associated with higher unchanged salicylate excretion in the urine of Reye patients. However, serum levels of salicylate were not increased [468]. Although a dose–response relationship has also been claimed occasionally [469], most available studies [468, 470, 471] do not suggest any direct relationship between salicylate levels in plasma and the severity of the disease (Section 1.2.2).

Inborn Errors of Metabolism Various inborn errors of metabolism can present clinically with Reye-like syndromes and become increasingly apparent with improved diagnostic facilities. Of particular interest in this context are inherent disorders of mitochondrial fatty acid metabolism [472–474]. These might cause recurrent Reye-like symptoms triggered by infections [475] and may be even fatal if exogenous alimentary supply is restricted for longer times than permitted by available (hepatic) glycogen stores [476]. Clinical differentiation is often difficult, and extensive laboratory testing is required. In countries where tandem mass spectrometry for inherited metabolic disorders is included in neonatal screening programs, for example, in Germany, fatty acid oxidation defects might be detected presymptomatically by acylcarnitine profiles [477]. The majority of these inherited disorders are specific enzyme defects in liver mitochondria with disturbed energy supply for cell functions via oxidative phosphorylation. Inborn mitochondrial failure usually becomes evident before the age of 3 years. It has been suggested that a proportion of about 10–20% of children diagnosed initially with Reye's syndrome suffer from a hitherto unknown inherited metabolic disorder (mainly fatty acid oxidation or urea cycle defects) [478, 479].

3.3.3.3 Clinical Studies

A Reye-like disease, except a few sporadic reports, did not exist (or was not reported) for any significant extent before 1950 and disappeared in the late 1980s [438, 480]. The reasons for this rise and fall are a matter of heavy controversies, and this in particular with respect to the role of aspirin.

Historically, an association between salicylates and Reye's syndrome arose from the recognition that the symptoms of salicylate poisoning are often similar to the clinical manifestations of Reye's syndrome [481]. It is now known that the typical symptoms of salicylate poisoning, in particular those in the central nervous system, have a different pathophysiological background. Thus, although some patients in the early epidemiological report of Linnemann *et al.* [482] had a history of excessive aspirin use, it was already evident at that time that aspirin intake alone could not explain all manifestations of Reye's syndrome [446, 483]. Today, with markedly advanced diagnostic facilities, including molecular biology, many Reye-like syndromes can be explained as inborn errors of metabolism without any causative role for aspirin, whereas the "real" aspirin-related Reye's syndrome apparently disappeared [438], if it had existed at all. Nevertheless, it were these and some early follow-up epidemiological studies, mainly in the United States and the United Kingdom, that were largely responsible for the removal of aspirin as an antipyretic analgesic in pediatrics – with the exception of Kawasaki disease (Section 4.2.3), and, therefore, require particular consideration.

Studies from the United States The first clinical data on a possible relationship between aspirin intake and Reye's syndrome came from epidemiological studies in the United States [484–487]. The four initial case–control studies, named after their respective localization as Arizona, Ohio, and Michigan 1 and 2, were widely cited by the lay and medical media and have stimulated much of the later concern about an association of Reye's syndrome with salicylates [438, 481].

In the first study from *Arizona*, a total of 7 children hospitalized with Reye's syndrome in 1978 were compared with 16 control individuals. All children with Reye's syndrome had influenza A and had taken aspirin, but only 50% of controls. No attempt was made to identify a viral background in the control individuals. Therefore, it is unknown whether the controls had the same prodromal illness as the cases [484].

The two studies from *Michigan* (1 and 2) had diagnostic weaknesses in terms of Reye's syndrome patient and control groups and also involved small numbers of patients. In addition, in the first study (Michigan 1) the final diagnosis was based upon interviews with the parents, conducted on average 6–8 weeks after the child had been diagnosed with Reye's syndrome. In these two studies, 30 of 46 patients with Reye's syndrome and 13 of 29 control individuals had received aspirin and did not develop Reye's syndrome; both groups had similar preceding illnesses, mostly, upper respiratory tract infections [486].

The *Ohio* study (1978–1980) was at this time both the largest and most controversial [438]. Ninety-four of the 97 patients with Reye's syndrome (97%) had taken aspirin and 110 of 156 controls (71%). However, 22% of Reye's syndrome patients compared with only 4% of controls had taken phenothiazines. Phenothiazines and other antiemetics could contribute to an escalation of the viral disease and specifically to extrapyramidal reactions [488] that might have an influence on clinical outcome. The questionnaire used in this study was consequently revised. Interestingly, 10% of the cases but 25% of the excluded patients had varicella and only 97 out of the total 227 cases of Reye's syndrome that were reported to the Ohio Department of Health were included in this study [485].

Each of these studies found a statistically significant association between salicylate use and Reye's syndrome, suggesting a "possible link" between the two. However, all of them have been subject of considerable criticism, regarding the way the study was published (only the Arizona study was originally published with a detailed description of methods), the results obtained, and their interpretation [438, 481, 489, 490]. All were retrospective case–control trials with small numbers of patients and three different definitions of cases. Thus, it was not proven whether the patients suffered from the "true" Reye's syndrome and biopsy confirmation of the diagnosis of Reye's syndrome was rare. There were significantly fewer individuals who took aspirin in the control groups of all of the studies. Insufficient, if any, data on salicylate levels were published with no data on the duration and dosage of aspirin treatment. Information about aspirin ingestion was usually obtained from interviews of the parents. The time interval between the presumed drug exposure and the interview varied from a few days to 3 months and, in the Ohio study, these time intervals were different between patients and controls. These and other limitations might have caused considerable bias, including selection bias in both the patients with Reye's syndrome and control individuals, information bias, and confounding bias [481, 489, 490].

The issue of possible information bias associated with retrospective epidemiological case–control studies in rare diseases, such as Reye's syndrome, was convincingly demonstrated by Heubi and colleagues.

Investigators at Cincinnati Children's Hospital, during periods when Reye's syndrome occurred commonly in their community, studied 85 children with a preceding viral upper respiratory tract infection or chicken pox and vomiting and an at least threefold elevated aspartate aminotransferase. The children did not show any of the neurological features of Reye's syndrome other than being quiet or withdrawn.

Liver biopsy demonstrated typical mitochondrial changes of Reye's syndrome with a severity no less than that observed in children with deep coma. On treatment with glucose and electrolyte infusion, only the five cases already more severe at diagnosis proceeded to a deeper grade of coma. One child survived with severe neurological deficits, all other recovered completely. There was no death.

Had these children not been presented to investigators who were studying this disorder prospectively, it is very unlikely that the diagnosis of Reye's syndrome would have been considered and if considered, it certainly would not have been confirmed. Such cases do not appear in most epidemiological surveys and may cause significant information bias [441, 491].

With reference to these epidemiological trials, a National Consensus Conference [492] in the United States stated that available studies have provided a strong statistical association between Reye's syndrome and aspirin use. However, other possible explanations for the disease should be considered and parents and physicians should be aware that most, if not all, medications have potential deleterious effects. Caution in the use of salicylates in children with influenza or varicella was suggested but finally concluded that the data also indicate that salicylates alone cannot be responsible for the development of Reye's syndrome [492]. As a consequence, the US Public Health Service (PHS) conducted two studies: a pilot study [493] and a main study [487]. The main study was among the largest epidemiological trials, searching for an interaction between salicylate intake and Reye's syndrome and will be discussed in more detail.

Throughout the United States, initially 50 and finally 70 pediatric centers participated. To be enrolled, all patients had to have a diagnosis of Reye's syndrome by a physician, an antecedent respiratory or gastrointestinal disease or chicken pox and an advanced (stage II or more) degree of encephalopathy.

Fifty-three patients were initially identified by attending physicians. Seven of them were subsequently reclassified to another diagnosis by the attending physician. Another 13 cases that had been enrolled by the attending physician were later excluded by the physician review panel because another diagnosis appeared more likely. Of the remaining 33 patients, 6 patients that had Reye's syndrome as confirmed by the expert panel were not included because an antecedent illness (as defined by the inclusion criteria) was not identified. Thus, 27 patients with Reye's syndrome and 140 matched control individuals were available for analysis – there were 30 patients in the pilot study [493]. Of these 27 patients, 3 (11%) died.

This mortality rate was suspiciously low. The number of Reye patients was also much lower than the "desired" 100–200 cases. However, a "strong association between salicylates and Reye's syndrome" in the midpoint analysis was found. As a result of this and the increasing rarity of the disease, the study was finished prematurely at this time point.

Fifteen (!) chemicals were given to at least 20% of the study participants, 26 out of 27 patients with Reye's syndrome (96%), and 53 out of 140 control individuals (38%) had taken salicylates, mostly aspirin, whereas 30% of cases and 86% of controls had taken acetaminophen. There was a highly significant difference in total and average doses of salicylates between patients with Reye's syndrome and controls, both being almost threefold higher in the former group. The total dose in patients with Reye's syndrome was 74 mg/kg, that is slightly more than 5 g per one 70 kg adult.

The conclusion was that more than 90% of patients enrolled with Reye's syndrome received salicylates, thereby suggesting an association between the two. Moreover, the risk of Reye's syndrome should also be related to the quantity of salicylates ingested. The recommendation was to limit the use of aspirin in children for the treatment of chicken pox or influenza-like illnesses [487].

Essential biases, however, remain in this study. This includes increasing public knowledge about a possible relation between aspirin and Reye's syndrome. Thus, the parents or other caregivers who were interviewed could have reported aspirin ingestion once they knew the diagnosis. There was no control of salicylate levels. Moreover, the diagnostic criteria were relatively nonspecific and histological support for the diagnosis was available only in 27% of patients [494].

In a further attempt to confirm the validity of the aspirin Reye's syndrome association, another even more thorough epidemiological study was undertaken to minimize possible sources of bias [495]. Eighty-eight percent of case subjects ($n = 24$) and only 17% of the matched controls ($n = 48$) had received aspirin prior to the onset of Reye's syndrome. Further analyses demonstrated that the association could not be attributed to the five potential sources of biases [495]. However, out of the total 24 cases, only 8 had liver biopsies and 12 had urine samples collected early during the disease, and it is unclear whether the latter 12 included any of the 8 who had liver biopsies [441].

Between 1980 and 1997, 1207 cases of Reye's syndrome were reported to the Center of Disease Control. The peak incidence occurred in 1980 with 555 cases reported and the number steadily declined thereafter (Figure 3.14). In 1994, the hospitalization

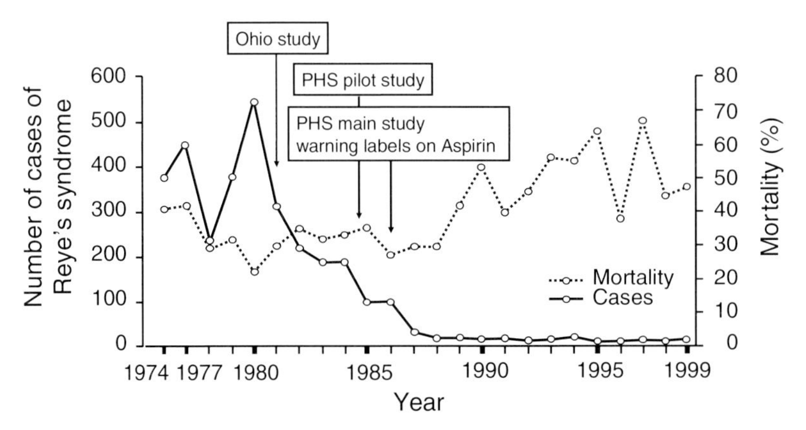

Figure 3.14 Incidence and mortality of Reye's syndrome in the United States from 1973 to 1999. The figure shows decline and apparent disappearance of the disease in relation to the first epidemiological studies and the placement of warning labels on aspirin (acetylsalicylic acid)-containing products (PHS study) (modified after [438]).

rate for Reye's syndrome in the United States was estimated at 0.06/100 000 persons aged less than 18 years and even this low number was considered to be an overestimate [452]. The annual peak correlated with the seasonal occurrence of viral upper respiratory tract infections, particularly influenza. Before 1990, the incidence of Reye's syndrome was higher in years with epidemics of influenza B than in years with influenza A. This association was not found subsequently. Two or fewer cases have been reported per year in the United States since 1994 [452]. The incidence is now considered to be <0.03–1/100 000 individuals aged less than 18 years [454].

Studies from the British Isles The epidemiology of Reye's syndrome in the British Isles differed from that in the United States. In the early 1980s, the median age of patients with Reye's syndrome in Britain was lower (14 months) than that in the United States (9 years) and there was no seasonal peak in winter, that is, no clear association with influenza waves [496]. However, an association between the disease and aspirin intake was also found.

In total, 264 cases were reported to the British Health Authorities between 1981 and 1985, when recruitment to the risk factors was terminated. A similar overall proportion of cases and controls (72% versus 68%) had taken antipyretics but 59 and 26%, respectively, had taken aspirin. It was concluded that there is an epidemiological association between the development of Reye's syndrome and aspirin [496].

Several publications discussed the incidence and possible causes for Reye's syndrome between 1982 and 1990 in the United Kingdom in more detail. Hardie *et al.* [456] divided their study into two parts: 4.5 years prior to and after the aspirin warnings by the British Health Authorities (June 1986). During this time, 445 cases of Reye's syndrome were reported; 91, that is, 20% of those, were misdiagnoses. Interestingly, 16% of diagnoses were revised in the first period but twice as much, that is, 34%, in the second. An explanation for this was possible misclassification and correct (re)identification as a "Reye-like" inherited metabolic disorder. According to the wide distribution of scores, the reported cases were considered a heterogeneous group of patients [446, 456, 497].

It has been suggested by expert panels in two other studies from the United States and Canada that one-third [498] and three-quarters [499] of cases, respectively, definitely or probably did not have real Reye's syndrome whereas Green and Hall [497] reported 10% of cases in the British Isles that were misdiagnosed because of inborn errors of metabolism. The highest scoring patients in the study from the British Isles by Hardie *et al.* [456] were those most likely exposed to preadmission aspirin. However, there was considerably less information about preadmission medications in the lower score cases (<50%) whereas this information was available to nearly all of the highest score cases. This was considered by the authors as a possible, though less likely source of bias. Only 33% of the highest scoring patients were reported as having received aspirin. This differs from the >90% of cases in the US case–control studies mentioned above and the 59% in the UK risk factor study [495, 496].

McGovern *et al.* [500] reported on five cases of Reye's syndrome in the Royal Belfast Hospital, Northern Ireland, over a 13-year period; two had a possible relationship to aspirin and three had no link to aspirin and were considered "atypical" by the authors. There was no positive identification of salicylates in the serum of the two aspirin-related cases after admission. The results of this study were heavily disputed and the possibility arose that the diagnosis of Reye's syndrome in these cases was put forward – in the best case – by default [501].

Studies from Continental Europe Reye's syndrome has always been very rare in Continental Europe. A survey of 99 children's hospitals indicated the incidence of Reye's syndrome to be 0.04–0.05 cases per 100 000 children aged below 18 years in West Germany between 1983 and 1985. The disease was fatal in about half of the cases [502]. Ten years later during a 1-year observation period (1997) of all severe complications of varicella infection in 485 German pediatric hospitals, there was not one case of Reye's syndrome [503].

In Denmark, all pediatric departments were asked for the year 1979, a year of epidemic influenza B (182 500 reported cases between January and April), if they had made a diagnosis of Reye's syndrome and none had. However, there was one case (autopsy report) of a girl with a Reye-like syndrome with measles but without any further information on treatment. From this one case, an incidence of 0.09/100 000 children up to the age of 14 years was estimated [504].

In France, 0.08/100 000 children aged below 15 years were hospitalized for Reye's syndrome in 1995/1996. Of the 46 suspected cases, 14 were classified as Reye's syndrome and 5 of them had a metabolic disorder. A total of eight children were exposed to aspirin, alone or in combination with other drugs [505].

In Switzerland, seven fatal cases of Reye's syndrome were diagnosed between 1971 and 1984, and aspirin intake was reported in one of them [506].

Thus, in European countries [502, 506–508] and many other countries worldwide [509–513], a significant number of children diagnosed with Reye's syndrome did not take aspirin: on average <30% (range: 0–71%) in 11 different countries as opposed to 94% in the main study of Hurwitz *et al.* [487] in the United States (Table 3.7).

Although epidemiological data might suggest a correlation between salicylate intake and Reye's syndrome, any causal relationship has never been established and probably will not establish in the future because of the rarity of the disease [495, 496, 516]. In the United States, there was a statistical relationship between reduced sales of "baby aspirin tablets" (81 mg aspirin per tablet) and decrease in the incidence of Reye's syndrome over 5 years (1980–1985) [517]. Similar relationships were also observed in other countries. Surprisingly, the disease is also disappearing in Australia, where it was originally detected, despite a total lack of association with salicylates, and in countries such as France and Belgium, which continued to use aspirin in children without change between 1970 and 1990 [438]. However, among the millions of children being infected with viruses of any kind every day, there might be a few cases of a Reye-like syndrome worldwide without any clear etiologic reason [438].

Taken together, epidemiologic and experimental studies are compatible with the hypothesis that

Table 3.7 Reye's syndrome and aspirin intake (alone or in combination with other drugs) throughout the world [447].

Country	Number of reported cases[a]	Number of cases taking aspirin[b]	Percentage	Reference
Australia	49	4	8	Orlowski *et al.* [513]
France	14	8	57	Autret-Lecat *et al.* [505]
(West) Germany	15	3	20	Gladtke and Schauseil-Zipf [502]
Hong Kong	27	3	11	Yu *et al.* [512]
India	71	None	—	Christo and Venkatesh [511]
Ireland	23	14	61	Glasgow [515]
Japan	30	7	23	Yamashita *et al.* [510]
Thailand	73	52	71[c]	Yamashita *et al.* [510]
South Africa	21	5	22	Hofman and Rosen [509]
Spain	57	23	40	Palomeque *et al.* [507]
Switzerland	7	1	14[d]	Sengupta *et al.* [506]
United States	27	26	94	Hurwitz *et al.* [487]

[a] Not all cases of originally considered to be "Reye's syndrome" were positively confirmed.
[b] Alone or in combination with other drugs.
[c] Possible relationship to aflatoxin intake with contaminated food.
[d] Only fetal cases.

"cryptogenic" Reye's syndrome may arise from an unusual response to viral infection, possibly determined by host genetic factors but modified by a range of exogenous agents [441]. If aspirin has any pathophysiological role, there must be an exceptional unpredictable combination of circumstances that act as a trigger [496]. There is not one single article that has ever established a causal relationship between Reye's syndrome and the intake of aspirin.

3.3.3.4 Actual Situation

Beginning in 1980, US American Health Authorities cautioned physicians and parents not to use salicylates in children with chicken pox or influenza-like illnesses. A warning label is required for all aspirin-containing medications since 1986. Other countries have followed these recommendations more or less completely.

Benefit–Risk Calculations Independent of the unclear relationship between salicylates and possible subsets of patients with Reye's syndrome sensitive to them, the question arises whether the benefits of salicylate removal – and replacement by other drugs, largely acetaminophen (paracetamol) – outweighs the possible benefits of aspirin [518].

Although both compounds can injure the liver, the adverse effect of aspirin is only seen on repeated administration of high doses in predisposed individuals, for example, in children with rheumatic diseases [461] (Section 3.2.3). Toxic liver injury with aspirin – in contrast to acetaminophen [519] – is rare and generally reversible and – in contrast to aspirin – not a typical symptom of acute overdosage (Section 3.3.1).

Inflammation of the Upper Airways and Larynx Aspirin possesses anti-inflammatory properties that may be useful for the prevention of bacterial superinfection and febrile inflammatory diseases of the upper airways. As outlined by Rivera-Penera *et al.* [520], acetaminophen at antipyretic doses does not have anti-inflammatory properties but may be hepatotoxic even at minor overdoses given during a few days. In an accompanying editorial to this paper, Heubi and Bien [521] stated that the current estimates of the occurrence of acetaminophen toxicity are the tip of the iceberg of the total number of cases seen in the United States. Thus, in infections, frequently occurring in children aged between 4 and 10 years, such as laryngitis/pharyngitis and otitis/sinusitis, replacement of aspirin by acetaminophen may be less effective [522]. Alternatively, overstating the risk of aspirin use may cause a compensatory increase in other NSAIDs that also have adverse effects and not been better tolerated [523].

Kawasaki Disease In Japan, up to 200 000 children have received aspirin for treatment of Kawasaki disease, the recommended initial dosage being between 30 and 100 mg/kg (Section 4.2.3). In Britain, during the acute phase of the illness, moderate doses of aspirin, 30–50 mg per day, are recommended. It is interesting to note that in a recent British guideline for practical therapy of Kawasaki's disease, Reye's syndrome is not even mentioned [524]. Only one case of Reye's syndrome associated with Kawasaki disease has ever been reported (in the Japanese literature), giving an incidence of <0.005% [525].

Asthma Another risk of replacing aspirin by acetaminophen is the possible facilitation of allergic sensitization asthma in genetically predisposed children [526]. This allergic sensitization has to be distinguished from nonallergic "aspirin-induced asthma" (Widal triad), which has no allergic background (Section 3.3.1).

The prevalence in childhood asthma in the United States increased by 23% from 1970 to 1980 but nearly twice as much, that is, 40%, from 1980 to 1986. Among other environmental factors, the nearly complete interruption of the use of aspirin in children with febrile respiratory infections and its replacement by acetaminophen have been discussed as a possible explanation [526]. There was a close linear correlation observed in the United States between the increasing use of acetaminophen and the prevalence of asthma in children and adolescents at the same time that aspirin use

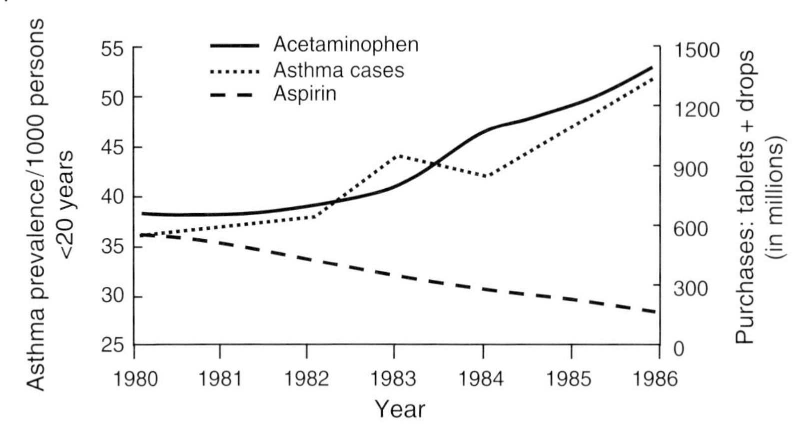

Figure 3.15 Prevalence of asthma in children and adolescents (aged <20 years) and total over-the-counter purchases of pediatric aspirin (acetylsalicylic acid) (tablets) and acetaminophen (paracetamol) (tablets + drops) in the United States from 1980 to 1986 [526].

declined (Figure 3.15). Frequent acetaminophen usage can cause asthma attacks [527].

The worldwide increase in the various allergic diseases, most notably asthma [528–530], might be due, at least in part, to the decreased use of aspirin [525]. Thus, the elimination of pediatric aspirin may be an unrecognized and important contributor to the increase in asthma prevalence.

Summary

Reye's syndrome is an extremely random but severe and often fatal hepatoencephalopathy. It presents clinically with protracted vomiting, a hepatopathy with signs of diverse hepatic dysfunctions, indicating mitochondrial failure. The consequences are particularly dramatic for the central nervous system and include several noninfectious neurological deficits. The disease is fatal in about 30–40% of cases because of brainstem dysfunction. The disease is typically preceded by a viral infection with an intermediate disease-free interval of 3–5 days before, in extremely rare cases, further progression into severe liver injury and CNS dysfunction occurs.

The etiology of the "cryptogenic" Reye's syndrome is unknown, but probably multifactorial. Hypothetically, the syndrome might result from an unusual response of the organism to a viral infection, which is determined by host genetic factors but modified by exogenous agents [441]. These include a number of pesticides, solvents, other chemicals, and at least 10 different drugs, aspirin being one of them.

The "rise and fall" of the Reye's syndrome pandemia is still poorly understood and unexplained. With a few exceptions, if any, there were no new Reye diseases during the past 10 years that could not be explained by an inherited disorder of metabolism or were just misdiagnoses. This may reflect scientific progress, that is, improved understanding of cellular and molecular dysfunctions as disease-determining factors. Alternatively, the immune response to and the virulence of a virus might have changed by altering its genetic code. Thus, there were no reports on Reye-like diseases prior to the early 1950s, though the use of aspirin in children was probably not much different during this time than at present. There was a similar decrease in the incidence in different countries, such as the United States, Belgium/France, and Australia in which aspirin prescription behavior in children differed.

Whether the benefit–risk ratio of drug treatment in febrile children is really improved by replacing aspirin with acetaminophen needs to be established. The missing anti-inflammatory component in the spectrum of actions of acetaminophen in the treatment of laryngitis/pharyngitis and otitis/sinusitis makes this compound less suitable for treatment of these disorders and, in addition, may favor allergies. In addition, acetaminophen bears a significant hepatotoxic potential even at slight overdoses.

It remains to be determined whether the restrictions in the use of pediatric aspirin did eliminate "real" Reye's syndrome or whether other reasons account for this phenomenon. Clearly, there is no drug treatment without a risk of side effects; this is also valid for aspirin. Thus, a carefully balanced decision whether a treatment with a certain drug justifies the risk of unwanted reactions is always required. Aspirin is no exception.

References

436 Johnson, G.M., Scurletis, T.D. and Carroll, N.B. (1963) A study of sixteen fatal cases of encephalitis-like disease in North Carolina children. *North Carolina Medical Journal*, **24**, 464–473.

437 Brain, W.R., Hunter, D. and Turnbull, H.M. (1929) Acute meningo-encephalomyelitis of childhood: report of 6 cases. *The Lancet*, **1**, 221–227.

438 Orlowski, J.P., Hanhan, U.A. and Fiallos, M.R. (2002) Is aspirin a cause of Reye's syndrome? A case against. *Drug Safety*, **25**, 225–231.

439 Reid, A.H., Fanning, T.G., Hultin, J.V. *et al.* (1999) Origin and evolution of the 1918 "Spanish" influenza virus hemagglutinin gene. *Proceedings of the National Academy of Sciences of the United States of America*, **96**, 1651–1656.

440 Gibbs, M.J., Armstrong, J.S. and Gibbs, A.J. (2001) Recombination in the hemagglutinin gene of the 1918 "Spanish flu". *Science*, **293**, 1842–1845.

441 Mowat, A.P. (1992) *Aspirin and Other Salicylates* (eds J.R. Vane and R.M. Botting), Chapman & Hall Medical, London, pp. 531–547.

442 Huang, R.T. and Dietsch, E. (1988) Anti-influenza viral activity of aspirin in cell culture. *The New England Journal of Medicine*, **319**, 797.

443 Mazur, I., Wurzer, W.J., Ehrhardt, C. *et al.* (2007) Acetylsalicylic acid (ASA) blocks influenza virus propagation via its NFκB-inhibiting activity. *Cellular Microbiology*, **9**, 1683–1694.

444 Heubi, J., Partin, E., Partin, J.C. *et al.* (1987) Reye's syndrome: current concepts. *Hepatology*, **7**, 155–164.

445 Visentin, M., Salmona, M. and Tacconi, M.T. (1995) Reye's and Reye-like syndromes, drug-related diseases? (Causative agents, etiology, pathogenesis, and therapeutic approaches). *Drug Metabolism Reviews*, **27**, 517–539.

446 Casteels-Van Daele, M., van Geet, C., Wouters, C. *et al.* (2000) Reye-syndrome revisited: a descriptive term covering a group of heterogeneous disorders. *European Journal of Pediatrics*, **159**, 641–648.

447 Schrör, K. (2007) Aspirin and Reye syndrome. A review of the evidence. *Pediatrics Drugs*, **9**, 195–204.

448 Van Coster, R.N., De Vivo, D.C., Blake, D. *et al.* (1991) Adult Reye's syndrome – a review with new evidence for a generalized defect in intramitochondrial enzyme processing. *Neurology*, **41**, 1815–1821.

449 Deschamps, D., Fisch, C., Fromenty, B. *et al.* (1991) Inhibition by salicylic acid of the activation and thus oxidation of long chain fatty acids. Possible role in the development of Reye's syndrome. *The Journal of Pharmacology and Experimental Therapeutics*, **259**, 894–904.

450 Corkey, B.E., Hale, D.E., Glennon, M.C. *et al.* (1988) Relationship between unusual hepatic acyl coenzyme A profiles and the pathogenesis of Reye syndrome. *The Journal of Clinical Investigation*, **82**, 782–788.

451 Reye, R.D.K., Morgan, G. and Baral, J. (1963) Encephalopathy and fatty degeneration of the viscera: a disease entity in childhood. *The Lancet*, **2**, 749–752.

452 Belay, E.D., Bresee, J.S., Holman, R.C. *et al.* (1999) Reye's syndrome in the United States from 1981 through 1997. *The New England Journal of Medicine*, **340**, 1377–1382.

453 Partin, J.S., Daugherty, C.C., McAdams, A.J. *et al.* (1984b) A comparison of liver ultra structure in salicylate intoxication and Reye's syndrome. *Hepatology*, **4**, 687–690.

454 Weiner, D.L. (2001) Pediatrics, Reye syndrome. eMedicine Specialities/Emergency Medicine/ Pediatric.

455 Glasgow, J.F.T. and Middleton, B. (2001) Reye-syndrome – insights on causation and prognosis. *Archives of Disease in Childhood*, **85**, 351–353.

456 Hardie, R.M., Newton, L.H., Bruce, J.C. *et al.* (1996) The changing clinical pattern of Reye's syndrome 1982–1990. *Archives of Disease in Childhood*, **74**, 400–405.

457 Larsen, S.U. (1997) Reye's syndrome. *Medicine Science and the Law*, **37**, 235–241.

458 Prandota, J. (2002) Important role of prodromal viral infections responsible for inhibition of xenobiotic metabolizing enzymes in the pathomechanisms of idiopathic Reye's syndrome. *American Journal of Therapeutics*, **9**, 149–156.

459 Glasgow, J.F.T., Middleton, B., Moore, R. *et al.* (1999) The mechanism of inhibition of β-oxidation by aspirin metabolites in skin fibroblasts from Reye's syndrome patients and controls. *Biochimica et Biophysica Acta*, **1454**, 115–125.

460 Cooperstock, M.S., Tucker, R.P. and Baublis, J.V. (1975) Possible pathogenic role of endotoxin in Reye's syndrome. *The Lancet*, **1**, 1272–1274.

461 Zimmermann, H.J. (1981) Effects of aspirin and acetaminophen on the liver. *Archives of Internal Medicine*, **141**, 333–342.

462 Giles, H.M.C. (1965) Encephalopathy and fatty degeneration of the viscera. *The Lancet*, **1**, 1075.

463 Yoshida, Y., Fujii, M., Brown, F.R. 3rd *et al.* (1988) Effect of salicylic acid on mitochondrial–peroxisomal fatty acid catabolism. *Pediatric Research* **23** 338–341.

464 Partin, J.S., Partin, J.C., Schubert, W.K. *et al.* (1982) Serum salicylate concentrations in Reye's disease. *The Lancet*, **1**, 191–194.

465 Trinder, P. (1954) Rapid determination of salicylate in biological fluids. *The Biochemical Journal*, **57**, 301–303.

466 Kang, E.S., Todd, T.A., Capaci, M.T. *et al.* (1983) Measurement of true salicylate concentrations in serum from patients with Reye's syndrome. *Clinical Chemistry*, **29**, 1012–1014.

467 Andresen, B.D., Alexander, M.S., Ng, K.J. *et al.* (1982) Aspirin and Reye's disease: a reinterpretation. *The Lancet*, **1**, 903.

468 Meert, K.L., Kauffman, R.E., Deshmukh, D.R. *et al.* (1990) Impaired oxidative metabolism of salicylate in Reye's syndrome. *Developmental Pharmacology and Therapeutics*, **15**, 57–60.

469 Pinsky, P.F., Hurwitz, E.S., Schonberger, L.B. *et al.* (1988) Reye's-syndrome and aspirin – evidence for a dose-response effect. *The Journal of the American Medical Association*, **260**, 657–661.

470 Clark, J.H., Nagamori, K. and Fitzgerald, J.F. (1985) Confirmation of serum salicylate levels in Reye's syndrome: a comparison between the Natelson colorimetric method and high performance liquid chromatography. *Clinica Chimica Acta*, **145**, 243–247.

471 Chu, A.B., Nerurkar, L.S., Witzel, N. *et al.* (1986) Reye's syndrome. Salicylate metabolism, viral antibody levels, and other factors in surviving patients and unaffected family members. *American Journal of Diseases of Children*, **140**, 1009–1012.

472 Ogawa, E., Kanazawa, M., Yamamoto, S. *et al.* (2002) Expression analysis of two mutations in carnitine palmitoyltransferase IA deficiency. *Journal of Human Genetics*, **47**, 342–347.

473 Rahbeeni, Z., Vaz, F.-M., Al-Hussein, K. *et al.* (2002) Identification of two novel mutations in OCNT2 from two Saudi patients with systemic carnitine deficiency. *Journal of Inherited Metabolic Disease*, **25**, 363–369.

474 Tamaoki, Y., Kimura, M., Hasegawa, Y. *et al.* (2002) A survey of Japanese patients with mitochondrial fatty acid beta-oxidation and related disorders as detected from 1985–2000. *Brain & Development*, **24**, 675–680.

475 Scaglia, F., Scheuerle, A.E. and Towbin, J.A. (2002) Neonatal presentation of ventricular tachycardia and a Reye-like syndrome episode associated with disturbed mitochondrial energy metabolism. *BMC Pediatrics*, **2**, 12.

476 Marsden, D., Nyhan, W.L. and Barshop, B.A. (2001) Creatine kinase and uric acid: early warning for metabolic imbalance resulting from disorders of fatty acid oxidation. *European Journal of Pediatrics*, **160**, 599–602.

477 Schulze, A., Lindner, M., Kohlmüller, D. *et al.* (2003) Expanded newborn screening for inborn errors of metabolism by electrospray ionization-tandem mass spectrometry: results, outcome, and implications. *Pediatrics*, **111**, 1399–1406.

478 Orlowski, J.P. (1999) What happened with Reye's syndrome? Did it ever really exist? *Critical Care Medicine*, **27**, 1582–1587.

479 Hall, S.M. and Lynn, R. (1998) Reye syndrome, in *British Paediatric Surveillance Unit, 12th Annual Report* (eds M. Guy, A. Nicoll and R. Lynn), RCPCH, London.

480 Sullivan-Bolyei, J.Z. and Corey, L. (1981) Epidemiology of Reye syndrome. *Epidemiologic Reviews*, **2**, 1–26.

481 Daniels, S.R., Greenberg, R.S. and Ibrahim, M.A. (1983) Scientific uncertainties in the studies of salicylate use and Reye's syndrome. *The Journal of the American Medical Association*, **249**, 1311–1316.

482 Linnemann, C.C., Jr, Shea, L., Partin, J.C. *et al.* (1975) Reye's syndrome: epidemiologic and viral studies, 1963–1974. *American Journal of Epidemiology*, **101**, 517–526.

483 Smith, T.C. (1996) Reye's syndrome and the use of aspirin. *Scottish Medical Journal*, **41**, 4–9.

484 Starko, K.M., Ray, C.G. and Dominguez, L.B. (1980) Reye's syndrome and salicylate use. *Pediatrics*, **66**, 854–864.

485 Halpin, T.J., Holtzhauer, F.J., Campbell, R.J. *et al.* (1982) Reye's syndrome and medication use. *The Journal of the American Medical Association*, **248**, 687–691.

486 Waldman, R.J., Hall, W.N. and van Amburg, G. (1982) Aspirin as a risk factor in Reye's syndrome. *The Journal of the American Medical Association*, **247**, 3089–3094.

487 Hurwitz, E.S., Barrett, M.J., Bregman, D. *et al.* (1987) Public Health Service study of Reye's syndrome and medications: report of the main study. *The Journal of the American Medical Association*, **257**, 1905–1911.

488 Casteels-Van Daele, M. (1991) Reye syndrome or side-effects of anti-emetics? *European Journal of Pediatrics*, **150**, 456–459.

489 Brown, A.K., Fikrig, S. and Findberg, L. (1983) Aspirin and Reye's syndrome. *The Journal of Pediatrics*, **102**, 157–178.

490 Hall, S.M. (1986) Reye's syndrome and aspirin: a review. *Journal of the Royal Society of Medicine*, **79**, 596–598.

491 Heubi, J.E., Daugherty, C.C., Partin, J.S. *et al.* (1984) Grade 1 Reye's syndrome – outcome and predictors of progression to deeper coma grades. *The New England Journal of Medicine*, **311**, 1539–1542.

492 Consensus Conference (1981) Diagnosis and treatment of Reye's syndrome. *The Journal of the American Medical Association* **246** 2441–2444.

493 Hurwitz, E.S., Barrett, M.J., Bregman, D. *et al.* (1985) Public Health Service study on Reye's syndrome and medications. Report of the pilot phase. *The New England Journal of Medicine*, **313**, 849–857.

494 Kang, A.S., Crocker, J.F.S. and Johnson, G.M. (1986) Reye's syndrome and salicylates. *The New England Journal of Medicine*, **314**, 920–921.

495 Forsyth, B.W., Horwitz, R.I., Acampora, D. *et al.* (1989) New epidemiologic evidence confirming that bias does not explain the aspirin/Reye's syndrome association. *The Journal of the American Medical Association*, **261**, 2517–2524.

496 Hall, S.M., Plaster, P.A., Glasgow, J.F.T. *et al.* (1988) Preadmission antipyretics in Reye's syndrome. *Archives of Disease in Childhood*, **63**, 857–866.

497 Green, A. and Hall, S. (1992) Investigation of metabolic disorders resembling's syndrome. *Archives of Disease in Childhood*, **67**, 1313–1317.

498 Forsyth, B.W., Shapiro, E.D., Horwitz, R.I. *et al.* (1991) Misdiagnosis of Reye's-like illness. *American Journal of Diseases of Children*, **145**, 964–966.

499 Gauthier, M., Guay, J., Lacroix, J. *et al.* (1989) Reye's syndrome: a reappraisal of diagnosis in 49 presumptive cases. *American Journal of Diseases of Children*, **143**, 1181–1185.

500 McGovern, M.C., Glasgow, J.F.T. and Stewart, M.C. (2001) Reye's syndrome and aspirin: least we forget. *British Medical Journal*, **322**, 1591–1592.

501 Casteels-Van Daele, M., Wouters, C. and van Geet, C. (2002) Reye's syndrome revisited. Commentary to the paper of McGovern *et al.*, Br Med J 2001;322:1591–1592]. *British Medical Journal*, **324**, 546.

502 Gladtke, E. and Schauseil-Zipf, U. (1987) Reye's syndrome. *Monatsschrift Kinderheilkunde*, **135**, 699–704.

503 Ziebold, C., von Kries, R., Lang, R. *et al.* (2001) Severe complications of varicella in previously healthy children in Germany: a 1-year survey. *Pediatrics*, **108**, E79.

504 Daugbjerg, P. and Ranek, L. (1986) Reye's syndrome in Denmark. *Acta Paediatrica Scandinavia*, **75**, 313–315.

505 Autret-Lecat, E., Jonville-Bera, A.P., Llau, M.E. *et al.* (2001) Incidence of Reye's syndrome in France: a hospital-based survey. *Journal of Clinical Epidemiology*, **54**, 857–862.

506 Sengupta, C., Steffen, R. and Schär, M. (1987) Das Reye-Syndrom in der Schweiz. *Schweizerische Rundschau fur Medizin*, **76**, 1114–1116.

507 Palomeque, A., Domenech, P., Martinez-Gutierrez, A. *et al.* (1986) Sindrome de Reye in España, 1980–1984 (Estudio cooperative Seccion de CIP de la AEP). *Anales Espanoles de Pediatria*, **24**, 285–289.

508 Mowat, A.P. (1988) Commentary [to the paper of Hall *et al.*, same issue]. *Archives of Disease in Childhood*, **63**, 857.

509 Hofman, K.J. and Rosen, E.U. (1982) Reye's syndrome in Johannesburg: epidemiology and clinical presentation. *South African Medical Journal*, **61**, 281–282.

510 Yamashita, F., Eiichiro, O., Kimura, A. *et al.* (1985) *Reye's Syndrome*, 4th edn (ed. J.D. Pollack), National Reye's Syndrome Foundation, Bryon, OH, pp. 47–60.

511 Christo, G.G. and Venkatesh, A. (1987) Reye syndrome: the Indian experience. *Indian Journal of Pediatrics*, **54**, 903–908.

512 Yu, E.C.L., Tang, P.S., Chow, C.B. *et al.* (1988) Reye's syndrome in Hong Kong. *Journal of the Medical Association (Hong Kong)*, **40**, 115–118.

513 Orlowski, J.P., Campbell, P. and Goldstein, S. (1990) Reye's syndrome: a case control study of medication use and associated viruses in Australia. *Cleveand Clinic Journal of Medicine*, **57**, 323–329.

514 Orlowski, J.P., Campbell, P. and Goldstein, S. (1990) Reye's syndrome: a case control study of medication use and associated viruses in Australia. *Cleveand Clinic Journal of Medicine*, **57**, 323–329.

515 Glasgow, J.F.T. (1984) Clinical features and prognosis of Reye's syndrome. *Archives of Disease in Childhood*, **59**, 230–235.

516 Hurwitz, E.S. (1989) Reye's syndrome. *Epidemiologic Reviews*, **11**, 249–253.

517 Arrowsmith, J.B., Kennedy, D.L., Kuritsky, J.N. *et al.* (1987) National patterns of aspirin use and Reye syndrome reporting, United States, 1980 to 1985. *Pediatrics*, **79**, 858–863.

518 Langford, N.J. (2002) Aspirin and Reye's syndrome: is the response appropriate? *Journal of Clinical Pharmacy and Therapeutics*, **27**, 157–160.

519 Larson, A.M., Polson, J., Fontana, R.J. *et al.* (2005) Acetaminophen-induced acute liver failure: results of a United States multicenter, prospective study. *Hepatology*, **42**, 1364–1372.

520 Rivera-Penera, T., Gugig, R., Davis, J. *et al.* (1997) Outcome of acetaminophen overdose in pediatric patients and factors contributing to hepatotoxicity. *The Journal of Pediatrics*, **130**, 300–304.

521 Heubi, J.E. and Bien, J.P. (1997) Acetaminophen use in children: more is not better [review]. *The Journal of Pediatrics*, **130**, 175–177.

522 Maison, P., Guillemot, D., Vauzelle-Kervroedan, F. *et al.* (1998) Trends in aspirin, paracetamol and non-steroidal anti-inflammatory drug use in children between 1981 and 1992 in France. *European Journal of Clinical Pharmacology*, **54**, 659–664.

523 Lindsley, C.B. (1993) Uses of nonsteroidal anti-inflammatory drugs in pediatrics. *American Journal of Diseases of Children*, **147**, 229–236.

524 Broken, P.A., Bose, A., Burgner, D. *et al.* (2002) Kawasaki disease: an evidence-based approach to diagnosis, treatment and proposals for future research. *Archives of Disease in Childhood*, **86**, 286–290.

525 van Bever, H.P., Quek, S.C. and Lim, T. (2004) Aspirin, Reye syndrome, Kawasaki disease, and allergies: a reconsideration of the links. *Archives of Disease in Childhood*, **89**, 1178.

526 Varner, A.E., Busse, W.W. and Lemasnke, R.F. Jr (1998) Hypothesis: decreased use of pediatric aspirin has contributed to the increasing prevalence of childhood asthma. *Annals of Allergy, Asthma & Immunology*, **81**, 347–351.

527 Shaheen, S.O., Sterne, A.C., Songhurst, C.E. *et al.* (2000) Frequent paracetamol use and asthma in adults. *Thorax*, **55**, 266–270.

528 Robertson, C.F., Heycock, E., Bishop, J. *et al.* (1991) Prevalence of asthma in Melbourne schoolchildren: changes over 26 years. *British Medical Journal*, **302**, 1116–1118.

529 Ninan, T.K. and Russell, G. (1992) Respiratory symptoms and atopy in Aberdeen schoolchildren: evidence from two surveys 25 years apart. *British Medical Journal*, **304**, 873–875.

530 ISAAC (1998) Worldwide variation in prevalence of symptoms of asthma, allergic rhino conjunctivitis and atopic eczema: ISAAC. The International Study of Asthma and Allergy in Childhood (ISAAC) Steering Committee. *The Lancet*, **351**, 1225–1232.

4 Clinical Applications of Aspirin

4.1 Thromboembolic Diseases
4.1.1 Coronary Vascular Disease
4.1.2 Cerebrovascular Diseases
4.1.3 Peripheral Arterial Disease
4.1.4 Venous Thrombosis
4.1.5 Preeclampsia
4.1.6 Aspirin "Resistance"
4.2 Pain, Fever, and Inflammatory Diseases
4.2.1 Aspirin as an Antipyretic Analgesic
4.2.2 Arthritis and Rheumatism
4.2.3 Kawasaki Disease
4.3 Further Clinical Indications
4.3.1 Colorectal Cancer
4.3.2 Alzheimer's Disease

Acetylsalicylic Acid. Karsten Schrör

ISBN: 978-3-527-32109-4

4
Clinical Applications of Aspirin

The potential clinical applications of aspirin are as broad as is its spectrum of biological activities. The reasons are the multiple pharmacological activities of the compound, including both the acetylation reactions and the interactions of the uncleaved compound and salicylate moiety with cellular signaling and metabolism.

Pharmacological Properties of Aspirin Versus Clinical Effects The inhibition of prostaglandin biosynthesis appears to be involved in all clinical indications and is the best understood and most thoroughly studied mode of action of aspirin. However, pleiotropic actions of salicylates may play a role in analgesic, anti-inflammatory, and antitumorigenic actions. This is reflected by the different doses that are effective in these indications: $\leq$100 mg for antiplatelet actions, 500–1000 mg for analgesic actions, and doses above 2 g for anti-inflammatory actions. The most effective doses for antitumour actions of salicylates are not yet known, but may be in the range of analgesic doses.

Not all of the multiple pharmacological actions of aspirin are currently used clinically. The reason is the availability of more effective and better tolerable therapeutic alternatives. This includes the hypoglycemic, uricosuric, and tocolytic actions of aspirin. One general pharmacological property that might be involved in most of them is the metabolic action of salicylates, that is, the reversible uncoupling of oxidative phosphorylation and the possibly associated kinase inhibition (Section 2.2.2). These actions might also cause a number of other clinical effects, for example, sweating, that is, removal of excess heat, in relation to the antipyretic action (Table 4.1). All of these actions with the exception of antiplatelet activities are largely or entirely mediated by the salicylate metabolite – including the symptoms of aspirin poisoning (Section 3.1.1).

Clinical Applications of Aspirin The current interest in aspirin as a therapeutic agent is focused on three areas: Prevention of thromboxane-dependent platelet activation, secretion, and aggregation and its consequences for thrombin generation, stimulation of leukocyte function, and arterial thrombogenesis. This is the main mechanism for antithrombotic actions (Section 4.1), that is, the prevention of myocardial infarction (MI), ischemic stroke, and peripheral arterial vessel occlusions. Historically, the first and still the most valuable clinical application is the analgesic/antipyretic activity (Section 4.2), practically used for the treatment of pain and febrile disorders. In this indication, aspirin is still among the most popular nonprescriptional drugs. The anti-inflammatory effect of the compound is also valuable. However, its use is restricted to particular inflammatory diseases, such as Kawasaki disease (Section 4.2.3). In other inflammatory conditions, therapeutic alternatives are available with an improved risk/benefit profile.

In addition to these "established" uses, basic research on aspirin is still ongoing to find new indications for the drug by improved understand-

Acetylsalicylic Acid. Karsten Schrör

ISBN: 978-3-527-32109-4

Table 4.1 Pharmacological actions of aspirin and their clinical application.

Action	Effective dose (g)	In clinical use
Antiplatelet	0.1[a]–0.3	Yes
Analgetic	0.5–2.0	Yes
Antipyretic	0.5–2.0	Yes
Anti-inflammatory[b]	2.0–4.0	(Yes)
Metabolic	>1	No
Antitumorigenic	?	No
Hypoglycemic	4.0–6.0	No
Tocolytic	1.0–2.0	No
Uricosuric	?	No

[a]Recommended maintenance dose for long-term use.
[b]30–60 mg/kg in children (Kawasaki disease) ~ 2–4 g/70 kg adults.

ing of its molecular mode(s) of action. This is facilitated by increased knowledge about molecular mechanisms of diseases, which will also help to understand the clinical actions of aspirin in more detail. Recent discoveries in this area involve actions of aspirin on the transcriptional regulation of genes, such as COX-2 and inducible NO synthase and also other "early response" genes that are involved in body defense mechanisms. Therapeutically interesting are also posttranscriptional actions, for example, tissue protection from oxidative stress and increased endothelial NO production as well as other nonprostaglandin-mediated activities. Some of these findings have already been translated into clinical research, such as the prevention of colorectal carcinoma or Alzheimer's disease (Section 4.3). New clinical indications will also have an impact on the pharmacological design and development of "second-generation" aspirin-like drugs, which, in contrast to the "first-generation" traditional NSAIDs, also might interact with the regulation of disease-modifying genes by modifying selected transcription factors and not solely with the enzymatic activity of a gene product, that is, cyclooxygenases. Clearly, targeted modification of gene regulation by modification of selected transcription factors comes much closer to the "intrinsic" activity of salicylates in plants, their natural origin, where they act as a transcriptionally regulated defense system in plant resistance.

Another disease-related factor that determines the use of aspirin in clinical settings is the goal and the duration of use. The appliance for relief in pain and fever is short-term, in many cases even single application. Treatment is done with the intention to cure either the disease or its symptoms. The appliance is independent of age and in most cases without paying too much attention to comorbidities and side effects, except (preexisting) subjective gastric intolerance. The patient groups in most cases are healthy (young) adults and treatment is usually performed by self-medication without prescription. Similarly, the use of aspirin for treatment of chronic pain, such as tension-related headache or migraine, is also done with the intention to cure the unpleasant symptoms of the disease, though higher doses and repeated administration may be necessary.

In contrast to analgesic use, the use of aspirin as an antiplatelet drug is entirely preventive. Treatment is done with the intention to avoid any "explosive" thromboxane formation by platelets as a catalyst for platelet recruitment, thrombus formation, and subsequent thrombotic artery occlusion. Thus, the goal is not to cure but to prevent diseases, and the administration is not a single but a long-term use, mostly in medium to older age patients who might have chronic diseases such as atherosclerosis or other comorbidities. Thus, side effects also become a very important issue for patients' compliance.

Types of Clinical Trials The clinical efficacy of aspirin – as that of all other drugs – is established by clinical trials and determined by the benefit/risk ratio. This also includes the consideration of available competitor drugs. The assessment of clinical efficacy is a dynamic self-correcting process in contrast to the pharmacological properties of the compound, which remain the same until new properties are detected. The clinical trials start with the estimation of safety (phase I) and feasibility (phase II), both having mechanism-based end

Table 4.2 Types and properties of clinical trials with clinical end points.

Type of trial	Important advantages	Important disadvantages	Examples
Randomized controlled (RCT)	Only study type with predictive information	Possible underestimation of risk because of patient selection	USPHS, WHS ISIS-2, CLASP
Epidemiological observational case–control	Real-life conditions	No randomly selected comparison group limited information about doses, duration of treatment, and comedications	NHS, GRACE, ARIC
Meta-analysis	Helps to address clinical questions in the ansence of data from large randomized trials	Pooled estimate of data, no inclusion of "missing studies", study bias passed on to the meta-analysis and might affect conclusions	Antiplatelet Trialists' CPS-II, PARIS

points, while phase III clinical trials determine the efficacy in a predefined, randomized, controlled patient population in terms of clinical end points. Clearly, after introduction of a drug to the market, clinical end points and the benefit/risk ratio are the important parameters to look at. For a drug-like aspirin with a long clinical experience, these clinical end points and the benefit/risk ratio in comparison with other drugs define its clinical use. In this respect, three types of clinical trials are available, all having their advantages and disadvantages, as shown in Table 4.2. It should be noted that it is the randomized controlled trial (RCT) that is the only one to provide defined information, based upon the (predetermined) differences in clinical end points between the treated and the control group. Non-randomized observational studies are also valuable, specifically because they are more close to real life than trials with predefined patient populations and many exclusion criteria. They are also more easy (and cheaper) to perform. In addition, they can be done in both prospective and retrospective manner and are well suited for the future design of a prospective randomized trial, for example, in the estimation of patient numbers. Finally, meta-analyses are available and helpful specifically for the detection of risk profiles and random events, as well as for generating hypotheses. However, they do not define any causality and their messages need to be confirmed in prospective randomized trials [1].

4.1 Thromboembolic Diseases

Arterial thromboembolism is usually due to the initial formation of platelet aggregates, for example, after tissue factor release from a ruptured atheromatous plaque with subsequent activation of the clotting system. There is also a circulation-dependent generation and release of inflammatory mediators. Thus, aspirin will modify the early platelet-dependent events in thrombus formation, though with different efficacy in different circulations.

While the overall efficacy of aspirin to prevent myocardial infarction (Section 4.1.1) and, to a lesser extent, ischemic stroke (Section 4.1.2) is now generally appreciated, it is also clear that its clinical use in these indications depend on the (individual) risk, that is, the occurrence of side effects. The most important side effect is bleeding, specifically in the cerebral and gastrointestinal circulations (Section 3.2.1). There is no convincing evidence that aspirin will prevent or improve peripheral arterial occlusive

Table 4.3 Thrombotic complications and recommended daily doses of (plain) aspirin to prevent thrombotic vessel occlusions in particular circulations.

Local circulation	Major thromboembolic complication	Recommended aspirin dosage (mg/day)
Coronary	Myocardial infarction	75–300
Cerebral	(Ischemic) stroke	100–300
Peripheral arterial	Critical leg ischemia	100–300[a]
Uteroplacental	Preeclampsia	100
Venous	Pulmonary embolism	Not recommended

[a]Because of increased cardiovascular risk.

diseases (Section 4.1.3). However, as these patients have a considerably elevated risk for myocardial infarctions because of the aggressiveness of their atherosclerosis, antiplatelet agents, such as aspirin, are strongly recommended. Aspirin is also effective in reducing the thrombotic risk in preeclampsia (Section 4.1.5), the pregnancy-induced hypertension. No significant benefit has been obtained with aspirin in venous thrombosis (Section 4.1.4). This is probably due to the small contribution of platelets, if any, to thrombus formation under low-pressure conditions in veins. An overview of these clinical indications of aspirin use for the prevention of thrombus formation is provided in Table 4.3.

Another clinically most relevant issue of aspirin use as an antiplatelet drug is so-called aspirin "resistance," more precisely a reduced or absent responsiveness of platelets to aspirin because of different reasons (Section 4.1.6). Unfortunately, the failure of clinical treatment and a pharmacological inability of the drug to act are often mixed in an inappropriate manner, which has caused much confusion. A disease-related increase in platelet reactivity and compliance problems, both primarily independent of aspirin treatment, are the most likely explanation for this phenomenon in many cases.

4.1.1 Coronary Vascular Disease

Etiology Coronary vascular disease is usually a consequence of generalized atherosclerosis. The major complication is myocardial infarction due to the thromboembolic occlusion of a large coronary artery.

The clinical use of aspirin for the prevention of thrombotic arterial vessel occlusions, such as myocardial infarction and stroke, started with the studies by a general practitioner, *Lawrence L. Craven*, in the early 1950s. He found that regular aspirin intake in middle-aged men prevented myocardial infarctions and strongly recommended its broader use in this indication (Section 1.1.6). The first randomized prospective trial on daily aspirin (300 mg) in male survivors of myocardial infarction reported a reduction in total mortality by 25% after 1 year, although this was not significant [2]. The Boston Collaborative Drug Surveillance Group [3] did a large though retrospective observational study in survivors of acute myocardial infarction (AMI). The results also suggested that regular aspirin intake may offer protection from another cardiovascular event. These data were the first to suggest a possible benefit of regular aspirin intake in the secondary prevention of myocardial infarction.

Epidemiology Complications of cardiovascular diseases, predominantly myocardial infarction, are the leading causes of death and disability in Western societies and contribute in large part to the escalating costs of health care in Europe and the United States [4, 5]. About 30–40% of deaths are caused by coronary artery disease, 30–40% of them are sudden and about half of them occur as a first manifestation of coronary heart disease. The estimated annual incidence of sudden cardiac deaths in persons aged between 20 and 75 years is 1 per 1000 inhabitants and the survival rate only 8% if the event occurs outside a hospital (80% of cases) [6]. In comparison, the event rate of sudden cardiac death in patients with known cardiac diseases is about 10-fold higher [4, 6].

There is a remarkable trend toward reduced infarction rates during the past two decades in Western societies, by 30–40%, in Western Europe [5] and the United States [7], whereas an opposite trend is now seen in the former communist countries, possibly caused by changes in socioeconomic factors [8]. Overall, this reflects the importance of myocardial ischemia as a triggering event as well as the necessity of appropriate prophylaxis, given the increasing age and morbidity of the population.

It is beyond the scope of this book to present and discuss all studies on aspirin and coronary prevention. This chapter is focused on a selection of historically important and clinically relevant investigations. Excellent reviews of the current status are available; for example, the large meta-analyses of the Antiplatelet/Antithrombotic Trialists' Collaboration [9, 10]. The latest available edition (2002) covers the effects of antiplatelet therapy in randomized trials for the prevention of death, myocardial infarction, and stroke in high-risk patients until 1997. Thus, with the exception of the CAPRIE trial on clopidogrel [11], these meta-analyses almost exclusively cover studies with aspirin. A critical review of the older studies, including benefit/risk evaluation, has been provided by Reilly and FitzGerald [12].

4.1.1.1 Thrombotic Risk and Mode of Aspirin Action

Platelets and Thromboxane A_2 The risk of acute thrombotic arterial vessel occlusion is mainly determined by the severity of atherosclerotic alterations of the vessel wall, preexisting risk factors, and comorbidities. All of them may cause platelet hyperreactivity [13] and can initiate an "explosion" of platelet-dependent thromboxane formation as an initial trigger of platelet activation and recruitment. Aspirin at antiplatelet doses inhibits the COX-1-dependent thromboxane formation. In this context, it is interesting to note that circulating plasma thromboxane levels are not increased in patients with stable angina compared to healthy controls but only in unstable angina and myocardial infarctions [14], that is, clinical situations with acute thromboembolic events. Similar changes are seen with PGI_2 (Figure 4.1).

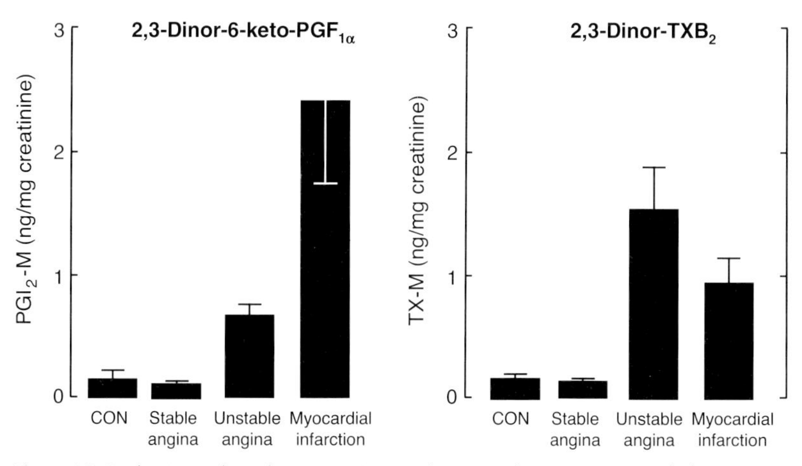

Figure 4.1 Peak urinary thromboxane (TX-M) and prostacyclin (PGI_2-M) metabolite excretion in controls as compared to patients after acute myocardial infarction and patients with unstable or stable angina. There is a marked increase of metabolite excretion in patients with unstable angina and myocardial infarction but no change in patients with stable angina and patients with noncardiac chest pain (not shown). CON: healthy controls, $n = 4$–16 (modified after [14]).

With exception of the Caerphilly study [15], in most if not all of the large clinical trials on primary and secondary prevention of cardiovascular events by aspirin, platelet function or thromboxane formation were either not measured or only determined in small subgroups of participants. It is not likely that platelets from apparently healthy males (USPHS) or females (WHS) show any increased thromboxane formation or platelet hyperreactivity *in vivo*. This raises the question how the beneficial effects of aspirin on prevention of myocardial infarction in males (USPHS) or ischemic stroke in females (WHS) can be explained.

Mechanistically, it is possible that continuous blockade of thromboxane formation by aspirin protects from sudden platelet activation and "explosion" of thromboxane biosynthesis. This may occur subsequent to plaque erosion or rupture. The (relatively) higher efficacy of aspirin with increasing atherothrombotic risk could then be explained by a higher probability of these "explosive" events. The analogy to β-receptor antagonists and their use for prevention of acute myocardial ischemias is obvious.

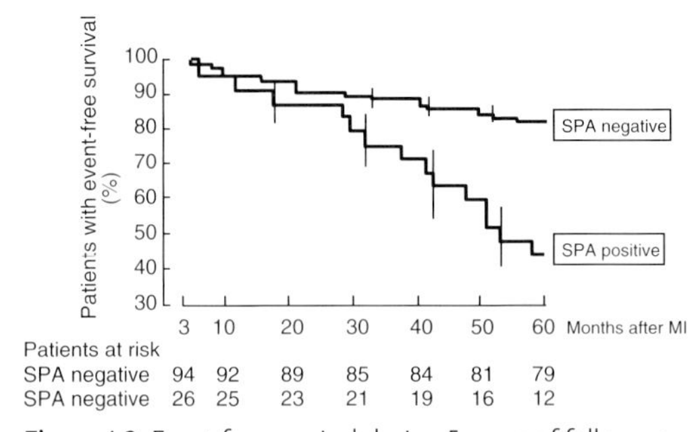

Figure 4.2 Event-free survival during 5 years of follow-up after acute MI, according to the platelet reactivity status at 3 months after the acute event. SPA: spontaneous platelet aggregation [18].

The Reactivity State of Platelets The "resting" state of circulating platelets is not a constant but might vary in relation to endothelial dysfunction in atherosclerotic patients. In general, platelets become more reactive with the progression of the disease, that is, they need lower concentrations of platelet agonists to become activated and, eventually, react with "exploding" thromboxane formation in acute coronary syndromes (ACSs) (Figure 4.1). The spontaneous platelet reactivity, as seen, for example, in patients who had already suffered a myocardial infarction, was suggested an independent predictor of long-term survival (Figure 4.2). Whether an enhanced agonist-induced, that is, ADP-induced, aggregation is able to predict the incidence of myocardial infarction in primary [16] or secondary prevention [17] is controversial. Nevertheless, there is no doubt that pharmacological prevention of platelet-dependent thromboxane formation protects from cardiovascular events. However, the efficacy of this measure depends on the role of thromboxane for platelet aggregate formation and this might differ in different clinical situations, that is, the contribution of thromboxane-independent factors and the initial state of platelet reactivity. This inherent variability in platelet function might also contribute to aspirin "resistance" or treatment failure, respectively, in clinical settings (Section 4.1.6).

Aspirin Dosing Although the general recommendation of aspirin dosing is 75–325 mg/day for cardiocoronary prevention, most guidelines recommend 100 mg/day plain or EC aspirin for long-term prevention. In case of acute events in aspirin-naive patients, a loading dose of 250–500 mg i.v. should be given.

Despite these generally accepted recommendations, several prospective, well-controlled though small randomized studies suggested that higher doses of aspirin may be more effective than lower ones in patients with vascular diseases [19]. However, in the acute situation (with upregulated COX-2), high-dose (500 mg) aspirin to aspirin-pretreated individuals may increase the size of periprocedural myonecrosis after percutaneous coronary interventions (PCI) [20], suggesting a deleterious effect. A recent large, though nonrandomized, post hoc analysis of more than 20 000 patients with acute coronary syndromes (GUSTO IIb and PURSUIT) has shown less frequent infarctions and more frequent strokes within 6 months among patients discharged while receiving higher (≥150 mg/day) aspirin doses. There was no change in mortality [21]. Kong *et al.* [22], analyzing all published randomized trials on aspirin versus placebo in acute myocardial infarction and unstable angina came to the conclusion that the most effective (individual) dose for secondary prevention has not yet been established. They also criticized that the Antiplatelet Trialists might not have sufficiently considered the heterogeneity between trials, for example, the greater efficacy of aspirin in unstable angina as opposed to postinfarction patients. Clearly, because of ethical reasons, no placebo-controlled studies with aspirin in secondary prevention will be done anymore; however, dose-range studies appear to be useful [22]. At this point, it should be noted that meta-analyses are hypothesis-generating statistical approaches but can never replace a prospective randomized trial with well-defined study protocols and patient selection. Those trials, comparing low-dose (i.e., 100 mg/day) aspirin with a higher dose (i. e., 325 mg/day) for cardiocoronary prevention in high-risk patients have not been done. It is also quite possible that the different disease-related (diabetes) or individual (comorbidities) risk is a relevant determinant also for dose selection. Numerically, the disease-related efficacy of aspirin in secondary prevention, varying between about 0 (aortocoronary bypass surgery, artificial heart valves) and 50% (unstable angina) [10], will give an average value of 25% protection – the frequently cited message from the Antithrombotic Trialists' Collaboration [10]. This is mathematically correct, but not very helpful for the individual case (and disease), and the lowest effective dose is not necessarily identical with the most effective dose in these particular patients.

Aspirin and Inflammation Platelets are also related to inflammatory processes in the vessel wall [23]. In addition to promoting vessel occlusion by platelet "plug" formation, platelets may also "prime" other cells, such as monocytes/macrophages or endothelial cells, to express adhesion molecules and to participate in the inflammatory and matrix-modifying processes of atherosclerosis and vascular remodeling.

This important function of inflammatory mediators in the progression of atherosclerosis and plaque stability becomes increasingly apparent. Several aspirin-sensitive markers of inflammation, such as C-reactive protein, leukocyte number, or fibrinogen levels are elevated in patients with coronary heart disease and are independent predictive markers of acute cardiovascular events [24–26], including sudden cardiac death [27, 28]. Similar results were obtained with inflammatory cytokines, such as interleukin 6 (IL-6) and macrophage colony stimulating factor (MCSF). These cytokines are elevated in patients with chronic stable angina and are significantly reduced by aspirin at antiplatelet doses [26] (Figure 4.3). A pathophysiological role of T and B leukocytes and their mediators is also suggested in acute myocardial ischemia. It is possible that the reduced incidence of "silent" ischemias in these patients also involves inhibition of thromboxane-induced leukocyte activation [29].

4.1.1.2 Clinical Trials: Primary Prevention in Individuals Without Risk Factors

Apparently Healthy Individuals Several large randomized trials on the benefit/risk ratio of long-term aspirin prophylaxis in apparently healthy subjects without overt risk factors are available. The US Physicians' Health Study (USPHS) and the British Medical Doctors' Study (BMDS) were the first prospective trials conducted in men. In addition to the prospective but observational US Nurses Study, the recently published Women's Health Study (WHS) now also provides information on primary prevention in apparently healthy women.

In the US Physician's Health Study (USPHS), 22 071 male physicians, 40–84 years of age at entry, were selected from a total the 60 000 doctors willing to participate. Exclusion criteria included a preexisting coronary heart disease, gastric intolerance to aspirin or preexisting ulcers, as well as missing compliance during an 18-week run-in period. Eligible participants received 325 mg aspirin, 50 mg β-carotene or placebo every other day in a 2×2 factorial design. Primary end point was cardiovascular death. The total study period was scheduled to 8 years.

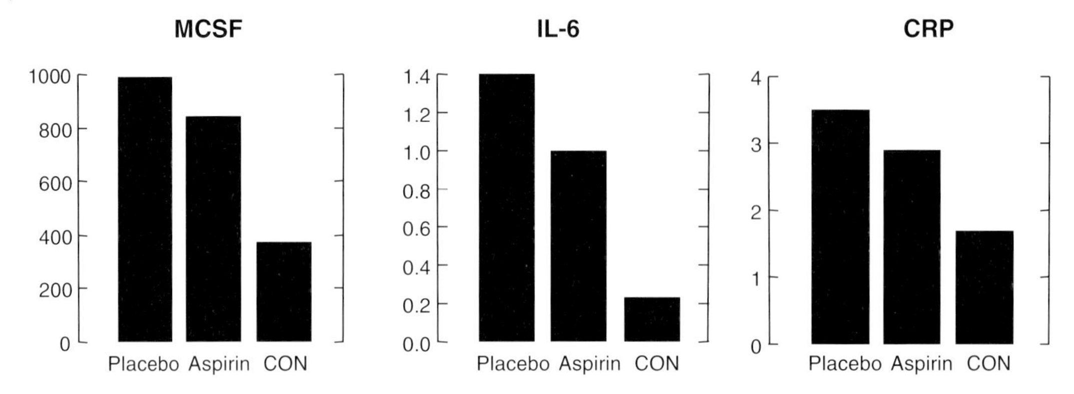

MCSF: Macrophage colony stimulating factor
CRP: C-reactive protein
IL-6: Interleukin 6

Figure 4.3 Inflammation markers in plasma of patients ($n = 40$) with ischemic heart disease treated with aspirin (300 mg/day) for 6 weeks or placebo in a randomized, double-blind crossover trial. CON: healthy untreated controls ($n = 24$) (modified after [26]).

The trial was stopped prematurely after an average treatment period of 5 years because of a significant reduction of the incidence of myocardial infarctions (which was not the original end point) in the aspirin group and the expectation that a sufficient number of cardiovascular deaths for statistical analysis would require to prolong the study beyond the year 2000. This was mainly due to the fact that the total number of cardiovascular events was much lower that originally anticipated.

The total incidence of a first myocardial infarction was reduced from 189 in the placebo group to 104 in the aspirin group ($p < 0.0001$). This corresponded to a relative risk reduction by 42%. In absolute numbers, this was equivalent to a reduction in the event rate from 0.4 to 0.2% per year. However, significant protection was only seen in men who were 50 years of age or older. The cardiovascular mortality results were inconclusive. Regarding side effects, there was an increase in hemorrhagic strokes and GI bleeding while the total stroke rate remained unaffected, due to a tendency in reduction of ischemic strokes.

The conclusion was that aspirin was an effective measure to prevent a first myocardial infarction in men and that the compound should be used for this purpose, if the benefit/risk ratio is appropriate [30–32].

The transfer of these study results to the general population was subjected to several critics and comments. Thus, the extremely healthy study population of participating American Physicians had an annual coronary event rate of only 0.4% as opposed to 2% in the general US population. This suggested an unrepresentative good health status, which was confirmed by the low overall mortality rate, amounting to only 15% of the general US American population. Explanations given by the investigators were as follows: high motivation of the participating doctors including personal lifestyle, educational status, and a higher individual acceptance rate of side effects. A significant influence of the personal motivation for clinical outcome was also suggested from other cardiocoronary prevention trials [33, 34].

Several post hoc analyses in subgroups of these participants provided additional information. For example, the beneficial effect of aspirin became apparent soon after initiation of treatment and did not change over time. This suggested the prevention of acute thrombotic events but not slowing the initiation and progression of atherosclerosis as an explanation for the positive results [35]. In agreement with this, the relative risk of angina pectoris (stable or unstable) in the 331 individuals of the original USPHS population who developed the disease was not changed by aspirin treatment [36].

Two days after the first report of the USPHS, the British Medical Doctors' trial (BMDS) was published, which, in contrast to the American counterpart, showed no beneficial effect of aspirin on the incidence of major cardiovascular events in primary prevention [37].

A total of 5139 apparently healthy male doctors were randomized to take 500 mg/day aspirin in different preparations (plain, soluble, effervescent) or no drug. The study design was open.

After a 6-year follow-up, there was no significant difference in the rates of fatal and nonfatal myocardial infarctions between the 3429 doctors who took aspirin and the corresponding controls that did not. The total mortality was 10% lower in the treatment group, possibly related to other diseases than atherothrombotic events. The reduction of a first myocardial infarction was 3%, that is, also nonsignificant ($p = 0.889$). Similar to the USPHS, the British study also showed an adverse effect on nonfatal stroke.

The conclusion was that the absolute benefits in primary prevention by aspirin, despite its established benefits in secondary prevention, remain uncertain [37].

This study also had a number of limitations. There was no placebo-treated control group and the definition of end points was less restrictive than in the American trial. A significant proportion of doctors – similar to the American trial – were judged ineligible for the study. Thus, the study population was also a selected sample [38]. The compliance to study treatment protocol was 70% in the aspirin group but 98% in the control group. In addition, the participants were allowed to change the medication if they wished to do so. At the end of the trial, 86% of physicians in the treatment group and 14% in the control group were taking aspirin or another antiplatelet medication.

The different results between the USPHS and the BMDS were extensively discussed, because, taken together, they did not answer the question whether regular aspirin intake will provide protection from cardiovascular events in a low-risk population. Both studies, however, showed an increased incidence of side effects, most notably hemorrhagic stroke [38, 39]. In addition, the studies also differed in other important parameters. There was apparently a different baseline health status: 2.1% of participants deceased in the USPHS trial period versus 8.8% in the BMDS; a large difference in the number of dropouts during the trial periods: about 5% in the USPHS versus 24% in the BMDS; a better compliance to drug therapy in the USPHS as opposed to multiple changes in the BMDS and marked differences in aspirin dosages, 325 mg each alternate day versus 500 mg daily [39].

Overall, these studies suggested that it appears to be unrealistic to conduct primary prevention trials in otherwise healthy individuals with cardiovascular mortality as a primary end point. Instead, occurrence of any major cardiovascular event (myocardial infarction, stroke, and sudden death) appears to be more appropriate. This policy was followed in subsequent studies.

Neither the USPHS nor BMDS did include women. However, coronary heart disease is not only the leading cause of death in men but also an equally important cause of death and disability in women above the age of 60 years [40]. The possible transfer of the data in men to women was studied for the first time in a large prospective though observational study in apparently healthy women, the American Health Professionals study [41].

The study cohort included 87678 American nurses (30–55 years of age at entry). The nurses were free from coronary heart disease, stroke, and cancer at the beginning of the study and were followed for 6 years. They were asked for use of aspirin and aspirin-containing medications and the frequency and reason for intake. The participants were also asked for personal lifestyle variables, including smoking and alcohol. Primary end points were fatal or nonfatal myocardial infarction and stroke.

After a 6-year follow-up, there was a highly significant, that is, 32% reduction of the risk of a first myocardial infarction in women taking 1–6 aspirin tablets per week (RR: 0.68; 95% CI: 0.52–0.89; $p = 0.005$). This was associated with an 11% reduction (RR: 0.89, $p = 0.56$; n.s.) in cardiovascular death. There was no change in the risk of stroke. The reduced risk of myocardial infarction was particularly evident in women above the age of 50 and in those with additional risk factors (smoking, history of hypertension, high serum cholesterol). The results were similar when 1–3 or 4–6 tablets were taken per week while the intake of 7 or more tablets per week did not cause risk reduction. The reasons for aspirin intake in the group with 1–6 tablets per week were headache (32%), arthritis or muscle pain (30%), or both (16%) but only in 9% of cases cardiocoronary prevention.

The conclusion was that the use of 1–6 aspirin tablets per week appears to be associated with a significantly reduced risk of a first myocardial infarction among women with little effect on the risk of ischemic stroke. However, any more decisive recommendations were not made and, according to the investigators, should only be made after an appropriately sized randomized, prospective trial was available [41].

According to the study design, there was no information about the exact aspirin dosage or the duration of use. This was investigated in the prospective randomized trial on primary prevention in women, the Women's Health Study [42].

A total of 39 876 initially healthy women (45 years of age or older) were randomly assigned to receive 100 mg aspirin each second day or placebo (vitamin E). In a foregoing pilot study in 22 (!) volunteers (males and females) for 2 weeks, it was found that this dosage and the alternate day treatment protocol will reduce the mean thromboxane and prostacyclin levels to 7.5 and 15% of baseline [43]. Primary end points were cardiovascular death, nonfatal myocardial infarction, and nonfatal stroke. The secondary end point was the individual risk in several subgroups. The total observation period was 10 years.

During the follow-up, 477 major cardiovascular events occurred in the aspirin group as opposed to 522 in the placebo group. This corresponded to a nonsignificant reduction of primary end point events by 9% (RR: 0.91; $p = 0.13$). Regarding the individual risk profiles, there was a significant reduction of ischemic stroke in the aspirin group by 17% (RR: 0.83, $p = 0.04$) but no change in the risk of fatal or nonfatal myocardial infarctions (RR: 1.02; $p = 0.83$) or cardiovascular mortality. However, subgroup analyses showed a 26% reduction of major cardiovascular events, including a 22% reduction in total stroke and a 30% reduction in ischemic strokes among women 65 years of age and older. There was a significantly increased risk of severe GI bleeding (RR: 1.40; $p = 0.02$).

The conclusion was that aspirin lowered the risk of stroke in women aged $\geq$45 years without affecting the risk of myocardial infarction or cardiovascular death [42].

These data were somehow contrary to the USPHS, although the study did also show a significant protection from an acute vascular ischemic event in women, here a 26% reduction in ischemic strokes, by regular long-term aspirin intake. In comparison to the USPHS, in which a 44% reduction of a first myocardial infarction was found, the rate of myocardial infarctions in the WHS was much smaller: only 97/100 000 patient years as opposed to 440/100 000 patient years at a comparable incidence of strokes in both groups [44]. The total rate of myocardial infarctions was extremely low, amounting only to one-third of the already low risk – 15% of that of the normal American population – in the USPHS (Figure 4.4). Thus, it will probably be difficult to further reduce the incidence of cardiovascular events by any preventive measure if the overall risk of cardiovascular disease is so low. If the vascular risk becomes increased with increasing age, beneficial effects become apparent. This was seen in the subgroup of elderly women ($\geq$65 years) in the WHS where both ischemic strokes and myocardial infarctions were significantly reduced.

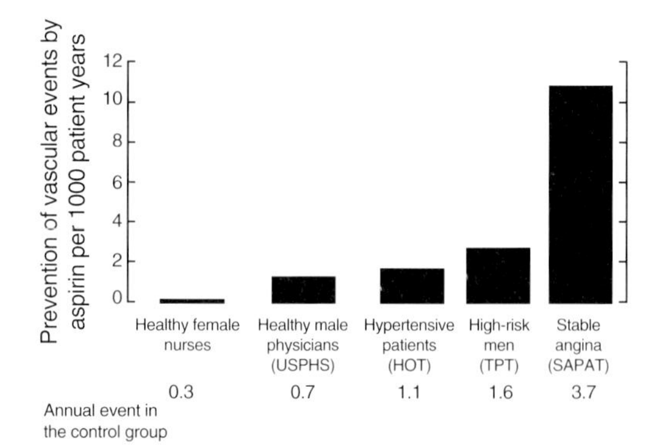

Figure 4.4 Absolute benefit of aspirin in primary prevention of cardiovascular events (nonfatal myocardial infarction, nonfatal stroke, and vascular death) in dependency on the statistical (individual) risk (modified after [45]).

The beneficial effects of aspirin prevention in women were clearly shown by the reduction of ischemic strokes. The relative proportion of strokes compared to myocardial infarctions in women was greater than that in men. Furthermore, women are more likely to die of stroke than men: 16% versus 8% [46]. The reasons for this are unknown but might be related to a different hormonal status [44] and/or a different function of vasoactive platelet-activating mediators such as serotonin.

Overall, available primary prevention studies with aspirin in healthy individuals show a statistically significant but – in absolute terms – low effect on the prevention of a first vascular event by aspirin: myocardial infarction in men and (ischemic) stroke in women [47]. There is no change in cardiovascular or overall mortality. One important variable especially in long-term use is the patient's adherence to the medication, that is, patient compliance. According to a subgroup analysis of the USPHS, the clinical efficacy of treatment largely disappears if the drug is taken less than on alternate days [48]. The possible beneficial effects of aspirin are limited by side effects, specifically, bleeding complications, including (hemorrhagic) strokes, which were similar in men and women. According to the benefit/risk ratio, aspirin prophylaxis is not recommended in healthy individuals without risk factors.

4.1.1.3 Clinical Trials: Primary Prevention in Individuals with Risk Factors

The use of antiplatelet agents, such as aspirin, for primary prevention is determined by the individual benefit/risk ratio. In addition to increasing age, this is determined by preexisting risk factors, such as hypercholesterolemia, hypertension, diabetes, cigarette smoking, and other environmental factors. The consequence is an increased cardiovascular mortality [4]. This suggests that prevention studies on subjects with preexisting risk factors might provide more convincing data than those in apparently healthy individuals, as also shown in Figure 4.4. Importantly, multiple risk factors potentiate the vascular risk rather than to act additive, making an appropriate prevention highly desirable.

Hypertension Hypertension is an independent risk factor for stroke (Section 4.1.2) and myocardial infarction. However, treating hypertensives with aspirin might increase the risk of cerebral hemorrhage, which can be reduced by adequate blood pressure control, for example by appropriate drug treatment with antihypertensives. This raises the question whether these individuals might benefit from the antiplatelet/antithrombotic actions of aspirin and whether the disease-related increased risk of (hemorrhagic) stroke further exists despite an adequate blood pressure control. This issue was addressed in the Hypertension Optimal Treatment (HOT) trial [49, 50].

This study included 18 790 patients with arterial hypertension, according to a diastolic blood pressure between 100 and 115 mmHg (mean: 105 mmHg). All patients were treated with antihypertensives. After adequate blood pressure control was obtained, they were randomly assigned to receive additionally either aspirin (75 mg/day) or placebo. The study was aimed to assess the optimum target diastolic blood pressure and the potential additional benefit of low-dose aspirin in adequately treated hypertensives. The average observation period was 3.8 years.

The decrease in diastolic blood pressure was comparable in all groups. In comparison to the antiplatelet-placebo group, there was a significant relative risk reduction of myocardial infarctions in the group subsequently treated with aspirin by 36% ($p = 0.002$) and of major cardiovascular events by 15% ($p = 0.03$) while the number of fatal and nonfatal strokes remained unchanged. Cardiovascular and total mortality rates were also unchanged. Similar effects were seen in diabetics and in elderly subjects. There was no increase in fatal bleeds (including cerebral hemorrhage) in the aspirin group but about twice as much nonfatal severe bleeds – 129 versus 70 – in the aspirin group, mainly from the GI tract. The estimated compliance rate for aspirin was 78%.

The conclusion was that intensive lowering of blood pressure in hypertensives is associated with a reduced rate of cardiovascular events. Aspirin treatment of these patients significantly reduces the incidence of myocardial infarctions but doubles nonfatal major bleeds. Aspirin does not change the incidence of strokes or fatal bleeds and, therefore, can be given daily to well-treated hypertensives, including elderly patients and diabetics [49, 50].

These findings agree with the Medical Research Council's Thrombosis Prevention Trial (TPT) [51] (see below) in which beneficial effects of aspirin were predominantly seen in individuals with the lowest blood pressure [52]. Interestingly, there was no significant reduction of myocardial infarctions in women ($p = 0.38$) [50]. Whether this can be explained by a different behavior of platelets in hypertensive women as opposed to men [53] remains to be determined. However, this finding is similar to that of the Womens' Health study, discussed above.

Similar results were obtained in the male hypertensives of the US Physicians' Health Study. No relative risk reduction was seen in about 2000 subjects with a diastolic blood pressure of more than 90 mmHg compared to the total study population [39]. However, the absolute risk reduction was twice as much, from 4.4 to 2.5% in hypertensives compared to 2.2% versus 1.3% risk reduction in the total study population [31].

Multiple Risk Factors There are two large prospective randomized trials on primary prevention with aspirin in individuals with multiple risk factors: the Thrombosis Prevention Trial [51] and the Primary Prevention Project (PPP). Another trial is the Clopidogrel for High Atherothrombotic Risk and Ischemic Stabilization Management and Avoidance (CHARISMA) study [54], using aspirin alone and in combination with clopidogrel. In addition to patients with preceding vascular events (secondary prevention), this trial also contained a group of patients with multiple cardiovascular risk factors without a preceding vascular event (see below).

The Thrombosis Prevention Trial [51] was a prospective comparison of aspirin and warfarin alone and in combination versus placebo in men at an enhanced cardiovascular risk because of preexisting risk factors.

A total of 5499 men aged between 45 and 69 years with elevated vascular risk because of preexisting risk factors (smoking, hypercholesterolemia, hypertension, a positive family history of ischemic coronary heart disease, elevated body mass index) were initially recruited. After a pilot study, the trial was expanded into a factorial comparison of low-intensity oral anticoagulation by warfarin (INR 1.5) and enteric-coated aspirin (75 mg/day). The four treatment groups were warfarin aspirin, warfarin + placebo, aspirin + placebo, and placebo + placebo. The median duration was 6.8 years. Primary end points were the total number of cardiovascular deaths and fatal and nonfatal myocardial infarctions.

The main effect of aspirin in comparison to placebo was a reduction of the primary end points by 20% ($p = 0.04$). This was almost entirely due to the 32% reduction in nonfatal cardiac events while the number of fatal events was modestly increased (n.s.). The main effect of warfarin was a reduction of primary end points by 21% ($p = 0.02$). This was mainly due to a 39% reduction in fatal events. Warfarin reduced the death rate from all causes by 17% ($p = 0.04$). There was a significant 20% increase of minor bleeds by aspirin and a tendency but no significant increase in intermediate or major bleeds. Combined treatment with warfarin and aspirin was more effective in reducing the risk of a cardiac event than either agent alone but also increased the risk of bleeding.

The conclusion was that aspirin reduces nonfatal coronary events, while warfarin reduces all cardiovascular events, chiefly because of an effect on fatal events. The combined treatment with both agents was more effective in the reduction of not only the cardiovascular events but also the bleeding risk than treatment with either agent alone [51].

The Primary Prevention Project [55] also studied the efficacy and safety of aspirin in patients with elevated cardiovascular risk. In contrast to the TPT, this study also contained women and the design was prospective but open.

In a controlled, randomized, but open trial, 4495 men and women (about half each, mean age 64 years) with multiple cardiovascular risk factors (older age, i.e., $\geq$65 years, hypertension, hypercholesterolemia, diabetes, obesity, family history of myocardial infarction) were randomly allocated to four treatment groups: enteric-coated aspirin (100 mg/day) or no aspirin and vitamin E (300 mg/day), or no vitamin E, according to a 2×2 factorial design. Efficacy end point was the cumulative rate of cardiovascular deaths, nonfatal myocardial infarctions, and nonfatal strokes.

After an interim analysis, the study was stopped prematurely at a mean follow-up of 3.6 years because of ethical grounds. At this time, there was a statistically significant benefit in the aspirin arm. In addition, two more primary prevention trials in similar risk groups (PPP [55], 1998; HOT, 1998; [50]) were published, demonstrating beneficial effects of aspirin. When the study was stopped, aspirin had significantly reduced the incidence of all end points. There was a reduction of cardiovascular deaths from 1.4 to 0.8% (RR: 0.56, 95% CI: 0.31–0.99) and of total cardiovascular events from 8.2 to 6.3% (RR: 0.77, 95% CI: 0.62–0.95). There were significantly more severe bleedings in the aspirin group. However, only one of the four deaths due to hemorrhage was in the aspirin group.

The conclusion was that the low-dose aspirin in men and women with at least one major risk factor, given in

1000 mg/day aspirin for the first 2 weeks. The 701 eligible patients were then randomized to treatment with either 30 (once or bid) or 1000 mg aspirin daily according to the location of the hospital (quasirandom design). The question was whether the efficacy of the 1000 and 30 (60) mg daily aspirin was comparable and whether there were less side effects in the low-dose group. There was no placebo control group.

After 2 years, there was a significant reduction of reinfarctions in the 30 mg group, by 58%, as compared to the group receiving 1000 mg aspirin. These beneficial effects of low-dose aspirin were still evident 4 and 6 years after the end of the controlled trial (the study was continued as an open trial). At this time, the incidence of nonlethal myocardial infarctions was reduced by 50% in patients previously treated with 30 mg aspirin as opposed to those who had received the 1000 mg dose.

The conclusion was that 30 mg aspirin per day is sufficient for the secondary prevention of myocardial infarction. Higher doses are not necessary and might even be dangerous because of the expected increased incidence of side effects [78].

This study was the first to suggest beneficial effects of lowest dose aspirin in secondary prevention of myocardial infarction. However, like many other innovative studies, the Cottbus trial has also caused a number of controversial discussions. These included the study design (initial treatment with high-dose aspirin), patient selection (no "random" randomization but hospital-specific distribution of patients to the treatment groups), heterogeneity in both the entrance criteria and concomitant treatments (linseed oil!), possible compliance problems, and the fact that no placebo group was included [79]. The absence of placebo at the time when the study was conducted was not a matter of ethical concerns since the efficacy of long-term aspirin treatment in secondary prevention was unknown. While the possibility exists that 30 mg aspirin is effective, as suggested by the authors, it can also not be excluded that the 1000 mg dose was less effective or not effective at all, as seen, for example, in a placebo-controlled trial in a similar group of postinfarction patients [80] (see below). It is also questioned whether 30 mg aspirin per day is sufficient to obtain an efficient inhibition of thromboxane formation [10]. Importantly, *all* patients, irrespective of their later randomization, had received 1000 mg aspirin daily during the initial postinfarction period (days 5–14 after the acute event). Thus, while the study had merits, it did not answer the question for an optimum low-dose aspirin regime for long-term reinfarction prophylaxis in postinfarction patients.

Two other trials have investigated the efficacy of low-dose (100 mg/day) aspirin versus placebo [81] or high-dose aspirin (1000 mg/day) [80] in postinfarction patients in a prospective, randomized, placebo-controlled manner. Treatment was started within 4–7 h after the acute event. In both studies, there was a significant reduction of the incidence of reinfarctions, by 44–55%, respectively, within 3 months with the 100 mg dosage but not with placebo or the 1000 mg dose. Infarct size at 3 days was unaffected as was the cardiovascular death rate at 3 months (Table 4.5) [81]. Husted *et al.* [80] addition-

Table 4.5 Secondary prevention in patients with acute myocardial infarction by low-dose aspirin (100 mg/day) as compared to placebo at 3 months.

Parameter	Aspirin ($n = 50$)	Placebo ($n = 50$)	Significance (aspirin versus placebo)
Mortality	10 (20%)	12 (24%)	n.s.
Re-infarction	2 (4%)	9 (18%)	<0.05
Unstable angina	14 (28%)	11 (22%)	n.s.
Infarct size (cumulative LDH release in 72 h: IU/I ± SD)	1431 ± 782	1492 ± 1082	n.s.

No change in infarct size at 3 days (LDH release) (modified after [81]). All patients were additionally treated with 5000 IU heparin until complete mobilization.

ally measured platelet aggregation and thromboxane formation in their patients and found a comparable inhibition of both parameters by the 100 and 1000 mg dose of aspirin. This suggested that the worse clinical outcome with 1000 mg aspirin cannot be simply explained by a different inhibition of platelet function. Other (negative) effects of aspirin may be involved that only become apparent at higher doses. One possible candidate is inhibition of prostacyclin formation (Section 2.3.1). In these infarcted patients, an advanced stage of atherosclerosis is likely. In this situation, a considerable proportion of prostacyclin is synthesized via COX-2 [82]. This enzyme is less sensitive to aspirin than COX-1. Clearly, anti-inflammatory actions of high-dose aspirin would rather suggest beneficial effects. Large randomized trials, comparing different doses of aspirin, are necessary to clarify this issue.

Aspirin and Fibrinolysis: The ISIS-2 Study Antiplatelet therapy of acute coronary syndromes prevents further thrombus growth and formation of new thrombi. Lysis of preexisting thrombi allows reopening of an occluded area and recovery of cardiac function if it is effective and occurs early enough. An unwanted side effect of thrombolysis is platelet activation. This is probably due to the formation and release of platelet activating compounds, specifically, thrombin, from the platelet–-fibrin clot. This is associated with a more than 20-fold increase in thromboxane formation, as seen from enhanced excretion rates of thromboxane metabolites [83]. Both lysis-induced hypercoagulability and platelet hyperreactivity can cause thrombotic reocclusion of the (reopened) coronary artery. The clinical significance of these events – inhibition of platelet function by aspirin in acute myocardial infarction with and without lysis – has been demonstrated for the first time in the ISIS-2 trial.

A total of 17 187 patients with clinical symptoms of acute myocardial infarction were randomized to receive enteric-coated aspirin (162 mg/day = half a 325 mg tablet), with the first dose crushed, sucked, or chewed after admission to the hospital for a rapid effect, i.v. infusion of streptokinase (SK) (1.5 million IU/1 h), both active treatments or neither treatment (placebo) within 5 h after the acute event. Fifty-six percent of patients had a transmural infarction (STEMI). The primary end point was cardiovascular mortality at 5 weeks.

At the end of this observation period, there were 804 cases of vascular deaths in the two aspirin (streptokinase) groups as opposed to 1016 fatality cases in the placebo group. This was equivalent to a reduction of vascular mortality from 13.2% in the placebo group to 8.0% in the group with combined treatment. Aspirin alone caused a decrease in mortality by 23%, while streptokinase caused a decrese by 25%. Both treatments were additive and reduced mortality more significantly by 38%. There was also a significant reduction of the number of reinfarctions and (ischemic) strokes by 50 and 40%, respectively. Aspirin alone reduced the number of reinfarctions from 284 in the placebo group to 156 in the treatment group, equivalent to an absolute reduction from 3.3 to 1.8%. There was a tendency for increased severe bleeds in both treatment groups, 0.4% and 0.5% compared to 0.2% in placebo, which, however, was only significant in the streptokinase group. Importantly, there was no additive effect on bleeding with the combined treatment.

The conclusion was that this study had shown for the first time the utility of the low-dose aspirin, given on patients' arrival at the hospital, to prevent reinfarction and improve survival after 5 weeks. The study also showed a synergistic effect of aspirin with streptokinase, that is, a doubling of efficacy, but no synergistic effects on bleeding [84] (Figure 4.7).

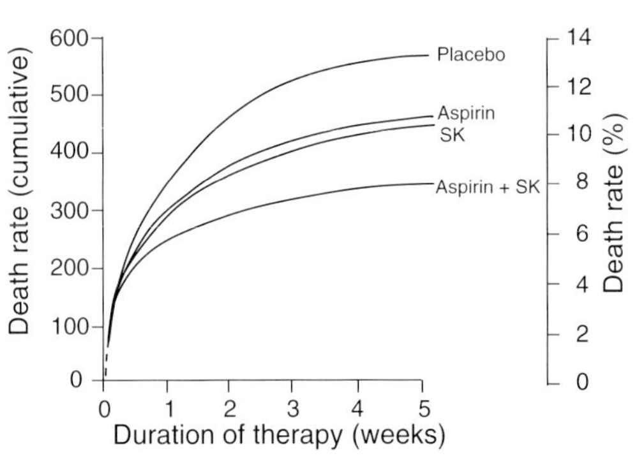

Figure 4.7 The ISIS-2 trial. Reduction of vascular death by aspirin (162 mg/day), streptokinase (SK) (1.5 million U/h), the combination of both or placebo in patients with acute myocardial infarction. Treatment was started within 5 h after the acute event [84].

These dramatic effects of aspirin were unexpected and, consequently, raised questions about the aspirin's mechanism of action, as the drug remains active for short time after the acute ischemic event but determines patients' survival after weeks. A post hoc analysis showed that the early survival advantage in the active treatment groups was maintained for at least 10 years [85]. Inhibition of enhanced thromboxane formation (Figure 4.1), which continues for hours within the already existing thrombus, and its potentiation by lysis is likely. However, other actions of aspirin may also have contributed, for example, acetylation of fibrinogen (Section 2.3.1), toward making polymerized fibrinogen more susceptible to fibrinolysis, as well as an inhibition of thrombin formation [86]. Whatever the final explanation might be, it was the ISIS-2 study that led to the introduction of aspirin as a first therapeutic measure in (suspected) acute myocardial infarction.

These positive data on aspirin in myocardial infarction alone and in combination with fibrinolytics were confirmed in a number of follow-up trials. According to a meta-analysis, the following results were obtained at 1 month after the acute event: fibrinolysis alone reduced the vascular mortality in comparison to placebo by 24% (placebo 3%), fibrinolysis plus aspirin by 40% (placebo 8%). This confirms the clinically highly relevant synergism between aspirin and fibrinolytics, originally found in ISIS-2. However, streptokinase in the meantime has been replaced by more selective, that is, thrombus-specific fibrinolytics, such as plasminogen activators.

4.1.1.6 Clinical Trials: Percutaneous Coronary Interventions

Treatment of ACS by percutaneous coronary interventions, that is, PTCA with or without subsequent implantation of an endoprosthesis (stent) is the therapeutic alternative to fibrinolysis to open an occluded coronary artery and currently the treatment of choice in appropriately equipped cardiocoronary units with skilled investigators. Neither PTCA nor stent implantation alters the natural history of coronary heart disease. However, these will improve symptoms and, eventually, extend the time interval until the next coronary intervention, but will not prolong survival.

Similar to fibrinolysis, PCI is also associated with platelet activation, here due to procedure-related loss of vascular endothelium and exposure of the thrombogenic subendothelium to the circulating blood. The clinical success of PCI is limited by two complications: in-stent thrombosis and restenosis of the reopened vessel by excessive neointima formation and matrix generation.

Thrombus Formation and In-Stent Thrombosis Thrombus removal and ballooning of the occluded area in ACS immediately restores blood flow and moves the dispersed thrombus particles downstream. Other particles, including calcified tissue, are translocated into the vessel wall, eventually causing a thrombotic/inflammatory reaction in the vessel wall with the activation of the platelet, activation of the clotting cascade, and formation of thrombus.

In-stent thrombosis is a random but acute life-threatening event. Recent evidence suggests that drug-eluting stents might have a higher potential for in-stent thrombosis than bare-metal stents. The combined use of different types of antiplatelet agents, that is, aspirin plus clopidogrel, will reduce this risk if treatment is performed for a sufficiently long time after stenting. According to randomized trials, the combination of antiplatelet compounds appears to be superior to the combination of aspirin with oral anticoagulants.

Restenosis Another aspect of PCI-induced injury of the vessel wall is migration and proliferation of vascular smooth muscle cells, natural processes during wound healing. However, in the absence of endothelial control, they may result in a critical lumen loss in the stent area, that is, restenosis. This lumen loss is due to the formation of a neointima from vascular smooth muscle cells, generating extracellular matrix that is finally covered by endothelium.

About 30–40% of patients subjected to PCI exhibit a clinically relevant late lumen loss due to neointima formation. This figure has continuously decreased with improved procedural technologies, including stent design and stent materials. The inhibition of cell proliferation by covering stents with growth inhibitory compounds, such as paclitaxel or sirolimus, will inhibit not only the proliferation of smooth muscle cells but also the endothelial coverage. This probably contributes to the increased thrombogenicity of currently used drug-eluting stents. The only available placebo-controlled PCI study with aspirin alone shows a significantly reduced restenosis rate: 30% versus 41% [87]. However, this effect is too small to become clinically relevant. According to current knowledge, aspirin and clopidogrel prevent thrombotic events associated with PCI but not restenosis. This suggests separate mechanisms for these events and agrees with early studies, which also found no correlation between inhibition of platelet function by aspirin and restenosis after PTCA within 3 months [88].

Both smooth muscle cell proliferation and migration are stimulated by the coagulation factors Xa and thrombin in a receptor-mediated manner [89, 90]. Both coagulation factors can also be generated by vascular smooth muscle cells from zymogens in circulating blood and might significantly contribute to intima proliferation and restenosis [90]. Therefore, inhibitors of thrombin and factor Xa formation, such as coumarins, may not only act as anticoagulants but may also reduce restenosis. Two recent studies have shown that coumarin pretreatment prior to PCI in addition to aspirin significantly improved the clinical outcome in PCI patients. There was a 67% risk reduction in late events and an angiographically significantly larger vessel lumen at 6 months after PCI in coumarin-treated patients [91, 92]. Thus, synergistic effects of inhibition of thrombin formation by direct anticoagulants and inhibitors of platelet activation, such as aspirin, might prevent not only thrombus formation but also restenosis. Specific, orally active low molecular weight thrombin inhibitors (dabigatran), inhibitors of factor Xa (rivaroxaban, apixaban) and thrombin receptor antagonists (SCH 530348, E5555) are in preclinical trials and might offer new options for combined anticoagulant/antiplatelet treatment in the prevention of PCI complications.

4.1.1.7 Clinical Trials: Coronary Artery Bypass Graft Surgery (CABG)

Surgical revascularization of coronary arteries by a bypass vessel is another option to restore sufficient blood supply to ischemic myocardium, but it is also associated with significant platelet activation. This is partially due to the surgical procedure itself, specifically the extracorporeal circulation. This will expose platelets to artificial surfaces, causing platelet adhesion, activation, secretion of storage products, and a significant drop of circulating platelet count. Thus, thrombotic reocclusion of successfully transplanted vessels is an inherent problem of bypass surgery and an appropriate antithrombotic treatment is strongly suggested.

Restitution of organ perfusion after opening of the bypass vessel is associated with an intense inflammatory response. This involves platelet activation and adherence to the injured vascular endothelium, intravascular thrombosis, and eventually organ ischemia and infarction [93]. At later times, migration and proliferation of vascular smooth muscle cells follows as atherosclerosis proceeds. Thus, similar to PCI, two thrombosis-related phenomena require conservative treatment: prevention of acute thrombotic vessel occlusion and prevention of late stenosis, probably initiated by the thrombotic/inflammatory process. However, antiplatelet treatment is associated with an increased risk of bleeding that may become a problem during surgical interventions.

Risk of Bleeding Versus Risk of Thrombosis Antiplatelet therapy enhances the risk of bleeding. This has caused concerns among surgeons, which may override the risk of thrombotic vessel occlusion. At least for aspirin, life-threatening bleedings are not to be expected in cardiovascular indications. A recent review on the effect of preoperative aspirin on bleeding risk in patients undergoing coronary artery bypass surgery came to the conclusion that an aspirin-related bleeding risk exists but may be avoided by the use of aspirin at low doses (<325 mg) [94] (Section 3.1.2). In contrast, there is definitely a risk for thrombotic graft occlusion, in particular, in cardiac surgery. In a series of 2606 consecutive patients who had to undergo a CABG procedure, 63% of them, that is, 1900 patients had taken aspirin within the past 12 h prior to surgery. Twenty-three percent of these patients required postoperative blood transfusion as opposed to 19% of patients without previous aspirin. This difference was only significant in a high-risk group [95]. In addition, aspirin withdrawal in these patients increased the risk for clinically relevant transplant occlusions, that is, myocardial infarctions [96].

Aspirin and Early Thrombotic Vessel Occlusions Treatment with antiplatelet drugs, such as aspirin or clopidogrel, is commonly employed in patients, subjected to coronary artery bypass surgery. First controlled trials showed protection of early thrombotic bypass occlusions of arterial bypasses by aspirin pretreatment while the results of venous graft were controversial [97–99]. In some studies, there was an increased risk of bleeding. A large case–control study confirmed a decreased mortality in CABG patients by preoperative aspirin treatment in the absence of any significant increase in hemorrhages or related morbidities [100].

According to the meta-analyses of the Antiplatelet/Antithrombotic Trialists' Collaboration, patients subjected to bypass surgery appear to be less well protected from a vascular atherothrombotic event than patients subjected to PTCA [10]. A possible explanation for this is aspirin "resistance" [101, 102]. However, treatment of these patients with antiplatelet agents is an essential procedure. A recent prospective trial has investigated the benefit/risk ratio of early aspirin treatment on the survival of CABG patients.

A total of 5022 patients undergoing coronary bypass surgery (cardiopulmonary bypass) who survived the first 48 h after surgery were included in a prospective multicenter study to discern the relation between early aspirin use and clinical outcome after 30 days. Patients received a total dose of 80–650 mg aspirin, according to the hospital recommendations, within the first 48 h after surgery or received no aspirin.

There were significantly more patients on aspirin on admission – 52% versus 39% ($p<0.001$) – however, only 1.3% of patients on aspirin (40 out of 2099) died 48 h after surgery or later as opposed to 5.0% (81 out of 2023) on placebo ($p<0.001$). In addition, in comparison to patients without aspirin, treatment caused a 48% reduction in the incidence of myocardial infarctions – 2.8% versus 5.4%; a 50% reduction in the incidence of strokes – 1.3% versus 2.6%; a 74% reduction in renal failure – 0.9% versus 3.4%; and a 62% reduction in bowel infarction – 0.3% versus 0.8%. There was no increased risk either for hemorrhage or for gastritis, infections and impaired wound healing by aspirin. There was also no dose dependency of these aspirin actions.

The conclusion was that early aspirin after CABG is safe and reduces the risk of death and ischemic complications, involving the heart, brain, kidneys, and gastrointestinal tract. There is no evidence of severe bleeding complications [96].

These are remarkable findings, suggesting at least twofold higher efficacy of aspirin in the prevention of thrombotic vessel occlusions than in "conventional" secondary or primary prevention. However, the study was also criticized for several reasons, including the fact that the treatment group assignment was nonrandomized. Since aspirin was used as a single therapy in this particular trial, these findings may have overestimated the efficacy of aspirin that might be achievable in clinical practice. The study also did not address the long-term outcomes for these patients, later than 1 month. Nevertheless, this trial suggests that early aspirin might be associated with a remarkable 68% (!) reduction in overall mortality and substantial reductions in the rates of ischemic complications, affecting the heart and other organs. Importantly, there was no increased risk of bleeding, possibly because of the marked inflammatory response associated with the procedure. Accordingly, there was no reason to increase the coagulation potential by the infusion of platelets, clotting factors, or antifibrinolytic drugs but rather there was evidence that all of these procedures might increase organ failure [96]. In a preface to this article, it was discussed whether the efficacy of aspirin might even have been underestimated because only

48 h survivors were included in the study. It was also suggested that in addition to antiplatelet effects, other pharmacological properties of aspirin might have been involved, for example, anti-inflammatory actions [103] or inhibition of platelet-induced activation of monocytes. Aspirin was shown to inhibit platelet/monocyte interactions in bypass patients by 25% [95]. Similar beneficial result, that is, a 27% reduction in mortality without increased bleeding was also previously shown in a large case–control study with preoperative aspirin administration [100].

Aspirin and Late Bypass Graft Occlusions Early studies have obtained some positive trends for the prevention of late transplant occlusion by aspirin [98]. A retrospective nonrandomized analysis showed an enhanced 5-year survival rate in regular aspirin users [104]. Interestingly, a recent meta-analysis reported that the efficacy of aspirin in this indication may depend on the dosage – medium doses being more effective than low doses during the first year after surgery [105]. Another prospective randomized placebo- and compliance-controlled double-blind trial in CABG patients also showed significant improvement of angiographic graft patency after aspirin intake compared to placebo: 1.6% versus 6.2% at one week and 5.8% versus 11.6% at one year for aspirin (324 mg/day, given within 1 h after surgery) as opposed to placebo [106].

Similar beneficial effects on long-term graft patency were not confirmed in all studies. Whether these different findings are due to the time point when aspirin was started, the aspirin-dosage or patient-related factors – there were no diabetics or patients with severe cardiac failure in the study of Gavaghan *et al.* [106] – remains to be determined. As mentioned above, nonplatelet-derived mitogenic and proinflammatory factors may be involved in late graft occlusion, including thrombin and other growth factors. Clearly, inhibition of platelet function by aspirin alone might not be sufficient for prevention of the progression of atherosclerosis, which is the ultimate cause of reocclusion of bypass vessels.

4.1.1.8 Aspirin and Other Drugs

Aspirin for cardiocoronary prophylaxis, the most important indication for long-term administration, is frequently used in combination with other drugs, including compounds that enhance the antiplatelet and antithrombotic actions of aspirin, such as thienopyridines (clopidogrel) and oral anticoagulants (coumarins) as well as statins because of a frequently coexisting hypercholesterolemia. Several new oral antithrombotics are currently subject to clinical trials and might soon enter the clinics, including new antiplatelet compounds such as ADP-receptor antagonists, inhibitors of factor Xa, and thrombin as well as thrombin receptor antagonists [107]. On the contrary, aspirin might cause unwanted side effects because of inhibition of prostaglandin biosynthesis if stimulation of prostaglandin biosynthesis is involved in comedications taken by the patients. This might be relevant to inhibitors of the angiotensin-converting enzyme (ACE) as well as traditional NSAIDs and coxibs.

Aspirin and Thienopyridines Aspirin inhibits platelet function by inhibiting the thromboxane formation, thienopyridines (clopidogrel, prasugrel) by blocking an ADP-receptor subtype (P_2Y_{12}) at the platelet surface that is involved in platelet aggregation and secretion. Both compounds act synergistically and independent of each other (Section 2.3.1). The efficacy of clopidogrel in cardiocoronary prevention was similar to that of aspirin according to the CAPRIE trial [11]. Thus, the combination of both drugs appears to be a useful strategy to enhance the efficacy of treatment, provided the benefit/risk ratio, that is, mainly the risk of bleedings, remains acceptable.

According to the CURE-Trial Investigators [72], combined use of aspirin with clopidogrel is superior to aspirin alone in percutaneous coronary interventions in ACS and reduced the absolute risk of a combined vascular end point (myocardial infarction, stroke, and cardiovascular death) by 2%. This change was significant for the combined end point but not for a single end point alone. At the same time, the rate of severe bleeding was increased by 1%.

A new discussion about combined use of clopidogrel came up, when it was seen that drug-eluting stents are associated with a markedly, four- to fivefold, increase of late in-stent thrombosis compared to bar metal stents. The median time for thrombus formation was 15–18 months. This suggests that the optimum duration of dual antiplatelet therapy probably needs to be extended to about 1 year [108]. It has also been shown in an observational study that early withdrawal of clopidogrel after stent implantation is associated with a significantly increased cardiovascular mortality after 1 year according to the PREMIER Registry [109].

Another question is whether the combined use of aspirin and clopidogrel also results in an improved clinical outcome during long-term prophylaxis of patients at elevated atherothrombotic risk. In other words, is it useful to generally replace the current monotherapy with aspirin in these patients by combining aspirin with clopidogrel? This issue was studied in the CHARISMA trial.

CHARISMA was a prospective double-blind randomized study in 15 603 patients (age $\geq$45 years) at high atherothrombotic risk. Three-quarters of the patients (12 153) already had suffered an atherothrombotic event (myocardial infarction, cerebrovascular ischemic event) or had a symptomatic peripheral arterial occlusive disease. One-quarter (3284) of the patients were asymptomatic but had multiple risk factors. Patients received aspirin (75–162 mg/day according to the discretion of the doctor) plus placebo or aspirin plus clopidogrel (75 mg/day). Primary efficacy end points were myocardial infarction, stroke, and cardiovascular death. Primary safety end points were severe bleedings. Secondary end points were similar but additionally included hospitalization because of vascular problems. The study end point was set at 1040 events.

The predefined primary end point was obtained at about 30 months in the total population. The cumulative event rate of primary events at this time was 7.3% in patients with aspirin and 6.8% in patients with the combined treatment. This was equivalent to a relative risk reduction by 7.1% with wide variations (95% CI: −4.5 to + 17.5) and not different from the treatment with aspirin alone ($p = 0.22$). There was a slightly improved efficacy in secondary end points in the group with combined treatment: 16% versus 17.9% ($p = 0.04$). The number of severe bleeds was not different. However, there were significantly more "moderate" bleeds (including those that required transfusion) in the combined group.

According to a subgroup analysis of primary efficacy end points of symptomatic and asymptomatic patients, there was a slight but significant ($p = 0.046$) benefit in favor of the symptomatic patients, that is, secondary prevention, but no benefit, or rather a tendency for a deleterious effect, in asymptomatic patients with multiple risk factors without preceding vascular event ($p = 0.20$). Of concern in this group was a significant increase in cardiovascular ($p = 0.01$) and total ($p = 0.04$) mortality after combined treatment.

The conclusion was that the combined use of clopidogrel and aspirin tends to improve the efficacy of secondary prevention in symptomatic patients who already had suffered a vascular event. The combined use of aspirin and clopidogrel cannot be recommended for primary prevention in patients with risk factors but without preceding vascular event. In these patients, addition of clopidogrel to aspirin might increase cardiovascular and total mortality as well as the risk of bleeding [54] (Figure 4.8).

Aspirin and Anticoagulants Exposure of the injured or dysfunctioning vessel wall to streaming blood promotes not only platelet adhesion and aggregate formation but also activation of the coagulation system. This occurs predominantly at the surface of the platelet–fibrin clot and results in increased thrombin formation and enhanced thrombin activity in plasma. Treatment with oral anticoagulants reduces the long-term (years) elevated thrombin levels in patients after acute myocardial infarction. Aspirin does not have this activity [110] and also does not inhibit thrombin-induced platelet activation. This is the reason for the consideration of anticoagulants as alternatives or comedication to antiplatelet compounds, such as aspirin.

Beneficial actions of oral anticoagulation at low INR (1.5) have been shown in primary prevention of high-risk patients. These effects of warfarin were amplified by comedication of aspirin [51]. However, data on secondary prevention were less convincing. Fixed low-dose warfarin (INR 1.2–1.5) was not found to provide any additional clinical benefit

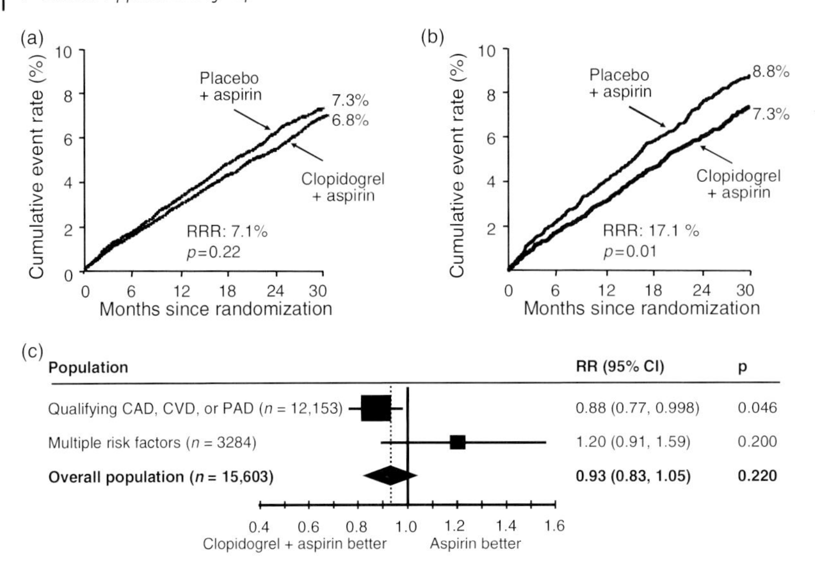

Figure 4.8 The CHARISMA trial. Cumulative event rate of primary end points (myocardial infarction, stroke, and cardiovascular death) in the overall population (a) and the subpopulation with a previous qualifying event (b) and in patients with multiple risk factors without a prior qualifying vascular event (c). There was no difference in the overall population between the two treatment groups and rather a tendency in favor of aspirin alone in the multirisk factor group, whereas the opposite was seen in patients with a previous qualifying event (for further explanation see text) adapted from data in [54].

when combined with aspirin (160 mg/day) over 14 months in patients after acute myocardial infarction but doubled the risk for spontaneous major hemorrhages [111]. Similar negative results were obtained in the OASIS trial, studying the efficacy of moderate-intensity oral anticoagulants in addition to aspirin in patients with unstable angina (NSTEMI). There was a small benefit after 5 months by the combined treatment; however, it also became clear that there were compliance problems with the oral anticoagulants and the investigators suggested that a good compliance to these compounds could potentially lead to clinically important reductions in major ischemic cardiovascular events [112]. A similar result was seen in the CHAMP study: low-INR (median 1.8) warfarin did not provide an additional benefit in postmyocardial infarction patients treated with aspirin (162 mg) [113]. In patients with acute coronary syndromes and prior CABG, combined therapy with aspirin and warfarin (INR 2.0–2.5) was not superior to low-dose aspirin alone in the prevention of recurrent ischemic events [114]. Thus, the majority of available studies indicates that the combined use of aspirin and low-INR anticoagulants does not provide additional therapeutic benefits but increases the risk of bleedings. A possible exception are lysis patients with a particular high risk of reocclusions of the infarcted coronary artery (APRICOT-2 Trial, [115]). In contrast, the efficacy of full-range anticoagulation (INR 2.8–4.2) in secondary prevention of cardiovascular events was clearly demonstrated in two other large randomized, double-blind, placebo-controlled trials. However, there was also a fourfold increased risk of major bleedings [116–118].

The available data suggest that in a setting of good compliance and well-organized INR monitoring, addition of oral anticoagulants (INR > 2) to aspirin seems to be beneficial. Safety of combined aspirin and dose-adjusted anticoagulation also appears to be acceptable. There is a two- to threefold increased risk of minor and major bleedings, without an increased risk of intracerebral hemorrhage [119]. Thus, there is evidence for a synergistic and clinically relevant effect of combined aspirin and coumarins. Possibly, new anticoagulants, such as (orally active) thrombin inhibitors (dabigatran) or inhibitors of factor Xa (rivaroxaban, apixaban), may overcome this problem and extend the range of the clinical useful combination of aspirin with (oral) anticoagulants.

Aspirin and ACE Inhibitors Aspirin and ACE inhibitors are frequently used in combination to treat coronary heart disease, hypertension, and chronic heart failure. ACE inhibitors stimulate prostacyclin formation via inhibition of bradykinin breakdown and this is probably part of their clinical efficacy. Thus, COX inhibition by aspirin might reduce or abolish this pharmacological action of ACE inhibitors. The clinical consequences would depend on the significance of (stimulated) prostaglandin biosynthesis for the particular disease and patient, respectively.

In agreement with this hypothesis, the available study data are variable and the results obviously influenced by the (severity of) basal disease. An antagonism of ACE inhibition by aspirin was seen in 1 out of 5 trials in hypertension, 1 out of 4 trials in coronary heart disease, and 9 out of 13 trials in congestive heart failure. These interactions were more likely to occur at higher doses of aspirin (>250 mg/day) [120]. However, all of these trials were retrospective with "weak" hemodynamic end points. The SOLVD trial was one of the first studies to suggest a possible reduction of efficacy of ACE inhibition by enalapril after aspirin cotreatment [121]. However, a more detailed analysis of this and other related studies on this issue has shown that this finding may be study specific and should not be extrapolated to ACE inhibitors in general [122]. Another recently published large epidemiological trial on more than 7300 patients with heart failure was also unable to show negative effects of aspirin cotreatment, also at higher doses, on the positive actions of ACE inhibitors, including the prolonged patients' survival [123].

Prospective randomized trials with hard end points (mortality) are definitely needed to answer the question of a possible ACE/aspirin interaction [120, 124]. Until these studies are available, lower doses of aspirin, that is, $\leq$100 mg/day should be used, specifically in patients with congestive heart failure, who are treated with ACE inhibitors. Here, sartans, that is, selective inhibitors of angiotensin II AT_1 receptors, are therapeutic alternatives because they do not interfere with kinin degradation and, therefore, will not modify PGI_2 biosynthesis.

Aspirin and Statins One of the most frequent and important cardiovascular risk factors is hypercholesterolemia. Therefore, appropriate treatment with lipid-lowering agents, such as statins, is a standard guideline recommendation with a well-established clinical benefit. A recent meta-analysis has shown an additive effect of aspirin and pravastatin in the secondary prevention of cardiovascular events [125]. Similar positive effects might be expected for other statins and would not be surprising because of the long-known antiplatelet effects of statins in hypercholesterolemic subjects [126].

Aspirin, Traditional NSAIDs, and Coxibs Recent investigations have provided evidence for a possible drug interaction between some NSAIDs, such as indomethacin and ibuprofen, with the antiplatelet effects of aspirin [127] (Section 2.2.1). The possible clinical significance of this finding was demonstrated in a small population-based study in patients with preexisting cardiovascular disease. There was a significant increase of cardiovascular events in users of combined aspirin (<325 mg/day) and ibuprofen as opposed to aspirin alone ($p = 0.001$). No such change was seen by comedication of diclofenac or other NSAIDs. This suggests that a combination of ibuprofen and aspirin may be deleterious, possibly by antagonizing the antiplatelet cardioprotective actions of aspirin in these patients [128]. A pharmacological inhibition of aspirin action during comedication of dipyrone (metamizol) has recently been reported and might be significant in case of postoperative analgesic treatment of patients who underwent cardiocoronary surgery [129]. However, a clinical trial with these patients is still missing.

Of particular concern are possible deleterious effects of selective COX-2 inhibitors (coxibs) on the incidence of cardiocoronary events in patients at elevated cardiovascular risk [130]. The VIGOR study [131] has shown a significant increase of myocardial infarctions by rofecoxib in patients at medium cardiovascular risk (rheumatoid arthritis) but not with naproxen. It has been suggested that naproxen might have a beneficial effect, possibly related to its relatively long half-life (13 h) [132] and combined COX-1/COX-2 inhibition.

A central issue is whether coxibs may affect the therapeutic benefits of aspirin in cardiocoronary prevention, in particular, if patients are at elevated risk. Vascular prostacyclin is mainly generated via COX-2 and this isoform is markedly induced in patients with atherosclerosis. This is associated with an increased formation of PGI_2 [82] (Section 2.3.1). Atherosclerosis is an inflammatory disease and an enhanced generation of vasodilatory prostaglandins, that is, PGI_2 and PGE_2, may reflect an increased demand of these mediators to maintain hemostatic balance, to keep blood pressure low, and to antagonize platelet hyperreactivity [133]. Importantly, coxibs will not block aspirin-sensitive thromboxane formation via COX-1 of platelets. Subsequent to VIGOR, several other recent trials have shown an increase in myocardial infarctions by coxibs and there appears to be a correlation with the preexisting cardiovascular risk [134]. A recent epidemiological Finnish trial found that all NSAIDs are associated with a modest increase in risk for a first myocardial infarction [135] while another reported a moderate increase in risk, comparable to high-dose ibuprofen and diclofenac but not naproxen [136]. However, the discussion on this issue is not closed yet [134, 137].

4.1.1.9 Actual Situation

Current recommendations on the use of aspirin in prevention of cardiocoronary events are found in actual guidelines. Risk calculators and tables are available at www.med-decisions.com or www.absoluterisk.com and were also published by Sanmuganathan *et al.* [138] and Lauer [139], respectively.

Primary Prevention Because of the generally low vascular risk, the expected benefit with aspirin is also low in primary prevention. Currently, the benefit/risk ratio is considered acceptable at an annual (statistical) event rate of 1%.

Providing regular use of aspirin for primary prevention in a population of 1000 individuals with a 5% risk for coronary heart disease events over 5 years would mean the following: prevention of 6–20 myocardial infarctions at the costs of 0–2 hemorrhagic strokes and 2–4 major gastrointestinal bleeding events. For patients with a risk of 1% over 5 years, this would mean prevention of 1–4 myocardial infarctions at the expense of the *same* number of severe side effects. Thus, the benefit/risk ratio would be considerably worse [140] (Figure 4.9).

In any case aspirin prophylaxis should only be used in addition to control cardiovascular risk factors, including dietary and lifestyle changes, smoking cessation, and control of blood pressure.

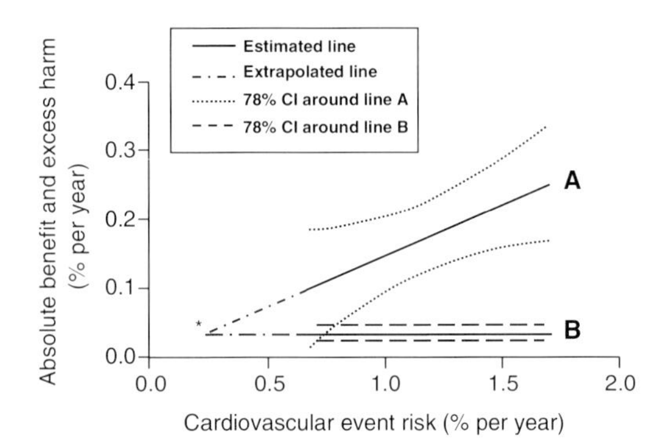

Figure 4.9 Absolute benefit (reduction in all cardiovascular events; line A) and absolute harm (increase in major bleeds; line B) by aspirin treatment in relation to the annual cardiovascular event risk. (*) Extrapolated point of equal benefit and harm: 0.22%/year. Suggested positive benefit/risk ratio for aspirin prevention: ≥1%/year [138].

Secondary Prevention Aspirin in high-risk patients, patients subjected to PCI, CABG, or those with an already preexisting coronary heart disease is recommended, if no contraindication exists. According to the European Expert Consensus document, in clinical situations when an immediate antithrombotic effect is required (such as in the case of acute coronary syndromes or in acute ischemic stroke), a loading dose of 250–500 mg (soluble) aspirin might be given at diagnosis to ensure a rapid and complete inhibition of TXA_2-dependent platelet aggregation if no aspirin prophylaxis was performed [141]. Similar recommendations were published by the American Heart Association [140, 142] and the National Heart Foundation of Australia [143]. The recommended maintenance dose of aspirin for the prevention of coronary events is in the range of 75–325 mg with Europeans tending to lower and others tending to higher dose recommendations.

Aspirin "Loading" Dose in ACS The immediate application of a "loading dose" of 250–500 mg aspirin as water-soluble lysine salt i.v. is now strongly recommended in many countries for initial, that is, prehospital, treatment of acute coronary syndromes in the absence of contraindications. This rather high initial dose is necessary for saturation of all COX-1 within the platelets. In case of acute myocardial infarctions (by ECG criteria), prehospital treatment with i.v. aspirin plus 5000 IU heparin has been found to markedly improve the clinical outcome, eventually resulting in an about 50% reduction of mortality at 30 days [144]. Similar results were obtained in two German Registries (MITRA and MIR), including 22 572 patients with STEMI; 92% of them receiving early (within 48 h) aspirin treatment. The 8%, who did not because of several reasons including relative contraindications or a critical clinical state, had an about three times higher in-hospital mortality [145]. Similar results were obtained in the GRACE-registry study, including 11 388 patients with and without a history of coronary artery disease [146]. Thus, initial aspirin treatment at sufficiently high doses is quite effective and reduces significantly the incidence of a (new) cardiovascular event, in-hospital complications, and mortality.

Stents According to current guidelines [147], PCI is a treatment of choice in patients presenting with STEMI if the appropriate technology and skilled personnel are available. Adjunctive antiplatelet treatment with aspirin plus clopidogrel is a therapeutic standard. Drug-eluting stents are associated with a higher risk of late in-stent thrombosis as opposed to bar metal stents and require a longer maintenance of dual antiplatelet therapy. The optimum time is still unknown, but may be 1 year and more.

Summary

Thromboembolic complications of coronary vascular disease are acute coronary syndromes, that is, instable angina, myocardial infarction, and sudden cardiac death. They result from a pathological interaction between platelets and the vessel wall and are caused by thrombus formation in a large coronary artery. Aspirin prevents platelet-dependent thromboxane formation and all thromboxane-related consequences for platelet function. This is the rationale for its prophylactic use in cardiocoronary prevention.

Aspirin is the drug of choice for secondary and primary prevention of cardiocoronary events at conventional antiplatelet doses (about 100–300 mg/day). Overall, there is a 15% risk reduction in primary and a 25% risk reduction in secondary prevention. These beneficial effects have to be balanced against side effects, most notably a bleeding tendency. Actual guidelines consider this risk as acceptable, if the (statistical) benefit/risk ratio is at least 2:1 or the average annual risk of a vascular event $\geq$1% or $\geq$10% in 10 years.

In primary prevention in men, the major effect is protection from cardiac events, and in women it provides protection from cerebrovascular events. Combined use of aspirin with clopidogrel is not recommended because of the increased risk of bleedings and an increased cardiovascular mortality. In secondary prevention, combined use is possible and depends on the risk profile. This is particularly valid for an antiplatelet therapy after implantation of drug-eluting stents with a possibly elevated risk of late in-stent thrombosis in risk populations.

Long-term treatment of patients with enhanced cardiovascular risk frequently requires comedication of other drugs, which might interact with aspirin. In addition to clopidogrel, synergistic (additive) interactions have been described for statins (pravastatin) and oral anticoagulants, although associated with an increased tendency for bleeding in both cases. Some retrospective trials suggest a reduced efficacy of ACE inhibitors by aspirin, specifically in congestive heart failure, although probably only at higher doses (≥250 mg/day).

There is no doubt that doses of aspirin between 100 and 300 mg/day are effective in coronary prevention. This does, however, not answer the question whether this dosing is optimal and the same for all clinical indications and patients or whether a more individualized treatment might be more effective. Because of the inflammatory nature of atherosclerosis, additional anti-inflammatory effects of aspirin may be useful and have been shown in some clinical trials. Whether these effects of higher dose aspirin are beneficial or deleterious because of inhibition of prostacyclin formation can only be decided by head-to-head comparisons of different doses in randomized trials, which have not been done so far. These studies might also include surrogate parameters of platelet function and mediators of inflammation, which might help define an individualized treatment.

References

1 Wang, H. and Bakhai, A. (Eds.) (2006) Clinical Trials, *A Practical Guide to Design, Analysis and Reporting*. Remedica, London.

2 Elwood, P.C., Cochrane, A.L., Burr, M.L. *et al.* (1974) A randomised controlled trial of acetylsalicylic acid in the secondary prevention of mortality from myocardial infarction. *British Medical Journal*, **1**, 436–440.

3 (?The) Boston Collaborative Drug Surveillance Group (1974) Regular aspirin intake and acute myocardial infarction. *British Medical Journal*, **1**, 440–443.

4 Kannel, W.B. and Thom, T.J. (1994) *The Heart*, 8th edn (eds R.C. Schlant and R.W. Alexander), McGraw-Hill, New York, pp. 185–197.

5 Sans, S., Kesteloo, H. and Kromhout, D. (1997) The burden of cardiovascular disease mortality in Europe. *European Heart Journal*, **18**, 1231–1248.

6 de Vreede-Swagemakers, J.J.M., Gorgels, A.P.M., Dubois-Arbouw, W.I. *et al.* (1997) Out-of-hospital cardiac arrest in the 1990s: a population-based study in the Maastricht area on incidence, characteristics and survival. *JACC*, **30**, 1500–1505.

7 Manson, J.E., Tosteson, H., Ridker, P.M. *et al.* (1992) The primary prevention of myocardial infarction. *The New England Journal of Medicine*, **326**, 1406–1416.

8 Ginter, E. (1997) The epidemic of cardiovascular disease in Eastern Europe. *The New England Journal of Medicine*, **336**, 1915–1916.

9 Antiplatelet Trialists' Collaboration (1994) Collaborative overview of randomized trials of antiplatelet therapy. I. Prevention of death, myocardial infarction, and stroke by prolonged antiplatelet therapy in various categories of patients. *British Medical Journal*, **308**, 81–106.

10 Antithrombotic Trialists' Collaboration (2002) Collaborative meta-analysis of randomised trials of antiplatelet therapy for prevention of death, myocardial infarction, and stroke in high risk patients. *British Medical Journal*, **324**, 71–86.

11 Gent, M. and CAPRIE Steering Committee (1996) A randomised, blinded trial of clopidogrel versus aspirin in patients at risk of ischaemic events (CAPRIE). *The Lancet*, **348**, 1329–1339.

12 Reilly, I.A.G. and FitzGerald, G.A. (1988) Aspirin in cardiovascular disease. *Drugs*, **35**, 154–176.

13 Gawaz, M. (2004) Role of platelets in coronary thrombosis and reperfusion of ischemic myocardium. *Cardiovascular Research*, **61**, 498–511.

14 Fitzgerald, D.J., Roy, L., Catella, F. *et al.* (1986) Platelet activation in unstable coronary disease. *The New England Journal of Medicine*, **315**, 983–989.

15 Elwood, P.C., Beswick, A., Pickering, J. *et al.* (2001) Platelet tests in the prediction of myocardial infarction and ischaemic stroke: evidence from the Caerphilly prospective study. *British Journal of Haematology*, **113**, 514–520.

16 Thaulow, E., Erikssen, I., Sandvik, L. *et al.* (1991) Blood platelet count and function are related to total and cardiovascular death in apparently healthy men. *Circulation*, **84**, 613–617.

17 Elwood, P.C., Renaud, S., Sharp, D.S. *et al.* (1991) Ischemic heart disease and platelet aggregation. The Caerphilly Collaborative Heart Disease Study. *Circulation*, **83**, 438–441.

18 Trip, M.D., Cats, V.M., van Capelle, F.J. *et al.* (1990) Platelet hyperreactivity and prognosis in survivors of myocardial infarction. *The New England Journal of Medicine*, **322**, 1549–1554.

19 Hart, R.G., Leonard, A.D., Talbert, R.L. *et al.* (2003) Aspirin dosage and thromboxane synthesis in patients with vascular diseases. *Pharmacotherapy*, **23**, 579–584.

20 Gulec, S., Ozdol, C., Vurgun, K. *et al.* (2008) The effect of high-dose aspirin pretreatment on the incidence of myonecrosis following elective coronary stenting. *Atherosclerosis*, **197**, 171–176.

21 Quinn, M.J., Aronow, H.D., Califf, R.M. *et al.* (2004) Aspirin dose and six-month outcome after an acute coronary syndrome. *Journal of the American College of Cardiology*, **43**, 972–978.

22 Kong, D.F., Hasselblad, V., Kandzari, D.E. *et al.* (2002) Seeking the optimal aspirin dose in acute coronary syndromes. *The American Journal of Cardiology*, **90**, 622–625.

23 Massberg, S., Schulz, C. and Gawaz, M. (2003) Role of platelets in the pathophysiology of acute coronary syndrome. *Seminars in Vascular Medicine*, **3**, 147–162.

24 Ridker, P.M., Cushman, M., Stampfer, M.J. *et al.* (1997) Inflammation, aspirin, and the risk of cardiovascular disease in apparently healthy men. *The New England Journal of Medicine*, **336**, 973–979.

25 Ridker, P.M., Buring, J.E., Shih, J. *et al.* (1998) Prospective study of C-reactive protein and the risk of future cardiovascular events among apparently healthy men. *Circulation*, **98**, 731–733.

26 Ikonomidis, I., Andreotti, F., Economou, E. *et al.* (1999) Increased proinflammatory cytokines in patients with chronic stable angina and their reduction by aspirin. *Circulation*, **100**, 793–79.

27 Held, C., Hjemdahl, P., Wallén, N.H. *et al.* (2000) Inflammatory and hemostatic markers in relation to cardiovascular prognosis in patients with stable angina pectoris. Results from the APSIS study. *Atherosclerosis*, **148**, 179–188.

28 Kennon, S., Price, C.P., Mills, P.G. *et al.* (2001) The effect of aspirin on C-reactive protein as a marker of risk in unstable angina. *Journal of the American College of Cardiology*, **37**, 1266–1270.

29 Ikonomidis, I., Andreotti, F. and Nihoyannopoulos, P. (2004) Reduction of daily life ischaemia by aspirin in patients with angina: underlying link between thromboxane A_2 and macrophage colony stimulating factor. *Heart (British Cardiac Society)*, **90**, 389–393.

30 Steering Committee of the Physicians' Health Study Research Group (1988) Preliminary report: findings from the aspirin component of the ongoing Physicians' Health Study. *The New England Journal of Medicine*, **318**, 262–264.

31 Steering Committee of the Physicians' Health Study Research Group (1989) Final report on the aspirin component of the ongoing Physicians' Health Study. *The New England Journal of Medicine*, **321**, 129–135.

32 Cook, N.R., Cole, S.R. and Hennekens, C.H. (2002) Use of a marginal structural model to determine the effect of aspirin on cardiovascular mortality in the physicians' health study. *American Journal of Epidemiology*, **155**, 1045–1053.

33 Tell, G.S., Fried, L.P., Hermanson, B. *et al.* (1993) Recruitment of adults 65 years and older as participants in the cardiovascular health study. *Annals of Epidemiology*, **3**, 358–366.

34 Reuben, D.B., Hirsch, S.H., Frank, J.C. *et al.* (1996) The Prevention for Elderly Persons (PEP) program: a model of municipal and academic partnership to meet the needs of older persons for preventive services. *Journal of the American Geriatrics Society*, **44**, 1394–1398.

35 Ridker, P.M., Manson, J.E., Buring, J.E. *et al.* (1991a) The effect of chronic platelet inhibition with low-dose aspirin on atherosclerosis progression and acute thrombosis: clinical evidence from the Physicians'

Health Study. *American Heart Journal*, **122**, 1588–1592.

36 Manson, J.E., Grobbee, D., Stampfer, M.J. *et al.* (1990) Aspirin in the primary prevention of angina pectoris in a randomized trial of United States Physicians. *The American Journal of Medicine*, **89**, 772–776.

37 Peto, R., Gray, R., Collins, R. *et al.* (1988) Randomized trial of prophylactic daily aspirin in British male doctors. *British Medical Journal*, **296**, 313–316.

38 De Gaetano, G. (1988) Primary prevention of vascular disease by aspirin. *The Lancet*, **1**, 1093–1094.

39 Bredie, S.J.H., Wollersheim, H., Verheugt, F.W.A. *et al.* (2003) Low-dose aspirin for primary prevention of cardiovascular disease. *Seminars in Vascular Medicine*, **3**, 177–183.

40 Rich-Edwards, J.W., Manson, J.E., Hennekens, Ch.H. *et al.* (1995) The primary prevention of coronary heart disease in women. *The New England Journal of Medicine*, **332**, 1758–1766.

41 Manson, J.E., Stampfer, M.J., Colditz, G.A. *et al.* (1991) A prospective study of aspirin use and primary prevention of cardiovascular disease in women. *The Journal of the American Medical Association*, **266**, 521–527.

42 Ridker, P.M., Cook, N.R., Lee, I.-M. *et al.* (2005) A randomized trial of low-dose aspirin in the primary prevention of cardiovascular disease in women. *The New England Journal of Medicine*, **352**, 1293–1304.

43 Ridker, P.M., Hennekens, C.H. and Tofler, G.H. (1996) Anti-platelet effects of 100 mg alternate day oral aspirin: a randomized, double-blind, placebo-controlled trial of regular and enteric-coated formulations in men and women. *Journal of Cardiovascular Risk*, **3**, 209–213.

44 Levin, R.I. (2005) The puzzle of aspirin and sex. *The New England Journal of Medicine*, **352**, 1366–1368.

45 Patrono, C., Coller, B., Dalen, J.E. *et al.* (2001) Platelet-active drugs. The relationship among dose, effectiveness, and side effects. *Chest*, **119**, 39S–63.

46 Bonita, R. (1992) Epidemiology of stroke. *The Lancet*, **339**, 342–344.

47 Berger, J.S., Roncaglioni, M.C., Avanzini, F. *et al.* (2006) Aspirin for the primary prevention of cardiovascular effects in women and men. A sex-specific meta-analysis of randomized controlled trials. *The Journal of the American Medical Association*, **295**, 306–313.

48 Glynn, R.J., Buring, J.E., Manson, J.E. *et al.* (1994) Adherence to aspirin in the prevention of myocardial infarction. *Archives of Internal Medicine*, **154**, 2649–2657.

49 Hansson, L., Zanchetti, A., Carruthers, S.G. *et al.* (1998) Effects of intensive blood-pressure lowering and low-dose aspirin in patients with hypertension: principal results of the Hypertension Optimal Treatment (HOT) randomised trial. HOT Study Group. *The Lancet*, **351**, 1755–1762.

50 Kjeldsen, S.E., Kolloch, R.E., Leonetti, G. *et al.* (2000) Influence of gender and age on preventing cardiovascular disease by antihypertensive treatment and acetylsalicylic acid. The HOT (Hypertension Optimal Treatment) study. *Journal of Hypertension*, **18**, 629–642.

51 (The) Medical Research Council's General Practice Research Framework (1998) Thrombosis prevention trial: randomised trial of low-intensity oral anticoagulation with warfarin and low-dose aspirin in the primary prevention of ischaemic heart disease in men at increased risk. *The Lancet*, **351**, 233–241.

52 Meade, T.W. and Brennan, P.J. (2000) Determination of who may derive most benefit from aspirin in primary prevention: subgroup results from a randomised controlled trial. *British Medical Journal*, **321**, 13–17.

53 Mundal, H.H., Nordby, G., Lande, K. *et al.* (1993) Effect of cold pressor test and awareness of hypertension on platelet function in normotensive and hypertensive women. *Scandinavian Journal of Clinical and Laboratory Investigation*, **53**, 585–591.

54 Bhatt, D.L., Fox, K.A., Werner, Ch.B. *et al.* (2006) Clopidogrel and aspirin versus aspirin alone for the prevention of atherothrombotic events. *The New England Journal of Medicine*, **354**, 1706–1717.

55 De Gaetano, G. on behalf of the Collaborative Group of the Primary Prevention Project (2001) Low-dose aspirin and vitamin E in people at cardiovascular risk: a randomized trial in general practice. Collaborative Group of the Primary Prevention Project. *The Lancet*, **357**, 89–95.

56 Juul-Möller, S., Edvardsson, N., Jahnmatz, B. *et al.*, for the Swedish Angina Pectoris Aspirin Trial (SAPAT) Group (1992) Double-blind trial of aspirin in primary prevention of myocardial infarction in patients with chronic stable angina pectoris. *The Lancet*, **340**, 1421–1425.

57 Ridker, P.M., Manson, J.E., Gaziano, M. *et al.* (1991b) Low-dose aspirin therapy for chronic stable angina: a randomized, placebo-controlled clinical trial. *Annals of Internal Medicine*, **114**, 835–839.

58 Willard, J.E., Lange, R.A. and Hillis, L.D. (1992) Current concepts: the use of aspirin in ischemic heart disease. *The New England Journal of Medicine*, **327**, 175–181.

59 Gum, P.A., Thamilarasan, M., Watanabe, J. *et al.* (2001) Aspirin use and all-cause mortality among patients being evaluated for known or suspected coronary artery disease. *The Journal of the American Medical Association*, **286**, 1187–1194.

60 Davies, M.J. and Thomas, A.C. (1985) Plaque fissuring – the cause of acute myocardial infarction, sudden ischaemic death and crescendo angina. *British Heart Journal*, **53**, 363–373.

61 Rentrop, K.P. (2000) Thrombi in acute coronary syndromes: revisited and revised. *Circulation*, **101**, 1619–1626.

62 Zaman, A.G., Herlft, G., Worthley, S.G. *et al.* (2000) The role of plaque rupture and thrombosis in coronary artery disease. *Atherosclerosis*, **149**, 251–266.

63 Lewis, H.D., Davis, J.W., Archibald, D.G. *et al.* (1983) Protective effects of aspirin against acute myocardial infarction and death in men with unstable angina. Results of a Veterans Administration Cooperative study. *The New England Journal of Medicine*, **309**, 396–403.

64 Cairns, J.A., Gent, M., Singer, J. *et al.* (1985) Aspirin, sulfinpyrazone, or both in unstable angina: results from a Canadian Multicenter Trial. *The New England Journal of Medicine*, **313**, 1369–1375.

65 Théroux, P., Ouimet, H., McCans, J. *et al.* (1988) Aspirin, heparin or both to treat acute unstable angina. *The New England Journal of Medicine*, **319**, 1105–1111.

66 (The) RISC Group (1990) Risk of myocardial infarction and death during treatment with low dose aspirin and intravenous heparin in men with unstable coronary artery disease. *The Lancet*, **336**, 827–830.

67 Wallentin, L.C. (1991) Aspirin (75 mg/day) after an episode of unstable coronary artery disease: long-term effects on the risk for myocardial infarction, occurrence of severe angina and the need for revascularization. Research group on instability in coronary artery disease in Southeast Sweden. *Journal of the American College of Cardiology*, **18**, 1587–1593.

68 Pedersen, A.K. and FitzGerald, G.A. (1985) Cyclooxygenase inhibition, platelet function, and metabolite formation during chronic sulfinpyrazone dosing. *Clinical Pharmacology and Therapeutics*, **37**, 36–42.

69 Forman, M.B., Uderman, H., Jackson, E.K. *et al.* (1985) Effects of indomethacin on systemic and coronary hemodynamics in patients with coronary artery disease. *American Heart Journal*, **110**, 311–318.

70 Vejar, M., Fragasso, G., Hackett, D. *et al.* (1990) Dissociation of platelet activation and spontaneous myocardial ischemia in unstable angina. *Thrombosis and Haemostasis*, **63**, 163–168.

71 Maseri, A., Davies, G., Hackett, D. *et al.* (1990) Coronary artery spasm and vasoconstriction. The case for a distinction. *Circulation*, **81**, 1983–1991.

72 CURE-Trial Investigators (2001) Effects of clopidogrel in addition to aspirin in patients with acute coronary syndromes without ST-segment elevation. *The New England Journal of Medicine*, **345**, 494–502.

73 Ronnevik, P.K., Folling, M., Pedersen, D. *et al.* (1991) Increased occurrence of exercise-induced silent ischemia after treatment with aspirin in patients admitted for suspected acute myocardial infarction. *International Journal of Cardiology*, **33**, 413–417.

74 Nyman, I., Larsson, H., Wallentin, L. *et al.* (1992) Prevention of serious cardiac events by low-dose aspirin in patients with silent myocardial ischaemia. *The Lancet*, **340**, 497–501.

75 Ferrari, E., Benhamou, M., Cerboni, P. *et al.* (2005) Coronary syndromes following aspirin withdrawal. *Journal of the American College of Cardiology*, **45**, 456–459.

76 Biondi-Zoccai, G.G., Lotrionte, M., Agostini, P. *et al.* (2006) A systematic review and meta-analysis on the hazards of discontinuing or not adhering to aspirin among 50279 patients at risk for coronary artery disease. *European Heart Journal*, **27**, 1667–1674.

77 Hoffmann, W. and Förster, W. (1991) Two year Cottbus reinfarction study with 30 mg aspirin per day. *Prostaglandins, Leukotrienes, and Essential Fatty Acids*, **44**, 159–169.

78 Hoffmann, W., Nitschke, M., Muche, J. *et al.* (1991) Reevaluation of the Cottbus reinfarction study with 30 mg aspirin per day, 4 years after the end of the study. *Prostaglandins, Leukotrienes, and Essential Fatty Acids*, **42**, 137–139.

79 Schrör, K. (1995) Reinfarct prophylaxis with 100 mg or 30 mg ASS daily [in German]. *Zeitschrift für Kardiologie*, **84**, 496–498.

80 Husted, S.E., Kraemmer-Nielsen, H., Krusell, L.R. *et al.* (1989) Acetylsalicylic acid 100 mg and 1000 mg daily in acute myocardial infarction suspects: a placebo-controlled trial. *Journal of Internal Medicine*, **226**, 303–310.

81 Verheugt, F.W.A., van der Laarse, A., Funke, A.J. *et al.* (1990) Effects of early intervention with low-dose aspirin (100 mg) on infarct size, reinfarction and mortality in anterior wall acute myocardial infarction. *The American Journal of Cardiology*, **66**, 267–270.

82 Belton, O., Byrne, D., Kearney, D. *et al.* (2000) Cyclooxygenase-1 and -2-dependent prostacyclin formation in patients with atherosclerosis. *Circulation*, **102**, 840–845.

83 Fitzgerald, D.J., Catella, F., Roy, L. *et al.* (1988) Marked platelet activation *in vivo* after intravenous streptokinase in patients with acute myocardial infarction. *Circulation*, **77**, 142–150.

84 ISIS-2 (1988) Randomized trial of intravenous streptokinase, oral aspirin, both or neither among 17,187 cases of suspected acute myocardial infarction. *The Lancet*, **2** 349–360.

85 Baigent, C., Collins, R., Appleby, P. *et al.* (1998) ISIS-2: 10-year survival among patients with suspected acute myocardial infarction in randomised comparison of intravenous streptokinase, oral aspirin, both, or neither. The ISIS-2 (Second International Study of Infarct Survival) Collaborative Group. *British Medical Journal*, **316**, 1337–1343.

86 Undas, A., Undas, R., Nusial, J. *et al.* (2000) A low dose of aspirin (75 mg/day) lowers thrombin generation to a similar extent as a high dose of aspirin (300 mg/day). *Blood Coagulation Fibrinolysis*, **11**, 231–234.

87 Savage, M.P., Goldberg, S., Bove, A.A. *et al.* (1995) Effect of thromboxane A_2 blockade on clinical outcome and restenosis after successful coronary angioplasty. Multi-Hospital Eastern Atlantic Restenosis Trial (M-HEART II). *Circulation*, **92**, 3194–3200.

88 Terres, W., Hamm, C.W., Ruchelka, A. *et al.* (1992) Residual platelet function under acetylsalicylic acid and the risk of restenosis after coronary angioplasty. *Journal of Cardiovascular Pharmacology*, **19**, 190–193.

89 Fager, G. (1995) Thrombin and proliferation of vascular smooth muscle cells. *Circulation Research*, **77**, 645–650.

90 Bretschneider, E., Braun, M., Fischer, A. *et al.* (2000) Factor Xa acts as a PDGF-independent mitogen in human vascular smooth muscle cells. *Thrombosis and Haemostasis*, **84**, 499–505.

91 ten Berg, J.M., Kelder, J.C., Suttorp, M.J. *et al.* (2000) Effect of coumarins started before coronary angioplasty on acute complications and long-term follow-up. A randomized trial. *Circulation*, **102**, 386–391.

92 ten Berg, J.M., Hutten, B.A., Kelder, J.C. *et al.* (2001) Oral anticoagulant therapy during and after coronary angioplasty. *Circulation*, **103**, 2042–2047.

93 Herskowitz, A. and Mangano, D.T. (1996) Inflammatory cascade: a final common pathway for perioperative injury? *Anesthesiology*, **85**, 957–960.

94 Sun, J.C.J., Whitlock, R., Cheng, J. *et al.* (2008) The effect of pre-operative aspirin on bleeding, transfusion, myocardial infarction, and mortality in coronary artery bypass surgery: a systematic review of randomized and observational studies. *European Heart Journal*, **29**, 1057–1071.

95 Ferraris, V.A., Ferraris, S.P., Joseph, O. *et al.* (2002) Aspirin and postoperative bleeding after coronary artery bypass grafting. *Annals of Surgery*, **235**, 820–827.

96 Mangano, D.T. for the Multicenter Study of Perioperative Ischemia Research Group (2002) Aspirin and mortality from coronary bypass surgery. *The New England Journal of Medicine*, **347**, 1309–1317.

97 Lorenz, R.L., von Schacky, C., Weber, M. *et al.* (1984) Improved aortocoronary bypass patency by low-dose (100 mg daily) aspirin. *The Lancet*, **1**, 1261–1264.

98 Goldman, S.J., Copeland, J., Moritz, T. *et al.* (1991) Starting aspirin therapy after operation. Effects on early graft patency. *Circulation*, **84**, 520–526.

99 Hockings, B.E.F., Ireland, M.A., Gorch-Martin, K. *et al.* (1993) Placebo-controlled trial of enteric coated aspirin in coronary bypass graft patients. *Medical Journal of Australia*, **1259**, 376–378.

100 Dacey, L.J., Munoz, J.J., Johnson, E.R. *et al.* (2000) Effect of preoperative aspirin use on mortality in coronary artery bypass grafting patients. *The Annals of Thoracic Surgery*, **70**, 1986–1990.

101 Zimmermann, N., Kienzle, P., Weber, A.A. *et al.* (2001) Aspirin resistance after coronary artery bypass grafting. *The Journal of Thoracic and Cardiovascular Surgery*, **121**, 982–984.

102 Zimmermann, N., Wenk, A., Kim, U. *et al.* (2003) Functional and biochemical evaluation of platelet aspirin resistance after coronary artery bypass surgery. *Circulation*, **108**, 542–547.

103 Topol, E.J. (2002) Aspirin with bypass surgery – from taboo to new standard of care. *The New England Journal of Medicine*, **347**, 1359–1360.

104 Johnson, W.D., Kayser, K.L., Jartz, A.J. *et al.* (1992) Aspirin use and survival after coronary bypass surgery. *American Heart Journal*, **123**, 603–608.

105 Lim, E., Ali, Z., Routledge, T. *et al.* (2003) Indirect comparison meta-analysis of aspirin therapy after coronary surgery. *British Medical Journal*, **327**, 1309–1311.

106 Gavaghan, T.P., Gebski, V. and Baron, D.W. (1991) Immediate postoperative aspirin improves vein graft patency early and late after coronary artery bypass surgery. A placebo-controlled, randomized study. *Circulation*, **83**, 1526–1533.

107 Meadows, T.A. and Bhatt, D.L. (2007) Clinical aspects of platelet inhibitors and thrombus formation. *Circulation Research*, **100**, 1261–1275.

108 Bavry, A.A., Kumbhani, D.J., Helton, T.J. *et al.* (2006) Late thrombosis of drug-eluting stents: a meta-analysis of randomized clinical trials. *The American Journal of Medicine*, **119**, 1056–1061.

109 Spertus, J.A., Kettelkamp, R., Vance, C. *et al.* (2006) Prevalence, predictors, and outcomes of premature discontinuation of thienopyridine therapy after drug-eluting stent placement. Results from the PREMIER Registry. *Circulation*, **113**, 2803–2809.

110 Seljeflot, I., Hurlen, M. and Arnesen, H. (2004) Increased levels of soluble tissue factor during long-term treatment with warfarin in patients after an acute myocardial infarction. *Journal of Thrombosis and Haemostasis*, **2**, 726–730.

111 CARS-Investigators (1997) Randomised double-blind trial of fixed low-dose warfarin with aspirin after myocardial infarction. *The Lancet*, **350**, 389–396.

112 (The) OASIS Investigators (2001) Effects of long-term, moderate intensity oral anticoagulation in addition to aspirin in unstable angina. *Journal of the American College of Cardiology*, **37**, 475–485.

113 Fiore, L.D., Ezekowitz, M.D., Brophy, M.T. *et al.* (2002) Department of Veterans Affairs Cooperative Studies Program Clinical Trial comparing combined warfarin and aspirin with aspirin alone in survivors of acute myocardial infarction: primary results of the CHAMP study. *Circulation*, **105**, 557–563.

114 Huynh, T., Theroux, P., Bogaty, P. *et al.* (2001) Aspirin, warfarin, or the combination for secondary prevention in patients with acute coronary syndromes and prior coronary artery bypass surgery. *Circulation*, **103**, 3069–3074.

115 Brouwer, M.A., van den Bergh, P.J., Aengevaeren, W.R. *et al.* (2002) Aspirin plus coumarin versus aspirin alone in the prevention of reocclusion after fibrinolysis for acute myocardial infarction. Results of the APRICOT-2 Trial. *Circulation*, **106**, 659–665.

116 Smith, P., Arnesen, H. and Holme, I. (1990) The effect of warfarin on mortality and reinfarction after myocardial infarction. *The New England Journal of Medicine*, **323**, 147–152.

117 (The) Anticoagulation in the Secondary Prevention of Events in Coronary Thrombosis (ASPECT) Research Group (1994) Effect of long-term oral anticoagulant treatment on mortality and cardiovascular morbidity after myocardial infarction. *The Lancet*, **343**, 499–503.

118 Anand, S.S. and Yusuf, S. (1999) Oral anticoagulant therapy in patients with coronary artery disease – a metaanalysis. *The Journal of the American Medical Association*, **282**, 2058–2067.

119 Brouwer, M.A. and Verheugt, F.W. (2002) Oral anticoagulation for acute coronary syndromes. *Circulation*, **105**, 1270–1274.

120 Stys, T., Lawson, W.E., Smaldone, G.C. *et al.* (2000) Does aspirin attenuate the beneficial effects of angiotensin-converting enzyme inhibition in heart failure? *Archives of Internal Medicine*, **160**, 1409–1413.

121 Al-Khadra, A.S., Salem, D.N., Rand, W.M. *et al.* (1998) Antiplatelet agents and survival: a cohort analysis from the studies of left ventricular dysfunction (SOLVD) trial. *Journal of the American College of Cardiology*, **31**, 419–442.

122 Teo, K.K., Yusuf, S., Pfeffer, M. *et al.* (2002) Effect of long-term treatment with angiotensin-converting enzyme inhibitors in the presence or absence of aspirin: a systematic review. *The Lancet*, **360**, 1037–1043.

123 McAlister, F.A., Ghali, W.A., Gong, Y. *et al.* (2006) Aspirin use and outcomes in a community-based cohort of 7352 patients discharged after first hospitalization for heart failure. *Circulation*, **113**, 2572–2578.

124 Meune, C., Mahe, I., Mourad, J.J. *et al.* (2000) Interaction between angiotensin-converting enzyme inhibitors and aspirin: a review. *European Journal of Clinical Pharmacology*, **56**, 609–620.

125 Hennekens, C.H., Sacks, F.M., Tonkin, A. *et al.* (2004) Additive benefits of pravastatin and aspirin to decrease risks of cardiovascular disease: randomized and observational comparisons of secondary prevention trials and their meta-analyses. *Archives of Internal Medicine*, **164**, 40–44.

126 Schrör, K., Löbel, P. and Steinhagen-Thiessen, E. (1989) Simvastatin reduces platelet thromboxane formation and restores normal platelet sensitivity against prostacyclin in type IIa hypercholesterolemia. *Eicosanoids*, **2**, 39–45.

127 Catella-Lawson, F., Reilly, M.P., Kapoor, S.C. *et al.* (2001a) Cyclooxygenase inhibitors and the antiplatelet effects of aspirin. *The New England Journal of Medicine*, **345**, 1809–1817.

128 MacDonald, T.M. and Wei, L. (2003) Effect of ibuprofen on cardioprotective effect of aspirin. *The Lancet*, **361**, 573–574.

129 Hohlfeld, T., Zimmermann, N., Weber, A.-A. *et al.* (2008) Pyrazolinone analgesics prevent the antiplatelet effect of aspirin and preserve human platelet thromboxane biosynthesis. *Journal of Thrombosis and Haemostasis*, **6**, 166–173.

130 Grosser, T., Fries, S. and FitzGerald, G.A. (2006) Biological basis for the cardiovascular consequences of COX-2 inhibition: therapeutic challenges and opportunities. *The Journal of Clinical Investigation*, **116**, 4–15.

131 Bombardier, C., Laine, L., Reicin, A. *et al.* (2000) Comparison of upper gastrointestinal toxicity of rofecoxib and naproxen in patients with rheumatoid arthritis. VIGOR Study Group. *The New England Journal of Medicine*, **343**, 1520–1528.

132 Capone, M.L., Tacconelli, S., Sciuli, M.G. *et al.* (2004) Clinical pharmacology of platelet, monocyte and vascular cyclooxygenase inhibition by naproxen and low-dose aspirin in healthy subjects. *Circulation*, **109**, 1468–1471.

133 Egan, K.M., Lawson, J.A., Fries, S. *et al.* (2004) COX-2-derived prostacyclin confers atheroprotection on female mice. *Science*, **306**, 1954–1957.

134 Schrör, K., Mehta, P. and Mehta, J.L. (2005) Cardiovascular risk of selective cyclooxygenase-2-inhibitors. *Journal of Cardiovascular Pharmacology and Therapeutics*, **10**, 95–101.

135 Helin-Salminvaara, A., Virtanen, A., Vesalainen, R. *et al.* (2006) NSAID use and the risk of hospitalization for first myocardial infarction in the general population: a nationwide case-control study from Finland. *European Heart Journal*, **27**, 1657–1663.

136 Kearney, P.M., Baigent, C., Godwin, J. *et al.* (2006) Do selective cyclooxygenase-2 inhibitors and traditional non-steroidal anti-inflammatory drugs increase the risk of atherothrombosis? Meta-analysis of randomised trials. *British Medical Journal*, **332**, 1302–1308.

137 Martinez-González, J. and Badimon, L. (2007) Mechanisms underlying the cardiovascular effects of COX inhibition: benefits and risks. *Current Pharmaceutical Design*, **13**, 2215–2227.

138 Sanmuganathan, P.S., Gharhamani, P., Jackson, P.R. *et al.* (2001) Aspirin for primary prevention of coronary heart disease: safety and absolute benefit related to coronary risk derived from meta-analysis of randomised trials. *Heart*, **85**, 265–271.

139 Lauer, M.S. (2002) Clinical practice. Aspirin for primary prevention of coronary events. *The New England Journal of Medicine*, **346**, 1468–1474.

140 Hayden, M., Pignone, M., Phillips, C. *et al.* (2002) Aspirin for the primary prevention of cardiovascular events: a summary of the evidence for the U.S. Preventive Services Task Force. *Annals of Internal Medicine*, **136**, 161–172.

141 Patrono, C., Bachmann, F., Baigent, C. *et al.* (2004) Expert Consensus Document on the Use of Antiplatelet Agents. The Task Force on the use of antiplatelet agents in patients with atherosclerotic cardiovascular disease of the European Society of Cardiology. *European Heart Journal*, **25**, 166–181.

142 Braunwald, E., Antman, E.M., Beasley, J.W. *et al.* (2002) ACC/AHA 2002 guideline update for the management of patients with unstable angina and non-ST-segment elevation myocardial infarction-summary article. A report of the American College of Cardiology/American Heart Association task force on practice guidelines (Committee on the Management of Patients with Unstable Angina). *Journal of the American College of Cardiology*, **40**, 1366–1374.

143 Hung, J. for the Medical Issues Committee of the National Heart Foundation of Australia (2003) Position statement. Aspirin for cardiovascular disease prevention. *Medical Journal of Australia*, **179**, 147–152.

144 Zijlstra, F., Ernst, N., de Boer, M.-J. *et al.* (2002) Influence of prehospital administration of aspirin and heparin on initial patency of the infarct-related artery in patients with acute ST-elevation myocardial infarction. *Journal of the American College of Cardiology*, **39**, 1733–1737.

145 Frilling, B., Schiele, R., Gitt, A.K. *et al.* (2001) Characterization and clinical course of patients not receiving aspirin for acute myocardial infarction: results from the MITRA and MIR studies. *American Heart Journal*, **141**, 200–205.

146 Spencer, F.A., Santopinto, J.J., Gore, J.M. *et al.* (2002) Impact of aspirin on presentation and hospital outcomes in patients with acute coronary syndromes (The Global Registry of acute Coronary Events [GRACE]). *The American Journal of Cardiology*, **90**, 1056–1061.

147 Silber, S., Albertsson, P., Avilés, F.F. *et al.* (2005) Guidelines for the percutaneous coronary interventions. The Task Force for Percutaneous Coronary Interventions of the European Society of Cardiology. *European Heart Journal*, **26**, 804–847.

4.1.2 Cerebrovascular Diseases

Etiology Cerebral ischemia is the consequence of impaired cerebral blood perfusion due to obstructions of the cerebral circulation. These obstructions may result from atherosclerotic narrowing of cerebral blood vessels, including plaque formation in carotid arteries. Embolic occlusions are caused by plaque rupture or thrombi from other sources, most notably cardiogenic thrombi in atrial fibrillation. Consequences are cerebral dysfunctions whose kind and severity are determined by the location and size of the obstruction as well as the duration of ischemia. Clinically, cerebral ischemia presents with transient ischemic attacks (TIAs), "minor," or major stroke.

Epidemiology and Types of Stroke Major stroke, if not resolved, is the most disabling disease with individual and social consequences that are much more significant than critical ischemia of the heart or lower extremities. The annual stroke rate increases markedly not only with increasing age but also with increasing numbers of risk factors, most notably hypertension and hypercholesterolemia. This is the reason for intense preventive measures, in particular, appropriate changes in lifestyle and medical treatment.

There are principally two different categories of stroke: intracranial bleeding (15% of strokes) and cerebral ischemia (85% of strokes). Both have a relationship with aspirin. Hemorrhagic stroke is an important aspirin-related side effect while the prevention of ischemic stroke is a major therapeutic goal of aspirin treatment. Importantly, ischemic stroke is also multifactorial. Most frequent reasons are large-artery atherosclerosis and atherothrombosis (20–30%), microatheroma and other small artery occlusive diseases (lacunar stroke) (20–25%), and cardiogenic embolism (15–20%). Atrial fibrillation is the most frequent cardiac reason of ischemic (embolic) cerebral infarctions and accounts for about 25% of strokes at the age of 75–84 years [148]. The remaining causes are of other (5%) or cryptogenic (30%) etiology [149].

The different etiologies of ischemic stroke – in contrast to the rather monocausal myocardial ischemia – largely determine the efficacy of treatment with antiplatelet compounds, including aspirin. Large artery atherothrombotic stroke is the only stroke subtype, which, as regard to its etiology, is comparable to myocardial infarction. However, there are important differences, which are also therapeutically relevant. Myocardial infarction usually arises from the rupture of an unstable atherosclerotic plaque. This exposes collagen to circulating platelets and eventually causes coronary artery occlusion by platelet-dependent plug formation. Stroke subsequent to macroangiopathy (carotid stenosis) is almost always caused by emboli originating from the stenosed artery. Very rarely does the occlusion of the carotid artery directly cause "hemodynamic" stroke.

The heterogeneity in the etiology of ischemic stroke is also reflected in the variable role of platelets for vessel occlusion and, consequently, the variable efficacy of antiplatelet drugs [150]. Low-dose aspirin can be effective only if platelet-dependent thromboxane formation is the critical event for the plaque formation and obstruction of the cerebral vasculature. In contrast to other circulation systems, human cerebral arteries are highly sensitive to thromboxane and serotonin, another vasoconstrictor [151], which is exclusively stored in platelet-dense granules and released during platelet secretion. Thus, the therapeutic benefit of aspirin in the prevention of cerebral ischemic insults might differ from that of the cardiac ischemia and there might also be gender [152] and racial [153] differences. In addition, the HOT study (Section 4.1.1) demonstrated that comedication of aspirin to antihypertensives protects from myocardial infarction but not from stroke. Because of this, the older age of stroke patients, and their frequent comorbidities as opposed to patients with myocardial ischemia, prospective randomized trials with cerebrovascular end points provide more relevant information than cardiovascular prevention studies in middle-aged individuals. However, there are only a few trials in the elderly (≥75 years) although these are the most stroke-prone population.

4.1.2.1 Thrombotic Risk and Mode of Aspirin Action

Platelet Hyperreactivity and Thromboxane A_2 Wu and Hoak [154] originally demonstrated platelet hyperreactivity in patients with cerebrovascular diseases (TIA, stroke), which was sensitive to oral aspirin treatment (Figure 4.10). Further studies indicated that platelets become activated during the passage of the cerebral circulation in TIA and stroke patients, which can be prevented by aspirin. This confirms an aspirin-sensitive mechanism of platelet activation in cerebral ischemia [155]. Another evidence for a role of aspirin-sensitive platelet activation is the occurrence of ischemic strokes as a "rebound" phenomenon in patients at elevated risk after aspirin withdrawal [156–158]. Finally, patients with cerebrovascular events are more likely to suffer another cerebrovascular event than a myocardial infarction. This points to different risk determinants between these two forms of atherothrombotic events.

Platelet hyperreactivity in cerebrovascular diseases is associated with elevated circulating levels of platelet activation markers [159–161], elevated platelet cytosolic Ca^{++}, and enhanced arachidonic acid metabolism in platelets [162]. All of these events can be largely normalized by treatment with aspirin at 100 mg/day [163]. However, this and many other studies in patients were experimental in nature and were not designed to provide information about the consequences of aspirin-sensitive platelet hyperreactivity for the clinical outcome, that is, primary or secondary prevention of stroke. Specifically, the question whether 100 mg/day aspirin is the maximum effective dose for cerebrovascular protection is not answered by this type of studies, and there is evidence to suggest that stronger antiplatelet treatment, for example, by combined use of aspirin and clopidogrel, might be more effective on circulating platelet activation markers

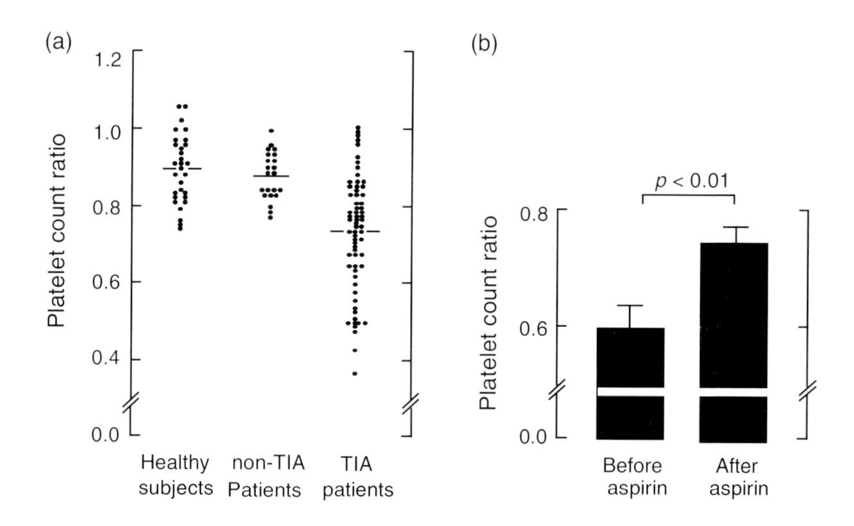

Figure 4.10 Platelet aggregate ratios in patients with TIA, healthy subjects and patients with TIA without thromboembolic disorders (a) and the effect of oral aspirin (+ dipyridamole) on platelet aggregates *ex vivo* (b). The platelet count ratio determines the proportion of nonactivated platelet in relation to total platelet count after maximum platelet stimulation (thrombin). Count ratio of 1 is equal to a complete inactivation, decrease indicates an increased proportion of activated platelets. Aspirin treatment reduces the number of activated platelets, as seen from an increased platelet count ratio (b) (modified after [154]).

than low-dose aspirin alone [164, 165]. However, it has been questioned whether more platelet inhibition is the answer for more effective (secondary) prevention or whether combination therapies, for example, with vasoprotective agents, are a more useful strategy for stroke prevention [166] (see below).

Variability in Aspirin Responses Platelets of patients with cerebral ischemia are not only hyperreactive but also less sensitive to inhibition by aspirin. Most notable is a large interindividual variability regarding doses of aspirin, necessary for antiplatelet effects [167–170]. This variability in stroke patients is seen in terms of both inhibition of platelet aggregation and thromboxane formation, suggesting an aspirin-insensitive component of platelet hyperreactivity [171, 172] (Figure 4.11). Several investigators have suggested that an incomplete inhibition of platelet function might be overcome by using combined antiplatelet treatment of aspirin and clopidogrel [164, 165]. However, Grau *et al.* [173] showed that the enhanced platelet activation in stroke patients was incompletely blocked by aspirin, clopidogrel, or the combination of both. Importantly, these studies as well as those by Grundmann *et al.* [174] were done in whole blood as opposed to the more conventional optical assay

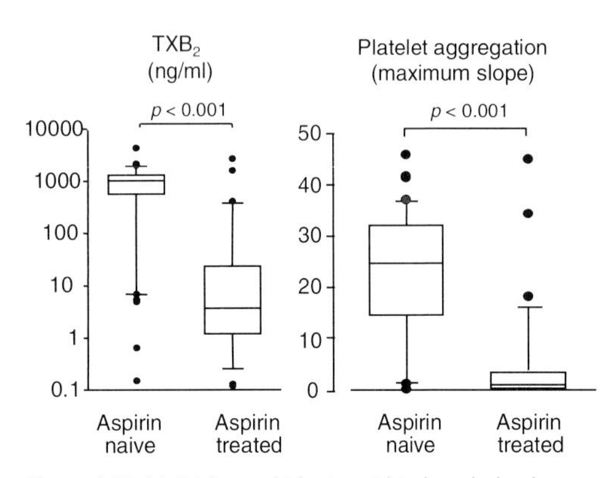

Figure 4.11 Variable arachidonic acid-induced platelet aggregation and thromboxane formation *ex vivo* in 90 patients with acute ischemic stroke. Patients were either "aspirin-naive" or treated with aspirin at antiplatelet doses prior to stroke [172].

techniques in platelet-rich plasma. A recent study in TIA and stroke patients, comparing three different technologies of measuring platelet function came to the conclusion that the results are largely test specific and the determination of their prognostic value for clinical outcome can only be estimated from appropriately sized randomized trials [175]. In addition, the subtype of stroke and its severity constitute another disease-related determinant of aspirin efficacy [159, 176].

Similar limitations apply to the measurement of antiplatelet actions of aspirin in terms of thromboxane (Section 4.1.6). The excretion of a thromboxane metabolite (11-DH-TXB_2) is elevated in TIA and stroke patients and significantly reduced by aspirin treatment [177]. However, the relation to outcome, that is, reoccurrence of cerebrovascular events, is uncertain. Another study on this issue [178], done in a subgroup of aspirin-treated patients in the HOPE [179] trial, found no increased risk for stroke, in contrast to an increased risk for myocardial infarction in patients less sensitive to aspirin in terms of 11-DH-TXB_2 excretion. The failure to show this association for stroke is unclear. The small number of strokes ($n = 80$) as well as the unbalanced groups and scarcely studied stroke etiology may be due to chance.

Aspirin Dosing The recommended aspirin dose for stroke prevention according to guideline recommendations is similar to that for cardiovascular prevention, that is, about 100–325 mg/day. The clinical efficacy might be enhanced by cotreatment with dipyridamole in extended-release formulations. However, the ideal antiplatelet agent for stroke prevention might still be elusive [180]. Some authors postulate that higher (i.e., 500 mg/day and more) doses of aspirin in (secondary) stroke prevention might be more effective than lower doses of aspirin [181–184]. This information is, however, derived from small experimental studies, frequently looking at surrogate parameters, and might not reflect sufficiently the real-life situation, which, in particular in (ischemic) stroke, requires an individualized therapy. Thus, researchers should rather look for the most effective treatment in well-defined high-risk populations in appropriately sized randomized trials and also consider the individual stroke etiology.

4.1.2.2 Clinical Trials: Primary Prevention

Stroke is a disease of the elderly. These individuals are frequently multimorbid and multidrug users. This will influence compliance and make the evaluation of the efficacy of one particular medication difficult. With increasing life expectancy in industrialized societies, both the percentage and the absolute number of elderly persons will increase, eventually resulting in an increased risk of stroke.

Epidemiological Trials In several large observational studies on primary prevention, self-selected use of aspirin was associated with higher rates of strokes [185–188]. This, at a first view surprising finding, is probably because of study-related factors such as ingestion habits and study protocols. For example, the observational cohort study of Kronmal *et al.* [185] in more than 5000 men and women above the age of 65 years suggested an increase in stroke (ischemic and hemorrhagic) after 4 years in "frequent" aspirin users. However, "frequent use" was defined as using aspirin for only 10 out of 14 days. Thus, many patients probably had taken aspirin not continuously on a long-term basis but probably rather episodically for reasons other than cardiovascular prevention. In treatment of pain and inflammation, higher doses of aspirin are required and aspirin might have been used together with traditional NSAIDs, which interfere with its antiplatelet action (Section 2.3.1). Alternatively, ischemic events might have occurred with the interruption of aspirin treatment [189]. Rebound phenomena, that is, the occurrence of TIA or stroke after cessation of aspirin use, have also been described [156–158]. Interestingly, regular use of aspirin, that is, 1–6 tablets per week in the prospective nurses' health study reduced not only cardiocoronary events (Section 4.1.1) but also markedly the risk of ischemic stroke due to thrombotic occlusions of large-to-medium-sized cerebral arteries (RR 0.50; 95% CI: 0.29–0.85; $p = 0.01$). Women who took 15 or more tablets per week had an increased risk of subarachnoid hemorrhage [190].

Prospective Trials in Apparently Healthy Individuals
The first large prospective randomized trials, that is, the US Physicians' Health Study [191] and the British Medical Doctors' Study [192] did not have stroke prevention as a separate study end point. Moreover, the study population consisted of middle-aged men with a very low risk of ischemic stroke. It is, therefore, not surprising that the overall stroke rate remained unchanged. There was, however, a trend for increased hemorrhagic strokes in both studies.

Somewhat different results were obtained in the recent Women's Health Study (see also Section 4.1.1).

A total of 39 876 initially healthy women (45 years of age or older) were randomly assigned to receive 100 mg aspirin each second day or placebo (vitamin E). Primary end points were cardiovascular death, nonfatal myocardial infarction, and nonfatal stroke, and the secondary end point was the individual risk in several subgroups. The total observation period was 10 years.

During follow-up, 477 major cardiovascular events occurred in the aspirin group as opposed to 522 in the placebo group. This was a nonsignificant reduction in events by 9% (RR 0.91; $p = 0.13$). Regarding the individual risks (secondary end point), there was a significant reduction in ischemic stroke in the aspirin group by 17% (RR: 0.83, $p = 0.04$) but no change in the risk of fatal or nonfatal myocardial infarctions (RR: 1.02; $p = 0.83$). However, subgroup analyses showed a significant reduction of risk of major cardiovascular events by 26%, ischemic stroke by 30%, and myocardial infarction among women 65 years of age and older. There was a no significant increase in hemorrhagic strokes.

The conclusion was that aspirin is effective in the primary prevention of stroke in women but does not provide protection from myocardial infarction [152].

The reasons for this surprising finding are not clear and the study as such has been discussed in more detail elsewhere in this book (Section 4.1.1). It, however, demonstrates that aspirin-sensitive risk factors that determine the incidence of stroke

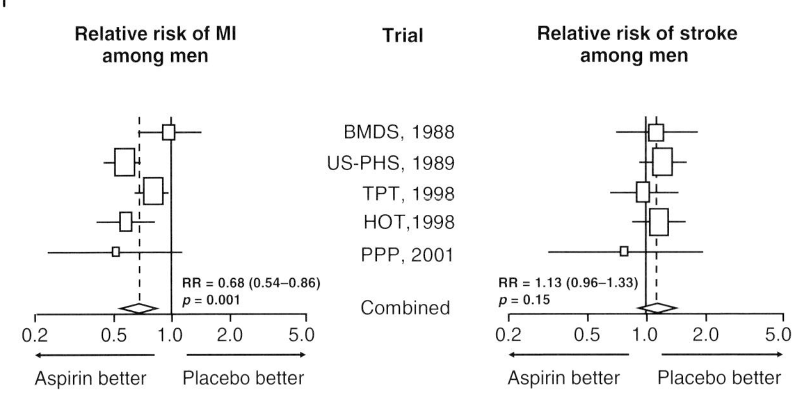

Figure 4.12 Aspirin in the primary prevention of stroke and MI in men [152].

in the primary prevention among apparently healthy women might be different from those for myocardial infarction and also different between men and women. An overview of published results from the large randomized primary prevention trials in men and women is summarized in Figures 4.12 and 4.13.

Hemorrhagic Stroke The USPHS and BMDS both showed an increased risk of hemorrhagic strokes at an unchanged number of ischemic strokes. In subjects with stable angina in the US Physicians' Health Study [191], there was a significant increase in overall stroke rates, specifically, hemorrhagic stroke, in the aspirin group [193]. This raises the question of the risk/benefit ratio in aspirin prophylaxis of stroke.

The risk of hemorrhagic stroke associated with aspirin treatment as opposed to its cardioprotective effects was analyzed in a meta-analysis of all 16 randomized controlled trials published until 1997. Only those studies were included where the treatment with aspirin (or control) was maintained at least for 1 month and the subtype of stroke indicated. The study population included 55 462 participants.

In this population, 108 hemorrhagic strokes were identified at a mean aspirin dosage of 273 mg/day and a mean duration of treatment of 37 months. There was an absolute *increase* in risk of hemorrhagic stroke by 12 events per 10 000 persons (95% CI: 5–20; $p < 0.001$) as opposed to an absolute risk *reduction* in myocardial

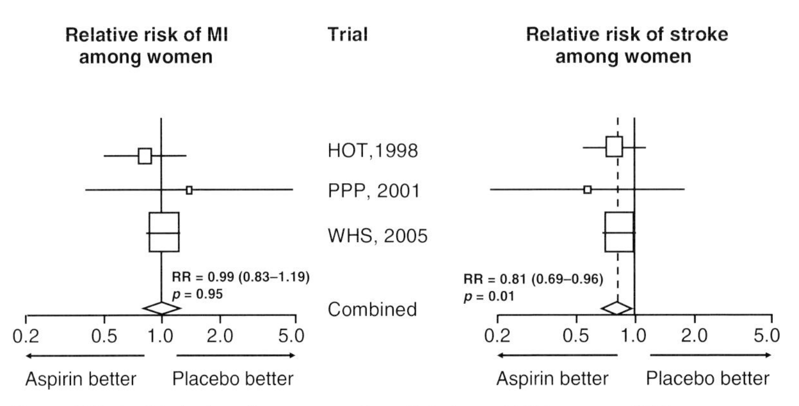

Figure 4.13 Aspirin in the primary prevention of stroke and MI in women [152].

infarctions by 137 events per 10 000 persons (95% CI: 107–167; $p < 0.001$) and in ischemic stroke of 39 events per 10 000 persons (95% CI: 17–61; $p < 0.001$).

The conclusion was that aspirin increases the risk of hemorrhagic stroke. However, the overall benefit of aspirin in the prevention of myocardial infarction and ischemic stroke by far outweighs its adverse effects on the risk of hemorrhagic stroke [194].

Taken together, current evidence suggests that aspirin has a significant but low preventive power in primary stroke prevention. This effect, however, is only significant in women.

4.1.2.3 Clinical Trials: Secondary Prevention

Meta-Analyses Up to 40% of patients with transient ischemic attacks or stroke will suffer new atherothrombotic events, in most cases stroke, within the next 5 years. A review of 11 randomized controlled trials of aspirin in about 10 000 patients with previous stroke or transient ischemic attack has shown that long-term aspirin reduced the total risk of combined vascular events (stroke, myocardial infarction, and vascular death) by 13%. This is equivalent to an absolute risk reduction by 1% per year or, in absolute terms, from 7 to 6% per year [195, 196]. Regarding an annual risk of stroke of 15–20% in high-risk populations, this indicates that most strokes even in high-risk populations are not prevented by aspirin prophylaxis.

The explanation might be the heterogeneity of strokes mentioned above. Cardiac emboli (about 15% of strokes) are the typical example of a platelet-poor (red) thrombus that might cause cerebral infarction but is not sensitive to aspirin. This subtype of stroke is the domain of oral anticoagulants (see below). However, patients with generalized atherosclerosis have not only an increased risk of stroke but also of myocardial infarctions and cardiac death. This additionally increases the protective power of aspirin prophylaxis.

Prospective Studies Early randomized placebo-controlled trials on aspirin prophylaxis in patients with TIA and cerebral infarction yielded different results. Only in one study there was a significant protective action of aspirin with reduced incidence of stroke and mortality by 13% [197]. In all studies, the aspirin doses were high (1.3–1.5 g/day) as were the dropout rates, in one study with 1.5 g aspirin/day 34% [198].

These trials had also a low number of patients, a low percentage of women and a low rate of (cerebral) events. All of these limitations might have caused statistical problems in detecting possible benefits of medical treatment. These weaknesses stimulated several large prospective, randomized multicenter trials. The first of them, comparing different doses of aspirin with placebo, was the UK-TIA trial; the second, comparing high and low-dose aspirin, was the Dutch TIA trial.

The UK-TIA trial included 2435 patients with transient ischemic attack or minor stroke. The patients were randomized to receive "blind" treatment with low-dose aspirin (300 mg/day), high-dose aspirin (600 mg bid), or placebo for a follow-up of 6 years.

Taking the two aspirin groups together, there was a significant reduction of strokes in comparison with placebo. However, there was no significant effect in either of the aspirin groups alone. This failure of aspirin was explained by a statistical type II error (too low patient numbers per group). There was a small, nonsignificant reduction of vascular events (death, nonlethal stroke, and nonfatal myocardial infarction) in the combined aspirin groups by 15%. However, significant dose-dependent differences existed with respect to bleeding: 600 mg aspirin bid caused 39 GI bleedings versus 25 bleedings at 300 mg aspirin and 9 bleedings with placebo. There was no difference in bleeding time that was measured in a subgroup of 120 patients [199], confirming the observation of several other studies that the antithrombotic effects of aspirin are not paralleled by an increase in bleeding time (Section 3.1.2).

The conclusion was that there is no significant reduction in major stroke or vascular death by aspirin in these high-risk, stroke-prone patients [200].

The Dutch TIA trial was also scheduled to compare two different aspirin doses: 30 mg/day versus 283 mg/day in a similar patient population. Inclusion criteria were previous TIA or minor stroke due to arterial thrombosis or thrombembolism (no atrial fibrillation!). The median observation period was 2.6 years.

The vascular death rate was reduced by 14.7% in the "low-dose" aspirin group and by 15.2% in the "high-dose" group. This difference was nonsignificant. Despite the significantly lower rate of minor bleedings with the lower aspirin dose, neither the number of major bleedings (40 versus 53) (nonsignificant trend in favor of 30 mg) nor GI intolerance (164 versus 179 patients) was reduced at low-dose aspirin.

It was concluded that 30 mg aspirin was not less effective than the 283 mg dose but had fewer adverse effects [201].

The end point stroke in each therapeutic arm in the Dutch TIA trial was rare, and the number of strokes was not different between the groups. Owing to the low number of strokes, the study had low statistical power, and more important, it was impossible to judge whether either dose was effective at all because there was no placebo arm. The reason for this study design were (positive) intermediate data of the aspirin group in the UK-TIA Study Group [202], suggesting positive effects of 300 mg aspirin versus placebo that were, however, not confirmed after the trial was completed (see above). The Dutch TIA study is often taken as an evidence for the efficacy of 30 mg/day aspirin in stroke prevention. This has *not* been shown and, in fact, it has never been demonstrated until now that daily doses of 30 mg aspirin are effective at all in primary or secondary prevention of stroke or myocardial infarction in a randomized, controlled trial (Section 4.1.1).

The Swedish Aspirin Low-Dose Trial (SALT) was the first to study low-dose aspirin (75 mg/day) in a prospective, placebo-controlled, randomized trial in patients with cerebrovascular disease [203]. This was also the first study to clearly demonstrate the efficacy of low-dose aspirin for stroke prevention in high-risk patients.

A total of 1360 patients (65% men, mean age 67 years) were randomized 1–4 months after TIA, "minor" stroke, or retina thrombosis and received 75 mg/day film-coated aspirin or placebo. Patients with a cardiac source of emboli, including those with atrial fibrillation or recent (within 3 months) myocardial infarction, were excluded. The observation period was 30 months. Primary end points were stroke or death from any cause, secondary end points other vascular events.

Aspirin significantly reduced the incidence of primary end points by 18% (RR: 0.82, 95% CI: 0.67–0.99; $p = 0.02$). Interestingly, the prevention of myocardial infarctions (one secondary end point) by aspirin was about twice as high: 36% (!). The rate of side effects was 22% in the aspirin and 18% in the placebo group. However, all five lethal hemorrhagic infarctions occurred in the aspirin group ($p = 0.03$) as well as 9 out of 13 severe GI bleedings, that required to stop the treatment. The compliance rate (pill count) was >90% in 99% of the patients. However, in terms of thromboxane levels, about 5% of patients in the aspirin and 10% in the placebo group had plasma (serum?) thromboxane levels of >100 ng/ml (!).

The conclusion was that 75 mg/day aspirin significantly reduces the risk of stroke and death in patients with preexisting cerebrovascular ischemia. However, it was also stated that a (total) risk reduction of vascular events by 17–25% is substantial but far from being the ultimate therapy since a large proportion of subsequent events were not prevented. It was not excluded that higher doses may be more efficient than the low dose used in this study [203].

The two probably largest trials on recurrent stroke prevention in patients with acute stroke were the "Chinese Acute Stroke Trial" [204] and the "International Stroke Trial" [205], both including about 20 000 patients. These trials studied the question whether aspirin protects from early recurrent stroke in real life, that is, a not particularly selected patient population if it is started immediately after a suspected stroke. In addition, the possibly increased risk of bleedings should be estimated.

The design of the two trials was similar. However, in CAST the control group was given placebo, whereas in IST they were not, that is, this was an open trial. A prospectively planned meta-analysis of these two studies showed that aspirin caused a significant reduction in recurrent ischemic stroke during the scheduled treatment period: 4 weeks in CAST, 2 weeks in IST: 1.6% versus 2.3% ($p < 0.00001$) and a modest reduction in death without further stroke: 5.0% versus 5.4% ($2p$ 0.05). There was an increased number of hemorrhagic strokes: 1.0% versus 0.8% ($2p = 0.07$). There was no heterogeneity between the 28 subgroups of patients studied.

The conclusion was that early aspirin is of benefit for a wide range of patients and its prompt use should be routinely considered for all patients with suspected acute ischemic stroke, mainly to reduce the risk of early recurrence [206].

These and another prospective cohort study [207] suggest beneficial effects of aspirin in the secondary prevention of stroke.

4.1.2.4 **Aspirin and Other Drugs**
Several attempts have been made to increase the efficacy of antiplatelet treatment for the prevention of stroke. One was the combination of aspirin with dipyridamole, another the combination of aspirin with antiplatelet drugs with a different mode of action, most notably ADP-receptor antagonists such as clopidogrel (Section 2.3.1). Finally, in patients with atrial fibrillation or other forms of primary noncerebral thromboembolism as a cause of stroke, oral anticoagulants, alone or in combination with aspirin, are the alternatives.

Dipyridamole Dipyridamole is a vasodilator and weak inhibitor of platelet aggregation: both actions are probably due to accumulation of cyclic GMP after the inhibition of phosphodiesterase V and subsequent activation of the NO/cGMP pathway [208]. Because of this mode of action, dipyridamole should synergize with aspirin with respect to the inhibition of platelet aggregation. The "European Stroke Prevention Study" [209] found a significant reduction of strokes by 33% for the combined use of aspirin (330 mg tid) plus dipyridamole. However, there was no aspirin-only treatment arm. Thus, no evaluation of the effects of single drugs was possible. In addition, a significant proportion of patients stopped treatment because of side effects, in particular headache, probably due to the vasodilator actions of dipyridamole [210].

A follow-up trial addressing in detail the issue of combined use of aspirin plus dipyridamole was the European Stroke Prevention Study-2 (ESPS-2) [211]. In this study, "lowest" dose aspirin was compared with placebo and dipyridamole both alone and in combination. Dipyridamole was applied in a new extended-release formulation, which allowed much higher total doses because of the slow release of the active compound.

A total of 6602 stroke/TIA patients were treated with aspirin (25 mg bid), extended-release dipyridamole (200 mg bid), the combination of both or placebo over 2 years in a double-blind, randomized manner. Primary end points were stroke and death, secondary end points TIA and other vascular events.

Strokes and vascular events after first ever stroke or TIA were significantly reduced by aspirin alone (18%), dipyridamole alone (16%), and the combination of the two (37%). All of these changes were highly significant versus placebo. There was no significant risk reduction for death alone, myocardial infarction alone or other vascular events alone. Side effects were bleeding (aspirin groups) and GI intolerance and headache (dipyridamole groups). The incidence of orthostatic hypotension was not reported.

The conclusion was that aspirin and dipyridamole at the dose and formulation used are equieffective in the secondary prevention of stroke and TIA in these patients. The combination of aspirin and dipyridamole acts as an additive and is significantly more effective than either treatment alone [211].

This study was the first to report a therapeutic benefit (as well as an increased bleeding tendency) for aspirin at the extremely low dose of 25 mg bid and extended-release dipyridamole alone as well as an additive effect for the combination. These were important findings; however, the therapeutic consequences for clinical reality study are still under discussion [180]. The ESPS-2 trial differed from earlier studies, using aspirin plus dipyridamole not only because of its larger size but also because of the relatively high dose of (extended-release) dipyridamole (200 mg versus 75 mg), resulting in a dipyridamole/aspirin ratio of 8 : 1. This excess of dipyridamole was previously reported in experimental settings to have an additive effect on human platelet aggregation in healthy volunteers [212]. The dose of aspirin (25 mg bid) used in ESPS-2 was extremely low and had not been shown before to be effective. However, it was effective in this study, although it cannot be excluded that higher doses of aspirin might have been even more effective. According to the Antithrombotic Trialists' Collaboration [196], <75 mg/day aspirin appears to be less effective than 75 mg/day or more and there was no evidence from any earlier trial that aspirin plus dipyridamole was superior to aspirin alone in the secondary prevention of vascular events [213]. In addition, aspirin alone at higher doses might have had the same effect as the combination had on stroke prevention but additional benefits for cardioprotection [214]. The dipyridamole component did not increase the risk of myocardial infarctions in this trial [215], although the number of patients with coronary arterial disease (35%) and peripheral arterial disease (PAD) (22%) was substantial. However, the benefit of aspirin for a simultaneous cardiocoronary protection may be reduced. On the contrary, significantly more patients in the two dipyridamole groups (29%) finished the study prematurely compared to the aspirin or placebo group (22%), probably because there were more GI events and headaches.

The results of the ESPS-2 study were for a long time the only results on the combination of low-dose aspirin with dipyridamole. Recently, the European/Australasian Stroke Prevention in Reversible Ischaemia Trial (ESPRIT) study was published, which was addressed to reconfirm the ESPS-2 data in a similar, randomized study population but with an open design.

Aspirin (30–325 mg/day, median dose: 75 mg/day) was given alone (1376 patients) or in combination with 200 mg bid dipyridamole (1363 patients), mostly (83%) in an extended-release formulation, to patients with a previous stroke of arterial origin. Primary end point was the composite of vascular death, nonfatal myocardial infarction, and stroke or major bleeding complication. The average duration of treatment was 3.5 years.

At least one primary outcome event was obtained in 13% of the patients with the combined treatment as opposed to 16% in the group who was treated with aspirin alone (RR: 0.80; 95% CI: 0.66–0.98) with the intention to treat analysis. This was equivalent to a reduction of the absolute risk by 1% per year. There were no differences between the two treatment groups in cerebral or cardiac events as single end points and no differences in the occurrence of severe bleedings. However, more than one-third of patients (34%) in the combined group stopped medication because of side effects, mainly (26%) headache, while only 13% of the aspirin group interrupted treatment, mostly because of medical reasons.

The conclusion was that this study – combined with the result of previous trials – provides sufficient evidence to prefer the combination of aspirin and dipyridamole over aspirin alone as antithrombotic treatment for the secondary prevention after cerebral ischemia of arterial origin [216].

The study was subjected to several criticisms and comments [180]. This included the open design and the change in the study protocol from a three-arm (including anticoagulation therapy) to a two-arm trial but also the fact that one-third of the patients discontinued treatment because of side effects of dipyridamole (similar to ESPS-2). The last effect was a particular issue of concern because of difficulties to maintain the combined treatment on long term in real life. Because of different aspirin dosing, the theoretically postulated optimum ratio of dipyridamole/aspirin (8 : 1) was not obtained in all patients and 17% of patients did not use the extended-release formulation of dipyridamole. It was also surprising and remained unexplained by the authors why the dipyridamole-related almost two- to threefold (34% versus 13%) higher dropout rate during treatment did not result in any difference between the "intention-to-treat" and "on-treatment" analysis of the study [217]. One possible explanation could have been that patients in the "intention-to-treat" group took an alternative antiplatelet drug, such as clopidogrel. It was also questioned whether the 1% absolute risk benefit of the combination over aspirin alone would justify a 40-times difference in costs [180].

The issue of secondary prevention of stroke by the combination of aspirin/extended-release dipyridamole versus clopidogrel was also studied in the most recent Prevention Regimen for Effectively Avoiding Second Strokes (PRoFESS) trial (2008).

Patients with a previous (< 120 days) ischemic stroke were randomized to aspirin (25 mg) plus extended-release dipyridamole (ERDP) (200 mg) bid or clopidogrel (75 mg daily). Primary endpoint was first recurrence of stroke of any type. Secondary endpoint was a composite of stroke, myocardial infarction or vascular death. Sequential statistical testing of noninferiority (margin of 1.075), followed by superiority testing was planned but not performed because of negative data.

A total of 20,332 patients were followed for a mean of 2.5 years. Recurrent stroke occurred in 9.0% of patients receiving aspirin + ERDP and in 8.8% of patients receiving clopidogrel (RR: 1.01; 95% CI: 0.92–1.11). Comparable results were obtained with the secondary endpoint. Regarding safety, 29% of patients in the aspirin + ERDP group but only 23% in the clopidogrel group discontinued treatment prematurely, frequently because of headache: 5.9% vs. 0.9%. These patients were also less compliant: 70% taking the study medication more than 75% of time. There were significantly more major hemorrhagic events (including hemorrhagic strokes) in patients receiving aspirin + ERDP: 4.1% vs. 3.6% for all events, 1.4% vs. 1.0% for intracranial hemorrhage ($P = 0.006$).

The conclusion was that the study did not meet the predefined criteria for inferiority, despite of similar rates of stroke in both treatment groups and, therefore, does not allow making a claim for noninferiority [218].

According to a recent meta-analysis, the combination of aspirin and dipyridamole is considered more effective than aspirin alone in the secondary prevention of cardiovascular events among patients with minor stroke and TIA [219]. Nevertheless, questions remain as outlined in an editorial to this publication [180].

Clopidogrel Thienopyridines, such as clopidogrel, appear to be slightly more effective than aspirin in patients at high vascular risk (CAPRIE trial; [220]). According to post hoc analyses, there is a nonsignificant trend toward a reduction in ischemic stroke [221]. More important than these marginal, if any, differences between aspirin and clopidogrel, are two other aspects: the availability of an oral alternative for aspirin in aspirin-intolerant individuals [222] and the option of a combined use of two antiplatelet compounds with a different mode of action and an expected additive effect.

No additional therapeutic benefits (but an increase in severe bleedings) were seen after combined treatment with aspirin and clopidogrel in the MATCH trial [223]. However, whether the data of this particular study can be transferred to all stroke patients is questionable [224]. Because of the different vascular risk profile in stroke versus myocardial infarction patients, data from patients with myocardial ischemia, for example, the CURE trial, should not be extrapolated to stroke patients [225]. CHARISMA has shown that comedication of clopidogrel to standard-dose aspirin does not reduce the number of ischemic vascular events, including stroke, in asymptomatic patients at elevated vascular risk, but it does increase the risk of bleeding and mortality. In contrast, some beneficial effect was seen in the secondary prevention of cardiovascular events (Section 4.1.1). Whether more advanced thienopyridines, such as prasugrel, will exhibit an improved benefit/risk ratio remains to be shown. The principal question that is raised by these data still remains: is more platelet inhibition the answer for improving the relatively low efficacy of antiplatelet drugs in stroke prevention [166]. The available data of the PROFESS trial do not suggest any benefit of dipyridamole plus aspirin versus clopidogrel in secondary stroke prevention and the study as a whole was negative because it failed to meet the prespecified noninferiority criteria.

Anticoagulants Another option to enhance the antithrombotic effect of aspirin in stroke prevention is the combined use with oral anticoagulants. Oral anticoagulants significantly decrease the risk of strokes and cardiovascular events in patients with nonvalvular chronic or paroxysmal atrial fibrillation and cardiogenic embolism. These compounds are significantly more effective than aspirin in this group of patients. However, they increase the risk of major bleedings. Thus, the individual benefit/risk ratio needs to be determined, for example, from available benefit/risk tables [148]; www.medscape.com). It has also to be considered that elderly patients are frequently multidrug users [226]. Thus, a tight control of target INR (2–3) is essential.

The incidence of recurrent strokes in atrial fibrillation is reduced from 12 to 11% per year by aspirin, while oral anticoagulants reduce the risk in this patient population to 4% [227]. Another meta-analysis showed that anticoagulation in patients with nonvalvular atrial fibrillation reduced the risk of stroke by 62% as opposed to only 22% with aspirin [228]. These and other data demonstrate a therapeutic superiority of anticoagulants versus antiplatelet drugs in patients with nonrheumatic atrial fibrillation [229]. However, so far, there are no prospective randomized trials on aspirin and oral anticoagulants in the elderly with atrial fibrillation. The recently started Birmingham Atrial Fibrillation Treatment of the Aged study (BAFTA; [230]) has addressed this issue and will compare warfarin (INR 2–3) with 75 mg/day aspirin in elderly patients with nonvalvular atrial fibrillation.

However, warfarin and other oral anticoagulants are not the first choice of treatment in other risk groups of stroke, including non-cardioembolic stroke and symptomatic intracranial arterial stenosis. The reason is the high risk of severe bleedings. Two studies [231, 232] have addressed this issue in stroke patients.

A total of 2206 nonselected patients (mean age 63 years, 41% women) who had suffered an ischemic stroke within the previous 30 days were randomized to aspirin (325 mg/day) or warfarin (INR 1.4–2.8) in a double-blind manner. The primary end point was recurrent ischemic stroke or death from any cause within 2 years.

The primary end point was reached by 17.8% of patients in the warfarin group and 16.0% in the aspirin group. This was not different ($p = 0.25$). Major hemorrhages occurred at a rate of 2.2 % per year in the warfarin group but only in 1.5% per year in the aspirin group ($p = 0.10$). The corresponding values for minor hemorrhages were 12.9% for aspirin but 20.8% for warfarin ($p < 0.001$).

The conclusion was that both warfarin and aspirin are reasonable therapeutic alternatives in the prevention of recurrent ischemic stroke. However, warfarin is associated with an increased risk of bleeding [233].

Another prospective trial was done in patients with transient ischemic attacks or nondisabling stroke. Inclusion criterion was an angiographically verified 50–99% stenosis of a major intracranial artery. Atrial fibrillation was an exclusion criterion. Patients were randomized to warfarin (INR 2–3) or aspirin (1300 mg/day) in a double-blind fashion. The primary end point was ischemic stroke, cerebral hemorrhage, or death from vascular causes other than stroke.

After 569 patients had undergone randomization, enrolment was stopped prematurely because of safety concerns in the warfarin group. After a mean follow-up of 1.8 years, there were significantly more severe adverse events including death in the warfarin group (9.7%) as compared to aspirin (4.3%) ($p = 0.02$). The numbers for major hemorrhage were 8.3% versus 3.2% ($p = 0.01$) and for myocardial infarction or sudden death: 7.3% versus 2.9% ($p = 0.02$). The rate of death from vascular sources was 5.9% versus 3.2% ($p = 0.16$) and from nonvascular sources 3.8% versus 1.1% ($p = 0.05$). There was no difference in efficacy. A primary end point occurred in 22.1% of patients in the aspirin group and 21.8% of patients in the warfarin group ($p = 0.83$).

The conclusion was that warfarin was associated with significantly higher rates of adverse events but produced no benefit above aspirin. Aspirin should be used in preference of warfarin for patients with intracranial arterial stenoses. In addition, the higher dose of aspirin used here may also protect patients from myocardial infarction and sudden death because of decreasing aspirin resistance and inhibiting the inflammatory component of atherosclerosis [232].

Therefore, warfarin is not the first-line treatment for stroke prevention in high-risk patients, except patients with atrial fibrillation [231, 234].

4.1.2.5 **Actual Situation**

In addition to reduction in blood pressure and lipid-lowering therapy, if appropriate, antiplatelet therapy is efficient in stroke prevention in high-risk populations. However, current prophylaxis by antiplatelet treatment does not prevent 80% of (ischemic) cerebral infarctions and even might cause (hemorrhagic) infarctions. Thus, high-dose aspirin, though possibly retarding the progression of atherosclerosis [182], might exhibit an unfavorable benefit/risk profile. The clinical problem is the complex etiology of (ischemic) stroke, containing subgroups that are less sensitive to antiplatelet treatment than others. There is a need for improved antithrombotic prevention and therapy of cerebral ischemia. Upcoming low molecular weight and more selective anticoagulants such as antithrombins or FXa inhibitors and thrombin receptor antagonists are currently in preclinical trials and might – alone or in combination with aspirin – open the door for improved therapeutic options in the future.

Primary Prevention Aspirin is recommended for the prevention of a first stroke in women whose risk is sufficiently high for the benefits to outweigh the risks but not for men. The use of aspirin for cardiovascular (including but not specific for stroke) prophylaxis in primary prevention is indicated at a 6–10% or higher risk of events per 10 years [235].

Secondary Prevention Although the efficacy is not very high, about 15–20% protection, the simultaneous protection from (also elevated) risk of myocardial infarction and sudden death is an additional benefit of aspirin, which disappears by combined treatment with dipyridamole. Comedication of dipyridamole with aspirin has been shown to be effective in secondary prevention of ischemic strokes. However, there was no positive effect on cardiocoronary prevention and no effect on mortality but an increased number of dipyridamole-related side effects in the combined treatment groups (headache), which resulted in patient-driven interruption of use in about each third patient (ESPS-2; [216]). In the case of aspirin intolerance or inefficacy, clopidogrel is an alternative [236]. Dual antiplatelet treatment with aspirin and clopidogrel for secondary stroke prevention has not been shown to be more effective than aspirin alone and could result in greater bleeding. Additional vascular protection by compounds such as dipyridamole or cilostazol might be useful [166, 237] and might open the door for a more disease-oriented therapy than inhibition of platelet hyperreactivity that is effective only in a minority of patients and frequently less sensitive to aspirin. Thromboxane antagonists might also have beneficial effects, including retardation of the progression of atherosclerosis, but the evidence is only based on small trials [238, 239]. Oral anticoagulants of the coumarin type at medium INR (2–3) are the treatment of choice in atrial fibrillation but not in other risk categories of cerebral ischemia.

Summary

Cerebrovascular ischemia, clinically appearing as transient ischemic attack or stroke, is a multicausal disease. About two-thirds of strokes are of vascular origin and therefore potentially sensitive to aspirin. Atrial fibrillation as a major reason for stroke in the elderly is less sensitive to aspirin and currently the domain of oral anticoagulants.

As with other manifestations of general atherothrombosis, platelets are hyperreactive in patients with TIA or stroke. Platelets of these patients also exhibit a large interindividual variability, also with respect to inhibition by aspirin and other antiplatelet agents. Whether an increased dose can overcome this problem is difficult to predict, also because there are not many studies with stroke as single study end point. Clopidogrel is the alternative in case of aspirin intolerance. Combined aspirin/extended-release dipyridamole appears to be slightly more effective (1% absolute risk reduction) than aspirin alone in stroke prevention but has also more side effects.

Overall, a reduction of ischemic strokes by 15–20% with about 5% increase of bleedings (cerebral and gastrointestinal) appears acceptable but may not be the final answer for stroke prevention by antiplatelet/antithrombotic agents. The future might be combined antiplatelet/antithrombotic plus vasoprotective treatment.

References

148 Dhond, A.J., Michelena, H.I. and Ezekowitz, M.D. (2003) Anticoagulation in the elderly. *The American Journal of Geriatric Cardiology*, **12**, 243–250.

149 Albers, G.W., Amarenco, P., Easton, J.D. *et al.* (2004) Antithrombotic and thrombolytic therapy for ischemic stroke. *Chest*, **126**, 483S–512S.

150 Chimowitz, M.I., Furlan, A.J., Navak, S. *et al.* (1990) Mechanism of stroke in patients taking aspirin. *Neurology*, **40**, 1682–1685.

151 Schrör, K. and Braun, M. (1990) Platelets as a source of vasoactive mediators. *Stroke*, **1** (Suppl. IV), 32–35.

152 Ridker, P.M., Cook, N.R., Lee, I.M. *et al.* (2005) A randomized trial of low-dose aspirin in the primary prevention of cardiovascular disease in women. *The New England Journal of Medicine*, **352**, 1293–1304.

153 Shinohara, Y. (2006) Regional differences in incidence and management of stroke – is there any difference between Western and Japanese guidelines on antiplatelet therapy? *Cerebrovascular Diseases*, **21** (Suppl. 1), 17–24.

154 Wu, K.K. and Hoak, J.C. (1975) Increased platelet aggregates in patients with transient ischemic attacks. *Stroke*, **6**, 521–524.

155 Kusunoki, M., Kimura, K., Nagatsuka, K. *et al.* (1982) Platelet hyperaggregability in ischemic cerebrovascular disease and effects of aspirin. *Thrombosis and Haemostasis*, **48**, 117–119.

156 Bachman, D.S. (2002) Discontinuing chronic aspirin therapy: another risk factor for stroke. *Annals of Neurology*, **51**, 137–138.

157 Sibon, I. and Orgogozo, J.M. (2004) Antiplatelet drug discontinuation is a risk factor for ischemic stroke. *Neurology*, **62**, 1187–1189.

158 Maulaz, A.B., Bezerra, D.C., Michel, P.O. *et al.* (2005) Effect of discontinuing aspirin therapy on the risk of brain ischemic stroke. *Archives of Neurology*, **62**, 1217–1220.

159 Shah, A.B., Beamer, N. and Coull, B.M. (1985) Enhanced *in vivo* platelet activation in subtypes of ischemic stroke. *Stroke*, **16**, 643–647.

160 Joseph, R., D'Andrea, G., Oster, S.B. *et al.* (1989) Whole blood platelet function in acute ischemic stroke. Importance of dense body secretion and effects of antithrombotic agents. *Stroke*, **20**, 38–44.

161 Grau, A., Ruf, A., Vogt, A. *et al.* (1998) Increased fraction of circulating activated platelets in acute and previous cerebrovascular ischemia. *Thrombosis and Haemostasis*, **80**, 298–301.

162 Joseph, R., Welch, K.M., Grunfeld, S. *et al.* (1988) Baseline and activated platelet cytoplasmic ionized calcium in acute ischemic stroke – effect of aspirin. *Stroke*, **19**, 1234–1238.

163 Lee, T.K., Chen, Y.C., Lien, I.N. *et al.* (1988) Inhibitory effect of acetylsalicylic acid on platelet function in patients with completed stroke or reversible ischemic neurological deficit. *Stroke*, **19**, 566–570.

164 Yip, H.-K., Chen, S.-S., Liu, J.S. *et al.* (2004) Serial changes in platelet activation in patients after ischemic stroke. *Stroke*, **35**, 1683–1687.

165 Serebruany, V.L., Malinin, A., Ziai, W. *et al.* (2005) Effects of clopidogrel and aspirin in combination versus aspirin alone on platelet activation and major receptor expression in patients after recent ischemic stroke. *Stroke*, **36**, 2289–2292.

166 Liao, J.K. (2007) Secondary prevention of stroke and transient ischemic attack. Is more platelet inhibition the answer? *Circulation*, **115**, 1615–1621.

167 Grotemeyer, K.-H. (1991a) Effects of acetylsalicylic acid in stroke patients. Evidence of nonresponders in a subpopulation of treated patients. *Thrombosis Research*, **63**, 587–593.

168 Grotemeyer, K.-H., Scharafinski, H.-W. and Husstedt, I.-W. (1993) Two-year follow up of ASA responder and ASA non-responder. A pilot study including 180 post-stroke patients. *Thrombosis Research*, **71**, 397–403.

169 Helgason, C.M., Tortorice, K.L., Winkler, S.R. *et al.* (1993) Aspirin response and failure in cerebral infarction. *Stroke*, **24**, 345–350.

170 Helgason, C.M., Bolin, K.M., Hoff, J.A. *et al.* (1994) Development of ASA resistance in persons with previous ischemic stroke. *Stroke*, **25**, 2331–2336.

171 Dippel, D.W.J., van Kooten, F., Leebeck, F.W.G. *et al.* (2004) What is the lowest dose of aspirin for maximum suppression of *in vivo* thromboxane production after a transient ischemic attack or ischemic stroke? *Cerebrovascular Diseases*, **17**, 296–302.

172 Hohlfeld, T., Weber, A.-A., Junghans, U. *et al.* (2007) Variable platelet response to aspirin in patients with ischemic stroke. *Cerebrovascular Diseases*, **24**, 43–50.

173 Grau, A.J., Reiners, S., Lichy, C. *et al.* (2003) Platelet function under aspirin, clopidogrel, and both after

ischemic stroke. A case-crossover study. *Stroke*, **34**, 849–855.

174 Grundmann, K., Jaschonek, K., Kleine, B. *et al.* (2003) ASA non-responder status in patients with recurrent cerebral ischemic attacks. *Journal of Neurology*, **250**, 63–66.

175 Harrison, P., Segal, H., Blasbery, K. *et al.* (2005) Screening for aspirin responsiveness after transient ischemic attack and stroke. Comparison of 2 point-of-care platelet function tests with optical aggregometry. *Stroke*, **36**, 1001–1005.

176 Englyst, N.A., Horsfield, G., Kwan, J.,*et al.* (2008) Aspirin resistance is more common in lacunar strokes than embolic strokes and is related to stroke severity. *Journal of Cerebral Blood Flow & Metabolism*, **28**: 1196–1203.

177 Koudstaal, P.J., Ciabattoni, G., van Gijn, J. *et al.* (1993) Increased thromboxane biosynthesis in patients with acute cerebral ischemia. *Stroke*, **24**, 219–223.

178 Eikelboom, J.W., Hirsh, J., Weitz, J.I. *et al.* (2002) ASA-resistant thromboxane biosynthesis and the risk of myocardial infarction, stroke, or cardiovascular death in patients at high risk for cardiovascular events. *Circulation*, **105**, 1650–1655.

179 (The) Heart Outcomes Prevention Evaluation Study Investigators (2000) Effects of an angiotensin-converting enzyme inhibitor, ramipril, on cardiovascular events in high-risk patients. *The New England Journal of Medicine*, **342**, 145–153.

180 Norris, J.W. (2008) The ideal antiplatelet drug for stroke prevention – still elusive. *Stroke*, **39**, 1076–1077.

181 Tohgi, H., Konno, S., Tamura, K. *et al.* (1992) The effects of low-to-high doses of aspirin on platelet aggregability and metabolites of thromboxane A_2 and prostacyclin. *Stroke*, **23**, 1400–1403.

182 Ranke, C., Hecker, H., Creutzig, A. *et al.* (1993) Dose-dependent effect of aspirin on carotid atherosclerosis. *Circulation*, **87**, 1873–1879.

183 Bornstein, N.M., Karepov, V.G., Aronovich, B.D. *et al.* (1994) Failure of aspirin treatment after stroke. *Stroke*, **25**, 275–277.

184 Barnett, H.J.M., Kaste, M., Meldrum, H. *et al.* (1996) ASA dose in stroke prevention. Beautiful hypotheses slain by ugly facts. *Stroke*, **27**, 588–592.

185 Kronmal, R.A., Hart, R.G., Manolio, T.A. *et al.* (1998) ASA use and incident stroke in the cardiovascular health study. CHS Collaborative Research Group. *Stroke*, **29**, 887–894.

186 Hart, R.G., Halperin, J.L., Mc Bride, R. *et al.* (2000) Aspirin for the primary prevention of stroke and other major vascular events: meta-analysis and hypotheses. *Archives of Neurology*, **57**, 326–332.

187 Hart, R.G., Pearce, L.A., Miller, V.T. *et al.* on behalf of the SPAF Investigators (2000) Cardioembolic vs. noncardioembolic strokes in atrial fibrillation: frequency and effect of antithrombotic agents in stroke prevention in atrial fibrillation studies. *Cerebrovascular Diseases*, **10**, 39–43.

188 Voko, Z., Koudstaal, P.J., Bots, M.L. *et al.* (2001) Aspirin use and risk of stroke in the elderly: the Rotterdam Study. *Neuroepidemiology*, **20**, 40–44.

189 Buring, J.E., Bogousslavsky, J. and Dyken, M. (1998) Aspirin and stroke. *Stroke*, **29**, 885–886.

190 Iso, H., Hennekens, C.H., Meir, J. *et al.* (1999) Prospective study of aspirin use and risk of stroke in women. *Stroke*, **30**, 1764–1771.

191 Steering Committee of the Physicians' Health Study Research Group (1989) Final report on the aspirin component of the ongoing Physicians' Health Study. *The New England Journal of Medicine*, **321**, 129–135.

192 Peto, R., Gray, R., Collins, R. *et al.* (1988) Randomized trial of prophylactic daily aspirin in British male doctors. *British Medical Journal*, **296**, 313–316.

193 Ridker, P.M., Manson, J.E., Gaziano, J.M. *et al.* (1991) Low-dose aspirin therapy for chronic stable angina. A randomized, placebo-controlled clinical trial. *Annals of Internal Medicine*, **114**, 835–839.

194 He, J., Whelton, P.K., Vu, B. *et al.* (1998) ASA and risk of hemorrhagic stroke: a metaanalysis of randomized controlled trials. *The Journal of the American Medical Association*, **280**, 1930–1935.

195 Algra, A. and van Gijn, J. (1999) Cumulative meta-analysis of aspirin efficacy after cerebral ischaemia of arterial origin. *Journal of Neurology, Neurosurgery, and Psychiatry*, **66**, 255 [letter].

196 Antithrombotic Trialists' Collaboration (2002) Collaborative meta-analysis of randomised trials of antiplatelet therapy for prevention of death, myocardial infarction, and stroke in high risk patients. *British Medical Journal*, **324**, 71–86.

197 Canadian Cooperative Study Group (1978) A randomized trial of aspirin and sulfinpyrazone in threatened stroke. *The New England Journal of Medicine*, **299**, 53–59.

198 Swedish Cooperative Study Group (1987) High-dose acetylsalicylic acid after cerebral infarction. *Stroke*, **18**, 325–334.

199 Frith, P.A. and Warlow, C.P. (1988) A study of bleeding time in 120 long-term aspirin trial patients. *Thrombosis Research*, **49**, 463–470.

200 UK-TIA Study Group (1991) The United Kingdom Transient Ischaemic Attack (UK-TIA) aspirin trial: final results. *Journal of Neurology, Neurosurgery, and Psychiatry*, **54**, 1044–1054.

201 (The) Dutch TIA Trial Study Group (1991) A comparison of two doses of aspirin (30 mg vs. 283 mg a day) in patients after a transient ischemic attack or minor ischemic stroke. *The New England Journal of Medicine*, **325**, 1261–1266.

202 UK-TIA Study Group (1988) The United Kingdom Transient Ischaemic Attack (UK-TIA) aspirin trial: interim results. *British Medical Journal*, **296**, 316–320.

203 SALT Collaborative Group (1991) Swedish Aspirin Low-Dose Trial (SALT) of 75 mg aspirin as secondary prophylaxis after cerebrovascular ischaemic events. *The Lancet*, **338**, 1345–1349.

204 CAST (Chinese Acute Stroke Trial) Collaborative Group (1997) CAST: a randomised, placebo-controlled trial of early aspirin use in 20,000 patients with acute ischaemic stroke. *The Lancet*, **349**, 1641–1649.

205 International Stroke Trial Collaborative Group (1997) The International Stroke Trial (IST): a randomised trial of aspirin, subcutaneous heparin, both, or neither among 19,435 patients with acute ischaemic stroke. *The Lancet*, **349**, 1569–1581.

206 Chen, Z.M., Sandercock, P., Pan, H. *et al.* on behalf of the CAST and IST Group (2000) Indications for early aspirin use in acute ischemic stroke. A combined analysis of 40,000 randomized patients from the Chinese Acute Stroke Trial and the International Stroke Trial. *Stroke*, **31**, 1240–1249.

207 Kalra, L., Perez, I., Smithard, D.G. *et al.* (2000) Does prior use of aspirin affect outcome in ischemic stroke? *The American Journal of Medicine*, **108**, 205–209.

208 Aktas, B., Utz, A., Hoenig-Liedl, P. *et al.* (2003) Dipyridamole enhances NO/cGMP-mediated vasodilator-stimulated phosphoprotein phosphorylation and signaling in human platelets. *In vitro* and *in vivo/ex vivo* studies. *Stroke*, **34**, 764–769.

209 (The) European Stroke Prevention Study (ESPS) (1987) Principal end-points. The ESPS Group. *The Lancet*, **2**, 1351–1354.

210 Sacco, R.L. and Elkind, M.S. (2000) Update on antiplatelet therapy for stroke prevention. *Archives of Internal Medicine*, **160**, 1579–1582.

211 Diener, H.C., Cunha, L., Forbes, C. *et al.* (1996) European Stroke Prevention Study 2. Dipyridamole and acetylsalicylic acid in the secondary prevention of stroke. *Journal of the Neurological Sciences*, **143**, 1–13.

212 Müller, T.H., Su, C.A.P.F., Weisenberger, H. *et al.* (1990) Dipyridamole alone or combined with low-dose acetylsalicylic acid inhibits platelet aggregation in human whole blood *ex vivo*. *British Journal of Clinical Pharmacology*, **30**, 179–186.

213 de Schryver, E.L.L.M., Algra, A. and van Gijn, J. (2003) Cochrane review: dipyridamole for preventing major vascular events in patients with vascular disease. *Stroke*, **43**, 2072–2080.

214 Dyken, M.L., Barnett, J.H.M., Easton, D. *et al.* (1992) Low-dose aspirin and stroke: "it ain't necessarily so". *Stroke*, **23**, 1395–1399.

215 Diener, H.C., Darius, H., Bertrand-Hardy, J.M. *et al.* for the European Stroke Prevention Study 2 (2001) Cardiac safety in the European Stroke Prevention Study 2 (ESPS 2). *International Journal of Clinical Practice*, **55**, 162–163.

216 (The) ESPRIT-Study Group (2006) Aspirin plus dipyridamole versus aspirin alone after cerebral ischemia of arterial origin (ESPRIT): randomised controlled trial. *The Lancet*, **367**, 1665–1673.

217 Norrving, B. (2006) Dipyridamole with aspirin for secondary stroke prevention. *The Lancet*, **367**, 1638–1639.

218 Sacco, R. L., Diener, H. C., Yusuf, S. *et al.* (2008) Aspirin and Extended-Release Dipyridamole versus Clopidogrel for Recurrent Stroke. *The New England Journal of Medicine*, **359**, e-pub ahead of print.

219 Verro, P., Gorelick, P.B. and Nguyen, D. (2008) Aspirin plus dipyridamole versus aspirin for prevention of vascular events after stroke or TIA: a meta-analysis. *Stroke*, **39**, 1358–1363.

220 CAPRIE Steering Committee (1996) A randomised, blinded, trial of clopidogrel versus aspirin in patients at risk of ischaemic events (CAPRIE). *The Lancet*, **348**, 1329–1339.

221 Hankey, G.J., Sudlow, C.L.M. and Dunbabin, D.W. (2000) Thienopyridines or aspirin to prevent stroke and other serious vascular events in patients at high risk of vascular disease. *Stroke*, **31**, 1779–1784.

222 Gorelick, P., Sechenova, O. and Hennekens, C.H. (2006) Evolving perspectives on clopidogrel in the treatment of ischemic stroke. *Journal of Cardiovascular Pharmacology and Therapeutics*, **11**, 245–248.

223 Diener, H.C., Bogousslavsky, J., Brass, L.M. *et al.* on behalf of the MATCH Investigators (2004a) Aspirin and clopidogrel compared with clopidogrel alone after recent ischemic stroke in high-risk patients (MATCH): randomized, double-blind, placebo-controlled trial. *The Lancet*, **364**, 331–337.

224 Amarenco, P. and Donnan, G.A. (2004) Should the MATCH results be extrapolated to all stroke patients and affect ongoing trials evaluating clopidogrel plus aspirin? *Stroke*, **35**, 2006–2008.

225 Albers, G.W. and Amarenco, P. (2001) Combination therapy with clopidogrel and aspirin. Can the CURE results be extrapolated to cerebrovascular patients? *Stroke*, **32**, 2948–2949.

226 van Walraven, C., Hart, R.G., Singer, D.E. *et al.* (2002) Oral anticoagulants vs aspirin on nonvalvular atrial fibrillation. *The Journal of the American Medical Association*, **288**, 2441–2450.

227 Atrial Fibrillation Investigators (1994) Risk factors for stroke and efficacy of antithrombotic therapy in atrial fibrillation. *Archives of Internal Medicine*, **154**, 1449–1457.

228 Hart, R.G., Benavente, O., McBride, R. *et al.* (1999) Antithrombotic therapy to prevent stroke in patients with atrial fibrillation: a meta-analysis. *Annals of Internal Medicine*, **131**, 492–501.

229 Saxena, R. and Koudstaal, P.J. (2004) Anticoagulants versus antiplatelet therapy for preventing stroke in patients with nonrheumatic atrial fibrillation and a history of stroke or transient ischemic attack (Cochrane review), in *The Cochrane Library*, Issue 2, John Wiley & Sons, Ltd, Chichester, UK.

230 Mant, J., Hobbs, F.D., Fletcher, K. *et al.* for the BAFTA Investigators; Midland Research Practices Network (MidReC) (2007) Warfarin versus aspirin for stroke prevention in an elderly community population with atrial fibrillation (the Birmingham Atrial Fibrillation Treatment of the Aged Study, BAFTA): a randomised controlled trial. *The Lancet*, **370**, 493–503.

231 Hankey, G.J. (2002) Warfarin-Aspirin Recurrent Stroke Study (WARSS) Trial. *Stroke*, **33**, 1723–1726.

232 Chimowitz, M.I., Lynn, M.J., Howlett-Smith, H. *et al.* (2005) Comparison of warfarin and aspirin for symptomatic intracranial arterial stenosis. *The New England Journal of Medicine*, **352**, 1305–1319.

233 Mohr, J.P., Thompson, J.L.P., Lazar, R.M. *et al.* for the WARSS Group (2001) A comparison of warfarin and aspirin for the prevention of recurrent ischemic stroke. *The New England Journal of Medicine*, **345**, 1444–1451.

234 Warlow, C.P. (2002) Aspirin should be first-line antiplatelet therapy in the secondary prevention of stroke. *Stroke*, **33**, 2137–2138.

235 Goldstein, L.B., Adams, R., Alberts, M.J. *et al.* (2006) Primary prevention of stroke. A guideline from the American Heart Association/American Stroke Association Stroke Council. *Circulation*, **113**, e873–e923.

236 Tran, H. and Anand, S.S. (2004) Oral antiplatelet therapy in cerebrovascular disease, coronary artery disease, and peripheral artery disease. *The Journal of the American Medical Association*, **292**, 1867–1874.

237 Huang, Y., Cheng, Y., Wu, J. *et al.* (2008) Cilostazol as an alternative to aspirin after ischaemic stroke: a randomised, double-blind pilot study. *Lancet Neurology*, 10.1016/S1474-4422(08)70094-2.

238 Cocozza, M., Picano, T., Oliviero, U. *et al.* (1995) Effects of picotamide, an antithromboxane agent, on carotid atherosclerotic evolution. *Stroke*, **26**, 597–601.

239 Neri Serneri, G.G., Coccheri, S., Marubini, E. *et al.*, Drug Evaluation in Atherosclerotic Vascular Disease in Diabetics (DAVID) Study Group (2004) Picotamide, a combined inhibitor of thromboxane A_2 synthase and receptor, reduces 2-year mortality in diabetics with peripheral arterial disease: the DAVID study. *European Heart Journal*, **25**, 1845–1852.

4.1.3 Peripheral Arterial Disease

Etiology Peripheral arterial disease is a debilitating atherosclerotic disease of the lower limbs, eventually resulting in thrombotic occlusions of the lower limb arteries. The clinical symptoms range from intermittent claudication during exercise (Fontaine stage II) to severe peripheral limb ischemia with pain at rest and ulcer formation, eventually requiring limb amputation. Intermittent claudication, the most common form, becomes symptomatic when blood flow is insufficient to meet the metabolic demands of leg muscles in ambulatory patients.

PAD is a systemic disease that reflects an aggressive type of atherosclerosis and thrombosis in multiple vascular beds [240, 241]. Consequently, PAD is associated with a several-fold increased cardiovascular morbidity and mortality [242] and this is the main reason for aspirin treatment.

Epidemiology PAD occurs in about 2–5% of individuals aged 60 years or more but may become more frequent as life expectancy increases [240, 243, 244]. The disease is frequently underdiagnosed, mainly because about only one-half of individuals with PAD are symptomatic [241, 245]. More important, PAD is also underestimated in its prognostic value for other thrombotic complications of generalized atherosclerosis, that is, myocardial infarction, stroke, and sudden cardiac death [246]. According to the GRACE registry, patients with PAD are less likely to be treated with effective medications, including aspirin, than patients without PAD [247].

According to a recent population-based study in Germany, about twice as much elderly patients with PAD had also a manifestation of cerebrovascular or cardiovascular disease than individuals without [246], a finding confirmed by other investigators [248]. Consequently, there is a two to four times greater risk of dying from the complications of generalized atherosclerosis in PAD patients compared to individuals without claudication [249, 250], that is, as much as 75% of PAD patients will die from a cardiocoronary or cerebrovascular event. Thus, the most serious problem of PAD is not the limitation of walking, even though this may be the only clinical symptom, but the coexistent coronary and cerebrovascular morbidities.

These different clinical features and symptoms of disturbed peripheral versus disturbed coronary or cerebral circulations also define the treatment goals in PAD. The first is relief of ischemic symptoms, in particular leg pain, and prevention or at least retardation of progression of the disease to critical arterial stenosis and finally occlusion, that is, critical limb ischemia. One option is percutaneous transluminal angioplasty (PTA), which requires appropriate antiplatelet treatment to prevent thrombotic reocclusions. The second is reducing systemic cardiovascular morbidity and mortality, that is, prevention of myocardial infarction, stroke, and vascular death. There are several attempts to reach these goals: treatment or avoidance of risk factors (diabetes, cigarette smoking, and immobility) and administration of antiplatelet drugs to reduce cardiovascular morbidity and mortality [240, 242, 244]. Among antiplatelet drugs, aspirin is still the first choice and should be given to every PAD patient in the absence of contraindications [244, 251].

4.1.3.1 Thrombotic Risk and Mode of Aspirin Action

Platelets PAD is associated with platelet hyperreactivity and also seen with other manifestations of generalized atherosclerosis in the coronary and cerebral circulations. In addition to their fundamental role in arterial thrombus formation, platelets are also a primary source of inflammatory mediators involved in the original injury to the vascular endothelium that promotes plaque formation [252]. Inflammation stimulates local thrombosis and vice versa. Consequently, antiplatelet therapy may be beneficial because of the inhibition of platelet-dependent plug formation and platelet-triggered inflammatory effects.

Platelet hyperreactivity in PAD is reflected by enhanced platelet secretion (serotonin, growth factors), thrombin formation [253], expression of adhesion molecules at the platelet surface (P-selectin) [254], and shortened platelet survival [255]. In addition, there is local shear stress induced stimulation of platelets in stenotic areas with nonlaminar blood flow [256]. Thus, there are multiple changes in platelet function in PAD, eventually resulting in inflammatory reactions and a synergistic impairment of platelet/vessel wall interaction. These changes in PAD are more complex and less aspirin sensitive to those in other circulations.

Aspirin is the most intensively studied antiplatelet drug in PAD. Several studies indicated that platelets in patients with PAD are relatively aspirin "resistant" [253, 256–259] (Section 4.1.6). For example, aspirin-treated platelets of PAD patients exhibited unchanged spontaneous or serotonin-induced platelet aggregation [260, 261], and are also not inhibited after stimulation by ADP [262, 263]. Consequently, selective platelet inhibition by aspirin at antiplatelet doses may be less effective in PAD patients than in patients with other manifestations of atherothrombosis [253, 264] and there is no convincing evidence that aspirin is beneficial in treatment of claudication per se [244]. However, the use of aspirin in combination with other antiplatelet drugs and cilostazol might cause synergistic activities and, eventually, cause actions that are not seen by one single drug alone, for example, inhibition of tissue factor related inflammatory/procoagulatory responses [265].

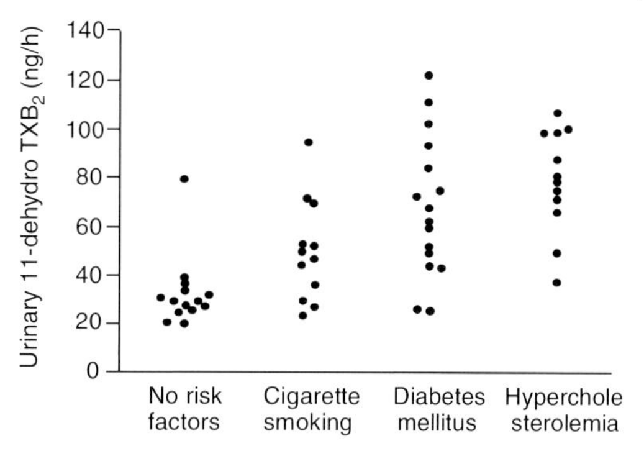

Figure 4.14 Urinary excretion of 11-DH-TXB_2 in subgroups of patients with PAD (stage II after Fontaine) [267].

Role of Thromboxane Since the antiplatelet actions of low-dose aspirin are determined by its effect on platelet-dependent thromboxane formation, the question arises whether thromboxane-dependent platelet activation also determines thrombotic vessel occlusions in PAD patients. Elevated plasma thromboxane levels and enhanced urinary excretion of the thromboxane metabolite, 11-Dehydro-TXB_2 (11-DH-TXB_2), have been repeatedly reported in PAD patients [255, 266]. Urinary excretion of 11-DH-TXB_2 was on average twice as much in PAD patients as in age- and sex-matched controls and was normalized by low-dose aspirin (50 mg/day), indicating that it was platelet driven [267]. However, these authors found an enhanced 11-DH-TXB_2 excretion in PAD patients only in association with coexistent cardiovascular risk factors, such as diabetes, hypercholesterolemia, or hypertension, suggesting that PAD per se is not a trigger of platelet activation *in vivo* and that the increased rate of thromboxane biosynthesis rather reflects the influence of coexisting disorders (Figure 4.14). This finding agrees with a population-based study in Finland, showing that after adjusting for symptoms and signs of coronary heart disease, claudication had no independent effect on mortality in men [249]. Thus, enhanced thromboxane formation may represent a common link between cardiovascular risk factors, specifically the activity of the process, and the thrombotic complications of PAD. Aspirin treatment will reduce the thromboxane-related elevated risk [267] but might be more effective in patients with coexisting atherosclerotic disorders. Since the probability of thrombotic occlusions of all arteries, including those of the limb, increases with the progression and severity of these coexisting diseases, aspirin as an antiplatelet/antithrombotic agent will be more effective at more advanced stages of PAD but primarily on coexisting disorder [268].

Coagulation and Fibrinolysis In addition to the disturbed platelet function, there are other hemostatic abnormalities that contribute to the atherothrombotic risk in PAD [254]. For example, blood viscosity and plasma fibrinogen levels are elevated and there is also evidence for enhanced thrombin formation [253]. High fibrinogen levels will also directly enhance platelet activity in PAD [256] in an aspirin-insensitive manner. This confirms an inflammatory component of the disease. The large mass of ischemic muscles of the leg might enhance generation and release of large amounts of inflammatory mediators from platelets, white cells, and the endothelium, not only during ischemia but also, in particular, during reperfusion after successful spontaneous or medical lysis of preformed clots.

Endothelial dysfunction In addition to abnormalities in platelet function, coagulation, and fibrinolysis in patients with PAD, there is evidence of severe endothelial dysfunction. This is best demonstrated in patients with type 2 diabetes who, according to the Framingham database, have a three- to fivefold increased risk of developing symptomatic PAD [269]. In these patients, there is markedly increased generation of reactive oxygen species. Diabetes is associated with an enhanced nonenzymatic generation of isoprostanes, which stimulate platelets via thromboxane receptors but in an aspirin-insensitive manner (aspirin "resistance") (Section 4.1.6). Inhibition of this platelet stimulatory effect of isoprostanes via the thromboxane receptor might be involved in the beneficial effects of picotamide, a thromboxane receptor antagonist, on mortality in PAD patients with diabetes [270].

4.1.3.2 Clinical Trials: Primary Prevention

There are no studies that have addressed symptoms and severity of PAD, such as critical limb ischemia or surgical interventions, as a primary or even only target of primary vascular prevention. Consequently, there is little evidence that antiplatelet medications, such as aspirin, will protect from claudication or its atherothrombotic complications. Nevertheless, development of atherosclerosis might also involve development of PAD with a considerably elevated risk of myocardial infarction and stroke and, therefore, needs protection by appropriate preventive measures.

The only available prospective randomized primary prevention trial on aspirin and PAD is a subgroup analysis of the US Physicians' Health Study (Section 4.1.1) [268]. The severity of the (developing) disease, by definition, was lower, the individuals much younger than the usual PAD patients, and the duration of the disease shorter. In addition, PAD complications were not a primary study end point but were calculated from a post hoc analysis of the data.

The study population were 22 000 apparently healthy US physicians, aged 40–84 years, who were treated for 5 years with aspirin (325 mg each other day) or placebo. Previous peripheral arterial surgery or preexisting claudication at baseline was exclusion criteria.

Within 5 years, 56 study participants had to undergo peripheral arterial surgery, 20 of them in the aspirin group and 36 in the placebo group. This was equivalent to a 46% relative risk reduction by aspirin ($p = 0.03$). No difference was seen in the incidence of claudication or the number of individuals that developed self-reported new claudication: 112 in the aspirin group and 109 in the placebo group ($p = 0.92$). Interestingly, from the nine claudicants who had to undergo peripheral arterial surgery, one was in the aspirin group and eight were in the placebo group ($p = 0.03$).

The conclusion was that long-term administration of aspirin will not protect from atherogenesis, that is, not retard the development or progression of atherosclerosis (in the limbs) in its early stages. However, aspirin will be beneficial in the more advanced stages of PAD when thrombosis within the narrowed vessels plays a critical role in disease complications and may require surgical treatment [268].

These data suggest that administration of aspirin will be effective in preventing thrombotic vessel occlusions but not in preventing the occurrence and progression of atherosclerosis, findings similar to the prevention of myocardial infarction (Section 4.1.1) and stroke (Section 4.1.2). Because of the higher risk with progression of atherosclerosis, the drug will be more effective at more advanced stages of the disease.

4.1.3.3 Clinical Trials: Secondary Prevention

Antiplatelet therapy in PAD patients is mainly done to prevent myocardial infarction and stroke because of the markedly increased vascular risk. These studies are summarized in Sections 4.1.1 and 4.1.2 and were most recently again confirmed in a new study and meta-analysis of the Critical Leg Ischemia Prevention Study (CLIPS) group (2007). The present section is focused on clinical trials with PAD complications as an end point. This includes progression of the disease to more severe stages as seen from angiography or sonography of the limb arteries, thrombotic vessel occlusions or narrowing, the requirement of surgical procedures (stenting), and, eventually, limb amputation.

Meta-Analyses The latest published meta-analysis of the Antithrombotic Trialists' Collaboration [271] on secondary prevention (including randomized studies published until 1997) contained 42 studies on 9214 patients with PAD. There was a proportional reduction by 23% in the risk of serious vascular events in PAD patients as compared to a 22% reduction in the total cardiovascular risk groups. This might have been expected because of the frequent association of PAD with other manifestations of atherosclerosis. However, the previous edition of the Antiplatelet Trialists' Collaboration did not report any significant benefit for PAD complications (Figure 4.15, suggesting that in this particular subgroup, the antiplatelet therapy, that is, aspirin, was not as effective as in other manifestations of atherosclerosis [272]. This difference to the 2002 edition came from one trial with picotamide [273], a mixed type thromboxane receptor antagonist/synthase inhibitor. This compound reduced fatal and nonfatal ischemic events by 19% (nonsignificant) but added more than 2300 patients with claudication to the Antithrombotic Trialists' investigation 2002 [245]. Because of its different spectrum of activities, most notably the thromboxane receptor blockade and its reversibility, picotamide may differ from aspirin and other inhibitors of platelet function. Other meta-analyses on antithrombotic drugs in the primary medical management of intermittent claudication also came to the conclusion that the beneficial effect of aspirin on claudication, that is, pain-free walking distance, is poor [274]. Thus, antiplatelet therapy with aspirin alone in these patients with respect to leg ischemia may be suboptimal [256].

Prospective Trials There are also few prospective trials on aspirin treatment in PAD with PAD-related complications as a primary study end point. Two of them used high doses of aspirin and may be less representative for the actual discussion on cardiovascular protection by antiplatelet drugs. However,

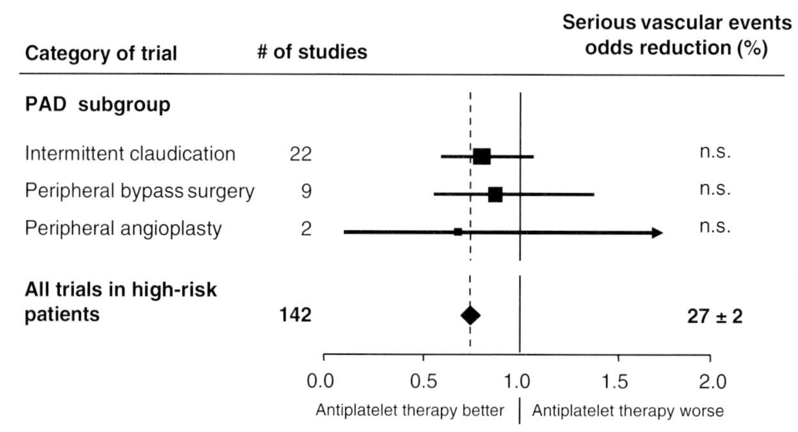

Figure 4.15 Proportional effects of antiplatelet therapy on serious vascular events in patients with symptomatic PAD as compared to the total high-risk population (modified after [272]).

they are interesting from a pharmacological point of view because they also included a placebo group. This, at the time, was not an ethical issue because the usefulness of aspirin for cardiocoronary prevention was not established.

The first large prospective trial on possible beneficial effects of antiplatelet treatment on the prevention of thrombotic femoral artery occlusions in patients with angiographically documented narrowing of the vessel was from the group of Schoop *et al.* [275].

The study included 300 men (age at entry 38–73 years, mean 56 years) who were randomized to one of the following treatment groups: aspirin (990 mg/day), aspirin 990 mg/day plus dipyridamole 225 mg/day, or placebo in a prospective, double-blind manner. Duration of treatment was 5 years. Efficacy parameter was the number of new occlusions of the affected femoral artery. Only patients were included that tolerated 1 g/day aspirin during a preceding 1-week run-in period. Angiographic control was done every 3 months.

At the end of the observation period there were 39 new peripheral vessel occlusions in the placebo group as opposed to 16 occlusions in the aspirin group and 23 occlusions in the aspirin + dipyridamole group (Figure 4.16). This difference was significant for the two treatment groups versus placebo but not within the two treatment groups. There were no deaths in the aspirin group, four in the placebo and eight in the combined aspirin dipyridamole group as well as five ulcers (no bleeds) with aspirin, one ulcer (one bleed) with placebo and three ulcers (three bleeds) in the combined treatment group.

The conclusion was that the thrombotic occlusion of stenosed femoral arteries can be retarded or even prevented by daily 1 g aspirin. The most significant effect is seen with regular smokers who are at the highest risk [275].

Similar results with the same doses of aspirin (330 mg tid) alone or in combination with dipyridamole (75 mg tid) or a matching placebo were reported in another early randomized, placebo-controlled trial. In 199 patients with occlusive arteriosclerosis of the lower extremities, serial angiograms were done before and after 2 years of treatment. Significantly more occlusions occurred in the placebo group than in the active treatment groups. However, significant differences were only obtained with the aspirin + dipyridamole combination but not with aspirin alone. It was concluded that long-term treatment with aspirin+ dipyridamole will retard the natural progression of atherosclerosis and that the combination with dipyridamole is superior to aspirin alone [276].

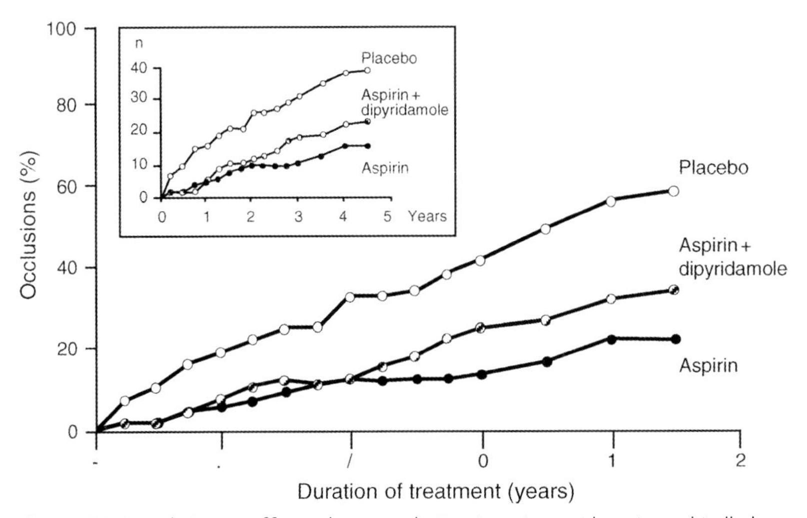

Figure 4.16 Cumulative rate of femoral artery occlusions in patients with angiographically documented stenosis (observation period: 4.5 years; $n = 100$ per treatment group) [275].

In addition to these studies, mostly involving patients at medium-to-moderate stages of PAD (claudicants), another prospective trial on the efficacy of aspirin in PAD was performed in patients at end stages of the disease, that is, diabetic patients with active limb gangrene or recent amputation.

A total of 231 patients were treated with aspirin (650 mg/day) and dipyridamole or placebo for 5 years. Primary end points were death from atherosclerotic vascular disease and amputation of the opposite extremity for gangrene.

In comparison to controls, there were no differences in treated patients in terms of vascular deaths: 22 and 19% or amputation of the opposite extremity: 20% versus 24%. There was no change in the incidence of myocardial infarctions but a 50% reduction in cerebrovascular events (stroke or TIA): 8% versus 19%, which, however, was not significant.

The conclusion was that antiplatelet agents have no effect on primary vascular end points but might affect the incidence of strokes that, however, was only a secondary end point in the study [277].

It is questioned whether any (conservative) treatment at this final stage of the atherosclerotic disease can still improve the symptoms or even retard the progression of the disease. However, the studies in claudicants do suggest that (high-dose) aspirin can reduce the incidence of thrombotic vessel occlusions in leg arteries although probably not retard the atherosclerotic process. This would be in line with the primary prevention data in the US Physicians' Health Study (see above) [268].

4.1.3.4 Clinical Trials: Peripheral Transluminal Angioplasty (PTA)

Meta-Analyses Meta-analyses of 14 randomized trials are available on antiplatelet and anticoagulant drugs for the prevention of restenosis/reocclusion following peripheral endovascular interventions in the Cochrane database. In patients with symptomatic PAD, a 60% reduction of recurrent obstruction of the dilated artery segment was found with aspirin (330 mg/day) combined with dipyridamole compared to placebo at 12 months. At 6 months after endovascular treatment, a positive effect on patency was seen with 50–100 mg/day aspirin, which, however, was not significant. It was concluded that aspirin with daily doses of 50–300 mg, starting prior to femoropopliteal endovascular interventions, appears to be effective and safe. Clopidogrel is considered as alternative, but sufficient data are lacking [278].

Prospective Trials A small prospective trial on PAD patients subjected to PTA indicated that enhanced levels of circulating surrogate markers, such as soluble adhesion molecules (P-selectin, VCAM-1) and markers of endothelial activation (thrombomodulin), had a predictive value for late restenosis. This suggests the significance of the activity of the ongoing atherosclerotic process for clinical outcome [279]. In general, prospective trials in PAD patients, subjected to PTA, failed to show convincing actions of aspirin [280] or any significant effect at all (CLIPS study, [281]) on leg ischemia. Interestingly, aspirin did significantly reduce the incidence of stroke [282] or myocardial infarctions [281] in the same studies. This confirms that antiplatelet effects of aspirin may not be equally important for peripheral as opposed to cerebral or coronary circulations.

A remarkable prospective, though small, study on the individual relationship between clinical outcomes of aspirin-treated PAD patients subjected to elective PTA and aspirin-induced inhibition of their platelet function as a prognostic surrogate parameter was published by Mueller *et al.* [283].

A total of 100 patients with intermittent claudication were subjected to elective percutaneous transluminal angioplasty. Clinical outcome (new vessel occlusion after successful intervention) and platelet function were monitored for 12 months in a prospective, compliance-controlled manner. All patients were treated with 100 mg aspirin/day. Whole-blood aggregometry was used to compare inhibition of collagen and ADP-induced platelet aggregation by aspirin with the clinical outcome.

All patients showed complete inhibition of arachidonic acid induced, that is, thromboxane-dependent platelet aggregation by aspirin (which was an inclusion criterion) at all study time points. Late reocclusions (from 8th to

52nd week) at the site of PTA occurred exclusively in male patients, for whom *in vitro* aggregometry failed to prove sufficient inhibition by collagen and ADP. In these patients, the risk of complications was increased by 87%.

It was concluded that only 40% of male patients showed the expected effect of aspirin on *in vitro* platelet aggregation and that whole-blood aggregometry was capable of predicting those patients at elevated risk of reocclusion following PTA [283].

One major difference between this and many other studies on antiplatelet agents in PAD patients is the use of whole-blood aggregometry. This technique also considers red cells and other plasma constituents as modifying factors for platelet function [284] (Section 2.3.1) and, therefore, may better reflect the complex situation of platelet activation in PAD patients *in vivo* than the measurement of platelet function in the conventional assays with platelet-rich plasma. Randomized trials in sufficiently sized patient populations are necessary to establish the relationship between reduced platelet sensitivity to aspirin and reduced therapeutic benefit in PAD patients subjected to PTA.

4.1.3.5 Aspirin and Other Drugs

As mentioned above, aspirin will probably not modify the claudication per se and the progression of the disease is mainly determined by aspirin-independent coexisting disorders such as diabetes, hypercholesterolemia, and hypertension. Thus, treatment with antihypertensives, antidiabetics, and lipid-lowering drugs is frequent in PAD patients. Neither of these compounds is considered to interfere with the antiplatelet effects of aspirin [242, 244].

Clopidogrel The currently most intensively studied alternative to aspirin in long-term prophylaxis of ischemic events with proven benefit in PAD patients is clopidogrel. In the CAPRIE trial, 6452 patients, that is, about one-third of the total study population, had PAD as a qualifying entry criterion. All patients were randomized to 325 mg/day aspirin or 75 mg/day clopidogrel and followed up for about 2 years.

In the subgroup with PAD as qualifying disease, clopidogrel (compared to aspirin) caused a 24% relative risk reduction in the combined vascular end point ischemic stroke/myocardial infarction/vascular death. However, the 95% confidence interval was wide (9–36%), and the group included not only patients with claudication but also patients with endovascular treatment [285]. The conclusion was that clopidogrel reduced the risk of vascular events in PAD patients, being slightly more effective than aspirin in a post hoc analysis [286]. However, the magnitude of this effect is uncertain and has to be balanced against the much higher costs [287]. A properly sized prospective randomized trial comparing aspirin with clopidogrel in PAD patients is still missing. There is also no information about the reocclusion rate of peripheral arteries after endovascular treatment by clopidogrel compared to aspirin.

Oral Anticoagulants Available studies with oral anticoagulants suggest that these compounds are of limited value in PAD [288]. This was recently confirmed in the randomized WAVE trial on more than 2100 PAD patients. The study compared a combination of warfarin (INR 2.0–3.0) with antiplatelet treatment (more than 90% aspirin at 81–325 mg/day) alone or in combination with the anticoagulant over 3 years. There was no reduced rate of myocardial infarctions, stroke, or cardiovascular death but a significant increase in moderate and life-threatening bleeding in the combined treatment group: 4.0% versus 1.2% (RR: 3.41, 95% CI: 1.84–6.35, $p < 0.001$) [289]. In one randomized but open trial over a period of 21 months, oral anticoagulation (INR 3.0–4.5) was found to be superior in preventing infrainguinal vein graft occlusion and lowering the rate of ischemic events whereas aspirin (80 mg/day) was more effective in preventing nonvenous graft occlusion and caused less bleeding [290]. Whether the benefits of oral anticoagulants outweigh the risk of severe and life-threatening bleeding at this high INR has to be confirmed in a second randomized trial.

Dipyridamole Whether the efficacy of aspirin treating PAD can be increased by comedication of dipyridamole is unproven. According to a recent Cochrane analysis, involving 29 trials with more than 23 000 participants, dipyridamole did not reduce the incidence of vascular death in the presence or absence of another antiplatelet drug. Dipyridamole appeared to reduce the risk of vascular events (RR: 0.88, 95% CI: 0.81–0.95). However, this benefit was only seen in patients with cerebral ischemia but not with other manifestations of atherothrombosis, including PAD [291].

Cilostazol Cilostazol is an orally active phosphodiesterase inhibitor (PDE III), and inhibits adenosine uptake, which resembles dipyridamole in several aspects [292]. Cilostazol has been shown in four randomized controlled trials to improve the function of ischemic legs, possibly by retarding the progression of atherosclerosis [242]. The compound appears not to interfere with platelet inhibitors, including aspirin [293] and does not prolong bleeding time. Cilostazol is the only compound that is approved for treatment of PAD in the United States and can be applied in combination with aspirin.

4.1.3.6 Actual Situation

In clinical practice, the optimum conservative treatment of PAD is still a matter of discussion as also seen from the multitude of agents that are recommended (by the manufacturers) for this purpose. In addition to treatment of basal diseases (diabetes, hypertension, and hypercholesterolemia) by appropriate drugs, aspirin is the drug of first choice in PAD to prevent thrombotic events. This includes the prevention of reocclusion after reopening of an occluded artery by PTA. In general, the efficacy of aspirin is low, regarding PAD-related problems, that is, the progression of limb atherosclerosis from claudication to critical limb ischemia. Aspirin is recommended in PAD patients because of the elevated risk of myocardial infarction and stroke, though the clinical evidence of its beneficial effects is low. Concomitant use of dipyridamole is an option but needs more clinical trials to become established. Clopidogrel is an alternative. Oral anticoagulants such as warfarin increase the bleeding risk and are of no therapeutic benefit. An interesting development with a significant therapeutic potential is cilostazol, a compound that not only acts mainly on vascular endothelium but also has tissue-protective, antiapoptotic, and platelet inhibitory properties [292]. Sufficiently sized randomized trials with PAD end points are required to determine the efficacy of these agents in patients with PAD [278].

Summary

Platelet hyperreactivity is part of the overall thrombotic/inflammatory syndrome in PAD patients. Further abnormalities of the hemostatic system include enhanced coagulation and disturbed fibrinolysis. Probably, because of this complexity, aspirin is less effective as an antiplatelet/antithrombotic drug in PAD patients than in patients suffering from other forms of generalized atherosclerosis. However, available data suggest a small effect of aspirin on the peripheral thrombotic complications of PAD.

PAD patients are at a two- to fourfold elevated risk of acute thromboembolic events in the coronary and cerebral circulations. These are the life-threatening events in these patients rather than critical limb ischemia or amputation. Thus, the use of antiplatelet drugs is strongly recommended. According to actual guidelines, all patients with PAD (whether symptomatic or not) should be considered for treatment with low-dose aspirin or with other approved antiplatelet drugs to reduce cardiovascular morbidity and mortality.

References

240 Weitz, J.I., Byrne, J., Clagett, G.P. *et al.* (1996) Diagnosis and treatment of chronic arterial insufficiency of the lower extremities: a critical review. *Circulation*, **94**, 3026–3049.

241 Hiatt, W.R. (2002) Preventing atherothrombotic events in peripheral arterial disease: the use of antiplatelet therapy. *Journal of Internal Medicine*, **251**, 193–206.

242 Hiatt, W.R. (2001) Medical treatment of peripheral arterial disease and claudication. *The New England Journal of Medicine*, **344**, 1608–1621.

243 Bradberry, J.-C. (2004) Peripheral arterial disease: pathophysiology, risk factors, and role of antithrombotic therapy. *Journal of the American Pharmaceutical Association*, **44** (Suppl. 1), S37–S44.

244 Duprez, D.A. (2007) Pharmacological interventions for peripheral artery disease. *Expert Opinion on Pharmacotherapy*, **8**, 1465–1477.

245 Duprez, D.A., De Buyzere, M.L. and Hirsch, A.T. (2003) Developing pharmaceutical treatments for peripheral arterial disease. *Expert Opinion on Investigational Drugs*, **12**, 101–108.

246 Diehm, C., Schuster, A., Allenberg, J.R. *et al.* (2004) High prevalence of peripheral arterial disease and co-morbidity in 6880 primary care patients: cross-sectional study. *Atherosclerosis*, **172**, 95–105.

247 Froehlich, J.B., Mukherjee, D., Avezum, A. *et al.* (2006) Association of peripheral artery disease with treatment and outcomes in acute coronary syndromes. The Global Registry of Acute Coronary Events (GRACE). *American Heart Journal*, **151**, 1123–1128.

248 Sukhija, R., Yalamanchili, K., Aronow, W.-S. *et al.* (2005) Clinical characteristics, risk factors, and medical treatment of 561 patients with peripheral arterial disease followed in an academic vascular surgery clinic. *Cardiology in Review*, **13**, 108–110.

249 Reunanen, A., Takkunen, H. and Aromaa, A. (1982) Prevalence of intermittent claudication and its effect on mortality. *Acta Medica Scandinavica*, **211**, 249–256.

250 Schoop, W. and Levy, H. (1982) Lebenserwartung bei Männern mit peripherer arterieller Verschlußkrankheit. *Lebensversicherungs Medizin*, **34**, 98–102.

251 Mannava, K. and Money, S.R. (2007) Current management of peripheral arterial occlusive disease: a review of pharmacological agents and other interventions. *American Journal of Cardiovascular Drugs*, **7**, 59–66.

252 Steinhubl, S.R., Newby, L.K., Sabatine, M. *et al.* (2005) Platelets and atherothrombosis: an essential role for inflammation in vascular disease – a review. *International Journal of Andrology*, **14**, 211–217.

253 Reininger, C.B., Graf, J., Reininger, A.J. *et al.* (1996) Increased platelet and coagulatory activity in peripheral atherosclerosis flow mediated platelet function is a sensitive and specific disease indicator. *International Angiology*, **15**, 335–343.

254 Koksch, M., Zeiger, F., Wittig, K. *et al.* (2001) Coagulation, fibrinolysis and platelet P-selectin expression in peripheral vascular disease. *European Journal of Vascular and Endovascular Surgery*, **21**, 147–154.

255 Zahavi, J. and Zahavi, M. (1985) Enhanced platelet release, shortened platelet survival time and increased platelet aggregation and plasma thromboxane B_2 in chronic obstructive arterial disease. *Thrombosis and Haemostasis*, **53**, 105–109.

256 Matsagas, M.I., Geroulakos, G., Mikhailidis, D.P. *et al.* (2002) The role of platelets in peripheral arterial disease: therapeutic implications. *Annals of Vascular Surgery*, **16**, 246–258.

257 Reininger, C.B., Graf, J., Reininger, A.J. *et al.* (1996) Increased platelet and coagulatory activity indicate ongoing thrombogenesis in peripheral arterial disease. *Thrombosis Research*, **82**, 523–532.

258 Robless, P., Mikhailidis, D. and Stansby, G. (2001) A systematic review of antiplatelet therapy in the prevention of myocardial infarction, stroke, or vascular death in peripheral vascular disease patients. *The British Journal of Surgery*, **88**, 787–800.

259 Roller, R.-E., Dorr, A., Ulrich, S. *et al.* (2002) Effect of aspirin treatment in patients with peripheral arterial disease monitored with the platelet function analyzer PFA-100. *Blood Coagulation & Fibrinolysis*, **13**, 277–281.

260 Barradas, M.A., Stansby, G., Hamilton, G. *et al.* (1993) Effect of naftidrofuryl and aspirin on platelet aggregation in peripheral vascular disease. *In Vivo*, **7**, 543–548.

261 Barradas, M.A., Stansby, G., Hamilton, G. *et al.* (1994) Diminished platelet yield and enhanced platelet aggregability in platelet-rich plasma of peripheral vascular disease patients. *International Angiology*, **13**, 202–207.

262 Moake, J.L., Turner, N.A., Stathopoulos, N.A. *et al.* (1988) Shear-induced platelet aggregation can be mediated by vWF released from platelets, as well as by exogenous large or unusually large vWF-multimers, requires adenosine diphosphate, and is resistant to aspirin. *Blood*, **71**, 1366–1374.

263 Jagroop, I.-A., Matsagas, M.-I., Geroulakos, G. *et al.* (2004) The effect of clopidogrel, aspirin and both antiplatelet drugs on platelet function in patients with peripheral arterial disease. *Platelets*, **15**, 117–125.

264 Walters, T.K., Mitchell, D.C. and Wood, R.F.M. (1993) Low-dose aspirin fails to inhibit increased platelet reactivity in patients with peripheral vascular disease. *The British Journal of Surgery*, **80**, 1266–1268.

265 Koneti Rao, A., Vaidyula, V.R., Bagga, S. *et al.* (2006) Effect of antiplatelet agents clopidogrel, aspirin, and cilostazol on circulating tissue factor procoagulant activity in patients with peripheral arterial disease. *Thrombosis and Haemostasis*, **96**, 738–743.

266 Gresele, P., Catalano, M., Giammaressi, C. *et al.* (1997) Platelet activation markers in patients with peripheral arterial disease. *Thrombosis and Haemostasis*, **78**, 1434–1437.

267 Davi, G., Gresele, P., Violi, F. *et al.* (1997) Diabetes mellitus, hypercholesterolemia, and hypertension but not vascular disease per se are associated with persistent platelet activation *in vivo*. Evidence derived from the study of peripheral arterial disease. *Circulation*, **96**, 69–75.

268 Goldhaber, S.Z., Manson, J.E., Stampfer, A.J. *et al.* (1992) Low-dose aspirin and subsequent peripheral arterial surgery in the Physicians' Health Study. *The Lancet*, **340**, 143–145.

269 Kannel, W.B. and Mc Gee, D.L. (1985) Update on some epidemiologic features of intermittent claudication: the Framingham Study. *Journal of the American Geriatrics Society*, **33**, 13–18.

270 Neri Serneri, G.-G., Coccheri, S., Marubini, E. *et al.* (2004) Picotamide, a combined inhibitor of thromboxane A_2 synthase and receptor, reduces 2-year mortality in diabetics with peripheral arterial disease: the DAVID study. *European Heart Journal*, **25**, 1845–1852.

271 Antithrombotic Trialists' Collaboration (2002) Collaborative meta-analysis of randomised trials of antiplatelet therapy for prevention of death, myocardial infarction, and stroke in high risk patients. *British Medical Journal*, **324**, 71–86.

272 Antiplatelet Trialists' Collaboration (1994) Collaborative overview of randomised trials of antiplatelet therapy. I. Prevention of death, myocardial infarction, and stroke by prolonged antiplatelet therapy in various categories of patients. *British Medical Journal*, **308**, 81–106.

273 Balsano, F. and Violi, F., ADEO Group (1993) Effect of picotamide on the clinical progression of peripheral vascular disease. A double-blind, placebo-controlled study. *Circulation*, **87**, 1563–1569.

274 Girolami, B., Bernardi, E., Prins, M. *et al.* (1999) Antithrombotic drugs in the primary medical management of intermittent claudication: a meta-analysis. *Thrombosis and Haemostasis*, **81**, 715–722.

275 Schoop, W., Levy, H., Schoop, B. *et al.* (1983) Experimentelle und klinische Studien zu der sekundären Prävention der peripheren Arteriosklerose, in *Thrombozytenfunktionshemmer* (eds A. Bollinger and K. Rhyner), Georg Thieme Verlag, Stuttgart, pp. 49–58.

276 Hess, H., Mietaschk, A. and Deichsel, G. (1985) Drug-induced inhibition of platelet function delays progression of PAD: a prospective double-blind arteriographically controlled trial. *The Lancet*, **1**, 415–419.

277 Colwell, J.A., Bingham, S.F., Abraira, C. *et al.* (1986) Veterans Administration Cooperative Study Group on antiplatelet agents in diabetic patients after amputation for gangrene. II. Effects of aspirin and dipyridamole on atherosclerotic vascular disease rates. *Diabetes Care*, **9**, 140–148.

278 Dörffler-Melly, J., Koopman, M.-M., Prins, M.-H. *et al.* (2005) Antiplatelet and anticoagulant drugs for prevention of restenosis/reocclusion following peripheral endovascular treatment. *Cochrane Database of Systematic Reviews*, **1**, CD002071.

279 Tsakiris, D.A., Tschöpl, M., Jäger, K. *et al.* (1999) Circulating cell adhesion molecules and endothelial markers before and after transluminal angioplasty in peripheral arterial occlusive disease. *Atherosclerosis*, **142**, 193–200.

280 Ranke, C., Creutzig, A., Luska, G. *et al.* (1994) Controlled trial of high-versus low-dose aspirin treatment after percutaneous transluminal angioplasty in patients with peripheral vascular disease. *The Clinical Investigator*, **72**, 673–680.

281 Catalano, M. for the Critical Leg Ischaemia Prevention Study (CLIPS) Group (2007) Prevention of serious vascular events by aspirin amongst patients with peripheral arterial disease: randomized, double-blind trial. *Journal of Internal Medicine*, **261**, 276–284.

282 Ranke, C., Hecker, H., Creutzig, A. *et al.* (1993) Dose-dependent effect of aspirin on carotid atherosclerosis. *Circulation*, **87**, 1873–1879.

283 Mueller, M.R., Salat, A., Stangl, P. *et al.* (1997) Variable platelet response to low-dose ASA and the risk of limb detoriation in patients submitted to peripheral arterial angioplasty. *Thrombosis and Haemostasis*, **78**, 1003–1007.

284 Santos, M.T., Valles, J., Marcus, A.J. *et al.* (1991) Enhancement of platelet reactivity and modulation of eicosanoid production by intact erythrocytes. *The Journal of Clinical Investigation*, **87**, 571–580.

285 CAPRIE Steering Committee (1996) A randomised, blinded, trial of clopidogrel versus aspirin in patients at risk of ischaemic events (CAPRIE). *The Lancet*, **348**, 1329–1339.

286 Davie, A.P. and Love, M.P. (1997) CAPRIE trial. *The Lancet*, **349**, 355.

287 Blinc, A. and Poredos, P. (2007) Pharmacological prevention of atherothrombotic events in patients with peripheral arterial disease. *European Journal of Clinical Investigation*, **37**, 157–164.

288 Cosmi, B. and Palareti, G. (2004) Is there a role for oral anticoagulant therapy in patients with peripheral arterial disease? *Current Drug Targets – Cardiovascular & Haematological Disorders*, **4**, 269–273.

289 (The) Warfarin Antiplatelet Vascular Evaluation (WAVE) Trial Investigators (2007) Oral anticoagulant and antiplatelet therapy and peripheral arterial disease. *The New England Journal of Medicine*, **357**, 217–227.

290 (The) Dutch Bypass Oral Anticoagulants or Aspirin (BOA) Study Group (2000) Efficacy of oral anticoagulants compared with aspirin after infrainguinal bypass surgery (The Dutch Bypass Oral Anticoagulants or Aspirin study): a randomized trial. *The Lancet*, **355**, 346–351.

291 De Schryver, E., Algra, A. and van Gijn, J. (2007) Dipyridamole for preventing stroke and other vascular events in patients with vascular disease. *Cochrane Database of Systematic Reviews*, Jul 18, CD001820.

292 Schrör, K. (2002) The pharmacology of cilostazol. *Diabetes, Obesity & Metabolism*, **4** (Suppl. 2), S14–S19.

293 Wilhite, D.-B., Comerota, A.-J., Schmieder, F.-A. *et al.* (2003) Managing PAD with multiple platelet inhibitors: the effect of combination therapy on bleeding time. *Journal of Vascular Surgery*, **38**, 710–713.

4.1.4
Venous Thrombosis

Etiology Venous thrombosis usually results from accumulation of activated clotting factors during local stasis in the low-pressure system of veins [294]. Activation of plasmatic clotting factors is stimulated by local inflammation, such as varicosis, or might have a genetic background. Blood stasis also retards the washout of activated clotting factors and their subsequent degradation. Quantitatively most important are deep vein thromboses (DVTs) after larger surgical interventions, such as hip or knee surgery. The prolonged postoperative immobilization of extremities associated with stimulation of the clotting cascade by the surgical intervention will facilitate the occurrence of thrombotic events. Another high-risk group of patients are bed-ridden persons and individuals, immobilized by sitting in an airplane during long-distance flights ($\geq$10 h). This will also facilitate venous stasis and activation of plasmatic coagulation without sufficient washout of locally accumulating clotting factors.

4.1.4.1 Thrombotic Risk and Mode of Aspirin Action

Venous thrombosis differs fundamentally from arterial thrombosis by the low if any contribution of platelets to the thrombotic event. This, however, does not mean that inhibition of platelet function cannot modify venous thrombosis. Statistically, among 100 patients who undergo surgical interventions of moderate severity, there are at least 30 cases of deep venous thrombosis, identifiable by plethysmography, if these patients are not given perioperative antithrombotic prophylaxis. Among these 100 patients, four cases of pulmonary embolism will be recognized; one or two of them may be fatal [295]. These figures illustrate the clinical significance of venous thrombosis and its most important complication – pulmonary embolism. They also indicate the importance of adequate thrombosis prophylaxis.

The goal of prophylaxis is to prevent complications and the methods used are both pharmacological and mechanical. Mechanical approaches reduce stasis by (early) mobilization or using compression devices, whereas pharmacological approaches bring about the the inhibition of clotting factors and stimulation of clot lysis.

Aspirin Versus Anticoagulants The venous thrombus is a fibrin-rich clot formed within the circulation in "dead waters," recirculating eddies, valve sinuses, and other areas of relative stasis [295]. Because of the low blood pressure in veins, there is little platelet activation by mechanical factors. Thus, direct inhibition of thrombin generation and action appear to be a more useful approach than inhibition of platelet function. However, because of the complex thrombotic risk in many patients, an individual risk stratification and multimodal management is necessary. In this context, aspirin treatment also has its place, although not primarily for the prevention of venous thrombembolism [296, 297]. Here antithrombins, that is, (low molecular weight) heparins and other anticoagulants are the treatment of choice.

4.1.4.2 Clinical Trials

Deep Vein Thrombosis Despite low significance of platelets as thrombogenic factors in venous thrombosis, several early studies have shown beneficial effects of aspirin in this indication [298, 299]. However, negative results have also been reported, and, according to McKenna *et al.* [300], beneficial actions of aspirin require doses of 3.9 g/day. This suggests platelet-independent actions of aspirin, such as anti-inflammatory effects, as an explanation for the therapeutic efficacy.

A prospective randomized study addressed the issue of efficacy and potency of aspirin compared to coumarins in the prevention of deep vein thrombosis after hip surgery (hip fracture). A total of 194 patients were randomized to aspirin (650 mg/bid), warfarin (10 mg/day), or placebo. Treatment lasted for 3 weeks after surgery.

Only 20% (13/65) of patients under warfarin developed DVT as opposed to 41% (27/66) and 46% (29/63) treated with aspirin or placebo, respectively.

The conclusion was that warfarin is significantly more potent than aspirin and that the efficacy of aspirin was not different from placebo [301].

Prophylactic treatment with aspirin in patients subjected to arthroplasty of the knee was also inferior to physical treatment by compression devices. The incidence of DVT was about twofold higher in the aspirin group compared to physical treatment [302]. Similar data, favoring compression treatment versus aspirin, were also reported by Westrich and Sculco [303]. However, beneficial effects of aspirin have been found in treatment of chronic venous leg ulcers. In a small, though placebo-controlled randomized trial, aspirin (300 mg/day) for 4 months plus standardized compression banding achieved ulcer healing in 38% of patients and significant reduction in ulcer size in another 52% of patients. These data suggest therapeutic efficacy for aspirin in this indication. However, no comparative treatment with another active drug was performed [304]. Overall, there is little evidence from these small studies for clinically significant beneficial antithrombotic effects of aspirin in DVT.

According to a meta-analysis, incorporating data from 50 trials in over 5000 patients, prophylactic aspirin reduced both proximal and distal DVT by 30–40% and pulmonary embolism by 60% in patients undergoing general surgical, orthopedic, and other medical procedures. Thus, aspirin might be beneficial. However, it provides less protection than anticoagulants and, therefore, cannot be recommended for the prevention of DVT in high-risk patients [305, 306].

DVT and Air Traveling The LONFLIT studies established that in high-risk subjects after long-distance (>10 h) flights, the incidence of DVT may be between 4 and 6%. The LONFLIT3 study randomized 300 high-risk subjects to aspirin (400 mg/day) for 3 days starting 12 h before the beginning of the flight, low molecular weight heparin (injection of one dose of enoxaparin 2–4 h before the flight), and a matching placebo.

In the placebo control group, 4.8% of subjects suffered a DVT, whereas in the aspirin group 3.6% of subjects suffered a DVT. No cases of DVT occurred in the low molecular weight heparin group. DVT was asymptomatic in 60% of subjects and 85% of events occurred in nonaisle seats.

It was concluded that one single dose of low molecular weight heparin is an effective measure to be considered in high-risk subjects during long-distance flights [307].

Thus, aspirin appears not to be an effective measure to prevent DVT in these conditions. Assuming that the rate of travel-related DVT is 20 per 100 000 travelers, one has to treat 17 000 people with aspirin to prevent one additional DVT [308].

Thrombosis in Venous Bypass Grafts A different situation is the prevention of thrombotic vessel occlusion after venous bypass grafting. In this indication, aspirin is first-line treatment (see Section 4.1.1).

4.1.4.3 Actual Situation

Aspirin is not recommended for the prevention of DVT. The reasons are the different etiologies of the disease as opposed to arterial thromboses. There is low if any protective action by aspirin. Alternatively, there are effective inhibitors of thrombin formation and action, that is, (low molecular weight) heparins and oral anticoagulants with proven efficacy in thrombosis prevention and treatment.

Summary

Venous thromboembolic disease, consisting primarily of deep vein thrombosis and pulmonary embolism is an important clinical issue. DVT occurs in conditions of venous stasis, predominantly not only after operative procedures but also after immobilization of extremities due to other reasons, such as long-distance air traveling.

In contrast to arterial thrombosis, platelets have no major role in venous thrombosis. Thrombus formation in veins is mainly due to (local) thrombin generation and depends less on platelets. Thus, anticoagulants, inhibiting thrombin formation and action, are the pharmacological treatment of choice rather than antiplatelet agents such as aspirin.

Although the therapeutic efficacy of aspirin in preventing venous thromboembolism in general is disappointing, aspirin is an effective measure by comparison with placebo. Nevertheless, anticoagulants, such as heparins, are the preferred option because of higher and more constant efficacy. In addition, physical treatment with compression devices will enhance the efficacy of medical treatment.

References

294 Hirsh, J., Dalen, J.E., Anderson, D.R. *et al.* (1998) Oral anticoagulants: mechanism of action, clinical effectiveness, and optimal therapeutic range. *Chest*, **114** (Suppl. 5), 445S–460.

295 Weinmann, E.E. and Salzman, E.W. (1994) Deep-vein thrombosis. *The New England Journal of Medicine*, **331**, 1630–1641.

296 Reitman, R.D., Emerson, R.H., Higgins, L.L. *et al.* (2003) A multimodality regimen for deep venous thrombosis prophylaxis in total knee arthroplasty. *Journal of Arthroplasty*, **18**, 161–168.

297 Berend, K.R. and Lombardi, A.V., Jr (2006) Multimodal venous thromboembolic disease prevention for patients undergoing primary or revision total joint arthroplasty: the role of aspirin. *American Journal of Orthopsychiatry*, **35**, 24–29.

298 Harris, W.H., Salzman, E.W., Athanasoulis, C.A. *et al.* (1977) Aspirin prophylaxis of venous thrombembolism after total hip replacement. *The New England Journal of Medicine*, **297**, 1246–1249.

299 Harris, W.H., Athanasoulis, C.A., Waltman, A.C. *et al.* (1985) Prophylaxis of deep-vein thrombosis after total hip replacement. Dextran and external pneumatic compression compared with 1. 2 or 0. 3 gram of aspirin daily. *The Journal of Bone and Joint Surgery. American Volume*, **67**, 57–62.

300 McKenna, R.J., Galante, J., Bachmann, F. *et al.* (1980) Prevention of venous thromboembolism after total knee replacement by high-dose aspirin or intermittent calf and thigh compression. *British Medical Journal*, **280**, 514–517.

301 Powers, P.J., Gent, M., Jay, R.M. *et al.* (1989) A randomized trial of less intense postoperative warfarin or aspirin therapy in the prevention of venous thrombembolism after surgery for fractured hip. *Archives of Internal Medicine*, **149**, 771–774.

302 Haas, S.B., Insall, J.N., Scuderi, R.E. *et al.* (1990) Pneumatic sequential compression boots compared with aspirin prophylaxis of deep vein thrombosis after total knee athroplasty. *The Journal of Bone and Joint Surgery. American Volume*, **72**, 27–31.

303 Westrich, G.H. and Sculco, T.P. (1996) Prophylaxis against deep venous thrombosis after total knee arthroplasty. *The Journal of Bone and Joint Surgery*, **78**, 826–834.

304 Layton, A.M., Ibbotson, S., Davies, J.A. *et al.* (1994) Randomised trial of oral aspirin for chronic venous leg ulcers. *The Lancet*, **344**, 164–165.

305 Antiplatelet Trialists' Collaboration (1994) Collaborative overview of randomised trials of antiplatelet therapy. I. Prevention of death, myocardial infarction, and stroke by prolonged antiplatelet therapy in various categories of patients. *British Medical Journal*, **308**, 81–106.

306 Antithrombotic Trialists' Collaboration (2002) Collaborative meta-analysis of randomised trials of antiplatelet therapy for prevention of death, myocardial infarction, and stroke in high risk patients. *British Medical Journal*, **324**, 71–86.

307 Cesarone, M.R., Belcaro, G., Nicolaides, A.N. *et al.* (2002) Venous thrombosis from air travel: the LONFLIT3 study prevention with aspirin vs. low-molecular-weight heparin (LMWH) in high-risk subjects: a randomized trial. *Angiology*, **53**, 1–6.

308 Loke, Y.K. and Derry, S. (2002) Air travel and venous thrombosis: how much help might aspirin be? *Medscape General Medicine*, **4**, 4.

4.1.5 Preeclampsia

Pregnancy-induced hypertension as isolated hypertension after 20 weeks of gestation or as hypertension with proteinuria (preeclampsia) is a leading cause of maternal and fetal morbidity and mortality. Preeclampsia is a multisystem disorder, unique to human pregnancy. The exact etiology and pathophysiology of the disease is still a matter of discussion [309].

Etiology The clinical symptoms of PIH appear late in pregnancy although the pathophysiological processes begin early, shortly after implantation of the trophoblast. Pathophysiological signals are generated in response to pathological interactions between the developing fetus and maternal tissues. This involves an anomaly in the invasion of the uterine spiral arteries by trophoblast cells, resulting in underperfusion of the placenta and, finally, placental ischemia and infarction. These processes involve immunogenic and inflammatory processes, eventually resulting in the release of inflammatory mediators from the placenta that causes the maternal syndrome called diffuse endothelial dysfunction [310–312]. This condition is associated with increased sensitivity of the vasculature to vasoconstrictor hormones (angiotensin II) and intense arterial vasoconstriction (hypertension) at reduced intravascular blood volume and an associated prothrombotic state [313]. There is not only underperfusion of the uteroplacental unit but also of maternal organs, specifically liver, brain, and heart. In addition, cytokines generated and released from the placenta cause a generalized inflammatory reaction in the maternal circulation. Clinically, this might result in abortion or preterm delivery, intrauterine growth retardation (IUGR), PIH, or death.

After placentation of the fetus, further fetal growth and development require development and growth of blood vessels, which come into connection with the maternal circulation. In certain immunological conditions, this might result in the generation of immunoregulatory cytokines and antiangiogenic factors. These placenta-derived proteins bind (and inactivate) growth factors [314, 315]. This is followed by insufficient vascularization and high vascular resistance in the uteroplacental circulation. These mediators also enter the maternal circulation and cause a generalized inflammatory response at the endothelium, eventually resulting in PIH [312]. Screening for antiangiogenic proteins or other soluble factors related to the pathology of these processes may allow early diagnosis of the disease, identification of high-risk groups and subsequently a more effective and causal prevention and treatment of PIH than currently possible [316].

Prothrombotic factors in plasma [317], for example, a prothrombotic genotype [318], or immunological/inflammatory disorders associated with placentation [312, 313] are possible pathophysiological explanations for preeclampsia. Preexisting medical illnesses may also contribute to PIH and aggravate the clinical symptoms and complications. A hypothetical scheme of etiology and pathophysiology of PIH with possible sites of action of aspirin is shown in Figure 4.17.

Treatment Goals Preeclampsia is a disease of the trophoblast and can currently only be cured by removing the placenta, that is, delivery or abortion. Otherwise, clinical symptoms can currently be treated only symptomatically. Central to many clinical symptoms of the disease is a pathology in arachidonic acid metabolism, possibly related to endothelial dysfunction, which results in platelet hyperreactivity and subsequently enhanced generation of platelet-derived thromboxane and other prothrombotic and vasoconstrictor factors. This is the rationale for the use of low-dose aspirin as an antiplatelet agent in treatment of PIH.

4.1.5.1 Thrombotic Risk and Mode of Aspirin Action

The generation of both PGI_2 and TXA_2 increases with increasing duration of the pregnancy. PGI_2 is mainly generated by vascular endothelium, trophoblastic cells and the placenta, whereas thromboxane A_2 is generated by platelets.

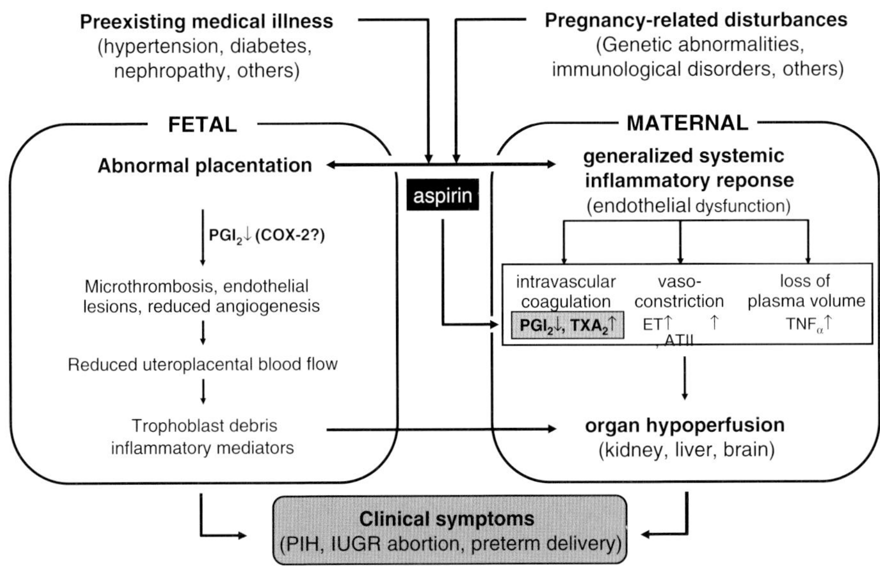

Figure 4.17 Etiology and pathophysiology of PIH and IUGR (adapted from Brown [329]).

Reduced Vascular PGI2 Generation Reduced generation of PGI_2 is one of the initial abnormalities in women at risk of developing PIH [319–322]. Significantly reduced endothelial PGI_2 generation, as seen from reduced urinary excretion of PGI_2 metabolites, becomes clinically detectable in 13–16 weeks of gestation, that is, before the clinical onset of the disease [320, 323]. One consequence of reduced prostacyclin formation is platelet hyperreactivity and enhanced platelet-derived thromboxane formation [324, 325], another increased sensitivity of vascular smooth muscle cells to endogenous vasoconstrictors, such as angiotensin II or endothelin. Attempts to correct PGI_2 deficiency by systemic administration of PGI_2 were not successful, probably because of its strong systemic vasodilatory activity and even fatalities [326, 327]. Selective inhibition of platelet thromboxane synthesis was effective in some cases [328] but has not been studied systematically. Thus, aspirin became the antiplatelet treatment of choice. This was based upon the assumption that the detrimental consequences of preeclampsia are platelet related and caused by an insufficient endothelial PGI_2 generation [329–331].

Enhanced Thromboxane Generation as Target of Aspirin Thromboxane levels are increased in normal gestation but furthermore two- to fourfold higher in preeclampsia [332–334]. This is probably due to platelet hyperreactivity. Accordingly, platelet activation markers are elevated and there is an increased platelet turnover rate [335]. Thrombocytopenia is the most common hemostatic abnormality in preeclampsia and probably related to intravascular clot formation. Platelet hyperreactivity and enhanced thromboxane formation precede the clinical onset of the disease and are associated with uteroplacental insufficiency and fetal and maternal hypoperfusion of vital organs. This makes prevention of excessive thromboxane formation a most useful tool for symptomatic treatment of vascular and hemostatic abnormalities in preeclampsia.

Serum thromboxane appears not to predict the outcome in high-risk pregnancies [336] and is also largely unchanged in normal pregnancy [337], despite markedly elevated plasma levels and urinary thromboxane excretion. This is not surprising since serum thromboxane only determines the capacity of thromboxane formation rather than endogenous thromboxane generation within the circulation *in vivo*. Thromboxane formation can be better determined by measuring excretion of thromboxane metabolites in urine. However, measurement of serum thromboxane is a useful compliance control if aspirin treatment is performed.

Mode of Aspirin Action The pharmacological reason for using (low-dose) aspirin to prevent preeclampsia is selective inhibition of thromboxane formation at undisturbed generation of PGI_2. Aspirin rapidly passes the placenta and enters the fetal circulation, approaching about 50% of the maternal plasma level [338]. In pregnant women, low-dose oral aspirin (100 mg/day) is completely cleared from maternal plasma within 4 h and is not detected in neonates. Neither renal thromboxane formation nor generation of PGI_2 or PGE_2 is affected by regular low-dose aspirin [339]. It is still uncertain, which proportion of the vascular PGI_2 is synthesized via COX-2 and whether the percentage remains the same during pregnancy and in women at risk for preeclampsia. Cytokine formation and release into the maternal and fetal circulations is most likely to stimulate COX-2 expression, including subsequently enhanced formation of PGI_2 and PGE_2, which is much less sensitive to aspirin than COX-1-dependent thromboxane formation in platelets. According to current knowledge and clinical experience, the goal of selective thromboxane suppression at (largely) undisturbed generation of PGI_2 can be obtained by daily doses of 100 mg aspirin (Figure 4.18) [340].

Possible Risks of Aspirin Use Although clinical data demonstrate a remarkable safety of aspirin even in high-risk pregnancies (see below), several possible side effects of aspirin should be considered. The bleeding risk is low in early pregnancy, although the clinical efficacy of aspirin in women at high risk for preeclampsia appears to be related to the prolongation of bleeding time [341] (Section 3.1.2). Another issue of concern is regional vasoconstriction subsequent to inhibition of vasodilator prostaglandin formation, in particular premature closure of the ductus arteriosus Botalli. Reversible constriction of fetal blood vessels was found with several NSAIDs and aspirin. However, these changes, if any, were small and without any negative effect on fetal development [342]. In addition, the sensitivity of the fetal and maternal circulation

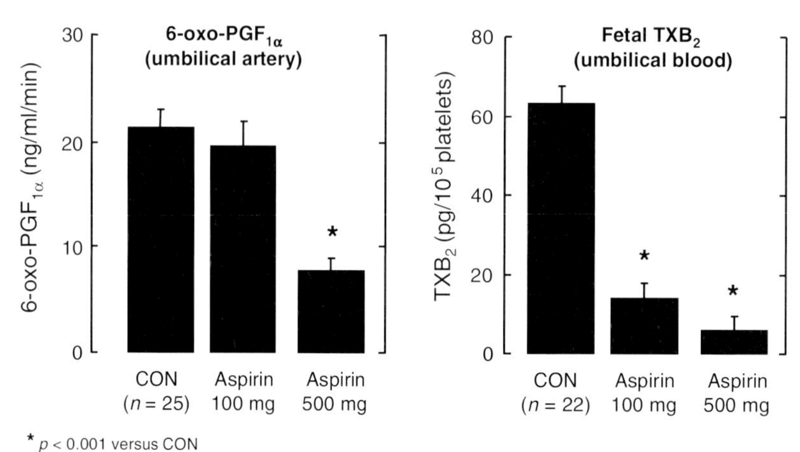

Figure 4.18 Dose-dependent inhibition of prostacyclin (6-oxo-PGF_{1a}) and thromboxane (TXB_2) formation by 100 and 500 mg oral aspirin in healthy women during labor at term. An aspirin dose of 100 mg causes a marked reduction of (platelet-dependent) TXB_2 in spontaneous clotting cord blood but leaves prostacyclin formation by umbilical artery (*in vitro*) unaffected. The high dose of 500 mg aspirin markedly reduced both thromboxane and prostacyclin formation [340].

to cyclooxygenase inhibitors might not be the same: aspirin doses that were sufficient to inhibit placental thromboxane formation did not block thromboxane or prostacyclin formation in isolated trophoblast cells [343]. Similarly, thromboxane synthesis of fetal platelets remained largely unaffected [322] under conditions when thromboxane generation by placental vessels was reduced [344]. The sensitivity of adult and neonatal platelets' thromboxane formation to aspirin and other NSAIDs *in vitro* appears to be similar [345]. The proportional contribution of COX-2 to PGE_2 and PGI_2 formation, which also plays a role in maintaining a functioning ductus arteriosus, remains to be determined.

4.1.5.2 Clinical Trials

Early Reports After a brief report by Goodlin *et al.* [346], Crandon and Isherwood [347] were the first to show that nulliparous women who took aspirin more than once each second week during pregnancy were at significantly lower risk of PIH than those who had taken no aspirin. The first randomized trial in early pregnancy (starting in the third month of gestational age) in women at high risk of preeclampsia showed that aspirin prevented PIH [348]. In this study, women were treated with aspirin (150 mg/day) plus dipyridamole (300 mg/day) until delivery. Preeclampsia and severe IUGR occurred in significant percentages of untreated but in none of the aspirin-treated patients. This suggests that regular low-dose aspirin may provide protection from preeclampsia in women at elevated risk for the disease.

Randomized Double-Blind Trials The first larger, placebo-controlled double-blind prospective randomized trial testing low-dose aspirin as a preventive measure of pregnancy-induced hypertension and preeclampsia in pregnant women at elevated risk for PIH was published by Wallenburg *et al.* [349].

Pregnant nulliparous, healthy women, normotensive at 26th week of gestation, were subjected to the angiotensin II sensitivity test to detect abnormal vasoconstriction as an index parameter for pathological vascular reactivity. From a total of 207 women, 46 were found to react with enhanced pressure responses. These patients at risk were enrolled into the study at 28th week of gestation and treated with aspirin (60 mg/day) or placebo until delivery. This aspirin dose reduced thromboxane generation by platelets (determined by measuring malondialdehyde) on average by 89%.

One intrauterine death (asphyxia?) and 2 cases of slight hypertension occurred in the aspirin group but 12 cases of a usually severe preeclampsia in the placebo group. No adverse effects of aspirin treatment on either mother or infants were observed. Specifically, there were no hemorrhages or any evidence for constriction of the ductus arteriosus Botalli.

The conclusion was that in these individuals at the high risk of PIH, treatment with low-dose aspirin is a useful protective measure [349].

Benigni *et al.* [322] confirmed the clinical benefit of protection of high-risk patients by low-dose (60 mg/day) aspirin. In this study, treatment was started already at the 12th week of gestation. The positive clinical outcome was associated with a significant reduction of thromboxane formation at unchanged levels of PGI_2. Other small studies with low-dose aspirin (60–150 mg/day) in pregnant women at elevated risk for preeclampsia also confirmed beneficial effects on PIH [350]. According to a meta-analysis of early trials, aspirin treatment reduced PIH by 65% and severe IUGR by 44% [351].

These positive results stimulated several larger, randomized placebo-controlled double-blind studies that, however, yielded different results and in general were unable to confirm the impressive beneficial effects of the earlier studies. The worldwide largest randomized, placebo-controlled trial so far was the "Collaborative Low-dose Aspirin Study in Pregnancy" [352].

A total of 9364 women were randomly assigned to film-coated aspirin (60 mg/day) or a matching placebo until delivery. Women were eligible if they were between 12th and 32nd weeks of gestation and were at sufficient preexisting risk of preeclampsia or IUGR. The majority of patients (74%) took aspirin for prophylactic purposes, that is, there was an increased risk of preeclampsia because of a history of preeclampsia or IUGR in a previous pregnancy, chronic hypertension, or renal diseases. Further risk factors were maternal age, family history, or multiple pregnancies. Main end points were proteinuric preeclampsia, duration of pregnancy, birth weight, stillbirth, and neonatal death ascribed to preeclampsia or IUGR.

Overall, 6.7% of women allocated to the aspirin group developed proteinuric preeclampsia compared to 7.6% of those allocated to placebo. This 12% relative risk reduction by aspirin was not significant. There was no significant effect of aspirin on the incidence of IUGR, stillbirth, or neonatal death. However, aspirin did significantly reduce the rate of preterm deliveries in comparison with placebo, 19.7% versus 22.2% ($2p = 0.003$), and there was a significant trend ($p = 0.004$) toward progressively greater reductions in proteinuric preeclampsia, the more preterm the delivery. Aspirin treatment was not associated with a significant increase in placental hemorrhages or bleedings during preparation for epidural anesthesia. However, there was a slight increase in the number of blood transfusion after delivery. Aspirin was generally safe for the fetus and newborn infant with no evidence of an increased likelihood of bleeding.

The main conclusion was that these findings do not support routine prophylactic or therapeutic administration of antiplatelet drugs such as aspirin in pregnancy to all women at an increased risk of preeclampsia or IUGR [352].

CLASP was one out of eight larger placebo-controlled randomized studies, that is, trials including at least 200 pregnant women, that were published between 1993 and 1998. The daily aspirin dose in these studies was 50–60 mg/day. The clinical outcomes were different, although the earlier studies tended to be more positive for aspirin than the later ones (Figure 4.19). According to a recent Cochrane analysis [353], considering results of all randomized trials until 2006, there was a moderate but significant reduction of preeclampsia (17%), preterm birth (8%), intrauterine growth retardation (10%), and perinatal mortality (14%) by antiplatelet agents, in most low-dose aspirin cases. These data were recently confirmed in a large meta-analysis of individual patient data – the perinatal antiplatelet review of international studies (PARIS) trial, also showing a moderate though significant 10% risk reduction for PIH and preterm birth in primary prevention of PIH in high-risk women. Importantly, the PARIS study did also show that treatment with antiplatelet agents is not associated with an increased bleeding risk for either the women or their babies [354].

4.1.5.3 Clinical Trials in Pregnancy-Induced Hypertension: Reasons for Data Variability

The different results on aspirin treatment and prevention of PIH in clinical trials are somehow unique to the disease and suggest additional variables for study outcome. Thus, it is possible that the less impressive results of the large trials compared

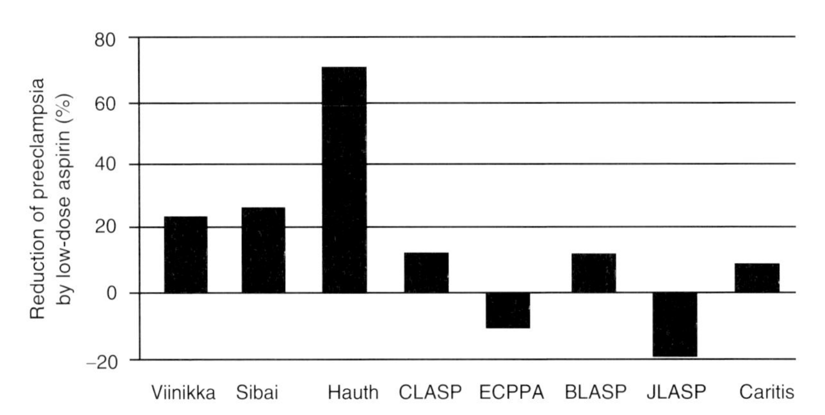

Figure 4.19 Effects of aspirin on the incidence of preeclampsia in eight larger trials, published between 1993 and 1998. Each trial involved more than 200 patients. The dose of aspirin in all studies was 60 mg/day or less (50 mg/day in the study of [355, 356].

to the smaller ones may be due to a too heterogeneous patient selection that dilutes any beneficial effect by inclusion of (relatively) more patients with a low-risk profile [357–359]. On the other hand, small-sized studies with negative results might not have been published, thus possibly causing a publication bias. Three variables with particular relevance to clinical outcome are (i) definition of women who are most likely to benefit from antiplatelet treatment; (ii) beginning (and end) of treatment, and (iii) selection of the optimum dose and control of adherence of the patient to regular drug intake (compliance!). The "perinatal antiplatelet review of international studies" retrospective meta-analysis for primary prevention of preeclampsia by antiplatelet treatment is a nice example to demonstrate the complexity of this issue.

The PARIS study was a meta-analysis of individual patient data from 32 217 women and their 32 819 babies, recruited from 31 randomized preeclampsia primary prevention trials. The prespecified main outcomes included preeclampsia, death *in utero* or death of the baby before discharge from the hospital, preterm birth at less than 34 weeks of gestation, and IUGR. Only randomized studies of antiplatelet agents (mostly aspirin) versus placebo or no treatment were included.

For women assigned to receive antiplatelet agents, the relative risk of developing preeclampsia was reduced by 10% (RR: 0.90, 95% CI: 0.84–0.97), delivering before 34 weeks by 10% (RR: 0.90; 95% CI: 0.83–0.98), and of having a pregnancy with a serious adverse outcome by 10% (RR: 0.90; 95% CI: 0.85–0.96). There was no effect on other parameters, including death of fetus or baby, IUGR, or bleeding events. The outcome was similar in several subgroups studied, including those with late start of treatment, dosing, and preexisting medical conditions.

The conclusion was that administration of antiplatelet agents during pregnancy causes moderate but consistent reductions in the risk of developing preeclampsia, preterm delivery (before 34 weeks' gestation), and pregnancy with serious adverse outcomes. There was no group of women in this study for whom there was evidence to justify withholding antiplatelet therapy [354].

While these data are important, in particular with respect to safety aspects of aspirin administration during pregnancy, they suffer from the principal problems of all meta-analyses, that is, the combination of data from different studies with different study designs, aspirin doses, and clinical outcome and preeclampsia definitions. Data only from 31 trials out of 115 trials that were considered by the authors as potentially eligible were presented, that is, less than one quarter from total. The aspirin dose varied between 50 and 150 mg/day. There was a 16% better outcome at doses ≥75 mg compared to doses <75 mg, which, however, was not significant. A planned ≥150 mg aspirin group could not be realized because of too small numbers of patients. Randomization and therapy at optimum time points, that is, before 20 weeks gestation could only be performed in 59% of women and women with preexisting medical conditions (i.e., renal disease, diabetes, and hypertension), that is, a possibly different pathophysiological background were also included. Finally, more than 37% of data extracted from the 31 studies came from *one* (principally negative) trial (CLASP) (see above). This indicates a rather mixed population that might dilute the real benefits of aspirin prophylaxis and might not transfer to all risk determinants to the same extent.

Similar beneficial findings were reported in a recent Cochrane analysis, involving 59 trials with 37 560 women [353]. There was a 17% reduction of preeclampsia by antiplatelet treatment, mostly aspirin. There was a significantly higher increase in risk reduction by antiplatelet treatment for high-risk (number needed to treat (NNT), 19, compared to moderate risk (NNT: 119) women. Thus, aspirin has benefits in the prevention of preeclampsia; however, further information defining subgroups of patients and conditions for optimum benefit are needed. Some of these aspects regarding patient selection, beginning and end of treatment, dosing and compliance issues are discussed below.

Patient Selection The pathophysiology of preeclampsia is related to early pregnancy, although clinical symptoms occur later in gestation and might be further aggravated by external risk factors and medical conditions. Thus, improved diagnostics, identifying patients with risk factors early, that is, before the appearance of clinical symptoms, is a most attractive approach to increase the success rate of treatment. For example, urinary screening for risk factors such as antiangiogenic proteins (sFlt-1, PlGF) (see above) is one of these promising procedures and may become available to more general use in the near future.

It has been shown that the clinical efficacy of antiplatelet treatment is not the same in all women with clinical symptoms of preeclampsia. Heyborne [360] qualified available studies on aspirin in the prevention of preeclampsia according to the individual risk profile of the patients. There was little if any benefit of aspirin in nonselected pregnancies but a significant protection in most if not all of the controlled trials when patients with a previous positive screening test for risk factors were included. Interestingly, no such benefit was seen in patients with a history of medical problems, such as hypertension, renal failure, diabetes, older age, and others (Table 4.6). This suggests that the individual risk profile is an important determinant for the efficacy of aspirin treatment and that patients at elevated risk by a medical condition might be less sensitive to aspirin protection.

Beginning and End of Treatment Because preeclampsia is a disease of early pregnancy, prevention of abnormalities of the uteroplacental circulation,

Table 4.6 Preeclampsia – effects of low-dose aspirin in different risk groups (adapted from [360]).

Risk group	Study	*n*	Risk determination	Incidence of preeclampsia (%)		Significance
				Control	Aspirin	
High risk because of diagnostic procedures. Nulliparous (no diabetes, no hypertension, and no other medical risk factors)	Wallenburg *et al.* [349]	46	Angiotensin II infusion	30	9.0	$p < 0.01$
	Schiff *et al.* [361]	65	Rollover test	35.5	11.8	$p = 0.024$
	McParland *et al.* [362]	100	Uterine artery Doppler scan	19	2.0	$p < 0.02$
	Brown *et al.* [363]	22	Angiotensin II infusion	63	27.0	$p < 0.01$
	Wenstrom *et al.* [364]	48	hCG > 2.0 multiples of the median	10.7	0.0	NS
	Bower *et al.* [365]	60	Uterine artery Doppler scan	41[a]	29.0[b]	NS ($p = 0.03$)
High risk because of medical conditions (hypertension, diabetes, chronic renal disease, and previous preeclampsia)	Italian Study of Aspirin in Pregnancy [366]	896	Presence of medical risk factors	2.7	2.9	NS
	CLASP Collaborative group [352]	6929	Presence of medical risk factors	6.7	7.6	NS
	ECPPA [367]	1900	Presence of medical risk factors	6.7	6.1	NS
	Sibai *et al.* [357]	2539	Presence of medical risk factors	22	18	NS

NS, not significant.
[a] Severe, $n = 38$.
[b] Severe, $n = 13$.

including early excess of placental thromboxane formation by aspirin, is probably most effective, when started before placentation is completed, that is, ≤13th week of gestation [368]. The efficacy might be lower with increasing duration of pregnancy beyond the 20th week of gestation. In several early trials, aspirin treatment was started up to the 32nd week of gestation and therefore might not be effective anymore [369]. Treatment should be stopped after the 37th week of gestation because of questionable benefit and a possible elevated bleeding risk, specifically with the concomitant intake of other antihemostatic agents, which was confirmed in the PARIS study.

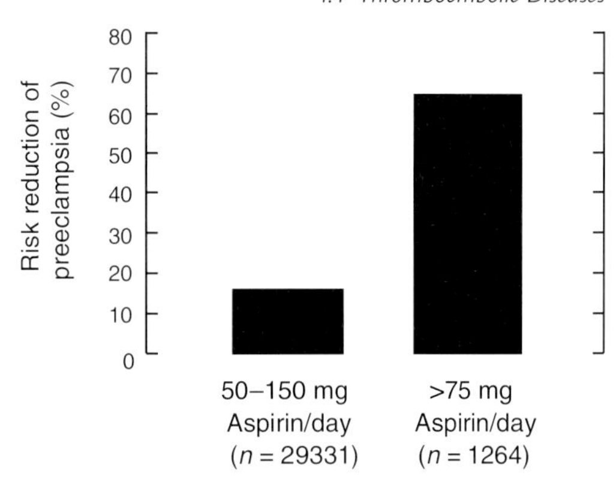

Figure 4.20 Evidence of a dose-dependent effect of aspirin on the incidence of preeclampsia according to Cochrane database (2001) [356].

Aspirin Dose Most of the studies that failed to show convincing beneficial effects of aspirin on preeclampsia were performed with 60 mg aspirin/day or less. This dose was chosen not to impair vascular prostacyclin production but it may not always be sufficient for inhibition of thromboxane formation, specifically in situations of increased platelet activation *in vivo*, such as preeclampsia. The least 95% inhibition of serum thromboxane levels [370], necessary for efficient inhibition of platelet function, was not obtained in all trials [349]. Doses below 75 mg/day appear also to be less effective in cardiocoronary prevention (Section 4.1.1). In agreement with this, recent Cochrane data indicate a risk reduction of preeclampsia by 65% (!) in women treated with higher doses of aspirin (>75 mg/day) as opposed to an overall only 15% risk reduction if all patients were included (Figure 4.20). Similar results with respect to IUGR were found in a meta-analysis by Leitich *et al.* [371]. According to data of 13 234 cases, aspirin antagonized IUGR and higher doses (100–150 mg/day) of the compound were more effective than the lower doses (50–80 mg/day).

Circadian Variations Another aspect of the efficacy of drug treatment, including antiplatelet agents, is possible circadian variations. Hermida *et al.* [372] showed in a placebo-controlled randomized trial in pregnant women at elevated risk of preeclampsia that the blood pressure lowering action of aspirin (100 mg/day) showed a circadian variation. Antihypertensive effects were most prominent when aspirin was given before bedtime as opposed to awakening time. It was concluded that the lack of positive results with aspirin in other trials might have partially been due to too low doses that do not affect placental thromboxane formation (<80 mg/day) and/or the lack of circadian timing of the drug application.

Compliance Insufficient patient compliance is a likely explanation for treatment failure. In addition to aspirin resistance (Section 4.1.6) missing adherence to drug treatment might also contribute to the treatment failure with aspirin in the prevention of gestation-related diseases. Specifically, pregnant women may not wish to take any drugs because of possible injury to the fetus. The compliance rate according to tablet counts was 57% in the JLASP trial [373] while only 42% of patients took more than 95% of prescribed aspirin tablets in the BLASP study [374]. Actually, compliance measurement by pill count might even overestimate the real intake [375, 376].

An impressive example for the validity of pill count as an estimate for patient compliance of aspirin use in the prevention of preeclampsia was published by Hauth *et al.* [376]. In this study, patient compliance was 94% according to tablet count but only 79% according to thromboxane determination as a more reliable parameter for aspirin intake.

Interestingly, an at least twofold reduction in thromboxane levels was seen in 33% of patients in the placebo group. Obviously, these patients had taken aspirin or aspirin-containing medications for other reasons. When all pregnancies included in the study, were divided into those with at least twofold reduction in thromboxane levels and those without – independent of randomization to the placebo or aspirin group –, there was a clear correlation between inhibition of thromboxane formation and clinical outcome, that is, rates of preeclampsia, IUGR, and preterm delivery [376].

Safety Aspects: Miscarriages and Malformations Regular low-dose aspirin to improve outcome of moderate- and high-risk pregnancies is generally well tolerated and is safe with respect to fetal and neonatal outcomes (Section 3.1.3). A recent meta-analysis of all published data on this issue until 2001 [377, 378] found no increased risk of miscarriage for women taking aspirin. However, these women exhibited a significantly lower risk of preterm delivery than did those on placebo (RR: 0.92, 95% CI: 0.86–0.98). There was no significant change in perinatal mortality and the rate of small-for-gestational-age infants. These data were confirmed in the PARIS meta-analysis on primary prevention [354]. There was also no evidence of an overall increase in the risk of congenital malformations that could be associated with aspirin if the compound was taken by the mother during the first trimester of pregnancy [377–379].

Safety Aspects: Placental Circulation and Bleeding Regular low-dose (100 mg/day) aspirin during the second and third trimester of pregnancy does not alter uteroplacental or fetoplacental hemodynamics and cause constriction of the ductus arteriosus [380]. There is also no overall increase in bleedings [354], in particular if aspirin is withdrawn prior to the end of the 37th week of gestation and there is no evidence for a significant bleeding tendency in the fetus or newborn, respectively (Section 3.1.2).

4.1.5.4 Actual Situation

According to the PARIS trial [354] and the most recent Cochrane data [353], aspirin is safe in pregnancy for both mother and child and exhibits a moderate, though significant, protective effect against preeclampsia. In addition to reducing the incidence of the disease in high-risk populations, it also retards the onset of clinical symptoms, improves clinical outcome, and reduces perinatal mortality. Aspirin prophylaxis and treatment is indicated in women, suffering from preeclampsia prior to 32nd gestational week, by familiar disposition and positive uterine Doppler flow in the 2nd trimester. Aspirin is less or not effective in preeclampsia associated with preexisting medical conditions, such as chronic hypertension, chronic renal disease, or diabetes. Treatment should be started at 12–16th week of gestation in women at elevated risk and should be terminated at 37th week of gestation.

Summary

Preeclampsia, that is, pregnancy-induced hypertension associated with proteinuria is a multisystem disorder and a leading cause of fetal and maternal morbidity and mortality. The clinical symptoms of the disease are possibly caused by a generalized endothelial dysfunction, initiated by a pathologic immune reaction between fetal and maternal tissues. This eventually results in disturbed placental development and function. This pathology develops with invasion of the uterine spiral arteries by the trophoblast but becomes clinical relevant only in later stages of pregnancy.

Vessel tone, permeability and platelet functions in preeclamptic women are significantly modified by a pathology in their arachidonic acid metabolism. The increased prostacyclin formation in normal pregnancy is reduced before symptoms of the disease occur and followed by a marked increase in thromboxane production. Platelet hyperreactivity and increased platelet-dependent thromboxane formation are seen before the onset of the disease.

Prevention of exaggerated platelet thromboxane formation is the rationale for prophylactic use of low-dose aspirin in the prevention and treatment of preeclampsia. A moderate but significant 10–15% improvement of clinical outcomes, specifically preeclampsia, intrauterine growth retardation, and perinatal mortality, are well established by several meta-analyses. The efficacy of aspirin is risk-dependent. The compound is rather ineffective if preeclampsia results from preexisting medical conditions of the mother, such as diabetes, renal failure, hypertension, or older age. It is expected that improved diagnostic procedures with early identification of women at risk may also improve the therapeutic efficacy of aspirin prophylaxis.

The aspirin doses used in most trials vary between 50 and 150 mg daily. Some data suggest that higher doses may be more effective than the lower ones and that early start of treatment is more useful than late. Randomized trials are required to assess which women benefit most from aspirin prophylaxis, when treatment is best started and at what dose. There is no evidence for significant side effects of aspirin at antiplatelet doses on vessel tone (premature closure of the ductus arteriosus, pulmonary hypertension), blood coagulation (bleeding), or fetal development (miscarriages, malformations).

References

309 Caritis, S., Sibai, B., Hauth, J., *et al.* (1998) Low-dose aspirin to prevent preeclampsia in women at high risk. National Institute of Child Health and Human Development Network of Maternal-Fetal Medicine Units. *The New England Journal of Medicine*, **338**: 701–705.

310 Roberts, J.M. and Cooper, D.W. (2001) Pathogenesis and genetics of preeclampsia. *The Lancet*, **357**, 53–60.

311 Sibai, B.M. (2004) Preeclampsia: an inflammatory syndrome? *American Journal of Obstetrics and Gynecology*, **191**, 1061–1062.

312 Redman, C.W. and Sargent, I.L. (2005) Latest advances in understanding preeclampsia. *Science*, **308**, 1592–1594.

313 Merviel, P., Carbillon, L., Challier, J.-C. *et al.* (2004) Pathophysiology of preeclampsia: links with implantation disorders. *European Journal of Obstetrics, Gynecology, and Reproductive Biology*, **115**, 134–147.

314 Maynard, S.E., Min, J.Y., Merchan, J. *et al.* (2003) Excess soluble fms-like tyrosine kinase 1 (sFlt 1) may contribute to endothelial dysfunction, hypertension, and proteinuria in preeclampsia. *The Journal of Clinical Investigation*, **111**, 649–658.

315 Levine, R.J., Thadani, R., Quian, C. *et al.* (2005) Urinary placental growth factor and risk of preeclampsia. *The Journal of the American Medical Association*, **293**, 77–85.

316 Lindheimer, M.D. (2005) Unraveling the mysteries of preeclampsia. *American Journal of Obstetrics and Gynecology*, **193**, 3–4.

317 Salomon, O., Seligsohn, U., Steinberg, D.M. *et al.* (2004) The common prothrombotic factors in nulliparous women do not compromise blood flow in the feto-maternal circulation and are not associated with preeclampsia or intrauterine growth restriction. *American Journal of Obstetrics and Gynecology*, **191**, 2002–2009.

318 Morrison, E.R., Miedzybrodzka, Z.H. and Campbell, D.M. (2002) Prothrombotic genotypes are not associated with pre-eclampsia and gestational hypertension: results from a large population based study and systematic review. *Thrombosis and Haemostasis*, **87**, 779–785.

319 Bussolino, F., Benedetto, C., Massobrio, M. *et al.* (1980) Maternal prostacyclin activity in pre-eclampsia. *The Lancet*, **2**, 702.

320 Fitzgerald, D.J., Entman, S.S., Mulloy, K. *et al.* (1987) Decreased prostacyclin biosynthesis preceding the clinical manifestation of pregnancy-induced hypertension. *Circulation*, **75**, 956–963.

321 Minuz, P., Paluani, F., Degan, M. *et al.* (1988) Altered excretion of prostaglandin and thromboxane metabolites in pregnancy-induced hypertension. *Hypertension*, **11**, 550–556.

322 Benigni, A., Gregorini, G., Frusca, T. *et al.* (1989) Effect of low-dose aspirin on fetal and maternal generation of thromboxane by platelets in women at risk for pregnancy-induced hypertension. *The New England Journal of Medicine*, **321**, 357–362.

323 Mills, J.L., DerSimonian, R., Raymond, E. *et al.* (1999) Prostacyclin and thromboxane changes predating clinical onset of preeclampsia: a multicenter prospective study. *The Journal of the American Medical Association*, **282**, 356–362.

324 Ballegeer, V.C., Spitz, B., De Baene, L.A. *et al.* (1992) Platelet activation and vascular damage in gestational hypertension. *American Journal of Obstetrics and Gynecology*, **166**, 629–633.

325 Klockenbusch, W., Somville, T., Hafner, D. *et al.* (1994) Excretion of prostacyclin and thromboxane metabolites before, during and after pregnancy induced hypertension. *European Journal of Obstetrics Gynecology and Reproductive Biology*, **57**, 47–50.

326 Steel, S.A. and Pearce, J.M. (1988) Specific therapy in severe fetal intrauterine growth retardation: failure of prostacyclin. *Journal of the Royal Society of Medicine*, **81**, 214–216.

327 Walsh, S.W. (1989) Low-dose aspirin: treatment for the imbalance of increased thromboxane and decreased prostacyclin in preeclampsia. *American Journal of Perinatology*, **6**, 124–132.

328 Van Assche, F.A., Spitz, B., Vermylen, J. *et al.* (1984) Preliminary observations on treatment of pregnancy-induced hypertension with a thromboxane synthetase inhibitor. *American Journal of Obstetrics and Gynecology*, **148**, 216–218.

329 Brown, M.A. (1991) Pregnancy-induced hypertension: pathogenesis and management. *Australian and New Zealand Journal of Medicine*, **21**, 257–273.

330 Uzan, S., Beaufils, M., Breast, G. *et al.* (1991) Prevention of fetal growth retardation with low-dose aspirin: findings of the EPREDA trial. *The Lancet*, **337**, 1427–1431.

331 Klockenbusch, W., Braun, M.S., Schröder, H. *et al.* (1992) Prostacyclin rather than nitric oxide lowers human umbilical artery tone *in vitro*. *European Journal of Obstetrics Gynecology and Reproductive Biology*, **47**, 109–115.

332 Walsh, S.W. (1985) Preeclampsia: an imbalance in placental prostacyclin and thromboxane production. *American Journal of Obstetrics and Gynecology*, **152**, 335–340.

333 Fitzgerald, D.J., Mayo, G., Catella, F. *et al.* (1987) Increased thromboxane biosynthesis in normal pregnancy is mainly derived from platelets. *American Journal of Obstetrics and Gynecology*, **157**, 325–330.

334 Fitzgerald, D.J., Rocki, W., Murray, R. *et al.* (1990) Thromboxane A_2 synthesis in pregnancy induced hypertension. *The Lancet*, **335**, 751–754.

335 Redman, C.W.G., Bonnar, J. and Beilin, L. (1978) Early platelet consumption in pre-eclampsia. *British Medical Journal*, **1**, 467–469.

336 Hauth, J.C., Sibai, B., Caritis, S. *et al.* (1998) Maternal serum thromboxane B_2 concentrations do not predict improved outcomes in high-risk pregnancies in a low-dose aspirin trial. The National Institute of Child Health and Human Development Network of Maternal–Fetal Medical Units. *American Journal of Obstetrics and Gynecology*, **179**, 1193–1199.

337 Louden, K.A., Broughton Pipkin, F., Heptinstall, S. *et al.* (1990) A longitudinal study of platelet behaviour and thromboxane production in whole blood in normal pregnancy and the puerperium. *British Journal of Obstetrics and Gynaecology*, **97**, 1108–1114.

338 Jacobson, R.L., Brewer, A. and Eis, A. (1991) Transfer of aspirin across the perfused human placental cotyledon. *American Journal of Obstetrics and Gynecology*, **165**, 939–944.

339 Leonhardt, A., Bernert, S., Watzer, B. *et al.* (2003) Low-dose aspirin in pregnancy: maternal and neonatal aspirin concentrations and neonatal prostanoid formation. *Pediatrics*, **111**, 77–81.

340 Ylikorkala, O., Mäkilä, U.M., Kääpä, P. *et al.* (1986) Maternal ingestion of acetylsalicylic acid inhibits fetal and neonatal prostacyclin and thromboxane in humans. *American Journal of Obstetrics and Gynecology*, **155**, 345–349.

341 Dumont, A., Flahault, A., Beaufils, M. *et al.* (1999) Effect of aspirin in pregnant women is dependent on increase in bleeding time. *American Journal of Obstetrics and Gynecology*, **180**, 135–140.

342 Huhta, J.C., Moise, K.J. Jr and Sharif, D.S. (1987) Human fetal ductus arteriosus constriction from nonsteroidal antiinflammatory drugs. *The American Journal of Cardiology*, **60**, 643 (Abstract 32).

343 Nelson, D.M. and Walsh, S.W. (1989) Aspirin differentially affects thromboxane and prostacyclin production by trophoblast and villous core compartments of human placental villi. *American Journal of Obstetrics and Gynecology*, **161**, 1593–1598.

344 Thorp, J.A., Walsh, S.W. and Brath, P.C. (1988) Low-dose aspirin inhibits thromboxane, but not prostacyclin production by human placental arteries. *American Journal of Obstetrics and Gynecology*, **159**, 1381–1384.

345 Stuart, M.J. and Dusse, J. (1985) *In vitro* comparison of the efficacy of cyclooxygenase inhibitors on the adult versus neonatal platelet. *Biology of the Neonate*, **47**, 265–269.

346 Goodlin, R.C., Haesslein, H.O. and Fleming, J. (1978) Aspirin for the treatment of recurrent toxaemia. *The Lancet*, **2**, 51.

347 Crandon, A.J. and Isherwood, D.M. (1979) Effect of aspirin on incidence of preeclampsia. *The Lancet*, **1**, 1356.

348 Beaufils, M., Uzan, S., Donsimoni, R. *et al.* (1985) Prevention of preeclampsia by early antiplatelet therapy. *The Lancet*, **1**, 840–842.

349 Wallenburg, H.C., Dekker, G.A., Makovitz, J.W. *et al.* (1986) Low-dose aspirin prevents pregnancy-induced hypertension and pre-eclampsia in angiotensin-sensitive primigravidae. *The Lancet*, **1**, 1–3.

350 Vainio, M., Kujansuu, E., Iso-Mustarjävi, M. *et al.* (2002) Low-dose acetylsalicylic acid in prevention of pregnancy-induced hypertension and intrauterine growth retardation in women with bilateral uterine artery notches. *British Journal of Obstetrics and Gynaecology*, **109**, 161–167.

351 Imperiale, T.F. and Petrulis, A.S. (1991) A meta-analysis of low-dose aspirin for the prevention of pregnancy-induced hypertensive disease. *The Journal of the American Medical Association*, **266**, 260–264.

352 CLASP (Collaborative Low-Dose Aspirin Study in Pregnancy) Collaborative Group (1994) CLASP: a randomized trial of low-dose aspirin for the prevention and treatment of pre-eclampsia among 9364 pregnant women. *The Lancet*, **343**, 619–629.

353 Duley, L., Henderson-Smart, D., Meher, S. *et al.* (2007) Antiplatelet agents for preventing pre-ecclampsia and its complications. *Cochrane Database of Systematic Reviews*, **2**, CD004659.

354 Askie, L.M., Duley, L., Henderson-Smart, D.J. *et al.* on behalf of the PARIS Collaborative Group (2007) Antiplatelet agents for prevention of preeclampsia: a meta-analysis of individual patient data. *The Lancet*, **369**, 1791–1798.

355 Viinikka, L., Hartikainen-Sorri, A.-L., Lumme, R. *et al.* (1993) Low dose aspirin in hypertensive pregnant women: effect on pregnancy outcome and prostacyclin-thromboxane balance in mother and newborn. *British Journal of Obstetrics and Gynaecology*, **100**, 809–815.

356 Klockenbusch, W. and Rath, W. (2002) Prävention der Präeklampsie mit Acetylsalizylsäure – eine kritische Analyse [Prevention of pre-eclampsia by low-dose acetylsalicylic acid – a critical appraisal]. *Zeitschrift fur Geburtshilfe und Neonatologie*, **206**, 125–130.

357 Sibai, B.M., Caritis, S.N., Thom, E. *et al.* (1993) Prevention of preeclampsia with low-dose aspirin in healthy nulliparous pregnant women. The National Institute of Child Health and Human Development Network of Maternal–Fetal Medicine Units. *The New England Journal of Medicine*, **329**, 1213–1218.

358 Darling, M. (1998) Low-dose aspirin not for pre-eclampsia. *The Lancet*, **352**, 342.

359 Subtil, D., Goeusse, P., Puech, F. *et al.* (2003) Aspirin (100 mg) used for prevention of pre-eclampsia in nulliparous women: the Essai Regional Aspirine Mere-Enfant study (Part 1). Essai-Regional-Aspirine-Mere-Enfant-ERASME-Collaborative-Group. *British Journal of Obstetrics and Gynaecology*, **110**, 475–484.

360 Heyborne, K.D. (2000) Preeclampsia prevention: lessons from the low-dose aspirin therapy trials. *American Journal of Obstetrics and Gynecology*, **183**, 523–528.

361 Schiff, E., Peleg, E., Goldenberg, M. *et al.* (1989) The use of aspirin to prevent pregnancy-induced hypertension and lower the ratio of thromboxane A_2 to prostacyclin in relatively high risk pregnancies. *The New England Journal of Medicine*, **321**, 351–356.

362 McParland, P., Pearce, J.M. and Chamberlain, G.V. (1990) Doppler ultrasound and aspirin in recognition and prevention of pregnancy-induced hypertension. *The Lancet*, **335**, 1552–1555.

363 Brown, C.E., Gant, N.T., Cox, K. *et al.* (1990) Low-dose aspirin. II. Relationship of angiotensin II pressor responses, circulating eicosanoids, and pregnancy outcome. *American Journal of Obstetrics and Gynecology*, **163**, 1853–1856.

364 Wenstrom, K.D., Hauth, J.C. and Goldenberg, R.L. (1995) The effect of low-dose aspirin on pregnancies complicated by elevated human chorionic gonadotropin levels. *American Journal of Obstetrics and Gynecology*, **173**, 1292–1296.

365 Bower, S.J., Harrington, K.F., Schuchter, K. *et al.* (1996) Prediction of pre-eclampsia by abnormal uterine ultra Doppler ultrasound and modification by aspirin. *British Journal of Obstetrics and Gynaecology*, **103**, 625–629.

366 Italian Study of Aspirin in Pregnancy (1993) Low-dose aspirin in prevention and treatment of intrauterine growth retardation and pregnancy-induced hypertension. *The Lancet*, **341**, 396–400.

367 ECPPA (Estudo colaborativo para prevencao da pre-eclampsia con aspirina) Collaborative (1996) ECPPA: randomized trial of low-dose aspirin for the prevention of maternal and fetal complications in high risk pregnant women. *British Journal of Obstetrics and Gynaecology*, **103**, 39–47.

368 Sullivan, M., Elder, M., de Swiet, M. *et al.* (1998) Titration of antiplatelet treatment in pregnant women at risk of pre-eclampsia. *Thrombosis and Haemostasis*, **79**, 743–746.

369 Carbillon, L. and Uzan, S. (2005) Early treatment with low dose aspirin is effective for the prevention of preeclampsia and related complications in high-risk patients selected by the analysis of their historic risk factors. *Blood*, **105**, 902–903.

370 FitzGerald, G.A., Reilly, I.A. and Pedersen, A.K. (1985) Inhibition of thromboxane formation *in vivo* and *ex vivo*: implications for therapy with platelet inhibitory drugs. *Circulation*, **72**, 1194–1201.

371 Leitich, H., Egarter, C., Husslein, P. *et al.* (1997) A meta-analysis of low dose aspirin for the prevention of intrauterine growth retardation. *British Journal of Obstetrics and Gynaecology*, **104**, 450–459.

372 Hermida, R.-C., Ayala, D.-E., Fernandez, J.-R. *et al.* (1999) Administration time-dependent effects of aspirin in women at differing risk for preeclampsia. *Hypertension*, **34**, 1016–1023.

373 Golding, J. (1998) A randomized trial of low-dose aspirin for primiparae in pregnancy. The Jamaica Low Dose Aspirin Study Group. *British Journal of Obstetrics and Gynaecology*, **105**, 293–299.

374 Rotchell, Y.E., Cruickshank, J.K., Gay, M.P. *et al.* (1998) Barbados low dose aspirin study in pregnancy (BLASP): a randomized trial for the prevention of pre-eclampsia and its complications. *British Journal of Obstetrics and Gynaecology*, **105**, 286–292.

375 Hauth, J., Goldenberg, R.L., Parker, C.R. Jr *et al.* (1993) Low-dose aspirin therapy to prevent preeclampsia. *American Journal of Obstetrics and Gynecology*, **168**, 1083–1091.

376 Hauth, J., Goldenberg, R.L., Parker, C.R. Jr *et al.* (1995) Maternal serum thromboxane B_2 reduction versus pregnancy outcome in a low-dose aspirin trial. *American Journal of Obstetrics and Gynecology*, **173**, 578–584.

377 Kozer, E., Nikfar, S., Costei, A. *et al.* (2002) Aspirin consumption during the first trimester of pregnancy and congenital anomalies: a meta-analysis. *American Journal of Obstetrics and Gynecology*, **187**, 1623–1630.

378 Kozer, E., Costei, A.M., Boskovic, R. *et al.* (2003) Effects of aspirin consumption during pregnancy on pregnancy outcomes: meta-analysis. *Birth Defects Research, Part B: Developmental and Reproductive Toxicology*, **68**, 70–84.

379 Werler, M.M., Mitchell, A.A. and Shapiro, S. (1989) The relation of aspirin use during the first trimester of pregnancy to congenital cardiac defects. *The New England Journal of Medicine*, **321**, 1639–1642.

380 Grab, D., Paulus, W.-E., Erdmann, M. *et al.* (2000) Effects of low-dose aspirin on uterine and fetal blood flow during pregnancy: results of a randomized, placebo-controlled, double-blind trial. *Ultrasound in Obstetrics & Gynecology*, **15**, 19–27.

4.1.6 Aspirin "Resistance"

Variability in Drug Responses Considerable heterogeneity exists in the way individuals respond to drugs in terms of both efficacy and safety. Antiplatelet drugs, such as aspirin, clopidogrel, GPIIb/IIIa antagonists and other compounds, which are used for prevention of myocardial infarction and stroke, are effective only in a certain proportion of patients but are ineffective in many others [381]. This variability to antiplatelet drugs is well documented in the large meta-analysis of the Antiplatelet/Antithrombotic Trialists' Collaboration [382]. The interindividual variability of platelet responses to aspirin and other antiplatelet agents clinically presents as treatment failure, that is, a vascular thromboembolic event despite of regular intake of the drug at recommended doses, that is, 75–325 mg/day. This phenomenon is frequently named "resistance" and is finding increasing clinical attention. However, treatment failures with drugs are neither new nor unusual nor is their incidence really increasing. There appears rather to be an increasing awareness of this phenomenon, which is also evident from an increasing detection of clopidogrel "resistance" (Figure 4.21). The large interindividual variability additionally suggests that there will be many more cases of treatment failures, including missing compliance but also an altered platelet sensitivity, than an insufficient pharmacological action of the drug [292]. Thus, clinical treatment failure and a failure of aspirin to work pharmacologically are *no* synonyms but, unfortunately, are frequently mixed up in an inappropriate way causing much confusion [383]. In addition, "resistance" is certainly not the optimum term to describe this phenomenon, which rather means a low or absent responsiveness. However, the term "resistance" will be used in this article because of its widespread acceptance.

Principal Pharmacological Mechanisms of Aspirin "Resistance" From a pharmacological point of view, drug treatment in general and antiplatelet treatment in particular may fail for three reasons: (i) the active (antiplatelet) compound is not available at the site of action at a sufficient concentration for a sufficient period of time; (ii) the active compound cannot work because of an impaired sensitivity of the cellular or molecular target, for example, a change in the phenotype of (platelet) COX-1 or other forms of platelet hyperreactivity, and (iii) the platelet-stimulating factors that initiate thrombus formation are not sensitive to aspirin.

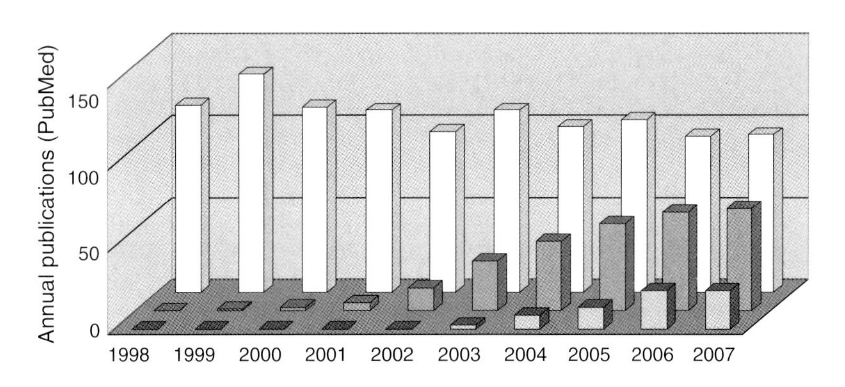

Figure 4.21 The increasing clinical interest in aspirin "resistance" during the past 10 years. Similar developments with clopidogrel.

Table 4.7 Mechanisms of pharmacological aspirin "resistance."

Drug related
Pharmacokinetics
Insufficient bioavailability
Prevention of binding of the Ser_{530} by traditional NSAIDs (ibuprofen, indomethacin, and naproxen)
Drug interaction with pyrazoles (dispyrone and others)
Pharmacodynamics
Impaired sensitivity of platelet COX-1 (CABG)
Gene polymorphism(s)
Disease related
Platelet hyperreactivity (diabetes)
Aspirin-insensitive mechanisms of platelet activation and aggregation
Facilitation of platelet adhesion (changes in the collagen receptor)
Platelet thromboxane receptor "sensitizing" (isoprostanes?)

An overview of possible mechanisms of aspirin "resistance" is shown in Table 4.7.

4.1.6.1 Definition and Types of Pharmacological Aspirin "Resistance"

Pharmacologically, the failure of aspirin to act is due to its inability to hit the molecular target, that is, to inhibit platelet function via inhibition of COX-1. However, there is neither a generally accepted definition of aspirin "resistance" nor any generally accepted procedure for its determination. In case of the antiplatelet action of aspirin, measurement of thromboxane formation in combination with measurement of platelet function appears to be a logical approach.

The Weber Assay Weber *et al.* [384] have proposed a typological approach to analyze aspirin "resistance" in pharmacological terms by studying the potency of aspirin to inhibit platelet-dependent thromboxane formation and thromboxane-dependent platelet aggregation in one and the same assay. This allows for a pharmacological separation of different types of resistance.

Aspirin "resistance" was measured pharmacologically by simultaneously determining the inhibition of collagen-induced thromboxane formation and platelet aggregation by aspirin *in vitro*. A low dose (1 μg/ml) of collagen was chosen that, under the conditions of this assay, required release of endogenous arachidonic acid for an aggregation response. In the case of incomplete or missing inhibition of platelet aggregation after oral treatment *ex vivo*, three different reaction profiles could be separated according to the alterations in thromboxane formation and platelet aggregation after *in vitro* addition of aspirin.

Type I (pharmacokinetic resistance): No inhibition of platelet aggregation after oral treatment *in vivo* but inhibition of aggregation and thromboxane formation after addition of aspirin *in vitro*. This suggests that aspirin does work in these platelets as expected but is not sufficiently bioavailable *in vivo*, most likely due to a missing compliance or drug competition with its binding to the hydrophobic channel of COX-1.

Type II (pharmacodynamic resistance): No inhibition of platelet aggregation after oral treatment *in vivo* but partial inhibition of platelet aggregation and thromboxane formation after addition of aspirin *in vitro*, which could be completed by increasing the aspirin dose. This "true" resistance suggests a pharmacodynamic failure of aspirin to act, for example, because of reduced sensitivity of the platelet COX-1 or residual platelet activity.

Type III (pseudoresistance): No inhibition of platelet aggregation after oral treatment *in vivo* and no inhibition of platelet aggregation *in vitro* despite of complete inhibition of thromboxane synthesis after addition of aspirin. Possible reasons are platelet activation by nonthromboxane-dependent pathways, for example, high-dose collagen (Figure 4.22).

This assay has also been successfully used for the classification of clinical syndrome of aspirin "resistance" [385].

4.1.6.2 Detection of Aspirin "Resistance"

As mentioned above, there are principally two mechanism-based methods to detect pharmacological platelet aspirin "resistance" – the measurement of one or more parameters of platelet function *ex vivo* or the determination of (platelet-dependent) thromboxane formation. Both approaches have limitations, the most important being (i) the absence of a generally accepted technology of measurement; (ii) the absence of generally accepted "normal" values to allow standardizations and

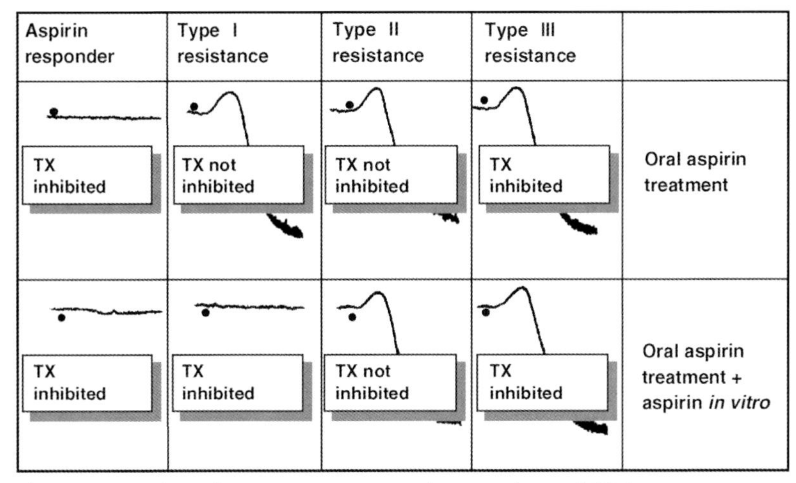

Figure 4.22 Typology of aspirin resistance according to Weber *et al.* [384].

comparisons between different laboratories, and (iii) the (still) poorly defined predictive value for clinical outcome, that is, the transfer of laboratory data into clinical reality.

A significant, aspirin-sensitive elevation of plasma and urinary 11-DH-TXB_2 associated with platelet activation syndromes, such as unstable angina, severe atherosclerosis, peripheral arterial disease, and pulmonary embolism is well known [386–389]. This confirms the usefulness of this approach.

Platelet Function and Aggregate Formation Light transmission aggregometry of platelets in citrated platelet-rich plasma is still the most popular technology to detect aspirin "resistance" [390, 391]. However, any study of blood platelets *ex vivo* or *in vitro* is done under conditions that differ fundamentally from the *in vivo* situation where circulating platelets do continuously interact with blood components and the vessel wall. In all of these assays, platelets are removed from this natural environment and the numerous platelet-active factors released from the endothelium or circulating in the blood. These factors control basal platelet activity and maintain its discoid shape under resting conditions. Several of these factors are either labile and are already inactivated (NO, prostacyclin) or not present *in vitro* for other reasons (endothelial ADPase), when the measurement of platelet function is performed. Mechanical stimulation of platelets by centrifugation or pipetting (shear stress) might result in *ex vivo* activation prior to addition of stimulating agents. This effect is probably more pronounced if platelets are already hyperreactive *in vivo*, for example, in patients with a history of previous atherothrombotic events. There are no red cells in many platelet assays. Red cells markedly increase platelet reactivity [392] and will reduce the antiplatelet actions of aspirin [393]. Finally, measurement of platelet function in platelet-rich plasma does not consider possible effects of aspirin on other cells circulating in blood that may influence the platelet-dependent hemostatic process. In addition to red cells, this includes white cells, which release inflammatory, that is, prothrombotic, cytokines. Thus, any analysis of platelet aspirin sensitivity that is solely based on the measurement of platelet inhibition *in vitro* after stimulation by one particular agonist would not capture the full anti-inflammatory benefit of antiplatelet therapy [394].

Different approaches are used to determine platelet activity. The most frequently used include measurement of platelet function, that is, aggregate formation (photometric aggregometry according to Born, PFA-100, and impedance aggregometry), expression of P-selectin, GPIIb/IIIa and other activation markers at the platelet surface (flow cytometry). The VerifyNow turbidimetric-based assay determines platelet aggregation in whole blood and currently becomes increasingly popular. Platelet reactivity tests originally described by Wu and Hoak [395] (Section 4.1.2) determine a platelet count ratio, for example, before and after standardized stimulation with an agonist. This will allow calculation of the fraction of platelets that become activated as percentage of the total.

In practical use, all of these platelet tests have limitations. Regarding the detection of aspirin "resistance," different assays yield different results [396]. The PFA-100, for instance, though a widely used approach appears to differ from other measurements of platelet function and provides significantly higher numbers of "resistant" individuals than other methods [397–401]. It even yielded different results in the same patients, that is, platelets, which were "resistant" in one assay were not "resistant" in another [402]. In the absence of appropriate controls, that is, platelet assays *prior* to aspirin ingestion, results may even disqualify patients [391]. Thus, these data should be interpreted with caution and large prospective studies will be required to determine the prospective value of each of the separate tests.

There are also technical issues regarding the measurement of platelet function in platelet-rich plasma *in vitro* that deserve documentation. For example, measurements should be started rapidly after blood withdrawal (because of the increase in pH with evaporating CO_2 that increases platelet reactivity). In addition, all platelet assays are markedly influenced by the type of anticoagulant and agonist used. For example, citrate will reduce the Ca^{2+} levels in the sample close to zero, while anticoagulants such as antithrombins (hirudin, factor Xa inhibitors) will maintain physiological Ca^{2+} levels, allowing a more natural activation of the platelet GPIIb/IIIa receptor. However, these compounds eliminate the most potent natural platelet activator, that is, thrombin, which might be an at least partially thromboxane-dependent target of aspirin and whose generation appears to be less inhibited in aspirin-resistant individuals [403, 404]. It should also be noted that the frequently used measurement of inhibition of ADP-induced aggregation by aspirin is an *in vitro* artifact (in non-aspirin pretreated individuals). In the presence of physiological Ca^{2+} concentrations, that is, *in vivo*, aspirin will not affect ADP-induced platelet responses because ADP-induced platelet secretion and aggregation are not aspirin sensitive. The reason is the absence of ADP-induced thromboxane formation [405, 406]. These limitations have prompted the European Heart Association not to recommend the measurement of platelet function as a parameter for aspirin "resistance" [407].

Thromboxane Formation Another frequently used procedure is the measurement of inhibition of thromboxane formation. Serum thromboxane corresponds to the capacity of thromboxane biosynthesis and is a useful surrogate parameter for the pharmacological aspirin action. The inhibition should amount to at least 95%, which is equivalent to thromboxane levels of about 25 ng/ml [408]. Less inhibition is clinically ineffective, that is, a 70% inhibition of serum thromboxane formation might be statistically significant but is biologically irrelevant [409]. Insufficient thromboxane inhibition appears also to be related to a poor clinical outcome, for example, after acute ST-elevation myocardial infarction [410]. Regarding the reliability of thromboxane measurements, one has to consider that this method is just a laboratory test of aspirin action without any natural correlate *in vivo*. Endogenous thromboxane levels, for example, in the area of thrombus formation, are considerably lower and also cannot accumulate with time (Section 3.1.2).

A more physiological approach to estimate the aspirin action on thromboxane formation in the cardiovascular system is the measurement of thromboxane metabolite excretion, such as 11-DH-TXB_2, in urine. However, 11-DH-TXB_2 is the only one out of 20 metabolites of $TXA(B)_2$ in urine [411] and the conversion rate of TXB_2 into this metabolite is only about 7% [412]. Importantly,

Table 4.8 Urinary levels of 11-dehydro-TXB_2 (11-DH-TXB_2) before and after aspirin treatment (selection).

	11-DH-TXB_2			
n	**Before aspirin**	**After aspirin**	**% Reduction**	**Reference**
24	75 ± 13 ng/mmol cr (SEM)	17 ± 3 ng/mmol cr	77	[388]
24	815 ± 183 ng/g cr (SEM)	266 ± 114 ng/g cr	67	[415]
64	About 450 pg/mg cr	About 160 pg/mg cr	72	[389]
16–71	1386 ng/g cr (176–3844, range)	783 (149–7415) ng/g cr	44	[416]
488	No data	22.8 ng/mmol cr (cases MI)	n.d.	[417]
488	No data	20.3 ng/mmol cr (control MI)	n.d.	

the proportional conversion into 11-DH-TXB_2 can be markedly influenced by environmental factors, such as smoking, being twice as much in nonsmokers (healthy or with cardiovascular disorders) as opposed to smokers [413]. A varying proportion of about 30% of 11-DH-TXB_2 in urine is probably not platelet derived, that is, COX-1-derived, but comes from other sources, such as monocytes/macrophages [414]. These cells are naturally "resistant" against aspirin at antiplatelet doses because of their COX-2-dependent thromboxane formation. This proportion of COX-2-dependent thromboxane formation will depend on the (acute) inflammatory state of atherosclerosis. Accordingly, there is a high variation in both published 11-DH-TXB_2 data and its reduction by aspirin (Table 4.8). Until now, there is no generally accepted definition of threshold or a normal range.

4.1.6.3 Pharmacological Mechanisms of Aspirin "Resistance"

An overview of drug- and disease-related mechanisms of aspirin resistance is shown in Table 4.7 [418]. This section discusses pharmacological mechanisms.

Insufficient Bioavailability of Low Aspirin Doses
The pharmacokinetics of aspirin should allow sufficient bioavailability of the active, that is, nonmetabolized compound, to its binding sites in the hydrophobic channel of platelet COX-1 protein (Section 2.2.1). Given that there are considerable differences in the presystemic and systemic bioavailability of aspirin, it can be predicted that there might be a fraction of individuals who do not show a sufficient inhibition of platelet COX-1 as a result of individual underdosing [419].

In order to study the reproducibility of antiplatelet effects of low-dose aspirin, 10 healthy volunteers were treated with three single doses of 80 mg/day, separated by a washout period of 2 weeks. The area under the aspirin concentration–time curve (AUC) in plasma was measured and correlated with the inhibition of arachidonic acid-induced platelet aggregation *ex vivo*.

There was a large intra- and interindividual variability in the AUC (0.4–1.7 μg h/ml), which, in some individuals and on some occasions, was related to an insufficient platelet response to aspirin, suggesting that an impaired bioavailability favors an insufficient platelet response to the compound. However, the overall correlation was also poor, possibly because the AUC in the systemic circulation was not the ideal pharmacokinetic parameter to determine the bioavailability of aspirin for the platelets [420] (Figure 4.23).

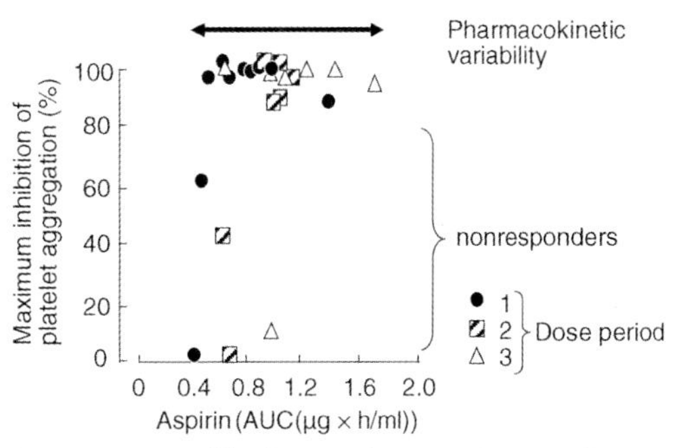

Figure 4.23 Variable pharmacokinetics of aspirin (for further explanation, see the text) (data from [420]).

A possibly too low bioavailability of low-dose aspirin (75 mg/day) was also described for enteric-coated preparations, which were less effective than a plain preparation at the same dose [421].

In addition to the drug, that is, aspirin, the platelets themselves may also become a variable because of kinetic reasons. Approximately, 10% of the existing platelet pool is renewed between the (daily) dosing intervals of aspirin. Because of the short half-life of aspirin (15–20 min), these platelets are largely unprotected and carry an active COX-1, capable of full thromboxane formation. An increased platelet turnover, for example, in myeloproliferative diseases or after platelet destruction by artificial surfaces, such as extracorporeal circulation [422] may result in a larger proportion of new, partially immature platelets that is sufficient to overcome the inhibition of thromboxane formation in pretreated platelets. This may explain, why low-dose aspirin has a limited efficacy in the reduction of large-vessel thrombosis in patients, suffering from essential thrombocythemia [423–425].

Competition with Other Drugs for Binding at COX-1 Another pharmacokinetic reason for insufficient antiplatelet effects of aspirin is an interference, not only with traditional NSAIDs but also with dipyrone [426] (Figure 4.24), with aspirin binding in the hydrophobic channel of COX-1. Binding of one of the reversible inhibitors to this site will prevent the initial (reversible) binding and subsequent (irreversible) acetylation of ser_{530} by aspirin (Figure 2.10 in Section 2.2.1). Because of the short half-life of aspirin (15–20 min), the active compound will be rapidly deacetylated to salicylic acid by aspirinases in the blood. When the reversible inhibitor leaves the binding site and becomes degraded, no active aspirin may be available anymore [426–428] (Section 2.2.1). This interaction is not to be expected with selective COX-2 inhibitors [429], which do not interact with COX-1. The possible clinical significance of this finding is underlined by the fact that three epidemiological studies provided evidence of a clinically important pharmacological interaction between aspirin and ibuprofen [430].

Changes in Platelet Cyclooxygenases Platelets of patients undergoing CABG surgery become largely resistant to conventional doses of oral aspirin, that is, 100 mg/day, shortly after the surgical procedure [422]. This "resistance" can be overcome *in vitro* by increasing the aspirin concentration [431]. Human platelets express an immunoreactive COX-2 [432], in particular in an immature state [433], which makes them less sensitive to aspirin [434]. Increased platelet turnover might result in an over-

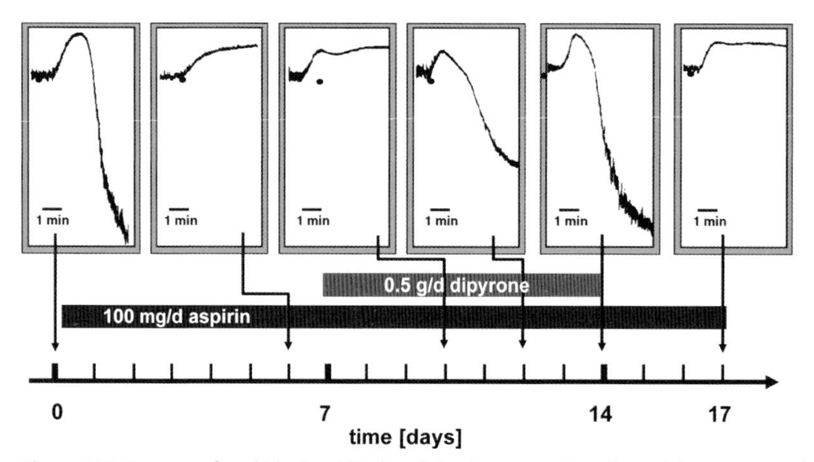

Figure 4.24 Recovery of arachidonic acid-induced platelet aggregation after oral dipyrone in one healthy subject (T.H.) during regular oral aspirin intake (Hohlfeld, personal communication).

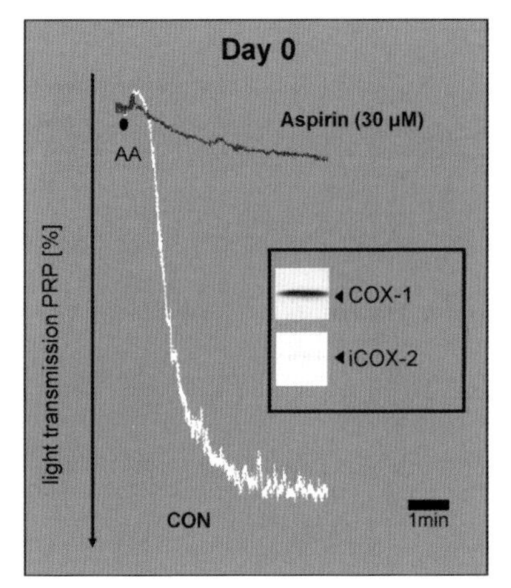

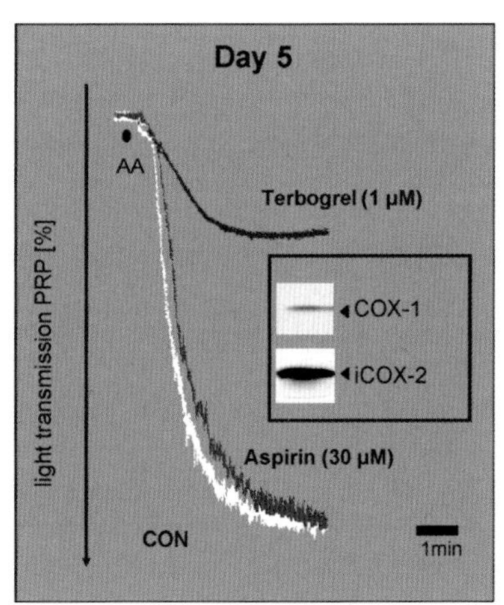

Figure 4.25 Expression of COX isoforms after coronary artery bypass grafting (CABG) and arachidonic acid (AA)-induced platelet aggregation before (day 0) and after (day 5) CABG in one patient. Different effects of aspirin and terbogrel (combined thromboxane synthase inhibitor and receptor antagonist) *in vitro*. Enhanced expression of immunoreactive COX-2 (i-COX-2, probably COX-2a [436]), no change in COX-1. (modified after data in [431]).

expression of COX-2 protein at unchanged COX-1 [432, 435]. Interestingly, this immunoreactive COX-2 was not involved in thromboxane production. The explanation was a new COX-2 isoform (COX-2a), which was about 50-fold upregulated in platelets of these patients [436] (Figure 4.25). Alternatively, the sensitivity of COX-1 against aspirin might be reduced because of gene polymorphisms (see below) or there might be an intrinsic change in platelet sensitivity that is not aspirin sensitive [437].

Gene Polymorphisms Several prothrombotic genetic variations, relevant to antiplatelet effects of aspirin, have been described [438–440]. Phenotypic alterations in the platelet GPIIIa receptor [441] have been found, which might be involved in residual platelet activity. Heritable factors are currently assumed to contribute prominently to the variability in residual platelet function after aspirin exposure [442].

Aspirin-Insensitive Platelet Activation In contrast to *in vitro* testing, platelets *in vivo* will regularly become activated by multiple stimuli that usually act in concert. Many of them are aspirin "resistant" because they do not require activation of the thromboxane pathway to become effective. This includes platelet activation by ADP [405] and higher concentrations of collagen as well as adrenaline [443], emotional stress [444, 445], and mechanical stimulation by shear stress [446, 447]. If these stimuli become critical to platelet activation, aspirin may not be effective despite sufficient inhibition of platelet-dependent thromboxane formation.

Isoprostanes Isoprostanes are arachidonic acid metabolites that are formed by nonenzymatic, free radical-catalyzed reactions [448]. Isoprostanes cause platelet activation via a mechanism that can be antagonized by blockade of the platelet thromboxane (TP) receptor [449]. Isoprostanes are increasingly formed *in vivo* when cyclooxygenase activation and oxidant stress coincide, for example, in stable and unstable angina [389] or type II diabetes [450]. The platelet activation by isoprostanes is not sensitive to aspirin [389] but can be prevented by thromboxane receptor antagonists [451]. Interestingly, the thromboxane antagonist picotamide was more effective than aspirin in type II diabetics with peripheral arterial disease, a situation with presumably increased oxidative stress (Section 4.1.3).

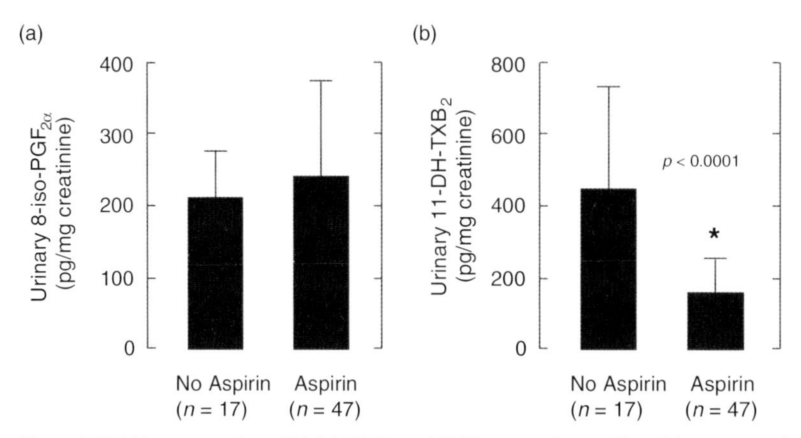

Figure 4.26 Urinary excretion of 11-DH-TXB_2 and $iPGF_{2\alpha}$ in patients with stable angina and its changes by aspirin treatment [389].

4.1.6.4 **Clinical Trials**

The clinical outcome is clearly the most important information that is to be expected from all laboratory measurements of aspirin "resistance." This includes disease-related factors, eventually resulting in a residual platelet activity even in the presence of a satisfying pharmacological action of aspirin, a fact that is not yet sufficiently appreciated [452, 453].

Depending on the method of measurement and the definition used, approximately 5–45% of cardiovascular and 5–65% of stroke patients [430, 454–458] are considered aspirin "resistant." So far, these numbers, elaborated by different techniques and protocols are hard to compare and rather suggest a large variability in assay procedures and clinical conditions because the "real" pharmacological inability of aspirin to work is much lower, probably <5% (see below). Therefore, these numbers should not be overestimated. Table 4.9 summarizes a selection of studies indicating the prevalence of aspirin "resistance" in various patient populations.

Compliance A frequent drug-related explanation of an insufficient efficacy of aspirin is missing compliance [466]. In long-term prevention studies, 2 days or more without aspirin may result in a too low percentage of platelets with sufficient COX-1 inhibition, specifically in situations with a more rapid platelet turnover [468]. In this respect, it is interesting to note that participants in the US Physicians' Health Study (Section 4.1.1) who took aspirin less than on alternate days [469] were not protected from cardiocoronary events (Section 4.1.1). The possible significance of one or more missed doses of aspirin during the 10-year (!) treatment period in the Women's Health Study, using also a two-day interval and only 100 mg aspirin each second day (Section 4.1.1), is unknown.

Aspirin "Resistance" and Cardiovascular Outcome In 2002, Eikelboom and colleagues were the first to present data from a subgroup of patients of the HOPE trial [179], which have generated concerns regarding the clinical significance of aspirin "resistance" in cardiocoronary prevention. This study is frequently cited to suggest that aspirin "resistance" is (i) associated with an increased risk of myocardial infarctions and cardiovascular death and that (ii) both can be predicted by measuring the excretion of the thromboxane metabolite 11-DH-TXB_2 in urine. Because of these important statements and the subsequently increasing use of urinary 11-DH-TXB_2 measurements to detect aspirin "resistance," this study and its limitations will be discussed in more detail.

The "Heart Outcomes Prevention Evaluation (HOPE) study" was designed to compare the ACE inhibitor

Table 4.9 Prevalence of aspirin "resistance" in various patient populations.

Patient population	*n*	Aspirin dose (mg/day)	Platelet function assay	% Aspirin resistance according to individual definition	Reference
Acute myocardial infarction	143	75–160	Platelet aggregation ratio	1.4–9.8	[459]
Stroke	180	1500	Aggregation	33.3	[460]
Stable coronary artery disease	325	325	Optical aggregation PFA-100	5.5 9.5	[402]
High risk of cardiovascular events	488	75–325	Urinary 11-DH TXB_2	25	[417]
Stable coronary artery disease	422	81–325	PFA-100	23	[455]
Stroke	53	100	PFA-100	34	[397]
Diabetes Type II	172	100	PFA-100	22	[461]
Nonurgent PCI	151	81–325	RPFA	19.2	[462]
Stable coronary artery disease	468	80–325	Ultegra	30 ($\leq$100 mg) 17 (150 mg) 0 (300 mg)	[463]
Diabetes	203	325	RPFA	18.7	[464]
PCI	223	325	LTA-1 mM AA TEG	<1	[465]
History of myocardial infarction	190	325	LTA plateletworks	<1	[466]
Stable coronary artery disease	100	81, 162 and 325	LTA (Collagen, ADP)	?16	[467]
Stable coronary artery disease	100	81, 162 and 325	LTA VerifyNow TEG	?1 ?5 ?5	[401]

ramipril and vitamin E with placebo in high-risk cardiac patients. A total of 5529 patients were enrolled and a retrospective subgroup analysis, using a nested case–control design was performed. All of the patients provided baseline urine specimens. Aspirin treatment was started at least 6 months before study entry. Thus, all (compliant) patients were on aspirin during the study and there was no urine sample available from non-aspirin-treated patients, that is, no untreated control group.

During the 5-year follow-up, 488 of these patients suffered a heart attack, stroke, or a fatal vascular event. The urinary 11-DH-TXB_2 level of these patients was compared with 488 matched controls of the same study who did not suffer an event. It was found that with increasing concentrations in urinary 11-DH-TXB_2, there was an increasing risk for cardiovascular events, being significantly higher in the highest quartile of urinary 11-DH-TXB_2 excretion compared to the lowest quartile. The median urinary 11-DH-TXB_2 level was 22.8 ng/mmol creatinine in the myocardial infarction cases and 20.3 ng/mmol in the controls ($p = 0.001$), the median in cases of cardiovascular death was 24.0 ng/mmol creatinine in cases and 19.9 in controls ($p < 0.001$). No differences were seen in stroke patients.

The conclusion was that (i) high-level 11-DH-TXB_2 in urine identifies patients who are resistant to aspirin and (ii) that these patients are at elevated risk for myocardial infarction and cardiac death [417].

This study was subjected to several comments and criticisms. First, measurement of 11-DH-TXB_2 in this and other studies is not a trivial procedure. This thromboxane metabolite exists in both open and closed forms, which are in a pH-dependent equilibrium. Complete conversion into one isoform requires about 3 h storage at pH 8 [470]. For the conventional enzyme immunoabsorption assay (EIA) this means that the sample pH is an important variable. The EIA used by Eikelboom and colleagues was based on an antibody that exclusively detected the open form that is spontaneously generated at alkaline pH. According to the producer, the assay buffer and protocol guaranteed complete conversion. Second, metabolite measurements might have been influenced by urinary contaminations. There is no information in the paper whether urinary samples were purified prior to the assay. Perneby *et al.* [471] strongly recommended efficient sample purification before analysis and, in addition, noted a doubling of sensitivity of their EIA for urinary 11-DH-TXB_2 if the samples were incubated and handled at alkaline pH. Although it has been suggested that urinary 11-DH-TXB_2 is the most appropriate analytical target to detect release of thromboxane into the circulation [386], it is also evident that the published urinary excretion rates of 11-DH-TXB_2 not only in healthy volunteers but also in patients with coronary heart disease are largely variable [442], making a standardization, that is, definition of normal values, extremely difficult.

The differences in urinary 11-DH-TXB_2 excretion reported by Eikelboom *et al.* [417] suggest a statistically significant trend for correlation of urinary 11-DH-TXB_2 excretion with the clinical outcome but could not establish any causal relationship according to the study design. In contrast to the total HOPE population, the cases in the study subgroup of Eikelboom also differed significantly from the controls in important entry criteria: for example, there were more patients with a history of myocardial infarctions, hypertension, diabetes, peripheral arterial disease, and current smokers among the cases. Finally, the about 10–20% difference in 11-DH-TXB_2 excretion between cases and controls is small compared to the highly variable literature data on excretion of this metabolite during aspirin treatment. Overall, simple conclusion from these data might be that patients who are more severely ill have a higher residual platelet sensitivity and are more likely to die earlier. In any case, the question whether or not urinary 11-DH-TXB_2 excretion is a suitable marker to predict increased vascular risk during aspirin treatment remains unanswered.

Aspirin Dosage A burning issue regarding aspirin resistance, its detection, and consequences is the question whether it can be overcome by higher doses. One of the first prospective randomized trials to study this issue in detail in patients with coronary heart disease was the Aspirin-induced Platelet Effect Study (ASPECT) by Gurbel and colleagues.

A total of 125 outpatients with stable coronary heart disease were randomized in a double-blind, double-crossover investigation to receive aspirin at 81, 162, and 325 mg daily for 4 weeks, each over a 12-week period. Platelet function and thromboxane metabolite excretion were measured by different laboratory procedures. The question was to determine the degree of platelet responsiveness to aspirin, to compare different techniques, and to evaluate the relation of aspirin doses to platelet inhibition.

At any one dose, resistance to aspirin was low – 0–6% – in the overall group when arachidonic acid was used as the stimulus but amounted to 1–27% when other methods were used, the figures seen with PFA-100 being highest (Table 4.10). Platelet responses to aspirin, as measured by collagen- and ADP-induced light transmission as well as the PFA-100, were dose related (81 mg/day

Table 4.10 Effects of assay and doses on measurement of aspirin resistance.

Technology for determination of aspirin resistance	Aspirin resistance (n)					
	81 mg	162 mg	325 mg	≥1 dose	2 doses	3 doses
LTA-AA	2	1	0	2	1	0
LTA-collagen	12	2	1	14	1	0
LTA-ADP	19	11	10	27	7	3
TEG-AA	5	3	5	11	2	0
PFA-100	32	14	21	42	15	5
Urinary 11-DH-TXB_2	31	22	14	42	16	5
VerifyNow	7	4	4	13	2	0

Abbreviations: LTA, light transmission aggregometry; TEG, thromboelastogramm (for further explanations see text) (modified after [401]).

versus 162 mg/day, $p \leq 0.05$) and there was also a dose-related inhibition of 11-DH-TXB_2 excretion. No carryover effects were observed.

The conclusion was that the assessment of aspirin resistance is highly assay dependent, arachidonic acid stimulation being the most sensitive stimulus. The dose-dependent effects despite nearly complete inhibition of arachidonic acid-induced aggregations suggest additional antiplatelet effects of aspirin that are COX-1 independent [401].

These data of Gurbel *et al.* [401] on dose dependency of aspirin-induced inhibition of platelet aggregation and "resistance" in cardiac patients confirm earlier studies on prevalence and variation of aspirin "resistance" in patients with previous ischemic stroke [472].

To evaluate the *ex vivo* effect of aspirin on platelet aggregation, 306 patients with previous ischemic stroke were treated over 33 months with aspirin. The efficacy of aspirin was studied by platelet aggregation testing. Increasing doses of aspirin (325–1300 mg/day) were administered and the extent of inhibition of platelet aggregation was determined 2 weeks after treatment and thereafter at 6 months intervals.

Of the 306 patients, 228 had complete and 78 partial inhibition of platelet aggregation at initial testing. At repeated testing, 39 of the 119 patients (33%) with complete inhibition at initial testing had lost part of the antiplatelet effect of aspirin without change in aspirin dosage. Of the 52 patients with partial inhibition at initial testing, 35 achieved complete inhibition by aspirin dosage escalation. Ultimately, 8% of patients exhibited aspirin "resistance" to 1300 mg/day.

The conclusion was that the antiplatelet (and possibly antithrombotic) effect of aspirin may not be constant over time in all individuals. There is evidence for increased dosage requirements or development of resistance over time, the mechanisms of which are undefined [472].

These studies and others confirm that high-dose aspirin might be more effective than low-dose in stroke patients [473], and the individual sensitivity of platelets in these patients may decrease with time after the acute event [460]. This conclusion is further supported by antiplatelet/antithrombotic actions of aspirin that are thromboxane-independent and detectable only at higher doses, such as inhibition of thrombin formation [474], increased fibrinolytic activity (Section 2.3.1) and the well-known anti-inflammatory effects of the compound [391].

Clinical Outcome Studies There are only a few trials, connecting the clinical outcome of patients at elevated vascular risk with a possible aspirin "resistance" (Table 4.11). Gum *et al.* [475] reported aspirin "resistance" as detected from light transmission aggregometry in 17 out of 326 patients with stable coronary disease. These patients were found to have a threefold higher risk of a cardiovascular event than those who were not "resistant" according to study criteria. However, this difference was only seen with optical aggregometry but not with

Table 4.11 Outcome data of aspirin resistance.

Patient population	n	Aspirin dose (mg/day)	Aspirin resistance assay	Follow-up duration	End points	Results	References
Prior cerebrovascular event	180	1500	Platelet reactivity	24 months	Stroke, MI, vascular death	OR: 14.5 95% CI: 5.2–40.9; $p < 0.001$	[460, 477]
Peripheral arterial occlusive disease	100	100	Whole blood aggregometry	18 months	Arterial reocclusion	87% higher risk in aspirin-resistant patients	[476]
Thromboembolic diseases (CAD, PAD, and cerebrovascular)	53	100	PFA-100	>60 months	Recurrent stroke/TIA	34% of patients with recurrence in aspirin nonresponder	[397]
Acute myocardial infarction (subgroup analysis of HOPE study)	976	75–325	Urinary 11-DH-TXB_2	5 years	MI/stroke, CV death	1.8 times higher risk of aspirin nonresponders	[417]
Stable CVD	326	325	Optical aggregometry	2–3 years	Death, MI, stroke	24% versus 10% (aspirin-resistant versus aspirin-sensitive)	[475]
Acute myocardial infarction (subgroup analysis of WARIS II trial)	202	75–160	PFA-100	4 years	Death, reinfarction, and stroke	36% versus 24% (aspirin-resistant versus aspirin-sensitive)	[478]
Nonurgent PCI	151	80–325 (+ clopidogrel)	Ultegra rapid platelet function assay	1 day	Risk of post-PCI myonecrosis (CCK-MB; TnT)	In the 19% resistant patients significant elevation of CKMB and TnT versus nonresistant	[462]
Patients on off-pump CABG	225	325	Thromboelastography whole blood aggregometry 11-DH-TXB_2 (30%)	30 days	Early graft thrombosis	30% aspirin-resistant, graft thrombosis associated with aspirin-resistant $p < 0.04$	[479]
Prior cerebral infarct or ischemic heart disease	136	81	Optical aggregometry PA-20 platelet analyzer (quartile comparison)	1 year	Risk of cardiovascular events in upper quartile	Patients in upper quartile (least sensitive): HR 7.98, $p = 0.008$; PA-20 HR 7.76, $p = 0.007$	[480]

the PFA-100 assay, and the group of patients was small. A correlation was also found by others between reduced aspirin sensitivity and coronary heart disease [455].

Chen and colleagues (2004) reported that patients undergoing elective percutaneous coronary interventions and combined treatment with aspirin and clopidogrel were more likely to have periprocedural myonecrosis if their platelets were aspirin-“resistant” according to the ULTEGRA-platelet assay. In a prospective trial, these authors also found a higher incidence of atherothrombotic complications in aspirin-treated (80–325 mg/day) patients with stable coronary artery disease, if they were “resistant” according to the ULTEGRA-platelet assay [481]. However, no increased appearance of myonecrosis in low-risk, aspirin-“resistant” patients, undergoing elective PCI was found by Buch *et al.* [482], and no difference in coronary patients with a history of myocardial infarction compared to those without was found by Dorsch *et al.* [483]. Both studies also used the ULTEGRA technique.

Residual Platelet Activity The inhibition of platelet function by aspirin can be overcome by increasing the dose of the agonist, the only exception being arachidonic acid or low-dose collagen that cause an entirely thromboxane-dependent response. This suggests that the variability in the inhibition of platelet function, that is, “resistance” or reduced responsiveness, may be due to factors independent of the platelet thromboxane pathway [401], ADP being a possible candidate [484]. In this respect, it should be remembered that platelets from patients at advanced stages of atherosclerosis circulate in an activated state and are hyperreactive to platelet agonists, such as ADP [485] without any increased levels of circulating thromboxane (Section 4.1.1). This suggests that a failure of aspirin to act might be due to the severity of the disease and/or the involvement of non-aspirin-sensitive, platelet-derived mechanisms, that is, a residual platelet activity.

Actual Situation According to recent data, pharmacological aspirin resistance is highly assay-dependent and a very random event affecting less than 5% of patients [401, 484, 486]. There are clinical data suggesting a relationship between clinical outcomes, that is, possible treatment failure in the case of pharmacological aspirin “resistance” [410, 462, 481]. However, large prospective randomized trials are still missing and it is entirely possible that the vast majority of treatment failure with aspirin are due to either disturbed platelet sensitivity (hyperreactivity) or missing compliance. According to data of Gurbel *et al.* [401], one might expect that increase in dosing may partially overcome a reduced responsiveness of the platelet. Again, there are data to suggest a lower incidence of aspirin “resistance” at higher doses [401, 463]. A recent meta-analysis of aspirin treatment after coronary surgery has shown that medium-dose aspirin (300–325 mg) is about twice as effective as low-dose aspirin (75–150 mg) – 45% versus 26% risk reduction in preventing graft occlusion in bypass patients [487]. However, more data are required, specifically regarding patient subgroups with particular risk profiles, such as diabetics, before general statements can be made. Specifically, a clear distinction should be made between monitoring patients on antiplatelet treatment to identify poor responders, that is, potentially “resistant” patients from those with a high residual platelet reactivity [452].

The general recommendation to date is that regarding the uncertainties of the transfer of measurements of platelet function *in vitro*, the absence of standardized procedures of measurement, the method-related differences in results, and the absence of proven effective alternative, patients should not be routinely tested for possible aspirin “resistance” [488]. Pharmacological aspirin “resistance” is a very random event [400, 401] and, in most cases, does not matter clinically [418].

Summary

A variable responsiveness to antiplatelet drugs is a phenomenon that does not principally differ from other kinds of drug treatment. The question is whether reduced responsiveness to aspirin ("resistance") is related to the clinical outcome of the individual patient and, therefore, needs to be controlled routinely and which method should be used to detect it. Pharmacological "resistance" is a very rare, highly assay-dependent event and occurs in less than 5% of patients.

Principally, there are two different methods of laboratory control for platelet sensitivity to aspirin treatment: Measurement of platelet activation (*ex vivo*) or measurement of inhibition of thromboxane formation, for example, in terms of thromboxane metabolite (11-DH-TXB_2) excretion. Both methods have limitations and did not yet result in a generally accepted definition of a pharmacological aspirin "resistance." Specifically, there is a wide variation in 11-DH-TXB_2 excretion and no definition of a normal range while assays of platelet function have a low predictability and may give different results according to the particular technique and protocol used.

The important issue of a possibly causal relationship between insufficient antiplatelet effects of aspirin and clinical outcome is still a matter of discussion and requires further sufficiently large prospective randomized trials. Nevertheless, only a very limited proportion of patients who exhibit pharmacological "resistance" to aspirin in any of these laboratory settings may experience thrombotic events and there may be many more in whom treatment fails without any evidence for a pharmacological reason. In this context, the residual platelet activity is another important variable and requires further analysis.

References

381 Mukherjee, D. and Topol, E.J. (2002) Pharmacogenomics in cardiovascular diseases. *Progress in Cardiovascular Diseases*, **44**, 479–498.

382 Antithrombotic Trialists' Collaboration (2002) Collaborative meta-analysis of randomised trials of antiplatelet therapy for prevention of death, myocardial infarction, and stroke in high risk patients. *British Medical Journal*, **324**, 71–86.

383 Hennekens, C.H., Schrör, K., Weisman, S. *et al.* (2004) Terms and conditions. Semantic complexity and aspirin resistance. *Circulation*, **110**, 1706–1708.

384 Weber, A.A., Przytulski, B., Schanz, A. *et al.* (2002) Towards a definition of aspirin resistance: a typological approach. *Platelets*, **13**, 37–40.

385 Macchi, L., Sorel, N. and Christiaens, L. (2006) Aspirin resistance: definitions, mechanisms, prevalence, and clinical significance. *Current Pharmaceutical Design*, **12**, 251–258.

386 Catella, F., Healy, D., Lawson, J.A. *et al.* (1986) 11-DH-Thromboxane B_2: a quantitative index of thromboxane A_2 formation in the human circulation. *Proceedings of the National Academy of Sciences of the United States of America*, **83**, 5861–5865.

387 Catella, F., Lawson, J.A., Fitzgerald, D.J. *et al.* (1987) Analysis of multiple thromboxane metabolites in plasma and urine. *Advances in Prostaglandin, Thromboxane, and Leukotriene Research*, **17**, 611–614.

388 Montalescot, G., Maclouf, J., Drobinski, G. *et al.* (1994) Eicosanoid biosynthesis in patients with stable angina: beneficial effects of very low dose aspirin. *Journal of the American College of Cardiology*, **24**, 33–38.

389 Cipollone, F., Ciabattoni, G., Patrignani, P. *et al.* (2000) Oxidant stress and aspirin-insensitive thromboxane biosynthesis in severe unstable angina. *Circulation*, **102**, 1007–1013.

390 De Gaetano, G. and Cerletti, C. (2003) Aspirin resistance: a revival of platelet aggregation tests? *Journal of Thrombosis and Haemostasis*, **1**, 2048–2050.

391 Cattaneo, M. (2004) Aspirin and clopidogrel. Efficacy, safety, and the issue of drug resistance. *Arteriosclerosis, Thrombosis, and Vascular Biology*, **24**, 1980–1987.

392 Santos, M.T., Valles, J., Aznar, J. *et al.* (1997) Prothrombotic effects of erythrocytes on platelet reactivity. Reduction by aspirin. *Circulation*, **95**, 63–68.

393 Valles, J., Santos, M.T., Aznar, J. *et al.* (1998) Erythrocyte promotion of platelet reactivity decreases the effectiveness of aspirin as an antithrombotic therapeutic modality: the effect of low-dose aspirin is less than optimal in patients with vascular disease due to prothrombotic effects of erythrocytes on platelet reactivity. *Circulation*, **97**, 350–355.

394 Bhatt, D.L. and Topol, E.J. (2003) Scientific and therapeutic advances in antiplatelet therapy. *Nature Reviews. Drug Discovery*, **2**, 15–28.

395 Wu, K.K. and Hoak, J.C. (1975) Increased platelet aggregates in patients with transient ischemic attacks. *Stroke*, **6**, 521–524.

396 Lordkipanidzé, M., Pharand, C., Schampaert, E. *et al.* (2007) A comparison of six major platelet function tests to determine the prevalence of aspirin resistance in patients with stable coronary artery disease. *European Heart Journal*, **28**, 1702–1708.

397 Grundmann, K., Jaschonek, K., Kleine, B. *et al.* (2003) Aspirin non-responder status in patients with recurrent cerebral ischemic attacks. *Journal of Neurology*, **250**, 63–66.

398 Gonzalez-Conejero, R., Rivera, J., Corral, J. *et al.* (2005) Biological assessment of aspirin efficacy on healthy individuals. *Stroke*, **36**, 276–280.

399 Harrison, P., Segal, H., Blasbery, K. *et al.* (2005) Screening for aspirin responsiveness after transient ischemic attack and stroke. *Stroke*, **36**, 1001–1005.

400 Fontana, P., Nolli, S., Reber, G. *et al.* (2006) Biological effects of aspirin and clopidogrel in a randomized cross-over study in 96 healthy volunteers. *Journal of Thrombosis and Haemostasis*, **4**, 813–819.

401 Gurbel, P., Bliden, K.P., DiChiara, J. *et al.* (2007) Evaluation of dose-related effects of aspirin on platelet function. Results from the aspirin-induced platelet effect (ASPECT) study. *Circulation*, **115**, 3156–3164.

402 Gum, P.A., Kottke-Marchant, K., Poggio, E.D. *et al.* (2001) Profile and prevalence of aspirin resistance in patients with cardiovascular disease. *The American Journal of Cardiology*, **88**, 230–235.

403 Undas, A., Placzkiewicz-Jankowska, E., Zielinski, L. and Tracz, W. (2007) Lack of aspirin-induced decrease in thrombin formation in subjects resistant to aspirin. *Thrombosis and Haemostasis*, **97**, 1056–1058.

404 Undas, A., Brummel-Ziedins, K.E. and Mann, K.G. (2007) Antithrombotic properties of aspirin and resistance to aspirin: beyond strictly antiplatelet actions. *Blood*, **109**, 2285–2292.

405 Packham, M.A., Bryant, N.L., Guccione, M.A. *et al.* (1989) Effect of the concentration of Ca^{2+} in the suspending medium on the responses of human and rabbit platelet to aggregating agents. *Thrombosis and Haemostasis*, **62**, 968–976.

406 Bretschneider, E., Glusa, E. and Schrör, K. (1994) ADP-, PAF- and adrenaline-induced platelet aggregation and thromboxane formation are not affected by a thromboxane receptor antagonist at physiological external Ca^{2+} concentrations. *Thrombosis Research*, **75**, 233–242.

407 Patrono, C., Bachmann, F., Baigent, C. *et al.* (2004) Expert Consensus Document on the use of antiplatelet agents. The Task Force on the use of antiplatelet agents in patients with atherosclerotic cardiovascular disease of the European Society of Cardiology. *European Heart Journal*, **25**, 166–181.

408 Mayeux, P.R., Morton, H.E., Gillard, J. *et al.* (1988) The affinities of prostaglandin H_2 and thromboxane A_2 for their receptor are similar in washed human platelets. *Biochemical and Biophysical Research Communications*, **157**, 733–739.

409 Reilly, I.A.G. and FitzGerald, G.A. (1987) Inhibition of thromboxane formation *in vivo* and *ex vivo*. *Blood*, **69**, 180–186.

410 Valles, J., Santos, M.T., Fuset, M.P. *et al.* (2007) Partial inhibition of platelet thromboxane A_2 synthesis by aspirin is associated with myonecrosis in patients with ST-segment elevation myocardial infarction. *The American Journal of Cardiology*, **99**, 19–25.

411 Roberts, L.J., Sweetman, B.J. and Oates, J.A. (1981) Metabolism of thromboxane B_2 in man. Identification of twenty urinary metabolites. *The Journal of Biological Chemistry*, **256**, 8384–8393.

412 Ciabattoni, G., Pugliese, F., Davi, G. *et al.* (1989) Fractional conversion of thromboxane B_2 to urinary 11-dehydro-thromboxane B_2 in man. *Biochimica et Biophysica Acta*, **992**, 66–70.

413 Uedelhoven, W.M., Rutzel, A., Meese, C.O. *et al.* (1991) Smoking alters thromboxane metabolism in man. *Biochimica et Biophysica Acta*, **1081**, 197–201.

414 Nüsing, R. and Ullrich, V. (1991) Immunoquantitation of thromboxane synthase in human tissues. *Advances in Prostaglandin, Thromboxane, and Leukotriene Research*, **21**, 307–310.

415 Uyama, O., Matsui, Y., Shimizu, S. *et al.* (1994) Risk factors for carotid atherosclerosis and platelet activation. *Japanese Circulation Journal*, **58**, 409–415.

416 Bruno, A., McConnell, J.P., Mansbach, H.H. *et al.* (2002) Aspirin and urinary 11-dehydrothromboxane B (2) in African American stroke patients. *Stroke*, **33**, 57–60.

417 Eikelboom, J.W., Hirsh, J., Weitz, J.I. *et al.* (2002) Aspirin-resistant thromboxane biosynthesis and the risk of myocardial infarction, stroke or cardiovascular death in patients at high risk for cardiovascular events. *Circulation*, **105**, 1650–1655.

418 Schrör, K., Hohlfeld, T. and Weber, A.-A. (2006) Aspirin resistance – does it clinically matter? *Clinical Research in Cardiology*, **95**, 505–510.

419 Hohlfeld, T. (2004) Pharmacology of aspirin resistance, in *Aspirin – "Resistance" or "Nonresponsiveness"* (ed. K. Schrör), Dr. Schrör Verlag, Frechen, pp. 17–32.

420 Benedek, I.H., Joshi, A.S., Pieniaszek, H.J. *et al.* (1995) Variability in the pharmacokinetics and pharmacodynamics of low dose aspirin in healthy male volunteers. *Journal of Clinical Pharmacology*, **35**, 1181–1186.

421 Cox, D., Maree, A.O., Dooley, M. *et al.* (2006) Effect of enteric coating on antiplatelet activity of low-dose aspirin in healthy volunteers. *Stroke*, **37**, 2153–2158.

422 Zimmermann, N., Kienzle, P., Weber, A.A. *et al.* (2001) Aspirin resistance after coronary artery bypass grafting. *The Journal of Thoracic and Cardiovascular Surgery*, **121**, 982–984.

423 Rocca, B., Ciabattoni, G., Tartaglione, R. *et al.* (1995) Increased thromboxane biosynthesis in essential thrombocythemia. *Thrombosis and Haemostasis*, **74**, 1225–1230.

424 Nurden, P., Bihour, C., Smith, M. *et al.* (1996) Platelet activation and thrombosis: studies in a patient with essential thrombocythemia. *American Journal of Hematology*, **51**, 79–84.

425 Pearson, T.C. (2002) The risk of thrombosis in essential thrombocythemia and polycythemia vera. *Seminars in Oncology*, **29**, 16–21.

426 Hohlfeld, T., Zimmermann, N., Weber, A.-A. *et al.* (2008) Pyrazolinone analgesics prevent the antiplatelet effect of aspirin and preserve human platelet thromboxane biosynthesis. *Journal of Thrombosis and Haemostasis*, **6**, 166–173.

427 Rao, G.H., Johnson, G.G., Reddy, K.R. *et al.* (1983) Ibuprofen protects platelet cyclooxygenase from irreversible inhibition by aspirin. *Arteriosclerosis*, **3**, 383–388.

428 Catella-Lawson, F., Muredach, P., Reilly, M.P. *et al.* (2001) Cyclooxygenase inhibitors and the antiplatelet effect of aspirin. *The New England Journal of Medicine*, **345**, 1809–1817.

429 Ouellet, M., Riendeau, D. and Percival, D. (2001) A high level of cyclooxygenase-2 inhibitor selectivity is associated with a reduced interference of platelet cyclooxygenase-1 inactivation by aspirin. *Proceedings of the National Academy of Sciences of the United States of America*, **98**, 14583–14588.

430 FitzGerald, G.A. (2003) Parsing an enigma: the pharmacodynamics of aspirin resistance. *The Lancet*, **361**, 542–544.

431 Zimmermann, N., Wenk, A., Kim, U. *et al.* (2003) Functional and biochemical evaluation of platelet aspirin resistance after coronary artery bypass surgery. *Circulation*, **108**, 542–547.

432 Weber, A.A., Zimmermann, K.C., Meyer-Kirchrath, J. *et al.* (1999) Cyclooxygenase-2 in human platelets as a possible factor in aspirin resistance. *The Lancet*, **353**, 900.

433 Rocca, B., Secchiero, P., Ciabattoni, G. *et al.* (2002) Cyclooxygenase-2 expression is induced during human megakaryopoiesis and characterizes newly formed platelets. *Proceedings of the National Academy of Sciences of the United States of America*, **99**, 7634–7639.

434 Guthikonda, S., Lev, E.I., Patel, R. *et al.* (2007) Reticulated platelets and uninhibited COX-1 and COX-2 decrease the antiplatelet effects of aspirin. *Journal of Thrombosis and Haemostasis*, **5**, 490–496.

435 Weber, A.A., Przytulski, B., Schumacher, M. *et al.* (2002) Flow cytometry analysis of platelet cyclooxygenase-2 expression. Induction of cyclooxygenase-2 in patients undergoing coronary artery bypass grafting. *British Journal of Haematology*, **117**, 424–426.

436 Censarek, P., Freidel, K., Udelhoven, M. *et al.* (2004) Cyclooxygenase COX-2a, a novel COX-2 mRNA variant, in platelets from patients after coronary artery bypass grafting. *Thrombosis and Haemostasis*, **92**, 925–928.

437 Pulcinelli, F.M., Riondino, S., Celestini, A. *et al.* (2005) Persistent production of platelet thromboxane A_2 in patients chronically treated with aspirin. *Journal of Thrombosis and Haemostasis*, **3**, 2784–2789.

438 Cambria-Kiely, J.A. and Gandhi, P.J. (2002) Aspirin resistance and genetic polymorphism. *Journal of Thrombosis and Thrombolysis*, **14**, 51–58.

439 Halushka, M.K., Walker, M.P. and Halushka, P.V. (2003) Genetic variation in cyclooxygenase 1: effects on response to aspirin. *Clinical Pharmacology and Therapeutics*, **73**, 122–130.

440 Hillarp, A., Palmqvist, B., Lethagen, S. *et al.* (2003) Mutations within the cyclooxygenase-1 gene in aspirin non-responders with recurrence of stroke. *Thrombosis Research*, **112**, 275–283.

441 Macchi, L., Christiaens, L., Brabant, S. *et al.* (2003) Resistance *in vitro* to low-dose aspirin is associated with platelet Pl^{A1} (GPIIIa) polymorphism but not with C807T (GPIa/IIa) and C-5T Kozak ($GPIb_{\alpha}$) polymorphisms. *Journal of the American College of Cardiology*, **42**, 1115–1119.

442 Faraday, N., Yanek, L.R., Mathias, R. *et al.* (2007) Heritability of platelet responsiveness to aspirin in activation pathways directly and indirectly related to cyclooxygenase-1. *Circulation*, **115**, 2490–2496.

443 Larsson, P.T., Wallén, N.H. and Hjemdahl, P. (1994) Norepinephrine-induced human platelet activation *in vivo* is only partly counteracted by aspirin. *Circulation*, **89**, 1951–1957.

444 Levine, S.P., Towell, B.L., Suarez, A.M. *et al.* (1985) Platelet activation and secretion associated with emotional stress. *Circulation*, **71**, 1129–1134.

445 Mittleman, M.A., Maclure, M., Sherwood, J.B. *et al.* for the Determinants of Myocardial Infarction Onset Study Investigators (1995) Triggering of acute myocardial infarction onset by episodes of anger. *Circulation*, **92**, 1720–1725.

446 Moake, J.L., Turner, N.A., Stathopoulos, N.A. *et al.* (1988) Shear-induced platelet aggregation can be mediated by vWF released from platelets, as well as by exogenous large or unusually large vWF-multimers, requires adenosine diphosphate, and is resistant to aspirin. *Blood*, **71**, 1366–1374.

447 Maalej, N. and Folts, J.D. (1996) Increased shear stress overcomes the antithrombotic platelet inhibitory effect of aspirin in stenosed dog coronary arteries. *Circulation*, **93**, 1201–1205.

448 Morrow, J.D., Hill, K.E., Burk, R.F. *et al.* (1990) A series of prostaglandin F_2-like compounds are produced *in vivo* in humans by non-cyclooxygenase, free radical-catalyzed mechanism. *Proceedings of the National Academy of Sciences*, **87**, 9383–9387.

449 Audoly, L.P., Rocca, B., Fabre, J.E. *et al.* (2000) Cardiovascular responses to the isoprostanes iPF (2alpha)-III and iPE(2)-III are mediated via the thromboxane A(2) receptor *in vivo*. *Circulation*, **101**, 2833–2840.

450 Véricel, E., Januel, C., Carreras, M. *et al.* (2004) Diabetic patients without vascular complications display enhanced basal platelet activation and decreased antioxidant status. *Diabetes*, **53**, 1046–1051.

451 Pratico, D., Smyth, E.M., Violi, F. *et al.* (1996) Local amplification of platelet function by 8-epi prostaglandin $F_{2\alpha}$ is not mediated by thromboxane receptor isoforms. *The Journal of Biological Chemistry*, **271**, 14916–14924.

452 Cattaneo, M. (2007) Laboratory detection of "aspirin resistance"; what test should we use (if any)? *European Heart Journal*, **28**, 1673–1675.

453 Zimmermann, N., Hohlfeld, T. (2008) Clinical implications of aspirin resistance. *Thrombosis and Haemostasis*, **100**, 379–390.

454 McKee, S.A., Sane, D.C. and Deliargyris, E.N. (2002) Aspirin resistance in cardiovascular disease: a review of prevalence, mechanisms, and clinical significance. *Thrombosis and Haemostasis*, **88**, 711–715.

455 Wang, J.C., Aucoin-Barry, D., Manuelian, D. *et al.* (2003) Incidence of aspirin nonresponsiveness using the Ultegra Rapid Platelet Function Assay-ASA. *The American Journal of Cardiology*, **92**, 1492–1494.

456 Alberts, M.J., Bergman, D.L., Molner, E. *et al.* (2004) Antiplatelet effect of aspirin in patients with cerebrovascular disease. *Stroke*, **35**, 175–178.

457 Mason, P.J., Jacobs, A.K. and Freedman, J.E. (2005) Aspirin resistance and atherothrombotic disease. *Journal of the American College of Cardiology*, **46**, 986–993.

458 Sztriha, L.K., Sas, K. and Vecsei, L. (2005) Aspirin resistance in stroke: 2004. *Journal of the Neurological Sciences*, **229/230**, 163–169.

459 Hurlen, M., Seljeflot, I. and Arnesen, H. (1998) The effect of different antithrombotic regimens on platelet aggregation after myocardial infarction. *Scandinavian Cardiovascular Journal*, **32**, 233–237.

460 Grotemeyer, K.H. (1991) Effects of acetylsalicylic acid in stroke patients. Evidence of nonresponders in a subpopulation of treated patients. *Thrombosis Research*, **63**, 587–593.

461 Fateh-Moghadam, S., Plockinger, U., Cabeza, N. *et al.* (2005) Prevalence of aspirin resistance in patients with type 2 diabetes. *Acta Diabetologica*, **42**, 99–103.

462 Chen, W.-H., Lee, P.Y., Ng, W. *et al.* (2004) Aspirin resistance is associated with a high incidence of myonecrosis after non-urgent percutaneous coronary intervention despite clopidogrel pretreatment. *Journal of the American College of Cardiology*, **43**, 1122–1126.

463 Lee, P.Y., Chen, W.H., Ng, W. *et al.* (2005) Low-dose aspirin increases aspirin resistance in patients with coronary artery disease. *The American Journal of Medicine*, **118**, 723–727.

464 Mehta, S.S., Silver, R.J., Aaronson, A. *et al.* (2006) Comparison of aspirin resistance in type 1 versus type 2 diabetes mellitus. *The American Journal of Cardiology*, **97**, 567–570.

465 Tantry, U.S., Bliden, K.P. and Gurbel, P.A. (2005) Overestimation of platelet aspirin resistance detection by thrombelastograph platelet mapping and validation by conventional aggregometry using arachidonic acid stimulation. *Journal of the American College of Cardiology*, **46**, 1705–1709.

466 Schwartz, K.A., Schwartz, D.E., Ghosheh, E. *et al.* (2005) Compliance as a critical consideration in patients who appear to be resistant to aspirin after healing of myocardial infarction. *The American Journal of Cardiology*, **95**, 973–975.

467 Tantry, U., Gurbel, P.A., Bliden, K.P. *et al.* (2006) Inconsistency in the prevalence of platelet aspirin resistance as measure by COX-1 non-specific assays in patients treated with 81, 162, and 325 mg aspirin. *Journal of the American College of Cardiology*, **47**, 290.

468 Weber, A.A., Liesener, S., Hohlfeld, T. *et al.* (2000) 40 mg of aspirin are not sufficient to inhibit platelet function under conditions of limited compliance. *Thrombosis Research*, **97**, 365–367.

469 Glynn, R.J., Buring, J.E., Manson, J.E. *et al.* (1994) Adherence to aspirin in the prevention of myocardial infarction. *Archives of Internal Medicine*, **154**, 2649–2657.

470 Granström, E. and Kumlin, M. (1987) Assay of thromboxane production in biological systems: reliability of TXB_2 versus 11-dehydro-TXB_2 for measurement. *Advances in Prostaglandin, Thromboxane, and Leukotriene Research*, **17**, 587–594.

471 Perneby, C., Granström, E., Beck, O. *et al.* (1999) Optimization of an enzyme immunoassay for 11-dehydro-thromboxane B_2 in urine. Comparison with GC–MS. *Thrombosis Research*, **96**, 427–436.

472 Helgason, C.M., Bolin, K.M., Hoff, J.A. *et al.* (1994) Development of aspirin resistance in persons with previous ischemic stroke. *Stroke*, **25**, 2331–2336.

473 Tohgi, H., Konno, S., Tamura, K. *et al.* (1992) Effects of low-to-high doses of aspirin on platelet aggregability and metabolites of thromboxane A_2 and prostacyclin. *Stroke*, **23**, 1400–1403.

474 Undas, A., Brummel, K., Musial, J. *et al.* (2001) Pl^{A2} polymorphism of β_3 integrins is associated with enhanced thrombin generation and impaired antithrombotic action of aspirin at the site of microvascular injury. *Circulation*, **104**, 2666–2672.

475 Gum, P.A., Kottke-Marchant, K., Welsh, P.A. *et al.* (2003) A prospective, blinded determination of the natural history of aspirin resistance among stable patients with cardiovascular disease. *Journal of the American College of Cardiology*, **41**, 961–965.

476 Mueller, M., Salat, A., Stangl, P. *et al.* (1997) Variable platelet response to low-dose ASA and the risk of limb detoriation in patients submitted to peripheral arterial angioplasty. *Thrombosis and Haemostasis*, **78**, 1003–1007.

477 Grotemeyer, K.H., Scharafinski, H.W. and Husstedt, I.W. (1993) Two-year follow-up of aspirin responder and aspirin nonresponder. A pilot study including 180 post-stroke patients. *Thrombosis Research*, **71**, 397–403.

478 Andersen, K., Hurlen, M., Arnesen, H. *et al.* (2002) Aspirin non-responsiveness as measured by PFA-100 in patients with coronary artery disease. *Thrombosis Research*, **108**, 37–42.

479 Poston, R.S., Gu, J., Brown, J.M. *et al.* (2006) Endothelial injury and acquired aspirin resistance as promoters of regional thrombin formation and early vein graft failure after coronary artery bypass grafting. *The Journal of Thoracic and Cardiovascular Surgery*, **131**, 122–130.

480 Ohmori, T., Yatomi, Y., Nonaka, T. *et al.* (2006) Aspirin resistance detected with aggregometry cannot be explained by cyclooxygenase activity: involvement of other signaling pathway(s) in cardiovascular events of aspirin-treated patients. *Journal of Thrombosis and Haemostasis*, **4**, 1271–1278.

481 Chen, W.-H., Cheng, X., Lee, P.-Y. *et al.* (2007) Aspirin resistance and adverse clinical events in patients with coronary artery disease. *The American Journal of Medicine*, **120**, 631–635.

482 Buch, A.N., Singh, S., Roy, P. *et al.* (2007) Measuring aspirin resistance, clopidogrel responsiveness and postprocedural markers of myonecrosis in patients undergoing percutaneous coronary intervention. *The American Journal of Cardiology*, **99**, 1518–1522.

483 Dorsch, M.P., Lee, J.S., Lynch, D.R. *et al.* (2007) Aspirin resistance in patients with stable coronary artery disease with and without a history of myocardial infarction. *Annals of Pharmacotherapy*, **41**, 737–741.

484 Frelinger, A.L., III, Furman, M.I., Linden, M.D. *et al.* (2006) Residual arachidonic acid-induced platelet activation via an adenosine diphosphate-dependent but cyclooxygenase-1- and cyclooxygenase-2-independent pathway: a 700-patient study of aspirin resistance. *Circulation*, **113** 2888–2896.

485 Weber, A.A., Meila, D., Jacobs, C. *et al.* (2002) Low incidence of paradoxical platelet activation by glycoprotein IIb/IIIa inhibitors. *Thrombosis Research*, **106**, 25–29.

486 Meen, O., Brosstad, F., Khiabani, H. *et al.* (2008) No case of COX-1 related aspirin resistance found in 289 patients with symptoms of stable CHD remitted for coronary angiography. *Scandinavian Journal of Clinical and Laboratory Investigation*, **68**, 185–191.

487 Lim, E., Ali, Z., Ali, A. *et al.* (2003) Indirect comparison meta-analysis of aspirin therapy after coronary surgery. *British Medical Journal*, **327**, 1309–1314.

488 Tran, H.A., Anand, S.S., Hankey, G.J. *et al.* (2007) Aspirin resistance. *Thrombosis Research*, **120**, 337–346.

4.2
Pain, Fever, and Inflammatory Diseases

Salicylate, manufactured as water-soluble sodium salt, was originally introduced into the clinics as an antipyretic analgesic. Synthetic salicylate soon replaced salicylates, isolated and prepared from natural sources and also became interesting because of its analgesic and anti-inflammatory properties. The combination of these three properties, although not independent of each other and at least partially mediated by the same mediators (Section 2.3.2), was highly desirable for treatment of flu and inflammatory pain, for example, in painful rheumatic diseases (Section 1.1.3). However, the unpleasant taste and gastric irritation frequently seen at the high doses that had to be taken by the patients limited the clinical acceptance. As a consequence, the compound was "updated" by acetylation, resulting in the synthesis of aspirin, to be used for the same indications but being better tolerable. Interestingly, aspirin was considered to be the prodrug of salicylate that was assumed as active principle, released by metabolic cleavage. This conclusion, originally made by Dreser in his first description of the pharmacological properties of aspirin a century ago (Section 1.1.3), is still basically correct, at least with respect to the indications, the compound was used for at the time.

For more than 70 years, antipyretic/analgesic actions were in the focus of aspirin use. The compound became soon an indispensable household remedy for almost every situation of distress, fever, pain, or just feeling bad, for example, during episodes of common cold. Despite the fact that aspirin in general was well tolerated, one should not ignore that both aspirin and salicylate are potent drugs. Since much of the aspirin is taken as an OTC preparation in self-medication ("take an aspirin"), there is always a chance of overdosing and appearance of unwanted side effects (Section 3.1.1). However, in daily practice, the real incidence of severe side effects is very low.

The first issue to be discussed here is the therapeutic use of aspirin as an antipyretic analgesic (Section 4.2.1). This is still the domain of its practical application in self-medication. In addition, aspirin is also used as an adjunct to NSAIDs or coxibs in the treatment of osteoarthritis and rheumatoid arthritis although this is not a primary indication anymore. Many of the patients with chronic inflammatory diseases might also be at an elevated risk for atherothrombotic events. In addition, patients with rheumatoid arthritis have a disease-related four- to fivefold elevated risk for myocardial infarctions and it is, therefore, important to know the individual benefit/risk ratio. These interactions between aspirin, NSAIDs, and coxibs are of particular interest in chronic inflammatory diseases and are discussed in Section 4.2.2.

Kawasaki's syndrome is an inflammatory disease in children where aspirin is still used as a standard medication together with immunoglobulin at anti-inflammatory doses. Here, both anti-inflammatory and antiplatelet effects of the compound might contribute to its clinical efficacy in preventing immune-vasculitis, coronary aneurysms, and myocardial infarctions in children at enhanced risk (Section 4.2.3).

4.2.1
Aspirin as an Antipyretic Analgesic

Aspirin has been successfully used for pain relief for many years. However, the field of applications has changed. In the beginning, it was mainly febrile states in children, in many cases due to flu, which were treated with the compound. However, after the intense discussions about aspirin and a possible relationship to Reye's syndrome, the compound was banned in many countries for this indication in children and young adults (Section 3.3.3). For treatment of inflammatory pain, aspirin was replaced by NSAIDs and coxibs (Section 4..2.2) although coxibs can only work if pain is related to (induced) COX-2 expression. Thus, migraine and tension-type headache along with flu-like symptoms are currently the domains of aspirin use.

4.2.1.1 Fever, Pain, and Antipyretic/Analgesic Actions of Aspirin

Fever and Mode of Antipyretic Actions of Aspirin Fever associated with upper respiratory tract infections is of suspected viral origin and, therefore mainly subjected to symptomatic treatment of the unpleasant symptoms, including fatigue, headache, and others. Mechanistically, aspirin interferes with endogenous pyrogens and ameliorates subsequent cytokine generation and action as well as their potentiation by prostaglandins. This includes upregulation of core temperature (Section 2.3.2). Interestingly, recent *in vivo* data from animal experiments suggest that aspirin and salicylate but not traditional NSAIDs will also inhibit replication of flu viruses and therefore might also interact with exogenous pyrogens (Section 2.2.2)

Aspirin does not interact with physiological temperature control. Therefore, there is no change in normal body temperature by aspirin. The antipyretic action of aspirin is mediated by salicylate (Figure 2.37) [489] and associated with sweating, that is, increased heat loss.

Pain and Mode of Analgesic Actions of Aspirin Pain can result from many reasons and is mediated by both peripheral and central mechanisms. The analgesic actions of aspirin include both peripheral and central sites of action, both involving prostaglandins (Section 2.3.2). A dose-dependent increase in pain threshold between 60 mg and 1.8 g of aspirin and maintenance of the analgesic effect after repeated application (300 mg in 2 h intervals) was already shown 60 years ago [490]. A dose dependency is seen for both the antipyretic and analgesic effects. Although it is likely that aspirin ultimately also interacts with other central mechanisms of pain control, that is, endocannabinoids and serotonin, there is currently little mechanism-based clinical research on this issue. Prostaglandins, the key mediators for pain sensitization arising from an area of tissue injury, are also involved in pain transmission and perception systems in peripheral nerves and the CNS via upregulated COX-2 (allodynia/hyperalgesia) [491].

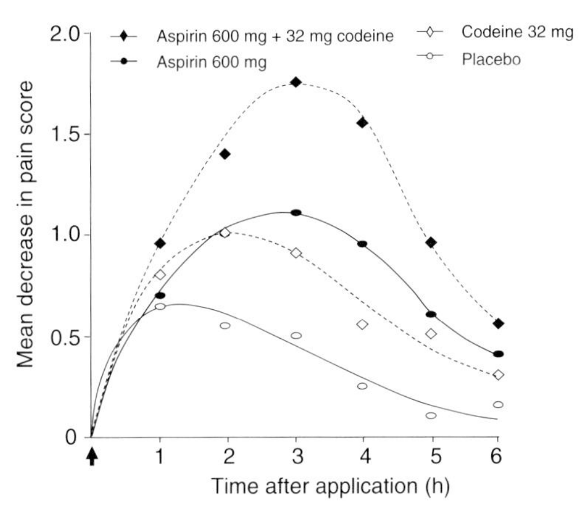

Figure 4.27 Time-effect curves for analgesic efficacy of placebo (lactose), aspirin, codeine and a combination of aspirin and codeine. Eleven patients with chronic pain (cancer) were treated with a total of 20 oral doses of each medication administered at random order modified after [492]).

Pain is a subjective experience. Therefore, the patient's self-report, for example, by a visual analogue scale (VAS), provides the most valid measurement of the intensity of pain sensitization. A 30% reduction of pain intensity on this scale is considered clinically significant. In addition, all mechanistic as well as clinical studies on pain have to consider the high placebo rate, underlining the significance of the subjective pain perception. In most studies, this placebo effect amounts to about 50% of the drug effect [492] and there is a clear time-dependency for both (Figure 4.27).

Clinical pain research with respect to established drugs and treatment protocols uses pain models. About two-thirds of experimental studies have used the extraction of the third molar (wisdom tooth) as a pain model. It might well be that findings on this postsurgical pain model differ from inflammatory, prostaglandin-mediated pain or the pain of headache, including tension-type headache and migraine.

Pain Research Models and Dose Dependency of Aspirin Action After an original report by Laska *et al.* [493], demonstrating remarkable differences in the analgesic potency of aspirin in different situations of clinical (gynecological/obstetric) pain, many follow-up studies have addressed the issue of possible influences of the pain model on the intensity of the analgesic effect of aspirin and other analgesics. According to current knowledge, the pain model used (postoperative, episiotomy, and dental pain, etc.), the kind of measurement and duration of observation have no effect on the magnitude of analgesia by aspirin [494]. A recent meta-analysis has added further evidence to this. The investigators calculated the "under the pain relief versus time curve" equivalent to at least 50% maximum pain relief over 6 h in dental and postsurgical pain. No major difference was obtained for the three analgesics studied: aspirin (600/650 mg), acetaminophen (600/650 mg; 1000 mg), and ibuprofen (400 mg) [495]. However, higher doses might be more effective (Table 4.12).

A frequently cited meta-analysis of randomized, placebo-controlled trials was performed to compare the analgesic efficacy of single-dose plain aspirin tablets in postoperative pain (post dental extraction, 68%), postsurgical or postpartum pain. Other pain conditions were not considered. Of the 169 publications found on this issue (containing at least 10 patients), 69 met the inclusion criteria. From these studies, pain intensity and pain relief data were extracted. These data were used to calculate the relative benefit (number needed to treat (NNT)) and relative harm (number needed to harm (NNH)) in dependency on the dose used. The median group size for all comparisons was 38 patients.

There was a significant benefit (estimated in terms of patients reporting at least 50% pain relief) with aspirin compared to placebo (Table 4.12). The effect was dose dependent and became significant at doses of 600 mg and more without significant differences between the causes of pain (Table 4.13). The number of patients needed to treat was four at a dose of 1000 mg. A total of 12% of patients on aspirin and 10% of patients on placebo reported adverse effects, most frequently drowsiness and gastric irritation. The total number needed to harm was 44, that is, 10-fold higher than the number to treat [494] (Table 4.14).

The conclusion was that aspirin is an effective, single-dose analgesic also in postoperative pain. The analgesic effect is dose dependent and the potency is comparable to that of acetaminophen (Figure 4.28) [494].

This type of studies has limitations, as also stated by authors, for example, by using a pain calculation model rather than an individual pain response. However, the dose dependency of analgesic actions of aspirin, acetaminophen, and ibuprofen was recently confirmed in a systematic review of randomized double-blind trials in acute pain [496]. More information has to be expected from prospective, double-blind randomized trials with a well-defined clinical end point, that is, definition of pain relief.

Table 4.12 Proportion of patients achieving at least 50% total pain relief (TOTPAR) according to a standard pain intensitiy scale after treatment with aspirin, paracetamol, or ibuprofen, relative benefit, and number needed to treat (NNT) (for further explanation, see the text) [495].

Drug	Dose (mg)	Pain model	#	TOTPAR relative benefit	NNT (95% CI)
Aspirin	600/650	Dental pain	3635	2.5 (2.2–2.8)	4.7 (4.2–5.4)
	600/650	Postsurgical pain	1427	2.3 (1.9–2.7)	3.9 (3.3–4.7)
Acetaminophen	600/650	Dental pain	1265	2.9 (2.9–3.7)	4.2 (3.6–5.2)
	600/650	Postsurgical pain	621	1.9 (1.5–2.4)	5.5 (3.9–9.1)
Acetaminophen	975/1000	Dental pain	1038	3.7 (2.7–5.1)	3.7 (3.1–4.7)
	1000	Postsurgical pain	1721	2.2 (1.9–2.5)	3.9 (3.3–4.7)
Ibuprofen	400	Dental pain	3402	5.2 (4.1–6.6)	2.2 (2.1–2.4)
	400	Postsurgical pain	1301	3.7 (2.6–5.1)	3.0 (2.6–3.4)

Table 4.13 Analgesic potency of aspirin in placebo-controlled trials.

Aspirin dose (mg)	# of trials	# of patients with at least 50% pain relief		TOTPAR relative benefit (98% CI)	NNT (95% CI)
		Aspirin	Placebo		
500	3	45/135	32/115	1.2 (0.8–1.8)	nc
600/650	68	960/2499	404/2562	2.0 (1.8–2.2)	4.4 (4.0–4.9)
1000	5	153/375	64/359	2.2 (1.4–3.4)	4.0 (3.2–5.4)
1200	5	85/140	27/139	3.3 (1.8–6.3)	2.4 (1.9–3.2)

Note the dose dependency of the analgesic effect, as shown by the decreasing number of patients needed to treat (NNT) with increasing aspirin doses [494].
nc: not calculated because relative risk not statistically significant.

Table 4.14 Relative risk (RR) of adverse effects to aspirin in analgesic doses (500–1200 mg) in placebo-controlled trials as documented by the number needed to harm.

Dose	# of trials	# of patients with adverse effects		RR (95% CI)	NNH (95% CI)
		Aspirin	Placebo		
All doses					
Total adverse effects	60	313/2619	261/2620	1.3 (0.0–1.5)	nc
Aspirin 600/650 mg					
Total adverse effects	53	157/1976	229/2088	1.2 (1.0–1.4)	44 (23–345)
Dizziness	30	41/1429	27/1557	1.6 (0.9–2.6)	nc
Drowsiness	33	103/1542	56/1672	1.9 (1.4–2.5)	28 (19–52)
Gastric irritation	11	20/546	6/562	2.5 (1.2–5.1)	38 (22–174)

Further adverse effects, including headache, nausea, and vomiting were on average lower in the aspirin group than with placebo (RR <1) [494].
nc: not calculated because relative risk not statistically significant.

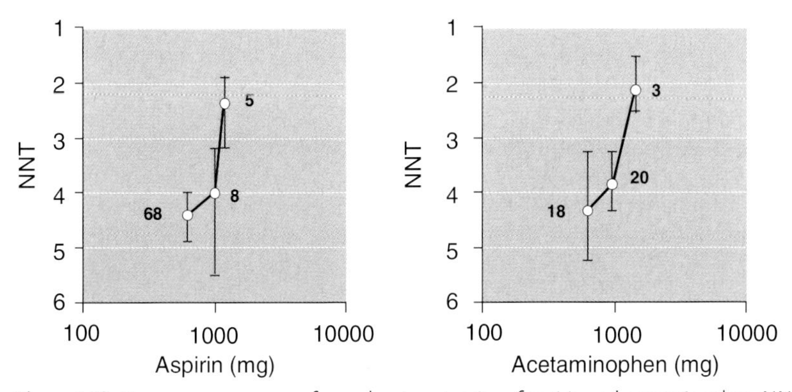

Figure 4.28 Dose–response curves for analgesic potencies of aspirin and acetaminophen. NNT: numbers needed to treat; numbers at the figure: number of studies for each dose [494].

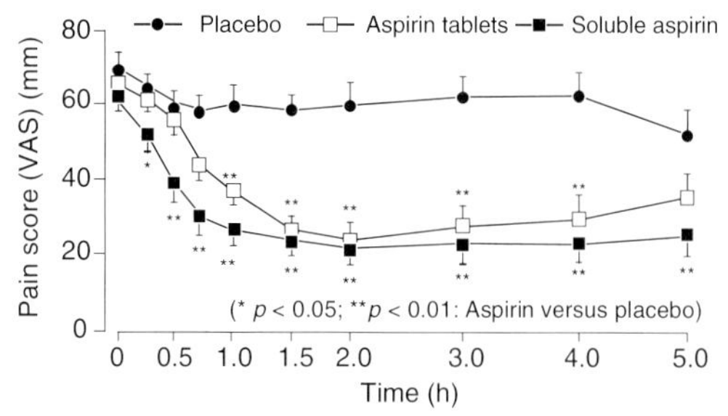

Figure 4.29 Analgesic potency of aspirin tablets and soluble aspirin (1.2 g each) in comparison to placebo as determined by individual pain sensitization (extraction of third molar) (VAS). Note the more rapid start of analgesia with soluble aspirin, a stronger action of soluble versus undissolved aspirin at 0.3–1 h and a potency similar to aspirin tablets at 2–5 h (modified after [497]).

Aspirin Formulations The analgesic efficacy of aspirin at a standard single dose of 0.5–1.0 g will be enhanced, if the compound is given in a predissolved or good water-soluble formulation [497]. This results in a more rapid increase in plasma levels and a parallel more rapid onset and efficacy of pain relief (Figure 4.29) [498, 499]. Soluble aspirin was also found to be more potent than solid acetaminophen in postsurgical pain [500]. A commercial effervescent preparation (solution) has been developed that allows to obtain the same peak plasma level, peak concentration, and half-life as plain aspirin tablets; however, the time to reach peak plasma concentrations is considerably shortened to about 30 min instead of 1 h in the plain preparation of plain aspirin tablets [501]. Similar benefits were obtained with a mouth-dispersible formulation [502].

Another possibility to even faster obtain effective plasma levels is the application of injectable aspirin water-soluble lysine salt. This is of particular advantage in migraine attacks when nausea and vomiting frequently occur (see below).

4.2.1.2 Clinical Trials

Flu and Other Feverish Diseases Aspirin as a household remedy is frequently used for treatment of influenza-like symptoms (headache, frontal and maxillary sinus sensitivity to percussion, sore throat, achiness, and feverish discomfort). These symptoms typically last for 3–5 days. They are not life threatening but markedly reduce the well-being. The maintenance or restoration of normal daily activity by reducing fever and influenza-like symptoms is the treatment goal and patients frequently use OTC antipyretics for this purpose. Among them, aspirin and acetaminophen are most commonly employed drugs [503, 504].

A recent placebo-controlled randomized double-blind trial has compared the antipyretic potency of aspirin and acetaminophen in adults. The patients suffered from an acute, noncomplicated infection of the upper airways that was likely to be of viral origin. Patients were treated with single doses of aspirin (500 or 1000 mg), acetaminophen (500 or 1000 mg), or placebo. Body temperature was controlled in regular intervals; feverish discomfort was evaluated on an interview basis. The total observation period was 6 h.

The average body temperature before treatment was 38.8 °C and remained essentially unchanged over the observation period of 6 h in the placebo group. Both aspirin and acetaminophen reduced the temperature to about 38.0 and 37.5 °C after single doses of 500 mg and 1 g, respectively. The antipyretic effect started 30 min after dosing and lasted for at least 6 h. The maximum effect was obtained 2.5–3 h after drug administration. Both compounds were about equipotent, also with respect to improvement of feverish discomfort, headache, and achiness. There was no significant difference in side effects between aspirin, acetaminophen, and placebo.

The conclusion was that aspirin and acetaminophen are equipotent antipyretics. This action is dose dependent and there are no differences in side effects [505] (Figure 4.30).

Migraine and Tension-Type Headache Most frequent pain is that of primary headache, that is, tension-type headache and migraine. In contrast to other diseases, the use of aspirin for treatment of these forms of pain is rather single dose or short term during attacks than regular, continuous intake by prescription. Patients are frequently young or middle aged and usually otherwise healthy. However, they need particular consideration for gastric tolerance because in situations such as migraine attacks, patients may experience nausea

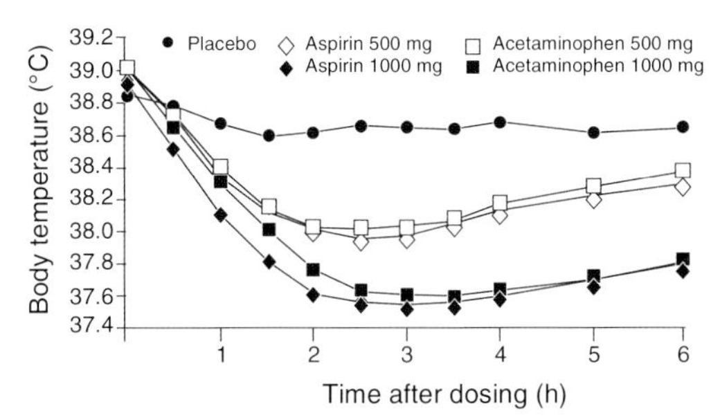

Figure 4.30 Time–course of orally measured body temperature in volunteers with acute uncomplicated febrile upper respiratory tract infection of suspected viral origin. Patients were treated with single dose of aspirin, acetaminophen, or placebo, as indicated. Data are a mean of 78–79 persons per treatment group (modeled after [505]).

or vomiting. Thus, rapid onset of action as well as less irritation or bypass of the stomach are desirable and, in case of aspirin, can be obtained by appropriate galenic formulations.

Migraine Aspirin is well established for the treatment of acute migraine attacks [506–508]. In a recent review, Diener *et al.* [501] summarized the evidence for aspirin as a drug of first choice in treatment of acute migraine attacks in randomized, double-blind controlled prospective trials. Overall, these studies showed not only significant beneficial effects of aspirin but also an increased efficacy and improved tolerability by a highly buffered effervescent preparation. This combination was found to be at least as effective as the combination of aspirin with metoclopramide in all but one [509] studies. Similar results were reported for the combination of lysine–aspirin plus metoclopramide versus ergotamine plus caffeine in relieving migraine attacks [510]. Aspirin (plus metoclopramide) was equieffective with triptanes (sumatriptan, zolmitriptan) [511–513] but with less side effects [501] (Table 4.15).

Tension-Type Headache Tension-type headache, also known as "normal" or "ordinary" headache, occurs in attack-like episodes but may also be chronic. This is the most frequent cause of pain and, probably, the best pain "model" in real life. It is a "featureless" disease, characterized by nothing but pain in the head. The pathophysiology is unknown but most likely complex. Tension-type headache may be episodic or chronic, that is, occurring at more than 15 days a month. The life-time prevalence of tension-type headache amounts to 79% with 3% suffering from the chronic form [517]. Psychical stress may be involved but also musculoskeletal functional or structural abnormalities, that is, tensions in the head and neck regions. Treatment is usually by OTC self-medication and – in contrast to migraine – triptanes do not work. From a pharmacological point of view this is a clear demonstration of a different pathophysiology of migraine and tension headache.

In a double-blind, placebo-controlled trial aspirin, 500 or 1000 mg, was compared with acetaminophen, 500 and 1000 mg, in a total of 572 compliant individuals. These persons suffered from episodic tension headache (not migraine). Treatment was single dose and the primary end point was subjective pain relief (total or worthwhile) after 2 h. In addition, individual severity of pain was measured by a visual analogue scale.

Aspirin at 1000 and 500 mg had a 76% and 70% responder rate, respectively. The response rate with acetaminophen was 71% at 1000 mg but only 64% at 500 mg. With the exception of acetaminophen 500 mg, all other treatments were significantly more effective than the placebo rate of 54%. Outcome was not affected by headache intensity at baseline. Adverse events were reported by 13–19% of subjects and were mild or moderate. No safety concerns arose.

The conclusion was that 1000 mg aspirin in moderate-to-severe headache is significantly more potent than placebo. Aspirin at 500 mg and acetaminophen at 1000 mg are also effective, but to less extent, whereas 500 mg acetaminophen are ineffective. There is a high placebo rate [518].

Aspirin (320 mg single dose) was also found to protect from headache at high altitude. This effect occurred at unchanged (low) oxygen saturation and was explained by interaction with sympathetic activity [519].

4.2.1.3 Aspirin and Other Drugs

Aspirin and acetaminophen are frequently used alternatives as antipyretic analgesics. A particular

Table 4.15 Prospective double-blind randomized trials with aspirin in migraine.

Type of study	G	#	Clinical end point	Outcome	Reference
Double-blind, parallel; aspirin 1000 mg versus PLA	p	485	% of patients with two-step improvement on a four-step scale after 2 h	pASA ≫ PLA	[508]
Double-blind, parallel; aspirin 900 mg versus PLA	p	101	% of patients with two-step improvement on a four-step scale after 2 h	pASA ≫ PLA	[502]
Double-blind, parallel; iLAS (=500 mg aspirin) versus PLA	i	40	Mean pain reduction on a 10-point VAS	iASA ≫ PLA	[506]
Double-blind, crossover; iLAS 1000 mg versus ERG 0.5 mg sc.	i	56	Pain reduction on a 10-point VAS	iASA = ERG	[507]
Double-blind, parallel; iLAS 1000 mg versus SUM 6 mg sc. versus parenteral PLA	i	279	% of patients with two-step improvement on a four-step scale after 2 h	SUM > iLAS ≫ PLA	[514]
Double-blind, parallel; eASA 1000 mg versus ePLA	e	343	% of patients with two-step improvement on a four-step scale agter 2 h	eASA ≫ PLA	[515]
Double-blind, crossover; eASA 1000 mg versus SUM 50 mg versus IBU 400 mg versus PLA	e	312	% of patients with two-step improvement on a four-step scale after 2 h	eASA = SUM = IBU ≫ PLA	[514]
Double-blind, parallel; eASA 1000 mg versus SUM 50 mg versus PLA	e	433	% of patients with complete remission of nausea, photo- and phonophobia after 2 h; % of patients with headache relief	eASA = SUM ≫ PLA	[516]

A comparison of three different aspirin formulations versus other analgesic monotherapy or placebo. G: Galenics; p: plain; e: effervescent; i: injectable (lysine salt); PLA: placebo; ERG: ergotamine; SUM: sumatriptan; IBU: ibuprofen (modeled after [501]).

issue with these compounds is possible misuse, in particular if the compounds are administered in combination with potentially habit-forming compounds, such as caffeine [520]. In general, the combined use of analgesics with a different mode of action, for example, aspirin + codeine, appears to be useful. However, fixed-dose combinations may not be as effective as patient-specific adjustment of doses [521]. In addition, the therapeutic efficacy may be diluted by addition of doubtful comedications, such as certain vitamins or other ingredients, which increase the price but do not add to the therapeutic efficacy. A notable exception is vitamin C that has been shown to protect the human stomach from aspirin-induced mucosal injury, possibly by its antioxidant properties [522].

4.2.1.4 Actual Situation

Aspirin, for example, in an effervescent formulation, is the OTC treatment of choice for tension headache. The compound is also effective in migraine and other forms of postsurgical inflammatory pain. The recommended single dose is 1 g. A meta-analysis of nine clinical trials with this single dose of aspirin in typical OTC medications, including tension-type headache showed that 6.3% of patients on aspirin and 3.9% of patients on placebo showed adverse effects. Only 3.1% of patients on aspirin and 2.0% of patients on placebo reported drug-related GI adverse effects [523]. In the case of aspirin intolerance or inefficacy, acetaminophen is the alternative. Whether ibuprofen at analgesic single doses of 400 mg can replace aspirin as an anti-inflammatory analgesic is not established yet. However, ibuprofen might interact with aspirin and antagonize its antiplatelet effects if regular intake is necessary for cardiovascular reasons (Section 2.3.1).

There is a clear need to develop new classes of analgesics, which more specifically interrupt individual pain perception and transmission pathways, allowing a more targeted treatment and prevention of pain. Whether coxibs are the final breakthrough for prostaglandin-associated pain perception can be disputed since these compounds will only work if (upregulated) COX-2 is significantly involved in prostaglandin production and the ultimate cause of pain. Unfortunately, current clinical analgesic trials select patients on the basis of disease and use crude global outcome measures instead to identify mechanism [524]. Regarding aspirin, its multiple pharmacological actions on gene regulation, signal transduction, and synaptic plasticity might also vary in different forms of pain. Unfortunately, none of these issues has been addressed sufficiently yet, although the US Congress has dedicated this decade the "Decade of Pain Control and Research."

Summary

Aspirin is an effective antipyretic analgesic for feverish conditions combined with influenza-like conditions. Standard doses between 1 and 2 g reduce the elevated body temperature shortly after intake. A maximum antipyretic effect is obtained at 2–3 h and is salicylate mediated.

Aspirin is an effective analgesic for control of acute pain of moderate-to-severe intensity and exhibits a clear dose–response relationship. Drowsiness and gastric irritation are the most frequent adverse effects, also at single-dose treatment. The benefit/risk ratio, that is, the ratio between number needed to treat and number needed to harm at the 1 g dose is 4 versus 30–40, which indicates that about 1 out of 10 patients will experience(subjective) side effects.

For treatment of migraine and tension-type headache, aspirin is a drug of first choice. The recommended single dose is 0.5–1 g and the efficacy can further be increased by using effervescent preparations or parenteral application as lysine salt. Another alternative in migraine is comedication of metoclopramide, which results in a clinical efficacy similar to triptanes.

References

489 Rosendorff, C. and Cranston, W.I. (1968) Effects of salicylate on human temperature regulation. *Clinical Science*, **35**, 81–91.

490 Schoen, R. (1952) Klinische Stellung zu den Analgetika. *Naunyn-Schmiedeberg's Archives of Experimental Pathology and Pharmackology*, **216**, 90–113.

491 Samad, T.A., Sapirstein, A. and Woolf, C.J. (2002) Prostanoids and pain: unraveling mechanisms and revealing therapeutic targets. *Trends in Molecular Medicine*, **8**, 390–396.

492 Houde, R.W., Wallenstein, S.L., Beaver, W.T. (1966) Evaluation of analgesics in patients with cancer pain. In *International Encyclopedia of Pharmacology and Therapeutics*, Vol. 1. Clinical Pharmacology, Oxford, Pergamon Press.

493 Laska, E.M., Sunshine, A., Wanderling, J.A. *et al.* (1982) Quantitative differences in aspirin analgesia in three models of clinical pain. *Journal of Clinical Pharmacology*, **22**, 531–542.

494 Edwards, J.E., Oldman, A.D., Smith, L.A. *et al.* (1999) Oral aspirin in postoperative pain: a quantitative systematic review. *Pain*, **81**, 289–297.

495 Barden, J., Edwards, J.E., McQuay, H.J. *et al.* (2004) Pain and analgesic response after third molar extraction and other postsurgical pain. *Pain*, **107**, 86–90.

496 McQuay, H.J. and Moore, R.A. (2007) Dose-response in direct comparisons of different doses of aspirin, ibuprofen and paracetamol (acetaminophen) in analgesic studies. *British Journal of Clinical Pharmacology*, **63**, 271–278.

497 Seymour, R.A., Williams, F.M., Luyk, N.M. *et al.* (1986) Comparative efficacy of soluble aspirin and aspirin tablets in postoperative dental pain. *European Journal of Clinical Pharmacology*, **30**, 495–498.

498 Holland, I.S., Seymour, R.A., Ward-Booth, R.P. *et al.* (1988) An evaluation of different doses of soluble aspirin and aspirin tablets in postoperative dental pain. *British Journal of Clinical Pharmacology*, **26**, 463–468.

499 Stillings, M., Havlik, I., Chetty, M. *et al.* (2000) Comparison of the pharmacokinetic profiles of soluble aspirin and solid paracetamol tablets in fed and fasted volunteers. *Current Medical Research and Opinion*, **16**, 115–124.

500 Seymour, R.A., Hawkesford, J.E., Sykes, J. *et al.* (2003) An investigation into the comparative efficacy of soluble aspirin and solid paracetamol in postoperative pain after third molar surgery. *British Dental Journal*, **194**, 153–157.

501 Diener, H.C., Lampl, C., Reimnitz, P. *et al.* (2006) Aspirin in the treatment of acute migraine attacks. *Expert Review of Neurotherapeutics*, **6**, 563–573.

502 MacGregor, E.A., Dowson, A. and Davies, P.T. (2002) Mouth-dispersible aspirin in the treatment of migraine: a placebo-controlled study. *Headache*, **42**, 249–255.

503 Hersh, E.V., Moore, P.A. and Ross, G.L. (2000) Over-the-counter analgesics and antipyretics: a critical assessment. *Clinical Therapeutics*, **22**, 500–548.

504 Mazur, I., Wurzer, W.J., Ehrhardt, C. *et al.* (2007) Acetylsalicylic acid (ASA) blocks influenza virus propagation via its NFκB-inhibiting activity. *Cellular Microbiology*, **9**, 1683–1694.

505 Bachert, C., Chuchalin, A.G., Eisebitt, R. *et al.* (2005) Aspirin compared with acetaminophen in the treatment of fever and other symptoms of upper respiratory tract infection in adults: a multicenter, randomized, double-blind, double-dummy, placebo-controlled, parallel group, single-dose, 6-hour dose-ranging study. *Clinical Therapeutics*, **27**, 993–1003.

506 Taneri, Z. and Petersen-Braun, M. (1995) Therapie des akuten Migräneanfalls mit intravenös applizierter Acetylsalicylsäure. Eine plazebokontrollierte Doppelblindstudie. *Der Schmerz*, **9**, 124–129.

507 Limmroth, V., May, A. and Diener, H.C. (1999) Lysine–acetylsalicylic acid in acute migraine attacks. *European Neurology*, **41**, 88–93.

508 Lipton, R.B., Goldstein, J., Baggish, J.S. *et al.* (2005) Aspirin is efficacious for the treatment of acute migraine. *Headache*, **45**, 283–292.

509 Diener, H.C. (1999) Efficacy and safety of intravenous acetylsalicylic acid lysinate compared to subcutaneous sumatriptan and parenteral placebo in the acute treatment of migraine. A double-blind, double-dummy, randomized, multicenter, parallel group study. The ASASUMAMIG Study Group. *Cephalalgia*, **19**, 581–588.

510 Titus, F., Escamilla, C., Gomes de la Costa palmeira, M.M. *et al.* (2001) A double-blind comparison of lysine acetylsalisylate plus metoclopramide vs. ergotamine plus caffeine in migraine. *Clinical Drug Investigation*, **21**, 87–94.

511 (The) Oral Sumatriptan Aspirin Plus Metoclopramide Comparative Study Group (1992) A study to compare oral sumatriptan with oral aspirin plus oral metoclopramide in the acute treatment of migraine. *European Neurology*, **32**, 177–184.

512 Tfelt-Hansen, P., Henry, P., Mulder, L.J. *et al.* (1995) The effectiveness of combined oral lysine acetylsalicylate and metoclopramide compared with oral sumatriptan for migraine. *The Lancet*, **346**, 923–926.

513 Geraud, G., Compagnon, A. and Rossi, A., COZAM Study Group (2002) Zolmitriptan versus a combination of acetylsalicylic acid and metoclopramide in the acute oral treatment of migraine: a double-blind, randomised, three-attack study. *European Neurology*, **47**, 88–98.

514 Diener, H.C., Bussone, G., de Liano, H. *et al.*, EMSASI Study Group (2004) Placebo-controlled comparison of effervescent aspirin and ibuprofen in the treatment of migraine attacks. *Cephalalgia*, **24**, 947–954.

515 Lange, R., Schwarz, J.A. and Hohn, M. (2000) Acetylsalicylic acid effervescent 1000 mg (Aspirin®) in acute migraine attacks; a multicentre, randomized, double-blind, single-dose, placebo-controlled parallel group study. *Cephalalgia*, **20**, 663–667.

516 Diener, H.C., Eikermann, A. and Gessner, U. (2004) Efficacy of 1000 mg effervescent acetylsalicylic acid and sumatriptan in treating associated migraine symptoms. *European Neurology.*, **52**, 50–56.

517 Rasmussen, B.K., Jensen, R., Schroll, M. *et al.* (1991) Epidemiology of headache in a general population. A prevalence study. *Journal of Clinical Epidemiology*, **44**, 1147–1157.

518 Steiner, T.J., Lange, R. and Voelker, M. (2003) Aspirin in episodic tension-type headache: placebo-controlled dose-ranging comparison with paracetamol. *Cephalalgia*, **23**, 59–66.

519 Burtscher, M., Likar, R., Nachbauer, W. *et al.* (1998) Aspirin for prophylaxis against headache at high altitudes: randomised, double-blind, placebo-controlled trial. *British Medical Journal*, **316**, 1057–1058.

520 Zhang, W.Y. (2001) A benefit–risk assessment of caffeine as an analgesic adjuvant. *Drug Safety*, **24**, 1127–1142.

521 Barkin, R.L. (2001) Acetaminophen, aspirin, or ibuprofen in combination analgesic products. *American Journal of Therapeutics*, **8**, 433–442.

522 Pohle, T., Brzozowski, T., Becker, J.C. *et al.* (2001) Role of reactive oxygen metabolites in aspirin-induced gastric damage in humans: gastroprotection by vitamin C. *Alimentary Pharmacology & Therapeutics*, **15**, 677–687.

523 Voelker, M. (2004) Safety and tolerability of aspirin in randomised controlled clinical trials. *Drug Safety*, **27**, 968.

524 Scholz, J. and Woolf, C.J. (2002) Can we conquer pain? *Nature Neuroscience*, **5** (Suppl.), 1062–1067.

4.2.2 Arthritis and Rheumatism

Aspirin was originally introduced into clinics for treatment of inflammatory pain and rheumatism as a better tolerable and more effective alternative to sodium salicylate. Although this was a remarkable step forward at the time, aspirin for this indication has meanwhile largely been replaced by other compounds, specifically those that were targeted to hit the suspected molecular target more precisely?: prostaglandins generated via upregulated COX-2. The first traditional NSAID, indomethacin, was introduced to the market in 1960 and was afterward followed by many successors, most important, diclofenac and ibuprofen.

These so-called "aspirin-like" drugs are only "aspirin-like" in one single aspect – inhibition of cyclooxygenase(s) by, again in contrast to aspirin, reversible inhibition of their enzymatic activity. This becomes clinically relevant, when aspirin comedication is necessary in patients with osteoarthritis or rheumatoid arthritis who frequently also suffer from (advanced) atherosclerosis – both chronic inflammatory diseases with an upregulated COX-2. The introduction of the first coxib, rofecoxib, in 1998, marked the beginning of a new prostaglandin-based era of symptomatic treatment of these diseases. However, it meanwhile became also clear that COX-2-derived prostaglandins, specifically prostacyclin and PGE_2, are not solely the "bad guys," which just enhance pain responses. They also may be useful or even essential for maintenance of an antithrombotic state in high-risk patients. In other words, COX-2-derived prostaglandins are mediators, made "on demand" and therefore, by definition, are neither good nor bad. They, however, do not specifically alter the natural history of the diseases, that is, the premature abrasion of cartilage in osteoarthritis or the pathologic immune reaction in rheumatoid arthritis. Nevertheless, they might be useful because of allowing physical exercise and improving the patient's quality of life. Meanwhile, more etiology-oriented drugs became available, including biologicals, such as anti-TNFα-antibodies (infliximab), tumor necrosis factor alpha (TNFα)-receptor blockers (etanercept), and other disease-modifying antirheumatic drugs (DMARDs). However, there is still need for improved analgesic treatment of the inflammatory/ ischemic process. Another clinically most important issue is the comedication of aspirin to traditional NSAIDs and its possible consequences for the efficacy of aspirin as an antiplatelet drug (Section 2.3.1).

4.2.2.1 Pathophysiology and Mode of Aspirin Action

Rheumatoid Arthritis Chronic rheumatoid arthritis is a multisystem disorder, caused by a pathologic immune reaction. The disease is associated with a significantly shortened (5–10 years) life expectancy. This is largely due to frequent comorbidities, which are associated with the systemic inflammatory process. This includes a 30–50% increased risk of atherothrombotic events, in particular myocardial infarction [525–527]. Recent studies with disease-modifying antirheumatic drugs," such as methotrexate, did not only retard the progression of the rheumatic disease [528] but also reduced the increased cardiovascular morbidity and mortality [529, 530]. Traditional NSAIDs and biologicals had no such effect [528, 530] (see below). Alternatively, appropriate treatment with antiplatelet/antithrombotic drugs might also reduce the enhanced cardiovascular mortality, for example, when given as adjunctive treatment to DMARDs.

In this context, it is interesting to note that patients with rheumatoid arthritis at a time when only aspirin was available for treatment of inflammatory pain suffered significantly less fatal myocardial infarctions than age-matched nonrheumatic controls [531]. Another early study in a small group of patients with rheumatoid arthritis reported a (nonsignificant) reduction of cardiovascular ischemic events by 30–50% after aspirin treatment [532] while Vandenbroucke *et al.* [533] found no significant difference in heart and vessel diseases – 10% lower risk in rheumatoid patients. Such studies cannot be performed anymore because of ethical reasons and the availability of more potent anti-inflammatory drugs. However, they are interesting from a pharmacological point of view.

COX-2 protein is upregulated in the synovia of patients with rheumatoid arthritis [534]. Aspirin at anti-inflammatory doses reduces pain and edema in the inflamed joint [535]. Because of the prompt effect, aspirin used to be the treatment of choice for arthritic pain. However, this has changed in the meantime for several reasons. One is the availability of NSAIDs as more effective analgesic/anti-inflammatory alternatives. Another is the high aspirin dose, that is, 3–4 g per day, which caused a number of side effects, in particular in the GI tract (Section 3.2.1) and inner ear (tinnitus) (Section 3.2.4). In addition, these high salicylate levels (>1 mM) might enhance apoptosis. This is wanted in tumor (Section 2.3.3) but not in human synovial cells [536]. In addition, aspirin suppresses cartilage proteoglycane synthesis, possibly via prostaglandin-independent mechanisms [537] but will not reduce the biosynthesis of hyaluronic acid, a most significant compound for maintenance of functional integrity of articular cartilage [538]. Another interesting finding was that aspirin might alter the disposition of methotrexate in patients with rheumatoid arthritis by altering its clearance and increasing the free, nonprotein-bound fraction. Although these actions required high-dose aspirin (3.9 g/day) and, therefore, might be not relevant for treatment of the disease today, they nevertheless suggest a potentially positive interaction between methotrexate and aspirin that might be valuable for the prevention of atherothrombotic events [539] (Figure 4.31).

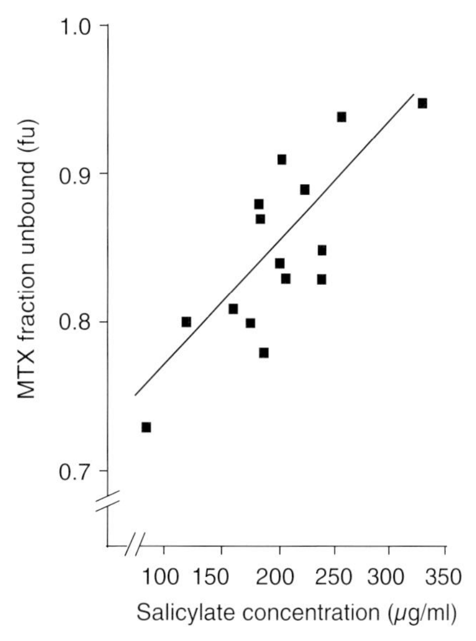

Figure 4.31 Correlation between free (unbound) methotrexate and the mean salicylate plasma level in 15 patients with rheumatoid arthritis treated with aspirin 975 mg qid for 1 week (modified after [539]).

Osteoarthritis In contrast to rheumatoid arthritis, osteoarthritis is not considered a primarily inflammatory, systemic disease but rather a local, degenerative disorder with an inflammatory component, mostly in joints under high "workload," such as the knee and hip. In contrast to rheumatoid arthritis, osteoarthritis is usually not associated with increased cardiovascular morbidity or a shortened life span [540]. Pain is the main devastating symptom and arises from different areas of the affected joint: bone (periostitis, subchondrial microfractures, and ischemia), synovia (synovitis), stimulation of nerve endings with neuronal inflammation, and release of inflammatory mediators as well as periarticular muscle spasms. Inflammatory pain at these sites might additionally induce expression of COX-2 in the spinal cord and cause neuropathic pain. Therefore, the intensity of the reactive, inflammatory reaction not only determines the intensity of pain but also might influence the progression of the degenerative processes in cartilage. Because osteoarthritis is an erosive disease of the cartilage without a chance for *restitutio ad integrum*, but only retardation of progression, the treatment is mainly symptomatic and focused on pain relief and improved mobility.

COX-2 protein is also upregulated in the synovia of patients with osteoarthritis as opposed to patients with traumatic knee injury [534]. This might be a target for aspirin, although the compound has been largely replaced by NSAIDs and acetaminophen. However, ibuprofen and other NSAIDs might interfere with the antiplatelet action of aspirin while acetaminophen at the recommended high dose might have a number of side effects in the kidney (Section 3.2.2) and liver (Section 3.2.3).

4.2.2.2 Clinical Trials

Rheumatoid Arthritis Aspirin is no longer a treatment of choice but still a possible adjunctive treatment for rheumatoid arthritis [541]. However, a significant proportion of these patients receive aspirin by prescription, for example, for cardiocoronary prevention, together with NSAIDs or coxibs for pain relief [542]. This might result in drug interactions, with regard to the COX-1-mediated antiplatelet effects of aspirin (Section 2.2.1) and the COX-2-dependent generation of antiplatelet, vasodilatory prostaglandins, such as PGI_2 and PGE_2. Another issue is gastric tolerance for both NSAIDs and coxibs (Section 3.2.1).

A recent meta-analysis on antirheumatic drug use and the risk of acute myocardial infarction investigated the possible risk of these compounds as opposed to DMARDs.

A nested case–control study was done within a cohort of subjects with rheumatoid arthritis to assess the risk of acute myocardial infarction associated with the use of DMARDs and anti-inflammatory medications in these patients.

The cohort included 107 908 patients, who were followed over an observation period of 5 years. End point was the first occurrence of myocardial infarction (cases). A total of 558 infarctions occurred during this follow-up (3.4% per year). This rate was significantly reduced with the current use of any DMARD, including methotrexate and leflunomide (RR: 0.80; 95% CI: 0.65–0.98) but not biologicals (infliximab, etanercept, and anakinra) (RR: 1.30; 95% CI: 0.92–1.83). The infarction rate was increased with glucocorticoids (RR: 1.32; 95% CI: 1.02–1.72) but not with traditional NSAIDs (RR: 1.05; 95% CI: 0.81–1.36) or coxibs (RR: 1.11; 95% CI: 0.87–1.43).

The conclusion was that DMARDs reduce the risk of myocardial infarctions in patients with rheumatoid arthritis. No risk increase was seen with traditional NSAIDs or coxibs [528].

Osteoarthritis Aspirin is still in use for treatment of osteoarthritis [541]. About 20% of patients suffering from osteoarthritis are prescribed aspirin and for about 50% of aspirin users, COX inhibitors, both NSAIDs and coxibs, are prescribed simultaneously. The prescription behavior of NSAIDs was independent from preexisting cardiovascular comorbidity, that is, a possible antagonism of aspirin-induced inhibition of COX-1 by coxibs or NSAIDs was not excluded [542].

Treatment of arthritic pain in osteoarthritis in most cases is by self-medication of OTC drugs. Among these drugs in addition to aspirin, acetaminophen and ibuprofen are the most frequently used. A recent meta-analysis of 23 randomized placebo-controlled trials has shown that these compounds can reduce short-term pain but because of serious side effects and low efficacy cannot be recommended for long-term use [543]. Another recent meta-analysis of clinical trials with single-dose aspirin treatment (1 g) in typical OTC medications showed that 6.3% of patients had side effects as opposed to 3.9% of patients on placebo [544]. A review on side effects of drugs, used in about 5700 patients with rheumatoid arthritis and 3100 patients with osteoarthritis, showed that OTC use of aspirin, acetaminophen, and ibuprofen in otherwise healthy individuals is not associated with a high risk of GI side effects. However, there is an enhanced intolerance in combined treatment with other NSAIDs, in particular, if corticosteroids are additionally prescribed [545] (Figure 4.32).

A certain reintroduction of aspirin for primary treatment of osteoarthritis and low-back pain was the remembrance of the fact that the compound was a chemical modification of the natural product salicylic acid, which in turn was the active principal of the prodrug salicin. Salicin is found in large amounts in the willow bark (Section 1.1.2). This stimulated studies on willow bark extracts for inflammatory pain relief.

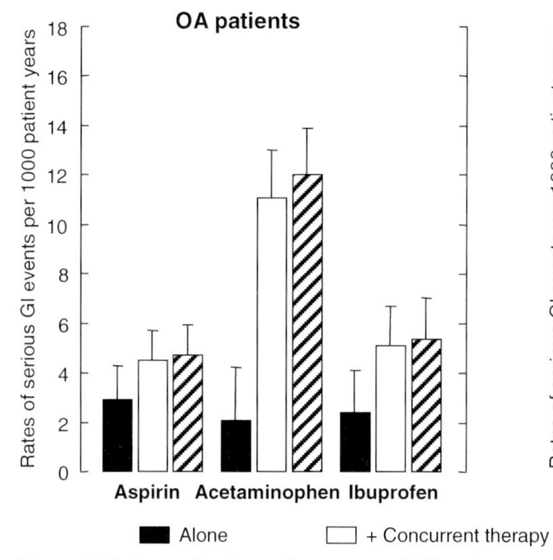

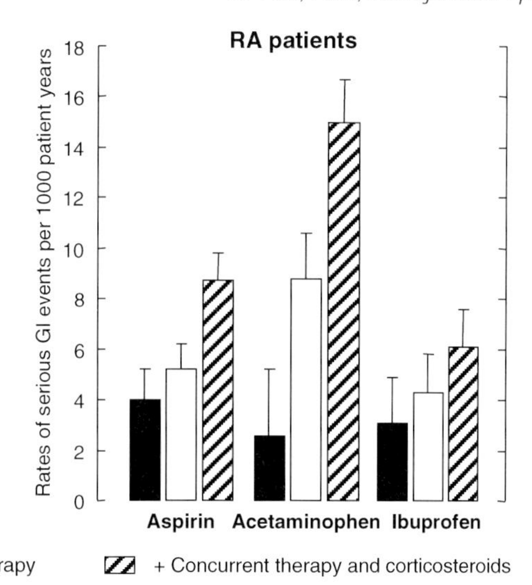

Alone + Concurrent therapy + Concurrent therapy and corticosteroids

Figure 4.32 Rates of serious GI events per 1000 patient-years in patients treated with aspirin, acetaminophen, or ibuprofen alone or concurrently with other drugs (mainly other NSAIDs). Patients who additionally received corticosteroids are shown separately. OA: osteoarthritis; RA: rheumatoid arthritis (modified after [545]).

Two randomized studies have been published recently, investigating the efficacy and tolerability of standardized willow bark extracts in low-back pain [546] and osteoarthritis [547]. Both studies reported beneficial effects. The active component in this extract was assumed to be salicin, the prodrug of salicylic acid. However, because of the different pharmacokinetics [548] and the much lower plasma salicylate levels than observed after analgesic doses of synthetic salicylate [549], the generation of salicylic acid alone was considered unlikely to explain analgesic or anti-inflammatory effects of willow bark.

4.2.2.3 Aspirin and Other Drugs

Acetaminophen Acetaminophen is recommended for treatment of symptoms of osteoarthritis in the United States. According to a study by Bradley *et al.* [550], acetaminophen was found to be as effective as ibuprofen. The acetaminophen dose in this particular study was 4 g (!) per day and the duration of treatment 4 weeks. Overall, systematic reviews of randomized controlled trials have not found important differences in efficacy between different NSAIDs. However, there is definitely need for large prospective randomized trials that compare a(ny) NSAID with acetaminophen [551]. Nevertheless, from a pharmacological point of view, it is questioned whether acetaminophen is really the first option for symptomatic treatment of an inflammatory complication of a degenerative disease, such as osteoarthritis: the analgesic potency is weak, the anti-inflammatory activity apparently absent, and, if any, of questionable clinical significance. A recent meta-analysis of randomized controlled trials, including about 6000 patients with osteoarthritis, suggested that NSAIDs are superior to acetaminophen for improving knee and hip pain [552].

Coxibs After the introduction of selective inhibitors of COX-2 and publication of the first long-term studies, a significantly increased risk of myocardial infarctions was found, specifically in patients with preexisting cardiovascular risk [553, 554]. One of these risk groups included patients with rheumatoid arthritis. In the VIGOR trial with rofecoxib, there was a significantly increased rate of myocardial infarctions in these patients. A total of 33% of myocardial infarctions were seen in a subgroup of patients (4%) with cardiovascular diseases who were not treated with aspirin although they met the general recommendations for aspirin use in cardiocoronary prevention [540, 555]. Comedication of aspirin was beneficial; however, it also caused more GI side effects than with coxibs alone [556]. On the other hand, coxibs do not interact with the antiplatelet action of aspirin. Therefore, in contrast to traditional NSAIDs, comedication of aspirin to coxibs might be useful with respect to efficacy and maintain the cardioprotective effects of aspirin. However, prospective randomized trials are necessary to establish this.

The other question is whether comedication of aspirin will worsen the GI tolerance to coxibs (Section 3.2.1). A recent register study in more than 4200 patients with rheumatoid arthritis taking coxibs with no ulcer at the beginning has shown only a small effect on the risk of developing dyspeptic syndromes in these patients [557].

Aspirin and Traditional NSAIDs One of the arguments against a possibly increased risk of myocardial infarctions with coxibs was that traditional NSAIDs have never been subjected to a comparative study. In addition, naproxen was suggested to exert a beneficial effect, for example, in the VIGOR trial, which might have been due to pharmacokinetic reasons, that is, its long half-life. Two recent trials have studied this issue in more detail. One epidemiological case–control study in Finland found an increased risk for a first myocardial infarction in adults after intake of traditional NSAIDs or coxibs. The risk was independent of the duration of treatment and similar with both classes of compounds (RR: 1.34; 95% CI: 1.26–1.43 versus RR: 1.31; 95% CI: 1.13–1.50) [558]. Similar results were obtained in a meta-analysis of randomized trials by Kearney *et al.* [559], comparing traditional NSAIDs with coxibs and placebo. In comparison to placebo, there was a comparable increase of atherothrombotic vessel occlusions by more than 40% for both traditional NSAIDs (RR: 1.45; 95% CI: 1.12–1.89) and coxibs (RR: 1.42; 95% CI: 1.13–1.78). Interestingly, in contrast to the Finnish study, there was no increased risk with naproxen [559]. However, the answer to the question whether traditional NSAIDs will interact with the antiplatelet effects of aspirin in a clinically relevant manner needs prospective randomized trials.

4.2.2.4 Actual Situation

Aspirin as an anti-inflammatory analgesic is still occasionally used for adjunctive treatment of osteoarthritis and rheumatoid arthritis. In rheumatoid arthritis, there is a markedly increased risk of atherothrombotic events, which might be reduced by aspirin. For pharmacological reasons coxibs plus aspirin might be the better alternative because of the absent interaction with the inhibition of platelet COX-1. In short-term use, aspirin appears to be well tolerated and not to differ from other analgesics, such as acetaminophen or ibuprofen as OTC medication. However, also in OTC use, reversible COX inhibitors such as ibuprofen or indomethacin and even dipyrone might interact with the antiplatelet effects of aspirin (Section 4.2.1). This could be relevant to patients requiring continuous aspirin intake for cardiocoronary prevention but needs prospective randomized trials to become firmly established.

Summary

Rheumatoid arthritis and osteoarthritis are frequent reasons for pain in musculoskeletal disorders. Rheumatoid arthritis is increasingly the domain of DMARDs while aspirin and NSAIDs may be (additionally) administered for pain relief. Effective analgesic treatment in osteoarthritis does not only reduce the symptoms of pain but might also indirectly retard the progression of the disease because of the improved mobility of patients.

Patients with rheumatoid arthritis have an increased risk of atherothrombotic events, in particular myocardial infarctions whereas patients with osteoarthritis have not. Most traditional NSAIDs – with a still unclear situation with naproxen – will not reduce the risk of atherothrombotic events but rather might increase it. Similar findings have been reported for coxibs. Inhibition of COX-2-dependent prostacyclin formation might be a common denominator for the elevated thrombotic risk in addition to the absence of antiplatelet effects by coxibs and a possible interaction of traditional NSAIDs with reversible inhibition of COX-1, that is, with the antiplatelet effects of aspirin. The clinical relevance of these findings is currently being studied in randomized trials.

Osteoarthritis is frequently treated by OTC drugs, that is, acetaminophen, aspirin or ibuprofen. Acetaminophen at the high doses recommended in the United States (4 g/day) not only might have a weak anti-inflammatory action but may also cause severe side effects, in particular during long-term use. OTC monotreatment with either of these compounds for acute or short-term periods appears to be well tolerated and is efficient for pain relief.

References

525 Wallberg-Jonsson, S., Ohmann, M.L. and Dahlqvist, S.R. (1997) Cardiovascular morbidity and mortality in patients with seropositive rheumatoid arthritis in Northern Sweden. *The Journal of Rheumatology*, **24**, 445–451.

526 Solomon, D.H., Karlson, E.W., Rimm, E.B. *et al.* (2003) Cardiovascular morbidity and mortality in women diagnosed with rheumatoid arthritis. *Circulation*, **107**, 1303–1307.

527 Turesson, C., Jarenros, A. and Jacobsson, L. (2004) Increased incidence of cardiovascular disease in patients with rheumatoid arthritis: results from a community-based study. *Annals of the Rheumatic Diseases*, **63**, 952–955.

528 Suissa, S., Bernatsky, S. and Hudson, M. (2006) Antirheumatic drug use and the risk of acute myocardial infarction. *Arthritis and Rheumatism*, **55**, 531–536.

529 Choi, H.K., Hernan, M.A., Seeger, J.D. *et al.* (2002) Methotrexate and mortality in patients with rheumatoid arthritis: a prospective study. *The Lancet*, **359**, 1173–1177.

530 Welsing, P.M., Kievit, W., Fransen, J. *et al.* (2005) The relation of disease activity and methotrexate with the risk of cardiovascular disease in patients with rheumatoid arthritis. *Annals of the Rheumatic Diseases*, **64** (Suppl. III), 90.

531 Cobb, S., Anderson, F. and Bauer, W. (1953) Length of life and cause of death in rheumatoid arthritis. *The New England Journal of Medicine*, **249**, 553–556.

532 Linos, A., Worthington, J.W., O'Fallon, W. *et al.* (1978) Effect of aspirin on prevention of coronary and cerebrovascular disease in patients with rheumatoid arthritis. A long-term follow-up study. *Mayo Clinic Proceedings*, **53**, 581–586.

533 Vandenbroucke, J.P., Hazefoet, H.M. and Cats, A. (1984) Survival and cause of death in rheumatoid arthritis: a 25-year prospective followup. *The Journal of Rheumatology*, **11**, 158–161.

534 Sano, H., Hla, T., Maier, J.A. *et al.* (1992) *In vivo* cyclooxygenase expression in synovial tissue of patients with rheumatoid arthritis and osteoarthritis and rats with adjuvant and streptococcal cell wall arthritis. *The Journal of Clinical Investigation*, **89**, 97–108.

535 Reid, J. (1960) Therapeutic properties of salicylates and its mode of action. *Annals of the New York Academy of Sciences*, **86**, 64–72.

536 Yamazaki, R., Kusunoki, N., Matsuzaki, T. *et al.* (2002) Aspirin and sodium salicylate inhibit proliferation and induce apoptosis in rheumatoid synovial cells. *The Journal of Pharmacy and Pharmacology*, **54**, 1675–1679.

537 Hugenberg, S.T., Brandt, K.D. and Cole, C.A. (1993) Effect of sodium salicylate, aspirin and ibuprofen on enzymes required by the chondrocyte for synthesis of chondroitin sulfate. *The Journal of Rheumatology*, **20**, 2128–2133.

538 Hugenberg, S.T., Kinch, M. and Brandt, K.D. (1987) The effect of salicylate on hyaluronic acid metabolism in articular cartilage. *Arthritis and Rheumatism*, **30** (Suppl.), S133.

539 Stewart, C.F., Fleming, R.A., Germain, B.F. *et al.* (1991) Aspirin alters methotrexate disposition in rheumatoid arthritis patients. *Arthritis and Rheumatism*, **34**, 1514–1520.

540 DeMaria, A.N. (2002) Relative risk of cardiovascular events in patients with rheumatoid arthritis. *The American Journal of Cardiology*, **89** (Suppl.): 33D–38D.

541 Fries, J.F., Ramey, D.R., Singh, G. *et al.* (1993) A reevaluation of aspirin therapy in rheumatoid arthritis. *Archives of Internal Medicine*, **153**, 2465–2471.

542 Greenberg, J.D., Bingham, C.O., Abramson, S.B. *et al.* (2005) Effect of cardiovascular comorbidities and concomitant aspirin use on selection of cyclooxygenase inhibitor among rheumatologists. *Arthritis and Rheumatism*, **53**, 12–17.

543 Bjordal, K.M., Ljunggren, A.E., Klovning, A. *et al.* (2004) Non-steroidal anti-inflammatory drugs, including cyclooxygenase-2 inhibitors, in osteoarthritic knee pain: meta-analysis of randomised placebo controlled trials. *British Medical Journal*, **329**, 1317–1322.

544 Voelker, M. (2004) Safety and tolerability of aspirin in randomised controlled clinical trials. *Drug Safety*, **27**, 968.

545 Fries, J.F. and Bruce, B. (2003) Rates of serious gastrointestinal events from low-dose use of acetylsalicylic acid, acetaminophen, and ibuprofen in patients with osteoarthritis and rheumatoid arthritis. *The Journal of Rheumatology*, **30**, 2226–2233.

546 Chrubasik, S., Künzel, O., Model, A. *et al.* (2001) Treatment of low-back pain with a herbal or synthetic anti-rheumatic: a randomized controlled study. Willow bark extract for low-back pain. *Rheumatology*, **40**, 1388–1393.

547 Schmid, B., Lüdtke, R., Selbmann, H.-K. *et al.* (2001) Efficacy and tolerability of a standardized willow bark extract in patients with osteoarthritis: randomized placebo-controlled, double blind clinical trial. *Phytotherapy Research*, **15**, 344–350.

548 Buss, T. (2005) Studie über die Einnahme von Weidenrinden-Extrakt, Salicin und Salcortin sowie Synthesen von Salicylsäure-Glykosiden und Salicin-Analoga. Inauguraldissertation, Marburg.

549 Schmid, B., Kötter, I. and Heide, L. (2001) Pharmacokinetics of salicin after oral administration of a standardised willow bark extract. *European Journal of Clinical Pharmacology*, **57**, 387–391.

550 Bradley, J.D., Brandt, K.D., Katz, B.P. *et al.* (1991) Comparison of an antiinflammatory dose of ibuprofen, an analgesic dose of ibuprofen and acetaminophen in the treatment of patients with osteoarthritis of the knee. *The New England Journal of Medicine*, **325**, 87–91.

551 Gøtzsche, P.C. (2000) Non-steroidal anti-inflammatory drugs. *British Medical Journal*, **320**, 1058–1061.

552 Towheed, T.E., Maxwell, L., Judd, M.G. *et al.* (2006) Acetaminophen for osteoarthritis. *Cochrane Database of Systematic Reviews*, **I**.

553 Ray, W.A., Stein, C.M., Daugherty, J.R. *et al.* (2002) COX-2 selective nonsteroidal anti-inflammatory drugs and risk of serious coronary heart disease. *The Lancet*, **360**, 1071–1073.

554 Schrör, K., Mehta, P. and Mehta, J.L. (2005) Cardiovascular risk of selective cyclooxygenase-2 inhibitors. *Journal of Cardiovascular Pharmacology and Therapeutics*, **10**, 95–101.

555 Bombardier, C., Laine, L., Reicin, A. *et al.* (2000) Comparison of upper gastrointestinal toxicity of rofecoxib and naproxen in patients with rheumatoid arthritis. *The New England Journal of Medicine*, **343**, 1520–1528.

556 Bolten, W.W. (2005) Problem of the atherothrombotic potential of non-steroidal anti-inflammatory drugs. *Annals of the Rheumatic Diseases*, **65**, 7–13.

557 Benito-Garcia, E., Michaud, K. and Wolfe, F. (2007) The effect of low-dose aspirin on the decreased risk of development of dyspepsia and gastrointestinal ulcers associated to cyclooxygenase-2 selective inhibitors. *The Journal of Rheumatology*, **34**, 1765–1769.

558 Helin-Salmivaara, A., Virtanen, A., Vesalainen, R. *et al.* (2006) NSAID use and the risk of hospitalization for first myocardial infarction in the general population: a nationwide case-control study from Finland. *European Heart Journal*, **27**, 1657–1663.

559 Kearney, P.M., Baigent, C., Godwin, J. *et al.* (2006) Do selective cyclooxygenase-2 inhibitors and traditional non-steroidal anti-inflammatory drugs increase the risk of atherothrombosis? Meta-analysis of randomised trials. *British Medical Journal*, **332**, 1302–1308.

4.2.3 Kawasaki Disease

Kawasaki disease (mucocutaneous lymph node syndrome) is a febrile disease in young children associated with an acute vasculitis of unknown etiology. The disease was originally described in Japan by *Tomisaku Kawasaki* [560] but was later also found in Europe and the United States. Kawasaki disease is a leading cause of acquired heart disease in small children. Cardiovascular symptoms are large vessel artheritis, myocarditis and aneurysms associated with a thrombosis tendency. Coronary artery aneurysms are of particular concern because patients with these abnormalities are at high risk of coronary thrombosis and myocardial infarction. Furthermore, these alterations may evolve into segmental stenoses in the chronic phase. Both incidence and mortality differ considerably throughout the world. The mortality rates vary between 1 and 3% in Europe but are less than 0.1% in Japan [561]. The incidence in Japan is 10-fold higher than in the US and 30-fold higher than in the United Kingdom and Australia. Thus, the disease among developed countries is most frequent in Japan and, consequently, object of intense research in Japan including timely diagnosis and appropriate treatment.

4.2.3.1 Pathophysiology and Mode of Aspirin Action

Clinical and Laboratory Findings Kawasaki disease is probably caused by an infectious agent in an immunologically susceptible subgroup of individuals. The causative agent remains elusive [562]. The patients present with fever, lasting for at least 5 days combined with signs of acute mucocutaneous inflammation and a pathologic immune reaction. This includes bulbar conjunctival injection, generalized erythema of skin and mucosae, and cervical unilateral lymph node enlargement [563]. About 20% of untreated children develop coronary artery aneurysms. Coronary artery aneurysms and myocardial infarctions most commonly occur after the second week of illness. They are paralleled by thrombocytosis and occurr at a time point when fever and mucocutaneous manifestations are subsiding [564]. Approximately half of these abnormalities regress within 5 years. The impressive decline in mortality in Japan from 0.1 to 0.01% was mainly due to avoidance of myocardial infarction or aneurysm rupture and underlines the prognostic significance of early diagnosis and treatment.

Mode of Aspirin Action The initial feverish phase is probably due to infection and then followed by an immune complex vasculitis that occurs when antibodies to the initiating agent appear in the circulation. The immune complexes activate and aggregate platelets, which in turn stimulates the release of platelet-derived vasoactive, proinflammatory mediators. This is associated with elevated plasma thromboxane levels at apparently unchanged PGE_2. However, there are not many data on the relationship between Kawasaki syndrome and prostaglandins and aspirin [565]. The somehow paradoxical thrombocytosis, occurring at this time point, is explained by a possible saturation of the reticuloendothelial system by immune complexes [564].

Overall, the laboratory findings are nonspecific and indicative of an immune complex vasculitis as pathophysiological explanation for Kawasaki syndrome. Acute-phase proteins and neutrophils are increased. There are elevated plasma levels of inflammatory cytokines, such as TNFα [566], and adhesion molecules, such as ICAM-1 [567]. Somehow indicative of an inflammatory immune reaction is the enhanced generation of cysteinyl leukotrienes during the acute phase of the disease [568] and circulating platelet-activating immune complexes in plasma after the 2nd week of disease [564]. These appear to be more frequent in those children who later develop coronary abnormalities [569]. Taken together, available data suggest a primarily pathologic immune reaction of unknown etiology with cytokine-induced generation of circulating cytotoxic antibodies for vascular cells and platelets.

Aspirin significantly reduces the elevated circulating thromboxane levels [565]. Interestingly, protein binding of aspirin (salicylate) is significantly lower in children during the acute phase of Kawasaki disease – 73% versus 90%. This results in on average a twofold higher level of free salicylate in these patients compared to normoalbuminemic controls [570] and also a significantly higher renal salicylate clearance during the febrile phase [571].

Kawasaki Disease and Reye's Syndrome Aspirin has been banned worldwide as an antipyretic analgesic in children because of a suggested risk of Reye's syndrome (Section 3.3.3). In this context, it is interesting to note that it is extremely difficult to find even one single case of Reye's syndrome in children with Kawasaki disease [572] despite the intense use of aspirin for treatment. In Japan, up to 200 000 children with Kawasaki disease have been treated with aspirin, the recommended initial dosage being between 30 and 100 mg/kg. Only one case of Reye's syndrome has been reported, giving an incidence of less than 0.005% [573]. In Britain, during the acute phase of the illness, moderate doses of aspirin, 30–50 mg per day, are recommended. However, in a recent British guideline for practical therapy of Kawasaki's disease, Reye's syndrome was not even mentioned [562]. This does not suggest any important relationship between the (virally induced?) febrile response in Kawasaki disease and Reye's syndrome and the use of aspirin, even in anti-inflammatory doses.

4.2.3.2 Clinical Trials

The therapeutic goal of treatment during the acute phase of the illness is to reduce inflammation and immune reactions in an effort to prevent clot formation and the later occurrence of coronary artery aneurysms. Early recognition and treatment with aspirin and intravenous immunoglobulin has been unequivocally shown to reduce the occurrence of coronary artery aneurysms. For this purpose, aspirin was originally given in anti-inflammatory doses of up to 100 mg/kg per day in the acute phase of the disease [570],573]. Because of the reduced bioavailability of salicylate in the febrile state as a consequence of increased renal clearance, these high doses of aspirin were considered necessary to obtain therapeutic salicylate plasma levels of 200 µg/ml. In one study, this was associated with a nearly complete prevention: 3% versus 39%, of coronary artery aneurysms [574].

In a retrospective, nonrandomized trial, a total of 162 patients with acute Kawasaki disease were treated with high-dose intravenous immunoglobulin (2 g/kg) as a single infusion without concomitant aspirin. Low-dose aspirin (3–5 mg/kg) was subsequently prescribed when fever subsided.

Patients who were nonresponsive to immunoglobulin according to study criteria had a significantly higher rate of coronary artery aneurysms: 25% versus 3%. The efficacy of immunoglobulin was not affected by the use of aspirin, which could not prevent the failure of immunoglobulin therapy. These results were similar to those of a previous trial [563] who had a similar >85% success rate in Kawasaki disease with high (80–100 mg/day) and low (3–5 mg/day) aspirin combined with high-dose intravenous immunoglobulin (2 g/kg) in the treatment of acute Kawasaki disease. The mean duration of fever was 47–8 h in high-dose and 34–5 h with low-dose aspirin. The duration of fever in the present study without aspirin was 48 h for 97% of patients in this study.

The conclusion was that the efficacy of immunoglobulin even without aspirin treatment raises questions concerning the use of aspirin in the acute phase of Kawasaki disease [575].

Despite these conclusions, it should be considered that the study of Hsieh *et al.* [575] was a retrospective, nonrandomized trial and, importantly, used historical controls for comparison of aspirin efficacy. In addition, the number of patients who were treated early was small and the long-term outcome of these children was unknown. Although aspirin may enhance the effects of immunoglobulin when given in the early phase of the disease, another earlier retrospective trial also reported no effect by aspirin on the response rate of intravenous immunoglobulin and the formation of coronary artery abnormalities and the duration of fever [576]. Randomized, prospective trials are urgently needed to clarify the possible benefits – and risks – of aspirin as an adjunctive to intravenous immunoglobulin in Kawasaki disease.

A meta-analysis on aspirin in Kawasaki disease indicated that 20–40% of children developed coronary artery aneurysms after treatment with aspirin alone. Combined therapy with aspirin and high-dose intravenous gammaglobulin given as a single infusion at high-dose (2 g/kg) reduced the occurrence of coronary artery aneurysms to 9% at 30 days and 4% at 60 days after the onset of the disease [577].

4.2.3.3 Aspirin and Other Drugs

There is no established antiplatelet or anti-inflammatory alternative to aspirin yet. However, intravenous gammaglobulin is the primary treatment of choice. Theoretically, anti-inflammatory glucocorticoids and TNFα antagonists might be used as well as clopidogrel for antiplatelet treatment [578].

4.2.3.4 Actual Situation

Intravenous immunoglobulin remains the mainstay of therapy of Kawasaki disease. Immunoglobulin treatment should be started early, preferably within the first 10 days of the illness. Aspirin is still considered a standard therapy as well, because of its anti-inflammatory and antiplatelet activities [579]. Aspirin is used initially in anti-inflammatory doses, which are followed by lower, antiplatelet dosages [580]. However, there are data to suggest that there are no major differences between high (75–100 mg/kg per day) and low-dose (1–74 mg/kg per day) aspirin in combination with immunoglobulin with respect to the duration of fever and the clinical outcome [563, 581]. Rescue therapies for immunoglobulin-resistant patients include corticosteroids as well as infliximab, an antagonist of tumor necrosis factor alpha.

There are only few randomized controlled trials of salicylate to treat Kawasaki disease in children. Until good quality randomized controlled trials are available, there is insufficient evidence to indicate whether children with Kawasaki disease should continue to receive salicylate as part of their treatment regimen [582]. However, administration of aspirin is a guideline recommendation in more severe cases, that is, cases with coronary artery aneurysms [583].

Summary

Kawasaki disease is an acute feverish disease that predominantly affects children below the age of 5 years. Fever is followed by a large-vessel immune vasculitis, myocarditis and coronary aneurysms associated with a thrombosis tendency. Patients with coronary artery abnormalities are at high risk of coronary thrombosis and myocardial infarction as well as accelerated atherosclerosis.

The pathogenesis and etiology of the disease are unknown. Possibly, the disease is initiated by infection, followed by an immune complex vasculitis with the appearance of antibodies in the circulation. These immune complexes cause a prothrombotic state with platelet aggregation, generation, release of inflammatory cytokines and expression of adhesion molecules for inflammatory cells at the vessel surface.

High-dose i.v. immunoglobulin combined with aspirin is the treatment of choice. Aspirin is initially given in high, anti-inflammatory doses of 30–60 mg/kg × day, followed by 3–5 mg/kg per day, that is, antiplatelet doses, in later phases if there is evidence for coronary abnormalities. Possibly, lower initial doses of aspirin are also effective, since both dose regimes are equipotent with respect to duration of fever. The combined treatment of immunoglobulin with aspirin might reduce the incidence of coronary artery aneurysms, myocardial infarctions, and vascular death. However, there are no sufficient data from prospective randomized trials to clearly establish this.

References

560 Kawasaki, T. (1967) Acute febrile mucocutaneous syndrome with lymphoid involvement with specific desquamation of the fingers and toes in children. *Arerugi*, **16**, 178–222.

561 Tizard, E.J. (1999) Recognition and management of Kawasaki disease. *Clinical Pediatrics*, **8**, 97–101.

562 Brogan, P.A., Bose, A., Burgner, D. *et al.* (2002) Kawasaki disease: an evidence based approach to diagnosis, treatment, and proposals for future research. *Archives of Disease in Childhood*, **86**, 286–290.

563 Saulsbury, F.T. (2002) Comparison of high-dose and low-dose aspirin plus intravenous immunoglobulin in the treatment of Kawasaki syndrome. *Clinical Pediatrics*, **41**, 597–601.

564 Levin, M., Holland, P., Nokes, T.J.C. *et al.* (1985) Platelet immune complex interactions in pathogenesis of Kawasaki disease and childhood polyarthritis. *British Medical Journal*, **290**, 1456–1460.

565 Fulton, D.R., Meissner, H.C. and Peterson, M.B. (1988) Effects of current therapy of Kawasaki disease on eicosanoid metabolism. *The American Journal of Cardiology*, **61**, 1323–1327.

566 Maury, C.P., Salo, E. and Pelkonen, P. (1989) Elevated circulating tumor necrosis factor alpha in patients with Kawasaki disease. *The Journal of Laboratory and Clinical Medicine*, **113**, 651–654.

567 Furukawa, S., Imai, K., Matsubara, T. *et al.* (1992) Increased levels of circulating intercellular adhesion molecule 1 in Kawasaki disease. *Arthritis and Rheumatism*, **35**, 672–677.

568 Mayatepek, E. and Lehmann, W.D. (1995) Increased generation of cysteinyl leukotrienes in Kawasaki disease. *Archives of Disease in Childhood*, **72**, 526–527.

569 Barron, K.S., Montalvo, J.F., Joseph, A.K. *et al.* (1990) Soluble interleukin-2 receptors in children with Kawasaki syndrome. *Arthritis and Rheumatism*, **33**, 1371–1377.

570 Koren, G., Silverman, E., Sundel, R. *et al.* (1991) Decreased protein binding of salicylates in Kawasaki disease. *The Journal of Pediatrics*, **118**, 456–459.

571 Koren, G., Schaffer, F., Silverman, E. *et al.* (1988) Determinations of low serum concentrations of salicylate in patients with Kawasaki disease. *The Journal of Pediatrics*, **112**, 663–667.

572 Lee, J.H., Hung, H.Y. and Huang, F.Y. (1992) Kawasaki disease with Reye syndrome: report of one case. *Zhonghua Min Guo Xiao Er Ke Yi Xue Hui Za Zhi*, **33**, 67–71.

573 van Bever, H.P., Quek, S.C. and Lim, T. (2004) Aspirin, Reye syndrome, Kawasaki disease, and allergies; a reconsideration of the links. *Archives of Disease in Childhood*, **89**, 1178.

574 Koren, G., Rose, V. and Lavi, S. (1985) Probable efficacy of high-dose salicylates in reducing coronary involvement in Kawasaki disease. *The Journal of the American Medical Association*, **254**, 767–769.

575 Hsieh, K.-S., Weng, K.-P., Lin, C.-C. *et al.* (2004) Treatment of acute Kawasaki disease: aspirin's role in the febrile stage revisited. *Pediatrics*, **114**, e689–e693.

576 Terai, M. and Shulman, S.T. (1997) Prevalence of coronary artery abnormalities in Kawasaki disease is highly dependent on gamma globulin dose but independent of salicylate dose. *The Journal of Pediatrics*, **131**, 888–893.

577 Durongpisitkul, K., Gururaj, V.J., Park, J.M. *et al.* (1995) The prevention of coronary artery aneurysm in Kawasaki disease: a meta-analysis on the efficacy of aspirin and immunoglobulin treatment. *Pediatrics*, **96**, 1057–1061.

578 Pinna, G.S., Kafetzis, D.A., Tselkas, O.I. *et al.* (2008) Kawasaki disease: an overview. *Current Opinion in Infectious Diseases*, **21**, 263–270.

579 Satou, G.M., Giamelli, J. and Gewitz, M.H. (2007) Kawasaki disease: diagnosis, management, and long-term implications. *Cardiology in Review*, **15**, 163–169.

580 Newburger, J.W. and Fulton, D.R. (2007) Kawasaki disease. *Current Treatment Options in Cardiovascular Medicine*, **9**, 148–158.

581 Simonsen, K.A., Rosenman, M.B., Conway, J.H. *et al.* (2005) Effectiveness of low-dose aspirin therapy in the management of Kawasaki disease. 8th International Kawasaki Disease Symposium, San Diego, CA, Abstract P124.

582 Baumer, J.H., Love, S.J.L., Gupta, A. *et al.* (2006) Salicylate for the treatment of Kawasaki disease in children. *Cochrane Database of Systematic Reviews*, **4**, CD004175, 10.1002/14651858.CD004175.pub.2.

583 Dajani, A., Taubert, K.A., Takahashi, M. *et al.* (1994) Guidelines for long-term management of patients with Kawasaki disease. *Circulation*, **89**, 916–922.

4.3
Further Clinical Indications

The clinical use of aspirin for prophylaxis of thromboembolic diseases and treatment of fever and pain is established since decades. This, however, covers only a small part of the broad spectrum of its pharmacological actions. Others, such as the anti-inflammatory/antirheumatic effects (with the exception of Kawasaki disease (Section 4.2.3), tokolytic and hypoglycemic activities, have not been considered therapeutically because of the availability of more effective and better tolerable alternatives. However, this does not mean that all of the possible therapeutic options provided by the unique pharmacological structure of aspirin are already utilized. One actual example with a considerable clinical impact is colorectal cancer. However, other carcinomas might also be affected, such as breast cancer [584], endometrial cancer [585], and nonsmall cell lung cancer among women [586].

Aspirin and salicylate are about equipotent inhibitors of the activation of several tumor-promoting and proinflammatory genes (Section 2.2.2). This effect is caused by an interaction of salicylate with the binding of transcription factors to the promoter region of these genes. Aspirin-induced upregulation (inflammatory cytokines) or down-regulation (tumor-suppressor genes) of disease-related genes is an example of another interesting pharmacological property of aspirin whose clinical significance has not been sufficiently appreciated yet. Indeed, new generation "aspirin-like" compounds could be synthesized that are targeted toward this regulation of gene expression. These compounds would not have anything in common with the conventional "aspirin-like" drugs, that is, traditional NSAIDs and coxibs. These compounds exclusively inhibit the generation of one product (prostaglandins) by an interaction with one single enzyme (cyclooxygenase). Clearly, both prostaglandin-dependent and prostaglandin-independent actions of aspirin may synergize and extend the spectrum of pharmacological actions and possible clinical applications.

Fresh insights into the etiology and pathophysiology of diseases in turn will also stimulate the development of new therapeutic strategies. In this respect, salicylate is unique because of its physicochemical properties that allow accumulation and enrichment in cell membranes. This, eventually results in an interference with transmembrane signal transduction and energy storage in the inner mitochondrial membrane (Section 2.2.3). These actions of salicylate, in strict sense nontoxic, are transient and principally reversible. Appropriate substitutions to the salicylate backbone structure – similar to penicillins or statins – may lead to the development of new derivatives with extended and improved pharmacological properties. Most attractive in this regard are salicylate-analogues, which upregulate defense genes or tumor-suppressor genes in case of an impending tissue injury or malignancy. Protection from injury by environmental factors is exactly the function of the (inducible) enzymatic generation of salicylates in plants. This wisdom from nature has so far not been transduced to animals or men. Clearly, increased resistance against injury, that is, inflammation and tumors, would be much closer to the intrinsic function of salicylate system than just the elimination of either signaling molecule, such as the prostaglandins.

4.3.1
Colorectal Cancer

Epidemiology Colorectal cancer is one of the most prevalent malignancies in Western societies. Any person older than 50 years has an about 5% chance of developing colorectal cancer with a 5-year survival rate among less than 40% patients [587, 588]. The vast majority (95%) of colorectal cancers occurs in individuals without a family history of cancer [589] and there is no gender difference [590]. However, there is a clear relationship to familial adenomatous polyposis coli (APC) and possibly also to the appearance of aberrant crypt foci (ACF) in the colon [591]. APC and ACF are considered premalignancies and are frequently found in patients with colorectal cancer – APC in about 90% of cases. Although only a minority of APC patients later develop cancer, the occurrence or recurrence of colorectal adenomas suggests a significant risk for later malignancy. This along with the long time from adenoma-to-carcinoma transition is the reason why APC in most of the clinical trials was taken as a surrogate parameter for the efficacy of chemoprevention.

Most data on cancer chemoprevention with aspirin are available for colorectal cancer. However, similar considerations may also apply to other GI malignancies. This includes carcinomas of the esophagus [592–594] and stomach [595, 596]. According to a meta-analysis, aspirin can reduce malignant tumors in these locations by about 50% if taken regularly at least over 3–5 years [597].

This section is focused on clinical aspects of the prevention of colorectal cancer. Basic pharmacological mechanisms of antitumor effects of aspirin are discussed in detail in Section 2.3.3.

4.3.1.1 Etiology and Pathophysiology of Intestinal Adenomas, Colorectal Cancer, and Mode of Aspirin Action

Etiology and Pathophysiology Colorectal cancer, like other epithelial tumors, is multifactorial in origin and a typical example of the multistep process of carcinogenesis. These include accumulation of mutations in specific genes controlling cell division, apoptosis, and DNA repair [598, 599]. Initially, germ line mutations of the APC gene cause familial polyposis coli where the vast majority of colorectal carcinomas arise from. The APC gene is believed to act as a "gatekeeper," ensuring that cell division is properly balanced by cell death. For a cell to become malignant, first the apoptosis mechanism has to be disturbed. This increases the number of adenomas in the intestinal mucosa, which are unable to undergo apoptosis because of mutations in apoptosis-related genes (p53, *k-Ras*). APC mutations alone probably do not cause the adenoma formation. This requires additional environmental factors, such as digestive secretion, dietary components (especially fat and fibers) and the intestinal flora. Overall, these factors determine both the progression of adenomatosis and the adenoma-to-carcinoma transition [600–602]. In this context, transcriptional induction of COX-2 is an important amplifying factor for both events. Accordingly, disruption of the COX-2 gene dramatically reduces the number of tumors in mice heterozygous for an APC mutation [603] (Table 4.16).

Heterozygous $APC^{\Delta716}$ mice, a model for human familial APC, develop hundreds of polyps per intestine within the first 10 weeks of age. In COX-2 heterozygous animals, generated by breading of $APC^{\Delta716}$ mice with COX-2 null mice, there was a 60% reduction in intestinal polyps

Table 4.16 Increase of number and size of intestinal polyps in a mouse model of human familial APC in dependency of COX-2 expression.

Parameter	COX-2 genotype		
	Homozygote (COX-2−/−)	Heterozygote (COX-2+/−)	Control (COX-2+/+)
Total # of polyps	93	224	652
% reduction	(86%)	(66%)	(0%)
# of larger polyps in colon (>2 mm diameter)	None	1.5 ± 1.9	6.8 ± 7.2
% reduction	(100%)	(78%)	(0%)

APC was induced by heterozygote deletion of the APC gene (APCΔ716). Heterozygote deletion of the COX-2 gene results in a significant homozygote deletion in a nearly complete disappearence of polyps in the colon as compared to animals with normal COX-2 expression (control) (modified after [603]).

compared to COX-$2^{+/+}$/APC$^{\Delta716}$ mice. In homozygous COX-2 knockouts, an 80–90% reduction was seen. In addition, tumor size was smaller in COX-2 deficient animals. Similar effects were obtained after the animals were treated with a selective COX-2 inhibitor.

These data provided first direct genetic evidence of a key role of COX-2 in colon tumorigenesis (adenomatosis) and suggest that induction of COX-2 is an early event in the sequence of polyp formation to colon carcinogenesis [603] (Table 4.16).

COX-2 and Tumor Promotion There is no immunoreactive COX-2 in mucosa biopsy samples taken from healthy colorectal mucosa but a marked upregulation in both adenomas and carcinomas. About 40–50% of mucosa biopsy samples taken from adenomas and 80–90% of cancer tissue show induced COX-2 expression [602]. No difference is seen in COX-1. COX-2 overexpression was also found in carcinomas of the stomach [604] and esophagus [605]. PGE_2 is generated as the dominating prostaglandin. PGE_2 stimulates tumor growth via EP receptors [606] and additionally acts as an immunosuppressive on monocytes/macrophages [607]. The tumorigenic action of PGE_2 might additionally become amplified by the metabolic activation of cocarcinogens via the peroxidase activity of the PGH-synthase complex (Section 2.3.3).

Upregulation of COX-2 and increase in PGE_2 synthesis are clinically correlated with the severity of the disease (lymph node metastases, tumor size) [608] and the survival rate of patients [609] (Figure 4.33). This makes the interaction with COX-2 expression and COX-2-dependent PGE_2 generation an attractive tool for therapeutic interventions.

Appropriate chemoprevention by suitable medications, including aspirin, is of particular therapeutic value in malignancies, such as colorectal cancer, where curative strategies for advanced or metastasizing disease have virtually no effect on life expectancy [610]. In the only available prospective, randomized, compliance-controlled trial, aspirin (600 mg bid for 2 years), given to patients with invasive colorectal cancer shortly after surgery, did neither prevent metastases nor improve the disease-free interval or survival time [611]. However, numerous studies indicate that the regular use of aspirin, non-selective NSAIDs and coxibs, reduces the incidence of colorectal carcinomas and/or appearance or reappearance of adenomas [612]. This points to enhanced COX-2 expression and aspirin-sensitive product formation, including prostaglandins, under the influence of tumor-promoting factors.

Mode of Aspirin Action Aspirin interferes with tumorigenesis at different levels; the most intensively studied is inhibition of PGE_2 production [613]. This, eventually, results in the inhibition

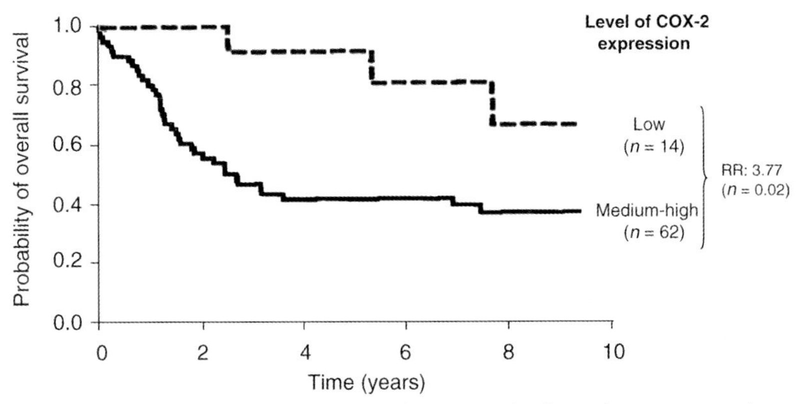

Figure 4.33 Kaplan–Meier survival estimates of patients with colorectal carcinomas in relation to histochemical expression of COX-2 in cancer tissue (modified after [609]).

of malignant cell proliferation by restoration of apoptosis. Similar to data in gene-manipulated mice mentioned above, clinical data also demonstrated chemoprotective actions of COX-2 selective inhibitors [614, 615]. This suggests that the antitumorigenic action of aspirin also involves the inhibition of COX-2-dependent prostaglandin formation. In addition, both aspirin and salicylate inhibit COX-2 expression and restore apoptosis via interactions with transcription factors, including p53 [610] and Bcl-2 [616]. Another variable for aspirin efficacy are genetic variations, for example, in the genotype of the uridine diphosphate glucuronyl transferase (UGT1A6), an enzyme involved in the metabolism of aspirin and some NSAIDs [617, 618]. Another gene of interest is ornithine decarboxylase. The G316A genotype is prognostic for colorectal adenoma occurrence and predictive in an enhanced protective action of aspirin [619]. The result of effective chemoprevention is restoration of apoptosis and inhibition of tumor growth. This hypothesis is shown in Figure 4.34.

4.3.1.2 Clinical Trials: Epidemiological Studies

First evidence for a chemopreventive action of aspirin on the risk of colon cancer was obtained in 1988 from the Melbourne Colorectal Cancer Study. *Gabriel Kune* and colleagues reported a 40% reduced risk of the incident colon cancer among regular aspirin users compared to those who did not regularly take the drug (Table 4.17). The results of the study were summarized by Dr Kune as follows:

> ... There was a statistically significant deficit of the use of aspirin and aspirin-containing compounds among cases and these differences remained statistically significant after adjustment for hypertension, heart disease, chronic arthritis, and diet in both males and females. This finding, whatever the mechanism may be, has potential significance in colorectal cancer chemoprevention and merits early confirmation. Aspirin is now widely used in the chemoprophylaxis of cardiovascular disease and may also be useful in a similar way in the prevention of colorectal cancer and perhaps also of other cancers [621].

These, at the time somewhat unexpected findings, were confirmed in another case–control trial, the Boston Collaborative Study Group. Regular intake of NSAIDs (usually aspirin-containing medications) reduced the incidence of colorectal carcinomas by 50%. The risk of colorectal carcinomas appeared to decrease with the duration of NSAID

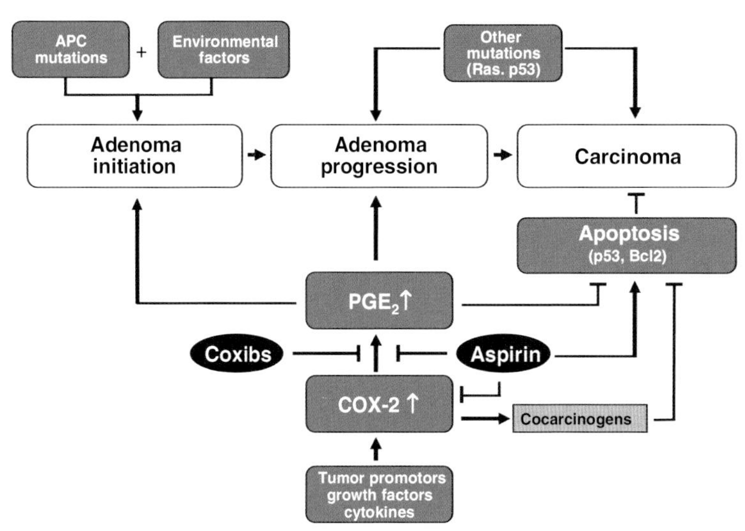

Figure 4.34 Pathophysiology of colorectal adenomas and carcinomas in patients with preexisting familial APC. Site of action of aspirin and coxibs (modified after [620]).

Table 4.17 Epidemiological and observational studies on the relationship between aspirin and risk of colorectal cancer (further explanation see text).

# of patients	Dose frequency	Study end point	RR (±95% CI)	Reference
715 cases 727 controls	Not stated	Colorectal cancer	0.53 (0.40–0.71)	[621]
1326 cases 4891 controls	>4 doses/week ≥3 months	Recurrent colorectal cancer	0.50 (0.20–0.90)	[622]
662 424 of either sex	≥16 times/month for >1 year	Death from colon cancer	0.60 (0.40–0.89) male	[623]
47 900 men	≥2 times/week	Colorectal cancer	0.68 (0.52–0.92) female	[624]
89 446 women	4–6 tablets/week for ≥20 year	Colorectal cancer	0.56 (0.36–0.90)	[624]
82 911 women 47 363 men	≥2 times/week	Colorectal cancer	0.64 (0.52–0.78) COX-2 (+) 0.96 (0.73–1.26) COX-2 (−)	[625]

(aspirin) use and to increase after withdrawal. However, none of these trends was significant [622].

An important impetus for the hypothesis of an anticancerogenic action of aspirin came from *Thun et al.* [623, 626]. These authors conducted a prospective mortality study in 662 424 adults above 60 years of age, the Cancer Prevention Study II (CPS-II, [627]). This is the largest epidemiological study on a possible relationship between aspirin and fatal colon cancer and one of the largest epidemiological studies at all.

Participants in the CPS-II study were asked two questions on aspirin, "how many times in the last month have you used the following [medication]" and "how long (years) have you used them."

The relative mortality from colon cancer among individuals who used aspirin 16 or more times per month for at least 1 year was 0.60 in men (CI: 0.40–0.89) and 0.58 in women (CI: 0.37–0.90), on average 0.58, compared to persons who did not take aspirin. There was also a trend of decreasing relative risk with more frequent and/or prolonged (at least 10 years) aspirin use, again similar in both sexes. Similar results were found with fatal rectum cancer with the relative risk reduced to 0.66 in men and women combined; however, greater reductions were obtained in men. No association was found between the use of acetaminophen and the risk of colon cancer.

The conclusion was that regular aspirin intake at low doses may reduce the risk of fatal colon cancer. In the study, published in 1993, there was a similar protective effect also for cancer of the stomach and esophagus. Whether this was due to a direct effect of aspirin, perhaps mediated by the inhibition of prostaglandin biosynthesis, or due to other factors, not associated with aspirin, remained unclear [623, 626].

The strength of this study is its size and prospective design, establishing dose–response trends with both the frequency and the duration of aspirin use in men and women (Figure 4.35). Its limitations include dependency upon a single brief, self-administered questionnaire, the absence of data on aspirin dosage (as opposed to frequency and duration of use), the possible intake of NSAIDs other than or in addition to aspirin, and, particularly, the study's reliance on cancer mortality rates rather than incidence [628]. There was no information whether or not the drug will also influence the development and progression of already existing tumors [590].

Two other large prospective cohort studies have addressed the question on duration and frequency of aspirin use and the incidence of colorectal carcinoma: The Health Professional Follow-up Study (HPFS) in male and the Nurses Health

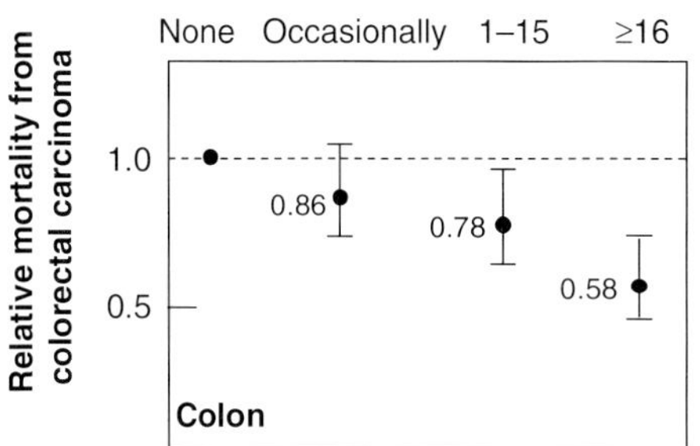

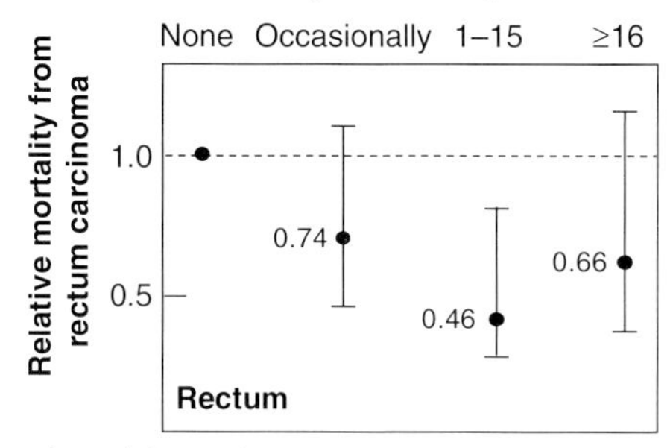

Figure 4.35 American Cancer Society Study. Relative mortality risk from colon or rectum carcinoma in dependency on the frequency of aspirin intake [628].

Study (NHS) in female health professionals (Table 4.17).

A total of 47 900 male health professionals, aged 40–75 years, were included in the HPFS study and asked every second year by a mailed questionnaire on aspirin and other NSAID use and on the history of cancer and other clinically diagnosed medical conditions.

A decrease in the number of colorectal adenomas was found in a subgroup of 10 521 men, subjected to endoscopy for reasons other than bleeding. Regular use of aspirin for two or more times a week, that is, a relatively infrequent use, reduced the relative risk of colorectal cancer compared to nonusers to 0.68 ($p = 0.008$). Decreased risk was noted for both colon (RR: 0.70) and rectum carcinomas (RR: 0.61). The inverse association between aspirin use and colorectal cancer became progressively stronger with evidence of more consistent use of aspirin. There was a strong inverse association between aspirin use and advanced (metastatic and fatal) cancer (RR: 0.51), suggesting that aspirin-related bleeding could further decrease mortality, for example, by allowing earlier diagnosis and (surgical) treatment.

According to a supplementary questionnaire to a randomly selected sample of 211 participants at the end of the study, the median duration of aspirin intake for these persons was 9 years. The data remained essentially the same after controlling for a number of variables (age, history of polyposis, previous endoscopy, family history of colorectal cancer, smoking, body mass, physical activity, intake of red meat, vitamin E, and alcohol).

The conclusion was that regular aspirin use is associated with a reduced risk of colorectal cancer in males [624].

This study additionally showed that regular screening for fecal occult blood loss, possibly combined with colonoscopy, will significantly reduce the mortality due to the disease. This was confirmed in a randomized controlled trial, showing a 33% cumulative decrease in colorectal cancer mortality at 13 years in the group of participants having annual screening compared to those who had not [629].

The latest edition of the HPFS study has confirmed the previous data but also provided some additional new information. During the now 18 years of follow-up there was a 21% reduction of the relative risk for colorectal cancer (RR: 0.79; 95% CI: 0.69–0.90) in men who regularly used aspirin ≥2 times a weeks. Maximum risk reduction was obtained at doses of more than 14 tablets per week and continuous use for at least 6–10 years was required. Interestingly, the protection disappeared within 4 years of discontinuing aspirin use. Thus, long-term treatment is necessary as well as a careful consideration of possible hazards, specifically bleeding at high aspirin doses [630].

A similar approach was used for women in the Nurses Health Study (NHS) who reported regular aspirin use on three consecutive questionnaires in a 2-year interval. The rates of colorectal cancer were determined according to the number of the consecutive years of regular aspirin use (defined as two or more standard aspirin tablets per week). The rates were compared with the rates among women who did not take aspirin. All cases of cancer over a period of 12 years were determined. The aim was to find out the effect of dose and duration of aspirin treatment on the risk of colorectal cancer.

During the observation period, 331 new cases of colorectal cancer were documented during 551 651 person-years of follow-up. Aspirin intake as defined above did not reduce the risk of colorectal cancer compared to nonusers after 4 years: RR: 1.05 (95% CI: 0.78–1.45) or after 5–9 years: RR: 0.84 (95% CI: 0.55–1.28). There was a slight, nonsignificant risk reduction after aspirin intake for 10–19 years: RR: 0.70 (95% CI: 0.41–1.20) but a significant reduction after 20 years of consistent use of aspirin: RR: 0.56 (95% CI: 0.36–0.90).

The conclusion was that regular aspirin use substantially reduces the risk of colorectal cancer in women. Four to six tablets per week appear to be optimal. However, this benefit may not be evident until after at least a decade of regular aspirin consumption [631].

A later study in this population addressed the issue of aspirin dosing and duration of treatment on the primary prevention of colorectal adenoma in women (34–77 years of age) without any particular risk, including familial polyposis, who underwent lower bowel endoscopy. The participants were considered regular users if they took the dose of two or more standard aspirin tablets (325 mg) per week. Self-reported data were obtained from biennial questionnaires.

The adjusted relative risk for adenoma of regular aspirin users compared to nonregular users was 0.75 (95% CI: 0.49–0.80). The risk decreased with increasing aspirin dosing from 0.80 in women who used less than 2 tablets per week to 0.74 with 2–5 tablets per week and 0.49 (95% CI: 0.36–0.65) in those who took more than 14 tablets per week ($p < 0.001$ for trend). Similar dose–response relationships were found among users for ≤5 years and >5 years.

The conclusion was that regular, short-term (≤5 years) aspirin use is inversely associated with the risk for colorectal adenoma. However, the greatest benefit is seen at substantially higher doses. This requires a more thorough benefit/risk evaluation before aspirin can be recommended for chemoprevention in the general adult population [632].

These observation and data of the Nurses Health Study agree with experimental findings from APC-deficient mice [603] and suggest that any protective action of aspirin is likely to take place early in the process of cancer development, that is, at the level of (asymptomatic) adenomas [633].

These findings so far suggested protective actions of regular aspirin at medium doses on the development of colorectal adenomas and their transition into carcinomas. However, there was no information about a possible relationship between aspirin and expression of COX-2. This issue was addressed in a recent update of the Nurses Health Study (NHS) combined with data from the (male) HPFS.

This study determined the level of (histochemical) COX-2 expression in colorectal cancer specimens from participants in the Nurses Health and HPFS studies. In 636 incident colorectal cancers that were accessible for determination of COX-2 expression, 423 (67%) had moderate to strong COX-2 expression. The preventive effect of aspirin was clearly related to COX-2 expression. Regular aspirin use conferred a significant reduction in the risk of colorectal cancers that overexpressed COX-2 (RR: 0.64, 95% CI: 0.52–0.78) but had no influence on tumors with weak or absent COX-2 overexpression (RR: 0.96; 95% CI: 0.73–1.26). There was also a higher incidence of cancers in individuals with COX-2 overexpression that could be reduced by aspirin: 56 versus 37 cases per 100 000 person years. No such effect was seen in individuals with weak or absent COX-2 overexpression: 28 versus 27 cases per 100 000 person years.

The conclusion was that aspirin appears to reduce the risk of colorectal cancers that overexpress COX-2 but not the risk of colorectal cancers with weak or absent expression of COX-2 [625] (Table 4.17; Figure 4.36).

With the exception of one trial [634], which has been criticized for several reasons [633, 635], and is discussed in detail elsewhere (Section 4.1.1), all of the more than 15 available epidemiological/observational studies uniformly demonstrated an about 30–50% risk reduction in colorectal adenomas and colorectal cancer, respectively, in individuals who regularly used aspirin or other NSAIDs over a certain period of time (Table 4.17). These studies can, however, by definition not establish any causal relationship between the two.

4.3.1.3 Clinical Trials: Randomized Prospective Prevention Trials

The first randomized, placebo-controlled prospective study on aspirin in primary prevention was the US Physicians' Health Study [636] (Section 4.1.1). Male physicians, taking 325 mg aspirin each other day had a slightly higher relative risk (RR) of invasive colorectal cancer (RR: 1.15) and a lower risk of *in situ* cancer or polyps (RR: 0.86) [637] during the original observation period of 5 years. However, transition time from adenoma to symptomatic

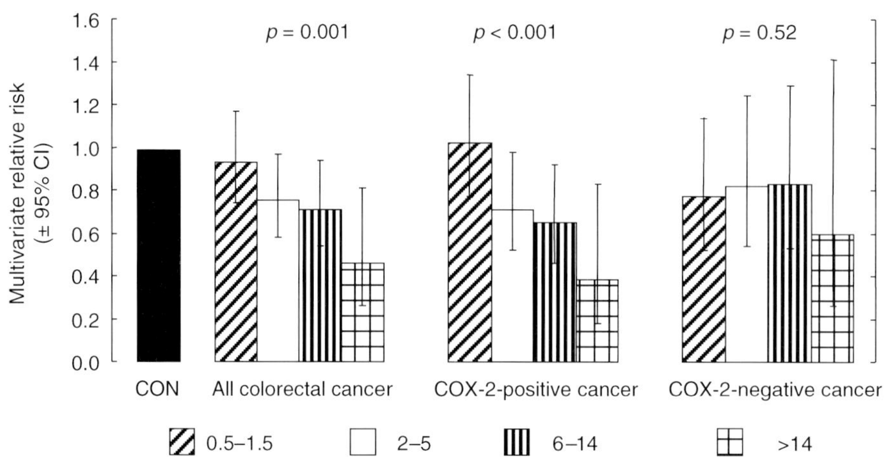

Figure 4.36 Relative risk of colorectal cancer in relation to COX-2 expression and aspirin dose according to data from the Nurses Health Study and the Health Professional Follow-up Study. Aspirin dose is classified according to the number of standard 325 mg tablets taken per week (modified after [625]).

cancer is probably at least 5–10 years [638] and a significant reduction in the risk of colorectal cancer is time-dependent and might require more than 5 years of regular aspirin use at higher doses [630]. No systematic screening for colorectal diseases was performed. There was also no relation between aspirin intake and the incidence of colorectal cancer in this population after another follow-up, excluding regular aspirin users from the randomized trial and studying only new users. Again, there was no reduction of colorectal carcinoma risk [639]. Similar, negative data were also reported from the women's health study [640] in which the subjects used 100 mg aspirin each alternate day over a period of 10 years [641].

The vast majority of colorectal carcinomas develop from colorectal adenomas (APC). Thus, patients with APC are at elevated risk for colorectal carcinomas and useful to determine the efficacy of preventive measures. Meanwhile, three more randomized prospective trials on aspirin in patients with colorectal adenomas became available that, overall, tend to confirm the positive data from nonrandomized observational trials.

The study of Baron *et al.* [642] included 1121 patients with a recent history of histological documented colorectal adenoma. The patients were randomized and received aspirin, 81 or 325 mg daily, or a matching placebo. All patients underwent a surveillance colonoscopy 34–40 months after the qualifying examination. A follow-up colonoscopy was performed at least 1 year after randomization. The primary outcome end point was the appearance of one or more colorectal adenomas.

The incidence for this event was 47% in the placebo group, 38% in the group given 81 mg/day aspirin, and 45% in the group given 325 mg/day aspirin ($p = 0.04$). The respective relative risks for advanced lesions in comparison to placebo were 0.59 (81 mg aspirin) and 0.83 (325 mg aspirin), respectively. During the treatment period, there were no differences in serious bleedings between the groups but seven (nonfatal) strokes in the aspirin groups as opposed to none with placebo (p for heterogeneity = 0.06).

The conclusion was that regular prophylactic use of aspirin had a moderate chemopreventive effect on recurrence of colorectal adenomas in these patients [642].

The Association pour la Prévention par L'Aspirine du Cancer Colorectal (APACC) intervention trial included 272 patients with a history of colorectal adenomas. The patients were randomly assigned to daily soluble (lysyl) aspirin (160 or 300 mg/day) or placebo for 4 years. All patients underwent colonoscopy at 1 year after enrollment. End point was adenoma recurrence (colonoscopy) at this time point. The 1-year data were published.

There was a relatively high adenoma recurrence, occurring in 38/126 patients (30%) in the aspirin groups and in 44 of the 112 patients (41%) on placebo. This was equivalent to a relative risk for at least one recurrent adenoma of 0.73 (95% CI: 0.52–1.04), which was not significant ($p = 0.08$). However, significant differences were seen with the adjusted relative risk for recurrent adenoma, amounting to 0.63 and 0.66 compared to placebo in the 300 and 160 mg aspirin groups, respectively.

The conclusion was that daily soluble aspirin is associated with a reduction in the risk of recurrent adenomas found by colonoscopy 1 year after starting treatment [643].

Similar positive results were obtained in a multicenter, randomized double-blind trial of aspirin versus folate supplements or placebo.

A total of 945 patients were included and observed over 3 years for recurrence of adenomas (colonoscopy) and treated with aspirin (300 mg/day), folate supplements (0.5 mg/day) or placebo. During this time 23% of the 434 patients receiving aspirin had a recurrent adenoma but 29% of the 419 patients on placebo. No effect was seen with folate [644].

These data also support the notion that regular aspirin inhibits the recurrence of colorectal adenoma and perhaps the development of advanced lesions. In addition, they show that folate is ineffective. This issue is discussed in more detail by Das *et al.* [645] reviewing available studies on a possible relationship between dietary intake of vegetables, fibers, and fruits and the risk of colorectal carcinoma.

A particular interesting study was by Sandler *et al.* [646] which appears to be the only published trial on secondary prevention of colorectal cancer.

A total of 517 patients with a previous history of colorectal cancer were included. All patients had curative resection of the primary tumor and colonoscopy with the removal of all polyps. They were randomized to enteric-coated aspirin (325 mg/day) or placebo in a double-blind fashion. The patients had at least one colonoscopic evaluation at 12.8 months (median) after randomization.

Because of significant differences between the treatment groups according to interim results, the study was terminated prematurely. One or more adenomas were found in 17% of patients in the aspirin group and in 27% of patients in the placebo group ($p = 0.004$). The number of adenomas was lower and the time to detection was longer in the aspirin group. This corresponded to a significant ($p = 0.022$) reduction of the relative risk for a new polyp of 0.64 (95% CI: 0.43–0.94). There were only few severe side effects, including one stroke in each group.

The conclusion was that daily intake of aspirin is associated with a significant reduction in the incidence of colorectal adenomas in patients with previous colorectal cancer [646].

The study of Sandler *et al.* [646] was the first to show significant secondary prevention (in terms of adenoma reappearance) in high-risk patients, taking 325 mg/day aspirin for at least 1 year. However, despite this clear protective effect of aspirin, adenomas still developed in some patients of the aspirin group. Thus, aspirin cannot be viewed as the replacement for surveillance colonoscopy [612]. This was at some variance with the study of Baron *et al.* [642] in adenoma patients, which provided less impressive results (Table 4.18). Whether these differences are due to the different patient populations, that is, the patients in the study of Sandler *et al.* [646] were at higher risk, or caused by other factors remains to be determined.

The frequency of colorectal cancer is too low to conduct large randomized clinical trials with cancer as an end point. In addition, low-dose aspirin is recommended as a cardioprophylactic measure for the same group of patients. This raises ethical concerns regarding placebo-controlled trials of long duration [587]. Because of these logistic (time) and practical (cost) problems associated with these studies, enumeration (recurrence) of adenomatous polyps was taken as an intermediate end point in the two studies mentioned above and in a number of others. Although this approach is reasonable, it is a compromise with several limitations [587]. The yearly conversion rate from adenoma to carcinoma varies between 0.25 and 37%, dependent on size and degree of dysplasia [647]. For this and other reasons, including the relatively short (usually 2–5 years or less) observation period (summarized by [612]), colorectal adenomas are not thought to be optimal surrogate parameters for cancer risk and the reduction (but not to complete prevention) of

Table 4.18 Placebo-controlled randomized interventional trials with aspirin in patients with colorectal adenomas [642, 643] or carcinomas [646].

# of patients	Aspirin dose (mg/day)	Treated	Study end point	Duration	RR versus placebo (±95% CI)	Reference
1121	325	(372)	Recurrent adenomas	3 years	0.96 (0.81–1.13)	[642]
	81	(377)			0.81 (0.69–0.96)	
	Placebo	(372)				
272	300	(67)	Recurrent adenomas	1 year	0.61 (0.37–0.99)	[643]
	160	(73)			0.85 (0.57–1.26)	
	Placebo	(132)				
635	325	(317)	Recurrent adenomas	3 years[a]	0.65 (0.46–0.91)	[646]
	Placebo	(318)				

Study end point was the appearance of recurrent adenomas.
[a] Premature termination after positive intermediate data in the aspirin group.

recurrent adenomas cannot be assumed to indicate the risk of colorectal cancer as the malignant potential of the lesions prevented is not known [612].

4.3.1.4 Aspirin and Other Drugs

Aspirin is the most intensively studied drug for chemoprevention of colorectal cancer. Therapeutic alternatives are NSAIDs and coxibs. In addition, sulindac, prodrug of an active NSAID, also reduces the growth of existing adenomas in APC. According to meta-analyses there were no larger differences between aspirin and NSAIDs [648]. However, there are no head-to-head comparisons of aspirin and NSAIDs to clearly answer this question. This is also valid for the main question in addition to efficacy: drug safety.

4.3.1.5 Actual Situation

The long-term use of any drug for preventive purposes, in particular for prevention of diseases with low incidence, makes high demands on safety. Drug safety is determined by dosage and duration of treatment but also the incidence and severity of side effects as well as costs and the cost/benefit ratio. There is a clear need for a randomized controlled prospective trial extending over 10 years or longer of regular daily aspirin consumption with cancer as an end point. It should be clarified whether the GI tolerance of aspirin can be improved by the inclusion of vitamin C [649]. Thus, several questions, specifically regarding the possible hazards of the long-term aspirin use have to be answered, before aspirin can be recommended clinically for the prevention of colorectal cancer [650].

Doses Observational and randomized trials on colorectal adenomas/carcinomas suggest that effective doses of aspirin are in the range of 300 mg/day. However, they do not answer the question for the most effective dose. Aspirin at doses of 81–650 mg per day for 28 days reduced PGE_2 levels in healthy human colorectal mucosa only incompletely, by 60–70% [651]. Interestingly, a similar incomplete inhibition was also seen in patients with preexisting colorectal adenomas or carcinomas [652, 653]. There was, however, a large interindividual variability. Accordingly, aspirin doses of less than 300 mg/day may be too low or too unsure for cancer chemoprevention, if PGE_2 was the key mediator of malignancy. Anti-inflammatory doses of aspirin are more likely to suppress COX-2-dependent prostaglandin formation, which appears to have a role in the disease. This is supported by a recent meta-analysis by Garcia Rodriguez and Huerta-Alvarez [654] suggesting that the efficacy of aspirin and non-aspirin-NSAIDs is similar and reduces the risk of colorectal cancer by about half. Interestingly, according to this meta-analysis, significant risk reduction by aspirin required doses of at least 300 mg daily while lower doses were ineffective. Thus, in addition to the inhibition of prostaglandin synthesis, further actions of aspirin on COX-2-dependent product formation, for example activation of cocarcinogens, or tumor-suppressor genes, might be involved (Section 2.2.3).

A population-based cohort study investigated the association between the risk of colorectal cancer and the use of aspirin (prescription only) and other NSAIDs, including the role of dose and duration of treatment.

The risk of colorectal cancer was reduced in long-term users of aspirin at doses of 300 mg/day but not at doses of 150 or 75 mg daily. Beneficial results with a comparable relative risk reduction by 40–60% were seen with a number of other NSAIDs but not by acetaminophen.

The conclusion was that continuous use of non-aspirin-NSAIDs for at least 6 months protects from colorectal cancer. The risk reduction is similar to aspirin at doses of at least 300 mg/day. One year treatment with NSAIDs would prevent one case of colorectal cancer in a population of 1000 persons 70–79 years of age [654].

Duration of Treatment A significant reduction in the incidence of colorectal carcinomas requires regular long-term intake of aspirin. There are no sufficient trials available to determine the optimum duration of treatment. According to available meta-analyses, administration of aspirin for 5 years or more reduces the incidence for colorectal carcinomas; however, this effect is significant only after a latency of 10 years or more and was greatest 10–14 years after randomization [631, 648, 655].

Environmental Factors Epidemiological data suggest that diets low in vegetables and fruits are related to an increased incidence of colorectal cancer [656, 657]. Many fruits (apples, apricots, cherries, grapes, peaches, and plums) and some vegetables (cucumber, pepper, and tomatoes) contain natural salicylates that may contribute to the reduced risk of colorectal cancer associated with fruit and vegetable consumption as observed in several epidemiological studies [628], and were found to increase the blood salicylate level to small but significant amounts [658] (Figure 2.4). In a survey on epidemiologic trials on the relationship between diet factors and colorectal adenoma risk, 11 out of 13 studies on vegetables and fruits showed a 50% or higher risk reduction for colorectal adenomas and carcinomas, respectively [659].

Benefit/Risk Ratio Any long-term regular use of any drug, including aspirin, carries an increased risk of side effects, which has to be balanced against possible benefits. If aspirin has to be given at an analgesic/anti-inflammatory dose for protection, that is, 500–1000 mg/day for many years, there is an estimated 2–4% risk of severe gastrointestinal complications (overt bleeding, ulcer) per year. Possibly, these side effects can be reduced by taking advantage from enteric-coated or predissolved preparations or comedication of proton-pump inhibitors (Section 3.2.1). Taking into account that the population to be treated for the prevention of colorectal cancer frequently is identical with the population at elevated risk of cardiovascular events, that is, myocardial infarctions, this will shift the benefit/risk ratio to more benefit with aspirin use. In contrast, coxibs were also efficient in colorectal prevention trials [648, 660] but showed no benefit and rather the opposite effect on atherothrombotic events, for example in the APPROVE trialy. Similar considerations apply to traditional NSAIDs, which also have no beneficial effect on the risk of myocardial infarctions during the long-term use (Section 4.2.2).

Cost-Effectiveness Analysis In addition to benefit/risk considerations in terms of health aspects, cost-effectiveness ratios have also to be considered, especially in long-term use for the prevention of low-incidence diseases. Regular aspirin intake for prophylaxis of colorectal carcinomas (325 mg/day) was compared in a hypothetical cohort of 100 000 asymptomatic subjects at 50 years of age (Markov model) with either no preventive measure or colonoscopy [661]. Compared to colonoscopy (one per 10 years at an estimated efficacy rate of 75%), the use of aspirin (estimated efficacy rate 50%) saved fewer lives at higher costs. Factors contributing to costs are mainly side effects of the drug, while the expenses for the drug itself are relatively low [661, 662]. However, another study, using the same statistical mode, came to opposite conclusions in favor of aspirin [663]. This study did consider the expected benefit not only in the prevention of colorectal carcinomas but also in the prevention of cardiovascular events. Thus, the combination of colonoscopy and aspirin in patients at elevated risk may represent a cost-effective option [666], particularly in those who are already taking aspirin regularly for other reasons, for example, for the prevention of myocardial infarction [664] and there might also be additional benefit of aspirin because of earlier diagnosis (bleeding). Finally, the efficacy of aspirin can clearly be improved if it is focused on individuals at high risk detected, for example, by routine screening for biomarkers, such as tumor-suppressor or apoptosis genes [602].

Summary

Colorectal cancer in most cases develops from adenomatous polyposis coli (APC). APC is probably initiated by genetic alterations of tumor-suppressor genes. Further defects in apoptosis genes and environmental factors, including an inappropriate diet, may eventually result in overt colorectal carcinomas. There is an overexpression of COX-2 in early stages of adenoma development and further increase with progression of the disease. The resulting PGE_2 formation facilitates polyp growth and multiplication as well as the malignant transition.

An interaction with COX-2-dependent product formation is probably related to the beneficial chemopreventive effects of aspirin. Additional actions on gene expression may contribute to this effect as well as a particular genotype, such as variants of the UGT1A6 or ornithine decarboxylase G316A genotypes.

Epidemiological data convincingly suggest that regular, long-term use of aspirin reduces the mortality from colorectal carcinoma and probably also the incidence and transition of colorectal adenomas to cancer. The risk reduction rate is in the range of 20–30% [665]. Prospective randomized trials in high-risk patients with adenoma recurrence as an end point show a comparable chemopreventive efficacy of the compound while the effects on cancer transition are less clear.

The optimum aspirin dose and duration of treatment are unknown. The majority of available information suggests that aspirin dose of 300 mg/day and a minimum duration of treatment of 10 years are necessary. Urgent questions to be answered by randomized controlled trials involve definition of the optimum dose and duration of treatment with cancer as an end point, identification of high-risk groups by appropriate biomarkers, and an estimation of the benefit/risk ratio on an individual, that is, risk-related background.

References

584 Mangiapane, S., Blettner, M. and Schlattmann, P. (2008) Aspirin use and breast cancer risk: a meta-analysis and meta-regression of observational studies from 2001 to 2005. *Pharmacoepidemiol Drug Safety*, **17**, 115–124.

585 Viswanathan, A.N., Feskanich, D., Schernhammer, E.S. *et al.* (2008) Aspirin, NSAID and acetaminophen use and the risk of endometrial cancer. *Cancer Research*, **68**, 2507–2513.

586 Van Dyke, A.L., Cote, M.L., Prysak, G. *et al.* (2008) Regular adult aspirin use decreases the risk of non-small cell lung cancer among women. *Cancer Epidemiology, Biomarkers & Prevention*, **17**, 148–157.

587 Garewal, H.S. (1994) Aspirin in the prevention of colorectal cancer. *Annals of Internal Medicine*, **121**, 303–304.

588 Hayne, D., Brown, R.S., McCormack, M. *et al.* (2001) Current trends in colorectal cancer: site, incidence, mortality and survival in England and Wales. *Clinical Oncology*, **13**, 448–452.

589 Ilyas, M., Straub, J., Tomlinson, I.P. *et al.* (1999) Genetic pathways in colorectal cancer and other cancers. *European Journal of Cancer*, **35**, 335–351.

590 Baron, J.A. and Greenberg, E.R. (1991) Could aspirin really prevent colon cancer? *The New England Journal of Medicine*, **325**, 1644–1646.

591 Shpitz, B., Klein, E., Buklan, G. *et al.* (2003) Suppressive effect of aspirin on aberrant crypt foci in patients with colorectal cancer. *Gut*, **52**, 1598–1601.

592 Funkhouser, E.M. and Sharp, G.B. (1995) Aspirin and reduced risk of esophageal carcinoma. *Cancer*, **76**, 1116–1119.

593 Jayaprakash, V., Menezes, R.J., Javie, M.M. *et al.* (2006) Regular aspirin use and esophageal cancer risk. *International Journal of Cancer*, **119**, 202–207.

594 Fortuny, J., Johnsson, C.C., Bohlke, K. *et al.* (2007) Use of anti-inflammatory drugs and lower esophageal sphincter-relaxing drugs and risk of esophageal and gastric cancers. *Clinical Gastroenterology and Hepatology*, **5**, 1154–1159.

595 Farrow, D.C., Vaughan, T.L., Hansten, P.D. *et al.* (1998) Use of aspirin and other nonsteroidal anti-inflammatory drugs and risk of esophageal and gastric cancer. *Cancer Epidemiol Biomarkers Prevention*, **7**, 97–102.

596 Zaridze, D., Borisova, E., Maximovitch, D. *et al.* (1999) Aspirin protects against gastric cancer: results of a case-control study from Moscow. *Russia International Journal of Cancer*, **82**, 473–476.

597 Bosetti, C., Talamini, R., Franceschi, S. *et al.* (2003) Aspirin use and cancers of the upper aerodigestive tract. *British Journal of Cancer*, **88**, 672–674.

598 Kinzler, K.W. and Vogelstein, B. (1996) Lessons from hereditary colorectal cancer. *Cell*, **87**, 159–170.

599 Morin, P.J., Vogelstein, B. and Kinzler, K.W. (1996) Apoptosis and APC in colorectal tumorigenesis. *Proceedings of the National Academy of Sciences of the United States of America*, **93**, 7950–7954.

600 DuBois, R.N., Giardiello, P.M. and Smalley, W.E. (1996) Nonsteroidal anti-inflammatory drugs, eicosanoids, and colorectal cancer prevention. *Gastroenterology Clinics of North America*, **25**, 773–791.

601 Prescott, S.M. and White, R.L. (1996) Self-promotion? Intimate connections between APC and prostaglandin H synthase-2. *Cell*, **87**, 783–786.

602 Williams, C.S., Smalley, W. and DuBois, R.N. (1997) Aspirin use and potential mechanisms for colorectal cancer prevention. *The Journal of Clinical Investigation*, **100**, 1325–1329.

603 Oshima, M., Dinchuk, J.E., Kargman, S.L. *et al.* (1996) Suppression of intestinal polyposis in $APC^{\Delta716}$ knockout mice by inhibition of cyclooxygenase-2 (COX-2). *Cell*, **87**, 803–809.

604 Ristimäki, A., Honkanen, N., Jankala, H. *et al.* (1997) Expression of cyclooxygenase-2 in human gastric carcinoma. *Cancer Research*, **57**, 1276–1280.

605 Zimmermann, K.C., Sarbia, M., Weber, A.-A. *et al.* (1999) Cyclooxygenase-2 expression in human esophageal carcinoma. *Cancer Research*, **59**, 198–204.

606 Sonoshita, M., Takaku, K., Sasaki, N. *et al.* (2001) Acceleration of intestinal polyposis through prostaglandin receptor EP_2 in $APC^{\Delta716}$ knockout mice. *Nature Medicine*, **7**, 1048–1051.

607 Marnett, L.J. (1992) Aspirin and the potential role of prostaglandins in colon cancer. *Cancer Research*, **52**, 5575–5589.

608 Yang, V.W., Sheilds, J.M., Hamilton, S.R. *et al.* (1998) Size-dependent increase in prostanoid levels in adenomas of patients with familial adenomatous polyposis. *Cancer Research*, **58**, 1750–1753.

609 Sheehan, K.M., Sheahan, K., O'Donoghue, D.P. *et al.* (1999) The relationship between cyclooxygenase-2 expression and colorectal cancer. *Journal of the American Medical Association*, **282**, 1254–1257.

610 Elder, D.J., Hague, A., Hicks, D.J. *et al.* (1996) Differential growth inhibition by the aspirin metabolite salicylate in human colorectal tumor cell lines: enhanced apoptosis in carcinoma and *in vitro*-transformed adenoma relative to adenoma cell lines. *Cancer Research*, **56**, 2273–2276.

611 Lipton, A., Scialla, S., Harvey, H. *et al.* (1982) Adjuvant antiplatelet therapy with aspirin in colo-rectal cancer. *Journal of Medicine*, **13**, 419–429.

612 Courtney, E.D.J., Melville, D.M. and Leicester, R.J. (2004) Review article: chemoprevention of colorectal cancer. *Alimentary Pharmacology & Therapeutics*, **19**, 1–24.

613 Barnes, C.J., Hamby-Mason, R.L., Hardman, W.E. *et al.* (1999) Effect of aspirin on prostaglandin E_2 formation and transforming growth factor alpha expression in human rectal mucosa from individuals with a history of adenomatous polyps of the colon. *Cancer Epidemiol Biomarkers Prevention*, **8**, 311–315.

614 Sheng, H., Shao, J., Kirkland, S.C. *et al.* (1997) Inhibition of human cancer cell growth by selective inhibition of cyclooxygenase-2. *The Journal of Clinical Investigation*, **99**, 2254–2259.

615 Kawamori, T., Uchiya, N., Sugimura, T. *et al.* (2003) Enhancement of colon carcinogenesis by prostaglandin E_2 administration. *Carcinogenesis*, **24**, 985–990.

616 Kim, K.M., Song, J.J., An, J.Y. *et al.* (2005) Pretreatment of acetylsalicylic acid promotes tumor necrosis factor-related apoptosis-inducing ligand-induced apoptosis by down-regulating BCL-2 gene expression. *The Journal of Biological Chemistry*, **280**, 41047–41056.

617 Bigler, J., Whitton, J., Lampe, J.W. *et al.* (2001) CYP2C9 and UGT1A6 genotypes modulate the protective effect of aspirin on colon adenoma risk. *Cancer Research*, **61**, 3566–3569.

618 Chan, A.T., Tranah, G.J., Giovanucci, E.L. *et al.* (2005) Genetic variants in the UGT1A6 enzyme, aspirin use, and the risk of colorectal adenoma. *Journal of the National Cancer Institute*, **97**, 457–460.

619 Hubner, R.A., Muir, R., Liu, J.-F. *et al.* (2008) Ornithine decarboxylase G316A genotype is prognostic for colorectal adenoma recurrence and predicts efficacy of aspirin chemoprevention. *Clinical Cancer Research*, **14**, 2303–2309.

620 Watson, A.J.M. and DuBois, R.N. (1997) Lipid metabolism and APC: implications for colorectal cancer prevention. *The Lancet*, **349**, 444–445.

621 Kune, G.A., Kune, S. and Watson, L.F. (1988) Colorectal cancer risk, chronic illnesses, operations, and medications: case control results from the Melbourne Colorectal Cancer Study. *Cancer Research*, **48**, 4399–4404.

622 Rosenberg, L., Palmer, J.R., Zauber, A.G. *et al.* (1991) A hypothesis: non-steroidal antiinflammatory drugs reduce the incidence of large-bowel cancer. *Journal of the National Cancer Institute*, **83**, 355–358.

623 Thun, M.J., Namboodiri, M.N. and Heath, C.W., Jr (1991) Aspirin use and reduced risk of fatal colon cancer. *The New England Journal of Medicine*, **325**, 1593–1596.

624 Giovannucci, E., Rimm, E.B., Stampfer, M.J. *et al.* (1994) Aspirin use and the risk for colorectal cancer and adenoma in male health professionals. *Annals of Internal Medicine*, **121**, 241–246.

625 Chan, A.T., Ogino, S. and Fuchs, C.S. (2007) Aspirin and the risk of colorectal cancer in relation to the expression of COX-2. *The New England Journal of Medicine*, **356**, 2131–2142.

626 Thun, M.J., Namboodiri, M.N., Calle, E.E. *et al.* (1993) Aspirin use and risk of fatal cancer. *Cancer Research*, **53**, 1322–1327.

627 Jacobs, E.J., Connell, C.J., Rodriguez, C. *et al.* (2004) Aspirin use and pancreatic cancer mortality in a large United States cohort (CPS-II). *Journal of the National Cancer Institute*, **96**, 524–528.

628 Thun, M.J. (1994) Aspirin, NSAIDs, and digestive tract cancers. *Cancer and Metastasis Reviews*, **13**, 269–277.

629 Mandel, J.S., Bond, J.H., Church, T.R. *et al.* (1993) Reducing mortality from colorectal cancer by screening for fetal occult blood. Minnesota Colon Cancer Control Study. *The New England Journal of Medicine*, **328**, 1365–1371.

630 Chan, A.T., Giovannucci, E.L., Meyerhardt, J.A. *et al.* (2008) Aspirin dose and duration of use and risk of colorectal cancer in men. *Gastroenterology*, **134**, 21–28.

631 Giovannucci, E., Egan, K.M., Hunter, D.J. *et al.* (1995) Aspirin and the risk of colorectal cancer in women. *The New England Journal of Medicine*, **333**, 609–614.

632 Chan, A.T., Giovannucci, E.L., Schernhammer, E.S. *et al.* (2004) A prospective study on aspirin use and the risk for colorectal adenoma. *Annals of Internal Medicine*, **140**, 157–166.

633 Muir, K.R. and Logan, R.F.A. (1999) Aspirin, NSAIDs and colorectal cancer – what do the epidemiological studies show and what do they tell us about the modus operandi? *Apoptosis*, **4**, 389–396.

634 Paganini-Hill, A., Chao, A., Ross, R.K. *et al.* (1989) Aspirin use and chronic diseases: a cohort study of the elderly. *British Medical Journal*, **299**, 1247–1250.

635 Weiss, H.A. and Forman, D. (1996) Aspirin, non-steroidal anti-inflammatory drugs and protection from colorectal cancer: a review of the epidemiological evidence. *Scandinavian Journal of Gastroenterology*, **31** (Suppl. 220), 137–141.

636 Goldhaber, S.Z., Manson, J.E., Stampfer, M.J. *et al.* (1992) Low-dose aspirin and subsequent peripheral arterial surgery in the Physicians' Health Study. *The Lancet*, **340**, 143–145.

637 Gann, P.H., Manson, J.E., Glynn, R.J. *et al.* (1993) Low-dose aspirin and incidence of colorectal tumors in a randomized trial. *Journal of the National Cancer Institute*, **85**, 1220–1224.

638 Stryker, S.J., Wolff, G.B., Culp, C.E. *et al.* (1987) Natural history of untreated colonic polyps. *Gastroenterology*, **93**, 1009–1013.

639 Stürmer, T., Glynn, R.J., Lee, I.-M. *et al.* (1998) Aspirin use and colorectal cancer: post-trial follow-up data from the Physicians' Health Study. *Annals of Internal Medicine*, **128**, 713–720.

640 Cook, N.R., Lee, I.M., Gaziano, J.M. *et al.* (2005) Low-dose aspirin in the primary prevention of cancer: the women's health study: a randomized controlled trial. *The Journal of the American Medical Association*, **294**, 47–55.

641 Zhang, S.M., Cook, N.R., Manson, J.E. *et al.* (2008) Low-dose aspirin and breast cancer risk: results by tumor characteristics from a randomized trial. *British Journal of Cancer*, **98**, 989–998.

642 Baron, J.A., Cole, B.F., Sandler, R.S. *et al.* (2003) A randomized trial on aspirin to prevent colorectal adenomas. *The New England Journal of Medicine*, **348**, 891–899.

643 Benamouzig, R., Deyra, J., Martin, A. *et al.* (2003) Daily soluble aspirin and prevention of colorectal adenoma recurrence: one-year results of the APACC trial. *Gastroenterology*, **125**, 328–336.

644 Logan, R.F.A., Grainge, M.J., Shepherd, V.C. *et al.* (2008) Aspirin and folic acid for the prevention of recurrent colorectal adenomas. *Gastroenterology*, **134**, 29–38.

645 Das, D., Arber, N. and Jankowski, J.A. (2007) Chemoprevention of colorectal cancer. *Digestion*, **76**, 51–67.

646 Sandler, R.S., Halabi, S., Baron, J.A. *et al.* (2003) A randomized trial of aspirin to prevent colorectal adenomas in patients with previous colorectal cancer. *The New England Journal of Medicine*, **348**, 883–890.

647 Eider, T.J. (1986) Risk of colorectal cancer in adenoma-bearing individuals within a defined population. *International Journal of Cancer*, **38**, 173–176.

648 Flossmann, E., Rothwell, P.M. *et al.*, British Doctors Aspirin Trial (2007) Effect of aspirin on long-term risk of colorectal cancer: consistent evidence from randomized and observational studies. *The Lancet*, **369**, 1603–1613.

649 Kune, G.A. (2007) Commentary: aspirin and cancer prevention. *International Journal of Epidemiology*, **36**, 957–959.

650 Imperiale, T.F. (2003) Aspirin and the prevention of colorectal cancer. *The New England Journal of Medicine*, **348**, 879–880.

651 Ruffin, M.T., IV, Krishnan, K., Rock, C.L. *et al.* (1997) Suppression of human colorectal mucosal prostaglandins: determining the lowest effective aspirin dose. *Journal of the National Cancer Institute*, **89**, 1152–1160.

652 Frommel, T.O., Dyavanapalli, M., Oldham, T. *et al.* (1997) Effect of aspirin on prostaglandin E2 and leukotriene B4 production in human colonic mucosa from cancer patients. *Clinical Cancer Research*, **3**, 209–213.

653 Krishnan, K., Ruffin, M.T., Normolle, D. *et al.* (2001) Colonic mucosal prostaglandin E_2 and cyclooxygenase expression before and after low aspirin doses in subjects at high risk or at normal risk for colorectal cancer. *Cancer Epidemiol Biomarkers Prevention*, **10**, 447–453.

654 Garcia Rodriguez, L.A. and Huerta-Alvarez, C. (2001) Reduced risk of colorectal cancer among long-term users of aspirin and non-aspirin nonsteroidal antiinflammatory drugs. *Epidemiology*, **12**, 88–93.

655 Kune, G.A. (2002) Colorectal cancer chemoprevention with aspirin. *Gastrointestinal Oncology*, **4**, 5–14.

656 Trock, B., Lanza, E. and Greenwald, P. (1990) Dietary fiber, vegetables, and colon cancer: critical review and meta-analyses of the epidemiologic evidence. *Journal of the National Cancer Institute*, **82**, 650–661.

657 Thun, M.J., Calle, E.E., Namboodiri, M.N. *et al.* (1992) Risk factors for fatal colon cancer in a large prospective study. *Journal of the National Cancer Institute*, **84**, 1491–1500.

658 Blalock, C.J., Lawrence, J.R., Wiles, D. *et al.* (2001) Salicylic acid in the serum of subjects not taking aspirin. Comparison of salicylic acid concentrations in the serum of vegetarians, non-vegetarians, and patients taking low-dose aspirin. *Journal of Clinical Pathology*, **54**, 553–555.

659 Kune, G.A. (1996) *Causes and Control of Colorectal Cancer. A Model for Cancer Prevention*, Kluwer Academic Publisher, Boston, MA.

660 Steinbach, G., Lynch, P.M., Phillips, R.K. *et al.* (2000) The effect of celecoxib, a cyclooxygenase-2-inhibitor, in familial adenomatous polyposis. *The New England Journal of Medicine*, **342**, 1946–1952.

661 Suleiman, S., Rex, D.K. and Sonnenberg, A. (2002) Chemoprevention of colorectal cancer by aspirin: a cost-effectiveness analysis. *Gastroenterology*, **122**, 78–84.

662 Ladabaum, U., Chopra, C.L., Huang, G. *et al.* (2001) Aspirin as an adjunct to screening for prevention of sporadic colorectal cancer. A cost-effectiveness analysis. *Annals of Internal Medicine*, **135**, 769–781.

663 Hur, C., Simon, L.S. and Gazelle, G.S. (2004) The cost-effectiveness of aspirin versus cyclooxygenase-2-selective inhibitors for colorectal carcinoma chemoprevention in healthy individuals. *Cancer*, **101**, 189–197.

664 Provenzale, D. (2002) The cost-effectiveness of aspirin for chemoprevention of colorectal cancer. *Gastroenterology*, **122**, 230–233.

665 Dubé, C., Rostom, A. and Lewin, G. (2007) The use of aspirin for primary prevention of colorectal cancer: a systematic review prepared for the U.S. preventive services task force. *Annals of Internal Medicine*, **146**, 365–375.

666 Dupont, A.W., Arguedas, M.R. and Wilcox, C.M. (2007) Aspirin chemoprevention in patients with increased risk for colorectal cancer: a cost-effectiveness analysis. *Alimentary Pharmacology & Therapeutics*, **26**, 431–441.

4.3.2 Alzheimer's Disease

Etiology Alzheimer's disease is a leading cause of progressive dementia in the elderly. Currently about 2% of the population is affected but the number is likely to increase with increasing life expectancy. Typical for the disease are a progressive loss of memory and higher cognitive functions. Alzheimer's dementia is usually diagnosed 2–4 years after the appearance of first symptoms and begins usually at the age of 70 with increasing incidence at increasing age. Thus, a delay in the onset by 2–3 years, for example, by identification of modifiable risk factors and their appropriate prevention or treatment is clinically most relevant. This would not only improve the quality of life of affected individuals but also help save costs for health care providers, institutionalization of the patient clearly being the most unwanted event for both sides. For these reasons, any effective preventive measure is much more desirable than solely the treatment of symptoms, which provides only marginal if any improvement in quality of life.

At present, several approaches for the therapy of Alzheimer's disease are available. These are focused on retardation of the development of the disease and the treatment of symptoms. However, since neuroinflammation and disturbed apoptosis are crucial to exacerbation of the disease, anti-inflammatory/antiapoptotic approaches might be useful, if they are applied *before* manifest brain injury emerges. Several classes of anti-inflammatory drugs, including traditional NSAIDs, selective COX-2 inhibitors (coxibs) and aspirin are of potential interest. Aspirin and NSAIDs are particularly attractive because the compounds might also be used for protection from and treatment of other disabilities of the elderly. This includes atherothrombotic events, such as myocardial infarction (Section 4.1.1) and stroke (Section 4.1.2) in case of aspirin and symptomatic treatment of chronic inflammatory diseases, such as osteoarthritis or rheumatoid arthritis (Section 4.2.2) in case of NSAIDs.

4.3.2.1 Pathophysiology and Mode of Aspirin Action

Pathophysiology In Alzheimer's disease, brain regions that are involved in learning and memory processes, are reduced in size as the result of degeneration of synapses and neurons [667]. Typical of the disease are senile plaques containing aggregated amyloid-β protein and neurofibrillary tangles. Neuronal overexpression of β-amyloid precursor and amyloid-β protein render the brain more vulnerable to ischemic injury [668]. Importantly, neurofibrillary tangle-containing neurons do not die of apoptosis but rather degenerate because of hyperphosphorylation of tau, the major protein subunit of neurofibrillary tangles [669]. Neuroinflammation is also intimately involved in the development and progression of functional disturbances of the disease, that is, loss of memory and cognitive functions. Biochemically, this includes activation of the complement cascade, generation of chemokines, cytokines, and reactive oxygen species [670]. Microglia, a macrophage-like cell population, congregates around amyloid plaques and degenerating neurons and releases toxins and inflammatory mediators that in turn promote neurodegeneration [667]. This activation of inflammation-associated signal transduction pathway is restricted not only to glial cells but is also found in neurons. It is likely to contribute to neuronal damage [671] and is an obvious target of anti-inflammatory drugs [668].

In contrast to many other tissues, COX-2 is expressed constitutively in the CNS. The neuronal expression of COX-2 is additionally regulated by synaptic activity, suggesting that in the CNS, this COX isoform and its enzymatic products might also be involved in activity-dependent neuronal plasticity. Brain microglia, which is crucial to inflammation and subsequent neurodegeneration, does not express high levels of COX-2 after stimulation with inflammatory cytokines or amyloid-β [672]. These findings suggest that selective inhibition of COX-2 may not be an optimum target if suppression of inflammation is the therapeutic goal and traditional NSAIDs, rather than coxibs, may be more useful to realize this approach [673].

Mode of Aspirin Action Extracellular deposition of amyloid-β protein as amyloid plaques and vascular amyloid is a typical feature of Alzheimer's disease and probably induced by chronic neuroinflammation as mentioned above. Formation of the amyloid-activated microglia complex is an early event in the disease [674]. Aspirin (1 mM) has been found *in vitro* to nearly completely prevent the precipitation of extracellular fibrils from dissolved β-amyloid in a cell-free system *in vitro* but not the amorphous, essentially nonfibrillar, insoluble association state of β-amyloid [675]. Antiamyloidogenic and fibril-destabilizing actions of aspirin and traditional NSAIDs can be seen at low, therapeutic concentrations (50 μM) of the compounds (Figure 4.37) [676]. If this mechanism works *in vivo*, it might result in reduced extracellular deposition of β-amyloid fibrils in brain tissue, thereby retarding the progression of the disease. In this context, it is interesting to note that aspirin was found to decrease tau phosphorylation [677]. Hyperphosphorylation of tau, the major protein subunit of neurofibrillar tangles in Alzheimer's disease, is thought to be responsible for the resistance against apoptosis of tangle-containing neurons [669].

Pharmacological, actions of aspirin with possible relevance to beneficial effects in Alzheimer's disease are already seen at antiplatelet doses. This does not exclude possible additional effects of higher doses. The clinical efficacy of NSAIDs in some observational trials (see below) is another, though indirect evidence of the involvement of prostaglandins in the (neuro)inflammatory process. In agreement with this, Alzheimer's patients have a three- to fourfold increased urinary excretion of thromboxane metabolites and of the lipid peroxidation marker 8-iso-$PGF_{2\alpha}$. The plasma levels of vitamin E, an antioxidant, are reduced and are inversely correlated with the excretion of these metabolites. These data suggest persistent platelet activation and reduced oxygen defense in Alzheimer's disease [678].

Low-dose aspirin (100 mg/day) markedly reduced urinary excretion of a thromboxane metabolite while the urinary excretion of the (nonenzymatically formed) stress marker 8-iso-$PGF_{2\alpha}$ was unchanged [678]. Platelets appear to be a primary source of β-amyloid in human blood [679]. Accordingly, amyloid precursor protein (APP) from platelets has recently been recommended as a biomarker for diagnosis, progression, and efficacy of treatment of Alzheimer's patients [680]. However, many questions still remain open. Specifically, Alzheimer's disease is probably multifactorial and heterogeneous, thus offering a large window of opportunities and a large number of therapeutic targets to inhibit it [681, 682], aspirin and the

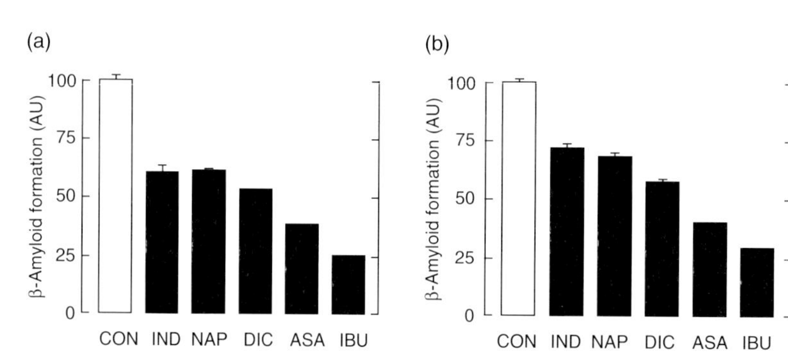

Figure 4.37 Effects of aspirin and traditional NSAIDs on the polymerization of amyloid-β peptide *in vitro*. The duration of incubation was 1 day (a) or 1 week (b), respectively, the concentrations of IND, NAP, DIC, IBU, and aspirin (ASA) were 50 μM. All drug treatments caused significant inhibition of β-amyloid polymerization (AU: arbitrary units) (modified after [676]).

antiplatelet/anti-inflammatory/antiapoptotic approach being one of them.

4.3.2.2 Clinical Trials

Observational Trials A number of retrospective and prospective observational trials have been conducted to study whether aspirin and/or traditional NSAIDs can prevent or retard the progression of the disease. One of the first trials showing a reduced prevalence of Alzheimer's in regular users of aspirin and NSAIDs was the prospective population-based Baltimore Longitudinal Study of Aging.

The Baltimore Longitudinal Study of Aging examined in 1686 participants whether the risk of Alzheimer's disease was reduced among users of aspirin and paracetamol compared to traditional NSAIDs. Information was collected by biennial examinations during 6 years. The question was whether self-reported medication with these drugs had any relation to the risk of Alzheimer's disease.

The relative risk of Alzheimer's disease was inversely correlated with increasing duration of NSAID use. The RR amounted to 0.40 in individuals with more than 2 years reported use and 0.65 in those with less than 2 years use. The overall risk for aspirin users was 0.74. This number was not significantly different from controls and no trend for decreasing risk with longer use was found. Acetaminophen had no effect at all (Figure 4.38).

It was concluded that regular intake of NSAIDs reduces the risk of Alzheimer's disease. This protective effect is more pronounced with longer use, suggesting that an inflammatory process might be involved. Aspirin was considered to confer some protection as well; however, an increasing proportion of elderly in the study used a prophylactic dose, that is, between 65 and 85 mg aspirin per day, for cardiocoronary prevention. This dose might have been too low for anti-inflammatory effects on the CNS [683].

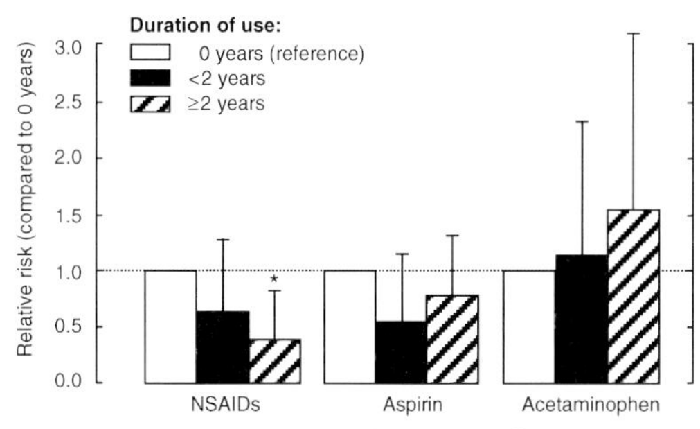

Figure 4.38 Baltimore Longitudinal Study of Aging. Relative risk of Alzheimer's disease by type and duration of medication use. There is a significant reduction by traditional NSAIDs, a tendency for reduction by aspirin, and no change by paracetamol [683].

In a comment to this study, the possibility was discussed whether aspirin might have been less effective because a significant number of Alzheimer's patients might have had vascular dementia because of a high incidence of vascular risk factors and that this might differ from its action in the general population [684]. Another population-based retrospective study confirmed that long-term use of aspirin and NSAIDs might decrease the risk of developing Alzheimer's disease [685] while no beneficial effect for aspirin was seen in a small Swedish population-based prospective trial [686]. However, no differentiation between aspirin use by prescription or self-medication was made in this study and no information about strength and dosages was provided. This study is also at variance with a retrospective observational trial from Australia [687].

The Sydney Older Persons Study was a retrospective case–control study in 647 recruited individuals, 75 years of age or older (average 81 years); 163 patients had been diagnosed for dementia (different categories) and were compared with 373 controls from the same population sample. The aim of the study was the detection of possible relationship between the used drugs, in particular, NSAIDs and aspirin, and the incidence of Alzheimer's disease.

There was an inverse association between the intake of NSAIDs and aspirin and the occurrence of Alzheimer's. No associations were seen with vascular dementia or any other diagnosis. There was no evidence of a dose dependency for either NSAIDs or aspirin at low (<175 mg/day) and medium doses (>175 mg/day).

The conclusion was that the regular use of NSAIDs or aspirin may protect from Alzheimer's disease. Since antiplatelet doses of aspirin were equieffective to higher doses, it was also assumed that the beneficial effects of aspirin were due to its antiplatelet activity and possibly related to inhibition of β-amyloid release from activated platelets [687].

A morphologic examination of postmortem brain tissue of Alzheimer's patients who were on long-term anti-inflammatory medications did not show reduced inflammatory microglia or neuropathological changes despite antemortem improved cognitive functions by drug treatment [688]. However, these findings have been controversial [689] and opposite results were obtained in an older subjects (84 years at inclusion) in a population-based cohort study in Sweden. Here, users of high-dose aspirin but not paracetamol, low-dose aspirin (75 mg) or other NSAIDs – even if given occasionally – had a significantly lower prevalence of Alzheimer's and better maintained cognitive functions than nonusers. No such effect was seen with paracetamol [690].

The largest available epidemiological study on the possible relation between NSAID and ASA intake and the risk of Alzheimer's disease was the Rotterdam study [691].

The Rotterdam study was a prospective, population-based cohort study in 6989 subjects 55 years of age or older (about 80% of the total cohort <75 years) who were free of dementia at baseline. On average, each participant was followed for about 7 years. End points were death, dementia, or the end of the study period. Only medications prescribed by a physician were considered. There was no control for cardiac or other vascular indications for prophylactic aspirin use. Complete information about prescriptions was available in an automated form from pharmacy records. A clinical diagnosis of dementia and its possible reason was done according to standard criteria.

A total of 394 subjects received a diagnosis of dementia. Out of these, 293 had Alzheimer's disease, 56 vascular dementia, and 45 other types of dementia. The use of NSAIDs at any time was associated with a reduced risk of Alzheimer's while no effect was seen with acetaminophen. The relative risk of Alzheimer's disease was 0.95 in subjects with short-term NSAID use, 0.83 in those with intermediate term use, and 0.20 (95% CI: 0.05–0.83) in those with long-term use. A total of 2314 individuals (33%) were on aspirin or other oral salicylates, almost all of them at antiplatelet doses, that is, less than 300 mg/day. There was a nonsignificant relative risk reduction to 0.76 (95% CI: 0.49–1.19) in long-term aspirin users. The risk of vascular dementia was not reduced by NSAIDs and significantly increased by aspirin.

The conclusion was that long-term use of NSAIDs, that is, for 2 years or more, is associated with a significantly reduced risk of Alzheimer's disease but does not protect from vascular dementia. Acetaminophen has no effect on any of these parameters.

These studies have been subjected to several meta-analyses and reviews. Etminan *et al.* [692] published a systematic review and a meta-analysis of observational studies that examined the role of NSAIDs and aspirin in preventing Alzheimer's disease. The nine studies, available at the time, included six cohort studies (13 211 participants) and three case–control studies (1443 participants). The pooled relative risk of Alzheimer's disease among NSAID users was 0.72 (95% CI: 0.56–0.94). The risk reduction depended on the duration of treatment and was highest in the long-term users (>2 years). The pooled relative risk in the eight studies of aspirin users was 0.87 (95% CI: 0.70–1.07). This was not significant. The conclusion was that NSAIDs offer some protection against Alzheimer's disease while the evidence behind the potential preventive use of aspirin is not robust. The optimum doses and duration of drug use as well as the benefit/risk ratio remain unclear. The meta-analysis of Szekely *et al.* [693], based upon 11 studies (out of 38), showed a combined risk estimate for development of Alzheimer's disease of 0.51 for NSAID exposure. In the prospective studies, the estimate was 0.74 for lifetime NSAID exposure and 0.42 for studies reporting duration of 2 years or more.

Overall, these studies suggest beneficial effects of NSAIDs on Alzheimer's disease. There might also be a similar, though less consistent beneficial effect of aspirin. Randomized primary prevention trials are necessary to validate these findings and to establish whether the treatment benefits outweigh the potential risks. In addition, treatment with these drugs appears to be effective only if it is started before neurological deficits become evident [691].

Randomized Trials The available prospective studies on aspirin and the CNS have mainly been focused on stroke. They showed a protective effect on secondary prevention and also some effect (in addition to a reduced number of myocardial infarctions) on primary prevention in patients at elevated risk, mainly in women (Section 4.1.2). Whether this protective effect can be extrapolated to nonvascular dementia, that is, Alzheimer's disease, can only be determined by randomized trials.

One randomized though open and small-sized trial in 310 community resident patients suffering from mild to moderate Alzheimer's disease were advised to take open-label aspirin (75 mg/day enteric-coated) or to avoid aspirin. After 2 years of treatment, there was no clinical benefit but an increased risk of bleedings, suggesting that in patients with existing Alzheimer's dementia, the use of low-dose aspirin has no beneficial effect [694].

The first prospective randomized, placebo-controlled trial on possible preventive effects of NSAIDs and coxibs, that is, anti-inflammatory treatment, the Alzheimer Disease Anti-inflammatory Prevention Trial (ADAPT) [695, 696] was started in 2001. This multicenter primary prevention trial compared celecoxib (200 mg/bid) and naproxen (220 mg/bid) in a total of 2528 patients. Another study was designed to investigate whether naproxen (220 mg/bid) or rofecoxib (25 mg/day) can prevent Alzheimer's dementia or delay cognitive decline [697]. At one year, there were no differences between the three groups regarding efficacy parameters. However, there was an increasing risk of cardio- and cerebrovascular morbidity with celecoxib, and the safety data in ADAPT suggested the possibility of similar increase with naproxen. The study was subsequently terminated prematurely after 3 years (of the planned 5–7 years). The significance of the study and its interpretation are still a matter of controversies [698, 699]. However, this study also demonstrates the inherent difficulties of large prospective randomized prevention trials in the elderly. In addition, one recent meta-analysis and a large retrospective case–control study on the cardiovascular risk of selective and nonselective COX inhibitors have shown that both classes of compounds carry a comparable risk of cardiovascular events [558, 136]. In addition, patients might receive aspirin simultaneously for atherothrombotic prevention and the possibility exists that some NSAIDs, including naproxen, might interact with aspirin for antiplatelet effects (Section 2.3.1) [700].

The most recent Cardiovascular Health Cognition Study investigated the association of Alzheimer's disease with the use of NSAIDs in dementia-free individuals. This study also confirmed a significant risk reduction by NSAIDs (HR: 0.63; 95% CI: 0.45–0.88) but did not find a significant effect for aspirin or acetaminophen. Interestingly, this risk reduction was seen only in patients, carrying an APOE ε4 allele (HR: 0.34, 95% CI: 0.18–0.65) and there was no advantage for $A_{\beta 42}$-lowering NSAIDs. One explanation for the failure of aspirin was that the dose that was taken by these individuals – mostly for cardiocoronary prevention – was possibly too low to provide the same neuroprotection as other NSAIDs [701].

4.3.2.3 Aspirin and Other Drugs

Many different classes of drugs are currently in use for the treatment of dementia and its behavioral disturbances, including Alzheimer's disease. Aspirin is one of them [681], although mainly from the aspect of stroke prevention. More drugs are currently under investigation in several primary prevention trials, including estrogen, selenium, ginkgo biloba extract, vitamin E, and others [698]. This multitude of therapeutic approaches documents the uncertainty about the pathophysiological target as well as the heterogeneity of cerebral dysfunctions in the elderly.

4.3.2.4 Actual Situation

An anti-inflammatory/antiplatelet approach is useful in symptomatic treatment of Alzheimer's disease. Aspirin is an option for patients with mild- to-moderate vascular or mixed Alzheimer's disease/vascular dementia according to guideline recommendations [681]. Unlike traditional NSAIDs and coxibs, aspirin is not expected to interfere with other therapeutic approaches but rather might have some additional beneficial effect with respect to the prevention of stroke and myocardial infarction.

Summary

Alzheimer's disease is a neurodegenerative disorder in the brain, associated with the loss of memory and cognitive functions. The early stages of the disease involve multiple neuro-inflammatory processes and disturbed apoptosis of affected neurons as well as platelet activation and peripheral signs of inflammation and reduced oxidative defense. These processes are subject to treatment with anti-inflammatory and/or antiplatelet drugs with the intention to retard the progression of the disease. Aspirin might affect these processes at different levels, including tau-related inhibition of apoptosis in degenerated neurofibrils.

The majority of available epidemiological trials suggest a marked reduction of the prevalence of Alzheimer's by NSAIDs, by about 50%, if treatment is started early and maintained for 2 or more years. Similar, though somewhat less impressive results were seen with aspirin in some but not all trials while acetaminophen appears to be inactive. There are no convincing data with coxibs but rather evidence of unwanted side effects. However, all of these findings need to be confirmed by appropriately sized, prospective randomized trials.

At present, no sufficiently large prospective randomized trials on aspirin for the prevention or treatment of Alzheimer's disease are available, with changes in cognitive defects or memory as primary end points. However, since Alzheimer's is a disease of the elderly, who are frequently multimorbid and might take aspirin for protection from atherothrombotic events, including stroke, any additional positive effect of the compound on cognitive functions is clearly desirable.

References

667 Mattson, M.P. (2004) Pathways towards and away from Alzheimer's disease. *Nature*, **430**, 631–639.

668 Koistinaho, M. and Koistinaho, J. (2005) Interactions between Alzheimer's disease and cerebral ischemia – focus on inflammation. *Brain Research. Brain Research Reviews*, **48**, 240–250.

669 Li, H.L., Wang, H.H., Liu, S.J. *et al.* (2007) Phosphorylation of tau antagonizes apoptosis by stabilizing beta-catenin, a mechanism involved in Alzheimer's neurodegeneration. *Proceedings of the National Academy of Sciences of the United States of America*, **104**, 3591–3596.

670 In 't Veld, B.A., Launer, L.J., Breteler, M.M. *et al.* (2002) Pharmacologic agents associated with a preventive effect on Alzheimer's disease: a review of the epidemiologic evidence. *Epidemiologic Reviews*, **24**, 248–268.

671 Hull, M., Lieb, K. and Fiebich, B.L. (2002) Pathways of inflammatory activation in Alzheimer's disease: potential targets for disease modifying drugs. *Current Medicinal Chemistry*, **9**, 83–88.

672 Hoozemans, J.J., Veerhuis, R., Rozemuller, J.M. *et al.* (2006) Neuroinflammation and regeneration in the early stages of Alzheimer's disease pathology. *International Journal of Developmental Neuroscience*, **24**, 157–165.

673 Firuzi, O. and Pratico, D. (2006) Coxibs and Alzheimer's disease: should they stay or should they go? *Annals of Neurology*, **59**, 219–228.

674 Eikelenboom, P., Rozemuller, A.J.M., Hoozemans, J.J.M. *et al.* (2000) Neuroinflammation and Alzheimer disease: clinical and therapeutic implications. *Alzheimer Disease & Associated Disorders*, **14** (Suppl. 1), S54–S61.

675 Harris, J.R. (2002) *In vitro* fibrillogenesis of the amyloid β_{1-42} peptide: cholesterol potentiation and aspirin inhibition. *Micron*, **33**, 609–626.

676 Hirohata, M., Ono, K., Naiki, H. *et al.* (2005) Non-steroidal anti-inflammatory drugs have anti-amyloidogenic effects for Alzheimer's beta-amyloid fibrils *in vitro*. *Neuropharmacology*, **49**, 1088–1099.

677 Tortosa, E., Avila, J. and Pérez, M. (2006) Acetylsalicylic acid decreases tau phosphorylation at serine 422. *Neuroscience Letters*, **396**, 77–80.

678 Ciabattoni, G., Porreca, E., Di Febbo, C. *et al.* (2007) Determinants of platelet activation in Alzheimer's disease. *Neurobiology of Aging*, **28**, 336–342.

679 Chen, M., Inestrosa, N.C., Ross, G.S. *et al.* (1995) Platelets are the primary source of amyloid beta-peptide in human blood. *Biochemical and Biophysical Research Communications*, **213**, 96–103.

680 Tang, K., Hynan, L.S., Baskin, F. *et al.* (2006) Platelet amyloid precursor protein processing: a bio-marker for Alzheimer disease. *Journal of the Neurological Sciences*, **240**, 53–58.

681 Alexopoulos, G.S., Jeste, D.V., Chung, H. *et al.* (2005) The expert consensus guideline series. Treatment of dementia and its behavioral disturbances. Introduction: methods, commentary, and summary. *Postgraduate Medicine*, Spec No. 6–22.

682 Iqbal, K. and Grundke-Iqbal, I. (2007) Developing pharmacological therapies for Alzheimer disease. *Cellular and Molecular Life Sciences*, **64**, 2234–2344.

683 Stewart, W.F., Kawas, C., Corrada, M. *et al.* (1997) Risk of Alzheimer's disease and duration of NSAID use. *Neurology*, **48**, 626–632.

684 Delanty, N. (1998) Risk of Alzheimer's disease and duration of NSAID use. *Neurology*, **51**, 652.

685 Anthony, J.C., Breitner, J.C., Zandi, P.P. *et al.* (2000) Reduced prevalence of AD of NSAIDs and H2 receptor antagonists: the Cache County study. *Neurology*, **54**, 2066–2071.

686 Cornelius, C., Fastbom, J., Winblad, B. *et al.* (2004) Aspirin, NSAIDs, risk of dementia, and influence of the apolipoprotein E epsilon 4 allele in an elderly population. *Neuroepidemiology*, **23**, 135–143.

687 Broe, G.A., Grayson, D.A., Creasey, H.M. *et al.* (2000) Anti-inflammatory drugs protect against Alzheimer disease at low doses. *Archives of Neurology*, **57**, 1586–1591.

688 Halliday, G.M., Shepherd, C.E., McCann, H. *et al.* (2000) Effect of anti-inflammatory medications on neuropathological findings in Alzheimer disease. *Archives of Neurology*, **57**, 831–836.

689 Mackenzie, I.R.A. (2001) Effect of anti-inflammatory medications on neuropathological findings in Alzheimer disease. *Archives of Neurology*, **58**, 517–518.

690 Nilsson, S.E., Johansson, B., Takkinen, S. *et al.* (2003) Does aspirin protect against Alzheimer's dementia? A study in a Swedish population-based sample aged ≥80 years. *European Journal of Clinical Pharmacology*, **59**, 313–319.

691 In 't Veld, B.A., Ruitenberg, A., Hofman, A. *et al.* (2001) Nonsteroidal antiinflammatory drugs and the risk of Alzheimer's disease. *The New England Journal of Medicine*, **345**, 1515–1521.

692 Etminan, M., Gill, S. and Samii, A. (2003) Effect of non-steroidal anti-inflammatory drugs on risk of Alzheimer's disease: systematic review and meta-analysis of observational studies. *British Medical Journal*, **327**, 128–131.

693 Szekely, C.A., Thorne, J.E., Zandi, P.P. *et al.* (2004) Nonsteroidal anti-inflammatory drugs for the prevention of Alzheimer's disease: a systematic review. *Neuroepidemiology*, **23**, 159–169.

694 AD2000 Collaborative Group , Bentham, P., Gray, R., Sellwood, E. *et al.* (2008) Aspirin in Alzheimer's disease (AD2000): a randomised open-label trial. *Lancet Neurology*, **7**, 41–49.

695 Martin, B.K., Meinert, C.L. and Breitner, J.C., ADAPT Research Group (2002) Double placebo design in a prevention trial for Alzheimer's disease. *Controlled Clinical Trials*, **23**, 93–99.

696 ADAPT Research Group , Lyketsos, C.G., Breitner, J.C., Green, R.C. *et al.* (2007) Naproxen and celecoxib do not prevent AD in early results from a randomized controlled trial. *Neurology*, **68**, 1800–1808.

697 Aisen, P.S., Schafer, K.A., Grundman, M. *et al.* (2003) Effects of rofecoxib or naproxen vs. placebo on Alzheimer disease progression. A randomized controlled trial. *Journal of the American Medical Association*, **289**, 2819–2826.

698 Green, R.C. and DeKosky, S.T. (2006) Primary prevention trials in Alzheimer disease. *Neurology*, **67** (Suppl. 3), S2–S5.

699 Martin, B.K., Breitner, J.C., Evans, D. *et al.* (2007) The trialist, meta-analyst, and journal editor: lessons from ADAPT. *The American Journal of Medicine*, **120**, 192–193.

700 Konstantinopoulos, P.A. and Lehmann, D.F. (2005) The cardiovascular toxicity of selective and nonselective cyclooxygenase inhibitors: comparisons, contrasts, and aspirin confounding. *Journal of Clinical Pharmacology*, **45**, 742–750.

701 Szekely, C.A., Breitner, J.C., Fitzpatrick, A.L. *et al.* (2008) NSAID use and dementia risk in the cardiovascular health study: role of APOE and NSAID type. *Neurology*, **70**, 17–24.

Appendix A. Abbreviations (No Acronyms for Clinical Studies)

AA	Arachidonic acid
ACE	Angiotensin converting enzyme
ACS	Acute coronary syndrome
ADP, ATP	Adenosine di/triphosphate
AP-1	Activating protein-1
ASA	Acetylsalicylic acid (aspirin)
APC	Adenomatous polyposi coli
ATL	Aspirin-triggered lipoxin
Bcl-2	B-cell lymphoma-2
BiP	Immunoglobulin-binding protein
CABG	Coronary artery bypass graft
cAMP	Cyclic adenosine monophosphate
CD39	5′-Nucleotidase
cEBPβ	CCAT/enhancer-binding protein β
CoA	Coenzyme A
COX	Cyclooxygenase
CRE	cAMP responsive element
CYP	Cytochrome
11-DH-TXB2	11-Dehydro-thromboxane B2
DMARD	Disease modifying antirheumatic drug
DNA	Deoxyribonucleic acid
ED_{50}	50% effective doses
EGF	Epidermal growth factor
eNOS	Endothelial nitric oxide synthase
ERK	Extracellular signal-regulated kinase
FFA	Free fatty acid
GA	Gentisic acid
GAPDH	Glycerolaldehyde-3-phosphate dehydrogenase
GPIIb/IIIa	Glycoprotein IIb/IIIa
HETE	Hydroxyeicosatetraenoic acid
HO	Heme oxygenase
HPETE	Hydroperoxyeicosatetraenoic acid
HSP	Heat shock protein
5-HT	5-Hydroxytryptamine (serotonin)
ICAM-1	Intercellular adhesion molecule 1 (CD54)
IKKβ	Inhibitor kinase β
IL	Interleukin
INFγ	Interferon-γ
iNOS	Inducible nitric oxide synthase
INR	International normalized ratio
IUGR	Intrauterine growth retardation
JNK	c-Jun N-terminal kinase
LD_{50}	50% lethal dose
5-LO	5-Lipoxygenase
LOX	Lipoxygenase
LPS	Lipopolysaccharide
LT	Leukotriene
LX	Lipoxin
MAPK	Mitogen-activated protein kinase (ERK 1/2)
NADP	Nicotinamide adenine nucleotide diphosphate
NFAT	Nuclear factor of activated T cells
NFκB	Nuclear factor κB

Acetylsalicylic Acid. Karsten Schrör

ISBN: 978-3-527-32109-4

NMDA	*N*-Methyl-D-aspartate
NNH	Number needed to harm
NNT	Number needed to treat
NOS	NO synthase
NSAID	Nonsteroidal antiinflammatory drug
NSTEMI	Non-ST-elevation myocardial infarct
OTC	Over the counter (drugs)
PAD	Peripheral arterial disease
PAI	Plasminogen activator inhibitor
PCI	Percutaneous coronary intervention
PDGF	Platelet-derived growth factor
PG	Prostaglandin
PHA	Phytohemagglutinin
PKC	Protein kinase C
PLC	Phospholipase C
PMA	Phorbol myristate acetate
PMN	Polymorphonuclear neutrophil
POX	Hydroperoxidase
PTA	Percutaneous transluminal angioplasty
PTCA	Percutaneous transluminal coronary angioplasty
RegI	Regenerative protein I
RSK	Ribosomal S6-kinase
SAG	Salicylic acid acyl glucuronide
SP	Substance P
SPG	Salicylic acid phenol glucuronide
src	Nonreceptor protein tyrosine kinases
SA	Salicylic acid
STAT	Signal transducers and activators of transcription
STEMI	ST-elevation myocardial infarct
SU	Salicyluric acid
SUPG	Salicyluric acid phenol glucuronide
TF	Tissue factor
TIMP	Tissue inhibitor of metalloproteinase
TNF	Tumor necrosis factor
tPA	Tissue plasminogen activator
TX	Thromboxane
VCAM	Vascular cell adhesion molecule
VEGF	Vascular endothelial growth factor
vWF	von Willebrand factor

Appendix B.
Acronyms for Clinical Studies

ADAPT	Alzheimer Disease Anti-Inflammatory Prevention Trial
AMIS	Aspirin-Myocardial Infarction Study
APRICOT	Antithrombotics in the Prevention of Reocclusion in Coronary Thrombolysis
ARIC	Atherosclerosis Risk in Communities Study
ASPECT	Aspirin Esomeprazole Chemoprevention Trial
BAFTA	Birmingham Atrial Fibrillation Treatment of the Aged Study
BLASP	Barbados Low-Dose Aspirin Study in Pregnancy
BMDS	British Medical Doctors' Trial
CAPRIE	Clopidogrel Versus Aspirin in Patients at Risk of Ischemic Events
CARS	Coumadin Aspirin Reinfarction Study
CAST	Chinese Acute Stroke Trial
CHAMP	Combination Hemotherapy and Mortality Prevention
CHARISMA	Clopidogrel for High Atherothrombotic Risk, Ischemic Stabilization, Management, and Avoidance
CLASP	Collaborative Low-Dose Aspirin Study in Pregnancy
CLASS	Celecoxib Long-Term Arthritis Safety Study
CLASSICS	Clopidogrel Aspirin Stent International Cooperative Study
CLIPS	Critical Leg Ischemia Prevention Study
CPS	Cancer Prevention Study
CREDO	Clopidogrel for Reduction of Events During Observation
CURE	Clopidogrel in Unstable Angina to Prevent Recurrent Ischemic Events
ESPRIT	European and Australian Stroke Prevention in Reversible Ischemia Trial
ESPS	European Stroke Prevention Study
GRACE	Global Registry of Acute Coronary Events
HOT	Hypertension Optimal Treatment
HPFS	Health Professionals Follow-Up Study
ISIS	International Study on Infarct Survival
IST	International Stroke Trial
JLASP	Jamaica Low-Dose Aspirin Study Project
MITRA	Maximal Individual Optimized Therapy for Acute Myocardial Infarction
NHS	Nurses' Health Study

Acetylsalicylic Acid. Karsten Schrör

ISBN: 978-3-527-32109-4

OASIS Orbital Atherectomy Study for Treatment of Peripheral Vascular Stenosis
PPP Primary Prevention Project
PROFESS Prevention Regimen for Effectively Avoiding Second Strokes
SAPAT Swedish Angina Pectoris Aspirin Trial
TOPPS Ticlid or Plavix Post-Stents
TPT Thrombosis Primary Prevention Trial
UK-HARP (First) United Kingdom Heart and Renal Protection Study
US-PHS US Physicians' Health Study
VIGOR VIOXX GI Clinical Outcomes Research
WAVE Warfarin Antiplatelet Vascular Evaluation
WHS Women's Health Study

Index

a
Aberrant crypt foci (ACF) 343
ACE inhibitors 184, 246, 249, 252
Acetaminophen (Paracetamol)
– Alzheimer's disease 361f.
– analgesic action 324ff.
– analgesic nephropathy 180, 182
– asthma 197ff., 217ff.
– hepatotoxicity 188
– osteoarthritis 334f.
– renal failure 181f.
Acetylsalicylic acid (ASA), *see* Aspirin
Acute coronary syndrome, *see* Coronary vascular disease
Adenomatous Polyposis Coli (APC) 118f., 343f., 346, 349, 354
Allodynia 114
Alzheimer's Disease Anti-inflammatory Prevention Trial (ADAPT) 363
Alzheimer's disease 18, 359ff.
– actual situation 363
– clinical trials 361ff.
– etiology 359
– mode of aspirin action 360
– pathophysiology 359
American Health Professionals' study 233f.
Angioedema 205
Anticoagulants
– cerebrovascular disease 269ff.
– coronary vascular disease 247ff.
– peripheral arterial disease 282
– venous thrombosis 287f.
Antiplatelet drugs 91f., 280, 289, 294
Antiplatelet/antithrombotic Trialists' Collaboration 229
Antipyretic analgesics 322ff., 329, 340
Antithrombotics in the Prevention of Reocclusion in Coronary Thrombolysis (APRICOT) 248
Antitumorigenic effects, *see* Malignancies
Apoptosis 69, 71, 118, 120ff., 354
Aprotinin 146f.
Arthritis and rheumatism 332ff.
– actual situation 336
– clinical trials 334f.
– mode of aspirin action 332ff.
– osteoarthritis 333ff.
– rheumatoid arthritis 332ff.
Aspirin, *see also* Salicylate
– absorption 34ff., 40
– actions on organs and tissues 86ff.
– alcohol 146, 168
– analgesic/antipyretic actions 14, 86, 105, 111, 113f., 323 ff.
– antiinflammatory actions 106ff., 232
– antiplatelet actions 15, 86, 88, 90, 262
– antitumor effects 120ff.
– antiulcer drugs 170
– bioavailability 38f.
– biotransformations 46ff.
– bleeding risk 15, 142ff., 152, 159f., 244f.
– chemical properties 25ff.
– clinical applications 225ff.
– coagulation 98
– COX-independent actions 66ff., 122 ff.
– COX-inhibition 13f., 54ff., 120ff., 163f.
– desensitization 201
– determination 30ff.
– dosing 41, 88, 92, 226ff., 251, 262f., 352f.
– Ductus arteriosus 151f.
– energy metabolism 79ff.
– esterases 46ff.
– excretion 49ff.
– fertility 150
– fibrinolysis 98f., 242f.
– formulations 27ff., 40, 326
– gene transcription 13f., 61ff., 69ff.
– hemostasis 89ff.
– hepatotoxicity 187f.

Acetylsalicylic Acid. Karsten Schrör

ISBN: 978-3-527-32109-4

– history 5ff., 19ff.
– hypersensitivity, *see also* Aspirin-induced asthma 197ff.
– inhibition of platelet function 90ff.
– inhibition of prostaglandin formation 12, 53ff.
– inhibition of kinases 67ff.
– interactions with NSAIDs 60f. 170, 249f. 308f.
– miscarriage 151
– modes of action 53ff.
– pharmacokinetics 33ff., 46ff.
– pharmacological properties 225
– pseudoallergic actions 197ff.
– renal failure 181ff.
– "resistance" 303ff.
– sphingosine-1-phosphate 58
– thrombin generation 143f.
– toxicity 131ff.
– transcription factors 69ff., 123f.
Aspirin Esomeprazole Chemoprevention Trial (ASPECT) 312f.
Aspirin-induced asthma 197ff., 217f.
– clinical studies 201f.
– clinical symptoms 198f.
– leukotrienes 200f.
– mode of aspirin action 199f.
– pathophysiology 197ff.
"Aspirin-like drugs" 18, 60
Aspirin-Myocardial Infarction Study (AMIS) 17
Aspirin-triggered lipoxin (ATL) 59, 108f., 163f.
Atherosclerosis Risk in Communities Study (ARIC) 227
Audiovestibular system, *see* Ototoxicity

b
Baltimore Longitudinal Study of Ageing 361
Barbados Low-dose Aspirin Study in Pregnancy (BLASP) 294, 297
Bioavailability 38ff.
Biotransformations 46ff.
Birmingham Atrial Fibrillation Treatment of the Aged Study (BAFTA) 270
Bleeding disorders 142ff.
– bleeding time 142ff.
– bleeding risk in surgical interventions 144ff.
– GI tract 159ff., 169f.
– hemorrhagic stroke 264
– pregnancy and labor 152
– role of platelets 143
British medical doctors' study (BMDS) 232f., 264

c
Cancer, *see* Malignancies
Cancer prevention study II (CPS-II) 227, 347f.
Cardiovascular diseases, *see also* Coronary vascular diseases
– ACE-inhibitors 249
– in patients with renal disease 182ff.
– statins 249
Celecoxib Long-term Arthritis Safety Study (CLASS) 171
Ceramide 121, 123
Cerebrovascular diseases 260ff.
– actual situation 270f.
– anticoagulants 269f.
– clopidogrel 269
– dipyridamole 267ff.
– etiology 260
– hemorrhagic stroke 264f.
– mode of aspirin action 261ff.
– primary prevention 263ff.
– secondary prevention 265ff.
Chinese Acute Stroke Trial (CAST) 266f.
Cilostazol 283
Clopidogrel
– cerebrovascular disease 269
– coronary vascular disease 246f.
– peripheral arterial disease 282
– "resistance" 303
Clopidogrel for High Atherothrombotic Risk, Ischemic Stabilization, Management and Avoidance (CHARISMA) 247f.
Clopidogrel in Unstable Angina to Prevent Recurrent Ischemic Events (CURE) 246, 269
Clopidogrel versus Aspirin in Patients at Risk of Ischemic Events (CAPRIE) 229, 240, 246, 269, 282
Collaborative low-dose aspirin study in pregnancy (CLASP) 227, 293ff.
Colorectal cancer 343ff.
– actual situation 352ff.
– clinical trials 346ff.
– COX-2 expression and aspirin 61f., 120f., 344ff.
– etiology 344f.
– mode of aspirin action 345f.
– NSAIDs 352
– pathophysiology 344ff.
Combination Hemotherapy and Mortality Prevention Study (CHAMP) 248
Coronary vascular diseases 411ff.
– actual situation 250ff.
– acute coronary syndrome (ACS) 238ff.
– aspirin historical use 15f.
– anticoagulants 247ff.
– Clopidogrel 246f.
– coronary artery bypass grafting 244ff.
– coxibs 249f.
– coronary artery bypass graft surgery (CABG) 244ff.
– etiology 228f.
– mode of aspirin action 229ff.
– NSAIDs 249
– percutaneous coronary interventions (PCI) 243f.
– primary prevention 231ff.

– secondary prevention 238ff.
Cottbus Reinfarction Trial 240f.
COX
– atherosclerosis 95f.
– gastric mucosa 160f.
– gene expression 14, 61, 359
– historical aspects 13ff.
– inhibition by NSAIDs 54ff., 171, 335ff., 363
– lipoxins 59
– role in asthma 199f.
– transcription factors 69ff.
COX-1
– acetylation by aspirin 56ff.
– active site 55f.
– aspirin-induced asthma 199f.
– 12-HPETE 62
– inhibition by aspirin 54ff.
– interaction with NSAIDs 308
– platelet function 90ff.
COX-2
– acetylation by aspirin 59ff.
– gene expression 61, 120f., 359
– inhibition by aspirin 56ff., 163
– lipoxins 59
– platelet function 90ff.
– tumor promotion 120ff., 345, 350
COX-3 111
Coxibs
– cardiovascular risk 335f.
– interactions with aspirin 108, 170f.
Critical Leg Ischemia Prevention Study (CLIPS) 279, 281
Cyclooxygenases, *see* COX

d

Deep vein thrombosis, *see* Venous thrombosis
Desmopressin 146
Diabetes 183
Dicarboxylic acids 81, 209
Dipyridamole 267ff., 280, 282f.
Dipyrone 308
Disease-modifying Antirheumatic Drugs (DMARDs) 332, 334
Ductus arteriosus 151f.
Dutch TIA Trial 265f.

e

Endocannabinoids 111
Endothelium
– antiplatelet factors 96f.
– atherosclerosis 95ff.
– dysfunction 278
– inflammation 108f.
– NO-synthase (eNOS) 74f.
– prostaglandin production 95ff.
Energy metabolism 79ff.
European and Australian Stroke Prevention in Reversible Ischaemia Trial (ESPRIT) 268
European Stroke Prevention Study-2 (ESPS-2) 267f.

f

Fatty acid metabolism 80ff.
Ferritin 74
Fertility 150ff.
Fetal development 150f.
Fever 113ff., 323, 326
Fibrinolysis 98f., 242f., 278

g

Gastrointestinal tract (GI) 157ff.
– aspirin and GI injury 161ff., 170 ff.
– clinical trials 166ff.
– coxibs 170f.
– gastric mucosa 358ff.
– gastrointestinal bleeding 159ff.
– *Helicobacter pylori* 164ff.
– mode of aspirin action 161ff.
– prostaglandins and GI injury 160f.
Global Registry of Acute Coronary Events (GRACE) 227, 251, 276

h

Habituation 139
Headache, *see* Pain
Health Professionals Follow-up Study (HPFS) 347ff.
Hearing disturbances, *see* Ototoxicity
Heart failure 183f.
Heart Outcomes Prevention Evaluation (HOPE) 310ff.
Helicobacter pylori 164ff.
Heme oxygenase 174f.
Hemorrhage, *see* Bleeding disorders
Hemorrhagic stroke, *see* Cerebrovascular diseases
Hemostasis 89ff.
Hepathoencephalopathy, *see* Reye's syndrome
Hypertension 183, 235
Hypertension in pregnancy, *see* Preeclampsia
Hypertension Optimal Treatment study (HOT) 234ff., 260, 264f.
Hyperthermia 137f.
Hyperventilation 79,134f.

i

Ibuprofen 56f., 60, 151, 171, 324
Indomethacin 56f., 143f., 151, 171
Inflammation 74, 105ff., 231
Inflammation marker 232
International Study on Infarct Survival (ISIS) 17, 99, 170, 227, 239, 242ff.
International Stroke Trial (IST) 266

Ischemia, *see individual organs*
Isoprostanes 94, 309

j
Jamaica Low-Dose Aspirin Study Project (JLASP) 294, 297

k
Kawasaki disease 339ff.
– clinical trials 340
– mode of aspirin action 339
– pathophysiology 339
– Reye's syndrome 217, 340
Kidney 179ff.
– analgesic nephropathy 179f.
– clinical studies 181f.
– diabetes 183
– mode of aspirin action 180
Kinases 66ff.
Leukotrienes
– asthma 198ff.
– inflammation 106, 108
– ototoxicity 193
Lipoxins, *see also* Aspirin-triggered Lipoxin 108
Liver 187ff.
– aspirin actions 80ff.,187
– drugs and liver injury 84, 187
– hepatotoxicity of salicylates 84f., 187, 210f.
– Reye's syndrome 84f.
Lyell syndrome 206

m
Malignancies 118ff., 343ff.
Malondialdehyde 122
Management of Atherothrombosis with Clopidogrel in High-risk Patients with recent Transient Ischemic Attack or Ischemic Stroke (MATCH) 269
Maximal Individual Optimized Therapy for Acute Myocardial Infarction (MITRA) 251
Melbourne Colorectal Cancer Study 346
Metamizol, *see* Dipyrone
Methylsalicylate 27
Migraine 326ff.
Myocardial infarction, *see* Coronary vascular diseases
Myocardial Infarction Registry (MIR) 251

n
Neuroprotection 72f., 360
Nitric oxide (NO) 73, 96f., 164
– endothelial NO-synthase (eNOS) 74, 143
– inducible NO-synthase (iNOS) 73f., 106
Nonsteroidal antiinflammatory drugs (NSAIDs)
– Alzheimer's disease 360ff.
– antiplatelet effects of aspirin 60, 249f., 308
– asthma 199f.
– colorectal carcinoma 346ff.
– COX-1/COX-2-selectivity 55f.
– Ductus arteriosus 151, 292f.
– GI-effects 160f., 170f.
– interactions with aspirin 60f.
– miscarriage 151
– mode of action 60, 108
– osteoarthritis 333, 336
– renal effects 181f.
Nuclear factor kB (NFkB) 70ff., 123
Nuclear factor of activated T-cells (NFAT) 70f.
Nurses' Health Study (NHS) 181, 227, 348ff.

o
Orbital Atherectomy Study for Treatment of Peripheral Vascular Stenosis (OASIS) 248
Organ toxicity 157ff.
Osteoarthritis 171, 333, 336
– acetaminophen 335
– clinical trials 334
Ototoxicity 191ff.
– clinical trials 193f.
– hearing disturbances 191f.
– mode of aspirin action 191f.
– pathophysiology 191
– tinnitus 192
Oxidative phosphorylation 82ff.

p
Pain 110ff., 322ff.
– acetaminophen 324ff.
– cannabinoids 111
– clinical studies 326ff.
– headache 326ff.
– mediators of pain 110ff.
– migraine 326ff.
– mode of aspirin action 111ff., 323ff.
– prostaglandins 110f.
– serotonin 112
Paracetamol, *see* Acetaminophen
Perinatal Antiplatelet Review of International Studies (PARIS) 295, 297ff.
Peripheral Arterial Disease (PAD) 276ff.
– actual situation 283
– clinical trials 278ff.
– etiology 276
– mode of aspirin action 276ff.
– other drug treatment 282f.
– peripheral transluminal angioplasty 281f.
Phenacetin 180
Physicians' Health Study, *see* US-Physicians' Health Study
Platelets 89ff.

– aspirin-insensitive inhibition 309, 315
– bleeding time 143
– cyclooxygenases 56ff., 308f.
– inhibition by aspirin 90ff.
– response variability, *see also* "resistance" 261f., 303ff.
– thrombotic risk 229f., 276f.
– thromboxane formation 91ff., 306f.
– time-dependent inhibition 91
Preeclampsia 290ff.
– actual situation 298
– clinical trials 293ff.
– etiology 290
– miscarriage 151, 298
– mode of aspirin action 290f.
Pregnancy 150ff.
Pregnancy-induced hypertension, *see* Preeclampsia
Prevention Regimen For Effectively avoiding Second Strokes (PROFESS) 268f.
Primary Prevention Project (PPP) 170, 236f., 264
Prostacyclin, *see also* Prostaglandins 95ff., 291f.
Prostaglandins
– biosynthesis 55ff.
– general aspects 12f.
– mode of aspirin action 11ff., 55
– pain mediators 110f.
– receptors 122
Prostaglandin synthases, *see* COX

r

"Resistance" 303ff.
– clinical trials 262, 310ff.
– definition 304
– measurement 304ff.
– mechanisms 303ff.
– role of thromboxane 306f.
Reye's syndrome 208ff.
– actual situation 217f.
– asthma 217f.
– clinical studies 212ff.
– etiology 209
– laboratory findings 209
– morphological findings 209
– pathophysiology 210
– salicylates 84f., 211ff.
Rheumatoid arthritis 332f.
– clinical studies 250, 334ff.
– mode of aspirin action 332f.
– myocardial infarction 332, 336
– pathophysiology 332
Rotterdam Study 362

s

Salicin 6, 26, 334f.
Salicylate
– antiinflammatory actions 107f.
– antipyretic actions 115
– biotransformations 47f.
– bleeding 98, 142f.
– chemical properties 25ff.
– determination 30f.
– energy metabolism 79ff.
– excretion 49f.
– fatty acid metabolism 80ff.
– gastric mucosal injury 162
– history 19ff.
– inhibition of kinases 67ff.
– inhibition of prostaglandin synthesis 56f.
– natural sources 6, 42f.
– pharmacokinetics 38, 48ff.
– preeclampsia 291f.
– protonophoric properties 83ff.
– thrombotic events 277
– toxicity 132ff.
– transcription factors 61f., 69ff.
– uncoupling of oxidative phosphorylation 82ff.
Salicylic acid, *see* Salicylate
Serotonin 112
Stroke, *see* Cerebrovascular diseases
Study of Left Ventricular Dysfunction (SOLVD) 249
Swedish Angina Pectoris Aspirin Trial (SAPAT) 237f.
Swedish Aspirin Low-Dose Trial (SALT) 266
Sydney Older Persons Study 361

t

Thrombin 89, 144
Thrombosis 90
Thrombosis Primary Prevention Trial (TPT) 170, 236ff., 264
Thrombotic risk 229f., 244ff., 261f., 276f., 287, 290f.
Thromboxane (A2)
– aspirin "resistance" 306f.
– biosynthesis 12f.
– inhibition 91, 97, 229, 261f., 292, 455
– urinary excretion 277, 306f.
Tinnitus, *see also* Ototoxicity 192
Toxicity, *see also individual organs* 131ff.
– clinical symptoms 131ff.
– dose-dependency 133ff.
– intoxication in children 135f.
– laboratory findings 135
– treatment 136ff.
Transcription factors
– colorectal carcinoma 123f.
– modulation by salicylates 69ff.

u

United Kingdom Transient Ischemic Attack Aspirin Trial (UK-TIA) 265

(First) United Kingdom Heart and Renal Protection Study (UK-HARP) 183
US-Physicians' Health Study (US-PHS) 169, 231f., 278, 349f.
US-Public Health Service Study (PHS) 213f.
Urticaria 205

v

Variability in antiplatelet responses, *see* "Resistance"
Venous thrombosis 287ff.
– actual situation 288
– air travelling 288
– clinical trials 287f.
– etiology 287
– mode of aspirin action 287
Veterans Administration Cooperative Trial 238f.
VIOXX GI Clinical Outcomes Research (VIGOR) 250f., 335f.

w

Warfarin, *see* Anticoagulants
Warfarin Antiplatelet Vascular Evaluation (WAVE) 270
Widal's Triad, *see* Aspirin-induced asthma
Women's Health Study (WHS) 227, 234f., 263f.